D1651752

Calculations in Chemistry: An Introduction

CALCULATIONS IN CHEMISTRY

An Introduction

SECOND EDITION

Donald J. Dahm
Rowan University

Eric A. Nelson
Falls Church, VA

W. W. NORTON & COMPANY
New York · London

If we did not make a complete analysis of the elements of the problem,
we should obtain an equation not homogeneous, and, a fortiori,
we should not be able to form the equations which express … more complex cases.

. . . every undetermined magnitude or constant has one dimension proper to itself,
and the terms of one and the same equation could not be compared
if they had not the same exponent of dimensions.

— Joseph Fourier, *The Analytical Theory of Heat* (1822)

W. W. Norton & Company has been independent since its founding in 1923, when William Warder Norton and Mary D. Herter Norton first published lectures delivered at the People's Institute, the adult education division of New York City's Cooper Union. The firm soon expanded its program beyond the Institute, publishing books by celebrated academics from America and abroad. By midcentury, the two major pillars of Norton's publishing program—trade books and college texts—were firmly established. In the 1950s, the Norton family transferred control of the company to its employees, and today—with a staff of four hundred and a comparable number of trade, college, and professional titles published each year—W. W. Norton & Company stands as the largest and oldest publishing house owned wholly by its employees.

Copyright © 2017, 2013 by W. W. Norton & Company, Inc.
Copyright © 2011 by ChemReview Publishing

All rights reserved
Printed in the United States of America

Editor: Erik Fahlgren
Project Editor: David Bradley
Developmental Editor: Andrew Sobel
Assistant Editor: Arielle Holstein
Managing Editor, College: Marian Johnson
Managing Editor, College Digital Media: Kim Yi
Production Manager: Eric Pier-Hocking
Media Editor: Christopher Rapp
Associate Media Editor: Julia Sammaritano
Media Project Editor: Jesse Newkirk
Media Editorial Assistant: Tori Reuter and Doris Chiu
Ebook Production Manager: Mateus Teixeira
Marketing Manager, Chemistry: Stacy Loyal
Designer: Anna Reich
Permissions Manager: Megan Schindel
Composition: Code Mantra
Manufacturing: Webcrafters

ISBN: 978-0-393-28420-1 (pbk.)

W. W. Norton & Company, Inc., 500 Fifth Avenue, New York, NY 10110
wwnorton.com
W. W. Norton & Company Ltd., 15 Carlisle Street, London W1D 3BS

4 5 6 7 8 9 0

Brief Contents

Contents

Preface for Instructors

Several years ago, at a meeting of first-year instructors, we noted broad agreement that each year, increasing numbers of students were entering our classes with gaps in their knowledge of the mathematics prerequisite for chemistry. Reviewing research that might explain these observations, we found that since 1995, K–12 math standards adopted in most states had significantly reduced the time devoted to computation topics, including exponential notation, simplification of fractions, and logarithms.[1]

To address these gaps while preserving class time for chemistry, we designed lessons to be completed during "study time" that reviewed prerequisite math, explained the chemistry that required the math, and then provided practice problems with worked-out solutions. The lessons were tested in both preparatory and general chemistry, and student success improved substantially in both courses.[2] Those lessons became the first edition of *Calculations in Chemistry: An Introduction*.

As a part of designing the instruction in *Calculations*, we reviewed cognitive research on how students solve problems. Recent studies have measured and verified that during problem solving, the brain can process all knowledge that can be quickly recalled from long-term memory, but only three to five small elements of knowledge that are not well memorized. During reasoning, you are limited in large measure to applying what you know—and know well.

Because of this dependence on memorized knowledge, cognitive science tells us that for each new topic in science, students must *practice* so that vocabulary definitions, fundamental relationships, and problem-solving procedures can be recalled quickly and accurately. Conceptual frameworks must then be constructed: the physical linkages in the brain that organize memory. The brain builds those frameworks in response to effort to solve problems that apply new knowledge in a variety of distinctive contexts.[3,4]

---

[1]Nelson, E. A. (2012). *Common Core Standards, Student Test Scores in Math Computation, and the Implications for Chemistry Instruction*. ACS Biennial Conference on Chemical Education (BCCE) 2012. Available at www.ChemReview.Net/BCCE2012Math.pdf.

[2]Hartman, J. R., Dahm, D. J., and Nelson, E. A. (2015). Time-Saving Resources Aligned with Cognitive Science to Help Instructors. *Journal of Chemical Education* 92(9): 1568–1569.

[3]Hartman, J. R., and Nelson, E. A. (2015). "Do We Need To Memorize That?" Or Cognitive Science for Chemists. *Foundations of Chemistry* 17(3): 263–274.

[4]Brown, P. C., Roediger, H. L., III, and McDaniel, M. A. (2014). *Make It Stick: The Science of Successful Learning*. Cambridge, MA: Harvard University Press.

To harness these findings of cognitive science, *Calculations* provides a systematic structure to help students increase and organize their knowledge.

- Each lesson highlights key facts and relationships. Students then complete exercises that move new knowledge into memory.

- Try It examples engage students in learning problem-solving procedures.

- Integrated within each lesson, Practice problems build conceptual frameworks by associating new and prior knowledge.

In this Second Edition of *Calculations*, we have added:

- A new lesson (Lesson 1.2, Numeracy) that helps students review and strengthen their mental arithmetic skills. In addition, throughout the text, new problems have been added with simplified numbers. Students are encouraged to solve these problems without a calculator, reinforcing skills in mental math and focusing attention on concepts.

- Chapter summaries that include study tips: tested, proven methods for students to learn the variety of strategies needed to solve problems in chemistry.

- End of chapter Review Quizzes, which supply the "interleaved practice" that science recommends to help students distinguish problem types, self-assess skills, and check understanding.[4]

In this edition, chapters on gas laws and equilibrium have each been divided into two chapters to provide more manageable length and additional scheduling flexibility.

Calculations is designed to serve as a primary text in preparatory chemistry or as a supplement in either general chemistry or allied health (GOB) chemistry. This Second Edition of *Calculations in Chemistry* is supported by Smartwork5 online homework with more than 325 questions spanning all 24 chapters. A modular design allows instructors to assign only the chapters their students need.

Learning studies stress the need for "distributed practice" (study sessions spaced over several days) rather than "massed practice" (cramming). Because most Smartwork5 questions have algorithmic variables, assignments in Smartwork5 can provide unique and frequent "homework quizzes" that encourage spaced study—and allow instructors to track student progress.

Posted reviews have reported that by filling in gaps in essential background knowledge, *Calculations* measurably increased the success of students in their courses. In our experience, the "study-time" assignments in *Calculations* opened additional time in lecture for inquiry, problem solving, demonstrations, and discussions. It is our hope that these lessons will assist in your work to improve student understanding and interest in chemistry.

—Donald J. Dahm and Eric A. Nelson (ToTheAuthors@ChemReview.Net)

Acknowledgments

We would like to thank our reviewers, who helped greatly in increasing the clarity and accuracy of this text.

Second Edition Reviewers

Charles Carraher, Florida Atlantic University

Jose Conceicao, Metropolitan Community College

Milagros Delgado, Florida International University

Donna Iannotti, Eastern Florida State College

Rajan Juniku, University of North Carolina, Wilmington

Laurie Lazinski, Fulton-Montgomery Community College

Jeremy Mason, Texas Tech University

Jeffrey Pribyl, Minnesota State University, Mankato

Dawne Taylor, College of Charleston

Steve Trail, Elgin Community College

Ingrid Ulbrich, Colorado State University

Melanie Veige, University of Florida

Jodi Wilker, Lakeland Community College

Mohannad Yousef, California State University, Dominguez Hills

Chad Yuen, Augustana College

Previous Edition's Reviewers

Doyle Barrow, Jr., Georgia State University

Nathan Barrows, Grand Valley State University

Ivana Bozidarevic, San Francisco City College

Claire Cohen, University of Toledo

Bernadette Corbett, Metropolitan Community College

Jennifer Coym, University of South Alabama

Steven Davis, University of Mississippi

Nell Freeman, St. Johns River State College

Aaron Fried, Los Angeles Community College

Daniel Groh, Grand Valley State University

Tamara Hanna, Texas Tech University

Donna Iannotti, Eastern Florida State College

Roy Kennedy, Massachusetts Bay Community College

Carol Martinez, Central New Mexico Community College

David Nachman, Mesa Community College

Randa Roland, University of California Santa Cruz

Mary Setzer, University of Alabama in Huntsville

Matthew Smith, Walters State Community College

Steve Trail, Elgin Community College

Marie Villarba, Seattle Central Community College

Note to Students

The goal of these lessons is to help you solve *calculations* in first-year chemistry. This is only one part of a course in chemistry, but it can be the most challenging.

We suggest purchasing a *spiral notebook* as a place to write your work when solving problems in these lessons. You will also need

- Two packs of 100 3 × 5-inch index cards (two or more colors are preferred) plus a small assortment of rubber bands, and
- A pack of long sticky notes (4 × 6-inch notes are recommended) for use as cover sheets

It is important that *you* use the *same* calculator to solve homework problems that you will be allowed to use during tests, in order to learn and practice the rules for that calculator before tests.

Many courses will not allow the use of a graphing calculator or other calculators with extensive memory during tests. If no type of calculator is specified for your course, any inexpensive calculator with a $\boxed{1/x}$ *or* $\boxed{x^{-1}}$, $\boxed{y^x}$ *or* $\boxed{\wedge}$, $\boxed{\log}$ *or* $\boxed{10^x}$, *and* $\boxed{\ln}$ functions will be sufficient for most calculations in first-year chemistry.

How to Use These Lessons

1. *Read* the lesson and *work* the ▭▭▭▭▷ **Try It questions.** Use this method:
   - As you start a new page, *if* you see a *stop* sign **STOP** on the page, *cover* the text *below* the *stop sign*. As a cover sheet, use either a sticky note or a folded sheet of paper.
   - In the space provided in the text or in your problem notebook, write your answer to the question (**Q.**) that is above the stop sign. Then move your cover sheet down to the next stop sign and check your answer. If you need a hint, read a *part* of the answer, then re-cover the answer and try the problem again.

2. *First* **learn the rules,** *then* **do the Practice.** The goal of learning is to move rules and concepts into *memory.* To begin, when working Try It questions, you may look back at the rules, but make an effort to commit the rules to memory before starting the Practice sets.

   Answers to the Practice problems are at the end of each chapter. If you need a hint, read a part of the answer and try again.

3. **How many Practice problems should you do?** It depends on your background. These lessons are intended to

   - Refresh your memory on topics you once knew, and
   - Fill in the gaps for topics that are less familiar

   If a topic is familiar, read the lesson for reminders and review, then do a *few* problems in each Practice set. Be sure to do the last problem (usually the most challenging).

   If a topic is unfamiliar, do more problems.

4. **Work Practice problems at least three days a week.** Chemistry is cumulative. What you learn in initial topics you will need in memory later. To retain what you learn, *space* your study of a topic out over several days.

   Begin lessons on new topics early, preferably before the topic is covered in a lecture.

5. **Memorize what must be memorized.** Use flashcards and other memory aids.

The key to success in chemistry is to *study* rules and concepts and *practice* solving problems at a *steady* pace.

About the Authors

Donald J. Dahm has retired from Rowan University after teaching first-year chemistry for 15 years. He was selected in 2011 by students as the professor they most wanted to give "The Last Lecture," an annual event based on the book by Randy Pausch. In 2014, Dr. Dahm received the Gerald S. Birth award from the Council on Near-Infrared Spectroscopy for the best recently published work. Don's most recent book, with his son Kevin, is *Interpreting Diffuse Reflectance and Transmittance: A Theoretical Introduction to Absorption Spectroscopy of Scattering Materials.*

Eric A. Nelson's career spanned work as a chemistry instructor and as an elected officer and legislative representative for his faculty organization. His professional interests include research on reading comprehension in the sciences and measures of student mathematics achievement in preparation for study in chemistry. Prior to his retirement, he served on committees at the state and regional levels working to raise academic standards and encourage the adoption of research-based curricula.

1

Numbers in Scientific Calculations

Lesson 1.1   Learning Chemistry

What does science tell us about the most effective ways to study?

1. **In the sciences, the goal of learning is to solve problems.** Your brain solves problems using information from your environment and from your memory. The brain contains different types of memory, including

   * **working memory (WM),** where you think by processing information; and

   * **long-term memory (LTM),** where the brain stores information you have learned.

2. **Problem solving depends on well-memorized information.** A key discovery of recent research in cognitive science is that working memory can hold and process essentially *unlimited* knowledge that one is able to recall quickly from long-term memory but can hold and process only a few small elements that have *not* previously been memorized. During initial learning, a primary goal is to move new knowledge into LTM so that it can then be linked to other knowledge.

3. **Reliably moving knowledge into LTM** requires *repeated thought* about the meaning of new information, *effort* at recalling new facts, and *practice* in applying new skills.

4. **"Automaticity" in recall of fundamentals** is the central strategy to overcome limitations in working memory. When knowledge can be recalled quickly and accurately, more space in WM is available both for processing and for noting the associations within knowledge that build conceptual understanding.

5. **Memorizing standard algorithms** (stepwise procedures) is another way around the "processing bottleneck" in WM.

6. **Concepts are crucial.** Your brain constructs "conceptual frameworks" to judge when information should be recalled. Frameworks are built by practice in applying new knowledge to a variety of problems.

7. **"You can always look it up" is a poor strategy for problem solving.** The more information you must stop to look up, the more difficult it is for WM to manage the steps needed to solve a complex problem.

Moving Knowledge into Memory

How can you promote the retention of needed fundamentals in LTM? The following strategies are recommended by cognitive scientists:

1. **Learn incrementally** (in small pieces). The brain is limited in how much new memory it can construct in a short amount of time. In learning, *steady* wins the race.

2. **Overlearn.** If you practice recalling new information only one time, it will tend to remain in memory for only a few days. *Repeated* practice to perfection (called *overlearning*) builds reliable recall.

3. **Space your learning.** To *retain* what you learn, 20 minutes of study spaced over 3 days ("distributed practice") is more effective than 1 hour of study for 1 day. Study by "massed practice" (cramming) tends not to "stick" in LTM.

4. **Focus on core skills.** The facts and processes you should practice most often are those needed most often in a discipline.

5. **Effort counts.** Experts in a field usually attribute their success to "hard work over an extended period of time" rather than talent.

6. **Self-testing builds memory.** Practicing recall (such as by use of flashcards) and *then* solving problems is more effective than highlighting or re-reading notes or texts.

7. **Use parallel processing.** To remember new knowledge, listen, observe, recite, write, and practice recall. Your brain stores multiple *types* of memory. Multiple cues help you to recall steps and facts needed to solve a problem.

8. **Get a good night's sleep.** While you sleep, your brain reviews the experience of your day to decide what to store in LTM. Sleep promotes retention of what you learn.

For more on the science of learning, see *Cognition: The Thinking Animal* by Daniel Willingham (Prentice Hall, 2007) and *Make It Stick: The Science of Successful*

Learning by Peter C. Brown, Henry L. Roediger III, and Mark A. McDaniel (Harvard University Press, 2014).

PRACTICE

Answer these questions in your notebook:

1. What is "overlearning"?

2. When is "working memory" limited?

3. Which better promotes long-term learning: "massed" or "distributed" practice?

Lesson 1.2   Numeracy

Before you begin this lesson, read "Note to Students" on page xvii.

Mental Arithmetic

Research has found that one of the best predictors of success in first-year college chemistry is the ability to solve simple mathematical calculations without a calculator. In part, this is because if you can recall "math facts" quickly from memory, the many relationships in science that are based on simple whole-number ratios make sense. In addition, speed matters in working memory. Fast recall of fundamentals leaves space for attention to the "science side" of demonstrations and calculations.

PRACTICE   A

Find a device that measures how long it takes to complete a task in seconds (such as a stopwatch or digital timer on a phone or computer). To learn to use the device, *practice* until you can reliably time "counting to 30 in your head." Then:

1. Time how long it takes, in seconds, to write answers to these eight **addition** problems. Go as *fast* as you can with accuracy.

$7 + 4 =$	$5 + 9 =$	$3 + 7 =$	$9 + 7 =$
$5 + 8 =$	$7 + 5 =$	$8 + 7 =$	$11 + 6 =$

Record your time: _____ seconds

2. Time how long it takes to write answers to these **subtraction** problems as fast as you can.

$9 - 4 =$	$12 - 9 =$	$11 - 5 =$	$10 - 3 =$
$12 - 7 =$	$9 - 5 =$	$12 - 3 =$	$15 - 4 =$

Record your time: _____ seconds

(continued)

3. Time answering these **multiplication** problems as fast as you can.

$8 \times 4 =$ $8 \times 9 =$ $3 \times 9 =$ $9 \times 5 =$

$6 \times 6 =$ $4 \times 3 =$ $8 \times 3 =$ $5 \times 8 =$

Record your time: _____ seconds

4. Time answering these **division** problems as fast as you can.

$24/4 =$ $36/9 =$ $56/8 =$ $45/9 =$

$42/6 =$ $49/7 =$ $60/12 =$ $40/5 =$

Record your time: _____ seconds

Analysis

Check your answers. Reading from left to right across the top row, then the bottom:

Addition: 11, 14, 10, 16, 13, 12, 15, 17

Subtraction: 5, 3, 6, 7, 5, 4, 9, 11

Multiplication: 32, 72, 27, 45, 36, 12, 24, 40

Division: 6, 4, 7, 5, 7, 7, 5, 8

With fast and accurate recall, you should be able to complete each numbered question above with 100% accuracy in 24 seconds. If you did so for all four questions, great!

If you missed that goal for one or more questions, search online for a "mental arithmetic flashcard" application for your smartphone or computer. Practice each "rusty" skill for 5–10 minutes a day until you can accurately solve at "3 seconds per problem" (20 correct in 1 minute).

> In your head, by *quick recall* not calculation, without a calculator or pencil, you must
> - be able to add and subtract numbers from 1 through 20, and
> - know your times tables (and corresponding division facts) through 12.

With practice, you will achieve the "automaticity" in math facts that is an essential foundation for work in the sciences.

Staying in Practice

The goal of study is learning that lasts, but we know that some of what we study is forgotten. What study practices build *long-term* memory?

Science has found that when we first learn something new, we best remember what we *practice recalling* for several days in a week, then again practice recalling about a week later, about a month later, and occasionally thereafter. This schedule for study seems to convey to the brain which new knowledge it especially needs to remember.

In these lessons, to keep your "math facts" sharp, we will ask you to occasionally perform simple multiplication and division without a calculator. There are several "standard algorithms" that can be used for multiplication and division, but to avoid confusion, research recommends that you learn one, then practice until its application becomes "automatic."

The multiplication algorithm usually taught in the United States includes these steps:

For $76 \times 42 = ?$

Step 1:	$^{1}76$	Steps 2 and 3:
	$\times\ 42$	
	152	

$$^{2}76$$
$$\times\ 42$$
$$152$$
$$\underline{304}\quad \leftarrow \text{(a } \mathbf{0} \text{ after the 4 is an option)}$$
$$\mathbf{3192}$$

The "long division" algorithm usually taught in the United States includes these steps:

For $2048 \div 8 = ?$

$$\begin{array}{r} 2 \\ 8\overline{)2048} \\ -16 \\ \hline \mathbf{44} \end{array}$$

$$\begin{array}{r} 256 \qquad = \mathbf{256}\\ 8\overline{)2048} \\ -16 \\ \hline 44 \\ -40 \\ \hline 48 \\ -48 \end{array}$$

In these lessons, using pencil and paper, you will be required to
- multiply two digits by two digits, and
- solve "long division" of an evenly divisible four-digit number by a single digit.

PRACTICE B

Complete the problems that follow using a standard algorithm you have practiced. If no standard algorithm is familiar, use the procedures given earlier. For additional help, search online for a "multiplication algorithm" or "long division algorithm" video.

Do one part of each problem if this is easy review but more if you need practice. Check answers as you go at the end of this chapter.

1. *Without* a calculator, working in your problem notebook, multiply these:

a.	93	b.	64	c.	49
	$\times\ 75$		$\times\ 82$		$\times\ 38$

(continued)

2. *Without* a calculator, in your notebook, solve these "evenly divisible" problems (answers will be a multidigit whole number—no decimals or remainders):

   a. 6)252   b. 9)801   c. 8)448

Decimal Equivalents

To divide 492 by 7.36, a calculator is useful, but when working in science, health care, or engineering, every calculation must be checked because every effort must be made to avoid errors.

One way to check your calculator use is estimation using mental math. Fast recall of the following **decimal equivalents** will help.

$1/2 = 0.50$	$1/3 \approx 0.33$	$1/4 = 0.25$	$1/5 = 0.20$
$2/3 \approx 0.67$	$3/4 = 0.75$	$1/8 = 0.125$	$3/2 = 1.5$

The squiggly $\approx$ sign means "approximately equals."

PRACTICE C

To practice conversion of fractions to decimal equivalents, try the following.

• On a sheet of paper, draw five columns and eight rows. List the fractions down the middle column.

		1/2		
		1/3		
		1/4		
		1/5		
		2/3		
		3/4		
		1/8		
		3/2		

• From memory, try writing the decimal equivalents of the fractions in the far right column. Then check your answers.

• Fold over those answers and repeat at the far left. Fold over those and repeat.

Using Mental Math to Simplify Fractions

Scientific calculations often involve fractions with several numbers. During estimation, these fractions can be simplified "on paper, with pencil," as a check on calculator use. Follow the steps across in this example, as "pencil arithmetic" is used to simplify the fraction.

$$\frac{8 \times 21 \times 3}{72} \quad \frac{8 \times 21 \times 3}{72_9} \quad \frac{8 \times 21 \times 3}{72_{9}{}_3} = \frac{21}{3} = 7$$

Other steps to simplify the arithmetic would be equally valid. In a problem with several cancellations, you may want to rewrite the fraction to summarize your progress, as in the "next to last" step above, to keep your work clear.

PRACTICE D

Without a calculator, use your mental math skills to reduce these fractions. Show your work on this paper. Hint: It usually helps to try to reduce larger numbers first.

1. Simplify each of the following to a one- or two-digit whole number:

    a. $\dfrac{56 \times 2 \times 3}{4 \times 7} =$

    b. $\dfrac{20 \times 4 \times 45}{8 \times 9} =$

    c. $\dfrac{42 \times 24 \times 5}{2 \times 6 \times 7} =$

    d. $\dfrac{16 \times 12 \times 7}{2 \times 96} =$

2. Convert each of the following to a decimal equivalent in the form **0.XXX**:

    a. $\dfrac{27 \times 4 \times 7}{6 \times 42 \times 9} =$

    b. $\dfrac{5 \times 3 \times 16}{8 \times 2 \times 30} =$

    c. $\dfrac{3 \times 2 \times 11}{22 \times 24} =$

    d. $\dfrac{8 \times 3 \times 63}{9 \times 7 \times 96} =$

Fixed Decimal Notation and Exponential Notation

In science, we often deal with very large and very small numbers. For example:

- A drop of water contains about **1,500,000,000,000,000,000,000** molecules.
- An atom of gold has a mass of **0.000 000 000 000 000 000 000 327** gram.

Values expressed as "regular numbers," such as 153 or 0.0024 or the two numbers above, are said to be in **fixed decimal notation** (also called **fixed notation**).

In science, very large and very small numbers are most often expressed in **base 10 exponential notation**: as a *number* multiplied by **10** to an *integer* (positive or negative whole number) power. In chemistry, unless otherwise noted, you should assume that "exponential notation" means *base 10* exponential notation.

For the measurements above, in exponential notation we can write:

- A drop of water contains about 1.5×10^{21} molecules.
- An atom of gold has a mass of 3.27×10^{-22} gram.

Values expressed in exponential notation can be described as having three parts.

Example: In -6.5×10^{-4}

- The $-$ in front is the **sign**.
- The **6.5** is a fixed decimal number that has a variety of names. In this text, we will call it the **significand**.
- The 10^{-4} is the **exponential** term: the **base** is **10** and the **exponent** (also called the **power**) is **−4**.

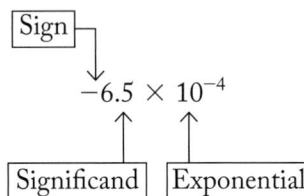

You will need to be able to label the parts of values in exponential notation using those six terms.

PRACTICE   E

1. Express the value of the fraction **1/4** in fixed decimal notation. _____

2. Circle the significand in 6.02×10^{23}

Lesson 1.3   Moving the Decimal

Powers of 10

Below are the numbers that correspond to powers of 10. Note the relationship between the exponent, the number of zeros, and the position of the decimal point in the fixed decimal numbers as you go down this sequence.

$$10^6 = 1,000,000 = 10 \times 10 \times 10 \times 10 \times 10 \times 10$$

$$10^3 = 1000 = 10 \times 10 \times 10$$

$$10^2 = 100 = 10 \times 10$$

$$10^1 = 10$$

$$10^0 = 1 \qquad \textit{Any positive number to the zero power equals one.}$$

$$10^{-1} = 0.1$$

$$10^{-2} = 0.01 = 1/10^2 = 1/100$$

$$10^{-3} = 0.001$$

When converting from powers of 10 to fixed decimal numbers, use these steps:

1. To convert a *positive* power of 10 to a fixed decimal number, write 1, then move the decimal point to the *right* by the number of places in the exponent.

Example: $10^2 = $ **100**

2. To convert a *negative* power of 10 to a fixed decimal number, write 1, then move the decimal to the *left* by the number of places equal to the number after the negative sign in the exponent.

▶ **TRY IT**

(See "How to Use These Lessons," point 1, p. xvii.)

Q. Convert to fixed decimal notation: $10^{-2} = $

Answer:

$$10^{-2} = \textbf{0.01}$$

PRACTICE   A

Convert these values to fixed decimal notation. Write your answers in the spaces below. Check answers at the end of this chapter.

1. $10^4 = $ 2. $10^{-4} = $ 3. $10^7 = $

4. $10^{-5} = $ 5. $10^0 = $

Multiplying and Dividing by 10, 100, and 1000

To multiply or divide by fixed decimal numbers that are positive whole-number powers of 10 (such as 100 or 10,000), use these rules:

1. When *multiplying* a fixed decimal number by 10, 100, 1000, and so forth, move the number's decimal to the *right* by the number of zeros in the 10, 100, or 1000.

Examples: $72 \times 100 = $ **7,200** $-0.0624 \times 1000 = $ **−62.4**

2. When *dividing* a number by 10, 100, 1000, and so forth, move the number's decimal to the *left* by the number of zeros in the 10, 100, or 1000.

TRY IT

Q. Write answers to these operations as fixed decimal numbers:

  a. 34.6/1000 =   b. −0.47/100 =

Answers:

  a. 34.6/1000 = **0.0346**   b. −0.47/100 = **−0.0047**

---

3. When writing a fixed decimal number between −1 and 1 (a number that "begins with a decimal point"), always place a *zero* in front of the decimal point.

    Example: Do not write .42 or −.74; do write 0.42 or −0.74.

During your written calculations, the zero in front helps in seeing your decimals.

PRACTICE **B**

1. When dividing by 1000, move the decimal to the _____ by _____ places.

2. Write answers to these operations in fixed decimal notation:

    a. 0.42 × 1000 =   b. 63/100 =   c. −74.6/10,000 =

Converting Exponential Notation to Fixed Decimal Notation

When working with numbers in science, we often need to convert between fixed decimal and exponential notation or to move the decimal in the significand of exponential notation.

A rule applies in all of these conversions: The sign in *front* never changes. The sign identifies whether the *value* is positive or negative. When moving the decimal, the sign of the *exponential* term *may* change, but positive values remain positive and negative values remain negative.

To convert from exponential notation (such as -4×10^3) to fixed decimal notation ($-4{,}000$), use these rules:

1. The sign in front never changes.

2. If the significand is multiplied by a *positive* power of 10, move the decimal point in the significand to the *right* by the number of places equal to the value of the exponent.

    Examples: $2 \times 10^2 = \textbf{200}$; $-0.0033 \times 10^3 = \textbf{−3.3}$

3. If the significand is multiplied by a *negative* power of 10, move the decimal point in the significand to the *left* by the number of places equal to the number *after* the negative sign in the exponent.

   Examples: $2 \times 10^{-2} = \mathbf{0.02}$; $-7,653.8 \times 10^{-3} = \mathbf{-7.6538}$

PRACTICE **C**

Convert these values to fixed decimal numbers:

1. $3 \times 10^3 =$

2. $5.5 \times 10^{-4} =$

3. $0.77 \times 10^6 =$

4. $-95 \times 10^{-4} =$

Converting Exponential Notation to Scientific Notation

In chemistry, when we work with very large or very small numbers, it is preferable to write the numbers in **scientific notation**, which is a special form of exponential notation that makes values easier to compare. There are many equivalent ways to write a value in exponential notation, but there is only one correct way to write a value in scientific notation.

The general rule is:

To convert a value from exponential notation to *scientific* notation, move the decimal so that the significand is a single number from 1 to 9, then adjust the power of 10 to keep the value the same.

Another way to say this is:

To convert from exponential notation to *scientific* notation,

- move the decimal point in the significand to be *after* the first digit that is *not* a zero, then

- adjust the exponent to keep the same numeric value.

To apply these rules, the specific steps are:

1. Do not change the + or − *sign* in *front* of the significand.

2. When moving the decimal *n* times to make the significand *larger*, make the power of 10 *smaller* by a count of *n*.

Example: Convert from exponential notation to scientific notation:

$$0.045 \times 10^5 = 4.5 \times 10^3$$

The decimal must be after the 4. Move the decimal two places to the right. This makes the significand 100 times larger. To keep the same numeric value, lower the exponent by 2, making the 10^x value 100 times smaller.

TRY IT

Q. Convert this value to scientific notation:

$$-0.0057 \times 10^{-2} =$$

Answer:

The value *must* be written as -5.7×10^{-5}

To convert to scientific notation, the decimal must be moved to be after the first number that is not a zero: the 5. That move of three places makes the significand 1000 times larger. To keep the same value, make the exponential term 1000 times smaller.

The logic of the math is:

$$-0.0057 \times 10^{-2} = -[(5.7 \times 10^{-3}) \times 10^{-2}] = \mathbf{-5.7 \times 10^{-5}}$$

---

3. When moving the decimal *n* times to make the significand *smaller*, make the power of 10 *larger* by a count of *n*.

TRY IT

Q. Convert this value to scientific notation:

$$-8{,}544 \times 10^{-7} =$$

Answer:

$$-8{,}544 \times 10^{-7} = \mathbf{-8.544 \times 10^{-4}}$$

In scientific notation, the decimal must be after the 8, so you must move the decimal three places to the left. This makes the significand 1000 times smaller. To keep the same numeric value, increase the exponent by 3, making the 10^x value 1000 times larger.

Remember, 10^{-4} is 1000 times larger than 10^{-7}.

The general rule is:

If you move the decimal *n* places, change the exponent by a count of *n*.

When you move a decimal, it is helpful to *recite*, for the significand and the exponential term *after* the sign in front:

If one gets *smaller*, the other gets *larger*. If one gets *larger*, the other gets *smaller*.

The Role of Practice

In the **Practice** sets, do as many problems as you need to feel "quiz ready."

- If the material in a lesson is easy review, do the *last* problem in each series of similar problems (which is usually the most challenging).
- If the lesson is not as easy, put a check mark (✓) by every other problem, then work that half of the problem set. If you miss one, do additional problems in the set.
- Save a *few* problems for your next study session or for later review.

During **Examples** and **Try Its**, you *may* look back at the rules, but write and practice recalling new rules from memory *before* starting the Practice set.

If you use Practice sets to learn the rules, it will be difficult to find time for all of the problems you will need to do. If you use Practice sets to *apply* rules that are in memory, you will need to solve fewer problems to be "quiz ready."

PRACTICE   D

Convert these values to scientific notation:

1. $5{,}423 \times 10^3 =$

2. $0.0067 \times 10^{-4} =$

3. $0.024 \times 10^3 =$

4. $-877 \times 10^{-4} =$

5. $0.00492 \times 10^{-12} =$

6. $-602 \times 10^{21} =$

Converting Fixed Decimal Notation to Scientific Notation

To convert fixed decimal numbers (that is, regular numbers) to *exponential* notation or *scientific* notation, we apply the following math rules:

- Any positive number to the zero power equals one: $2^0 = 1; 42^0 = 1$.
  Exponential notation most often uses $10^0 = 1$.
- Because any number can be multiplied by 1 without changing its value, any number can be multiplied by 10^0 without changing its value.

   Example:

   $$42 = 42 \times 1 = 42 \times 10^0 \text{ in } \textit{exponential} \text{ notation}$$

   $$= 4.2 \times 10^1 \text{ in } \textit{scientific} \text{ notation}$$

To convert fixed notation to *scientific* notation, the steps are:

1. Add "$\times 10^0$" after the number.

2. Apply the steps that convert exponential notation to scientific notation:
   - Do not change the sign in *front*.
   - Move the decimal in the significand to be *after* the *first* digit that is not a zero.
   - Adjust the power of 10 to compensate for moving the decimal.

➤ TRY IT

Q. Convert these fixed decimal numbers to scientific notation:

a. 943 =   b. −0.00036 =

Answers:

a. $943 = 943 \times 10^0 = \mathbf{9.43 \times 10^2}$ b. $-0.00036 = -0.00036 \times 10^0 = \mathbf{-3.6 \times 10^{-4}}$

When converting between fixed notation and scientific notation, the sign in front never changes. For the terms *after* the sign in front:

- When converted to scientific notation, a fixed decimal number with a value *larger than one* will have a *positive* whole-number power of 10 (a power zero or greater).
- When converted to scientific notation, a fixed decimal number with a value *between zero and one* (such as 0.25) will have a *negative* whole-number power of 10.
- Comparing the position of the decimal in a fixed decimal number and in the significand of the value written in scientific notation, the number of *places* that the decimal in a number moves is the *number after* the sign of the scientific notation *exponent*.

Note how these rules apply to the two parts of the Try It above.

Note that in both exponential notation and scientific notation, whether the sign in *front* is positive or negative has no relation to the sign of the *exponential* term. The sign in front states whether the value is positive or negative. The sign of the exponential term indicates how many places the decimal moved in converting from fixed notation to exponential or scientific notation.

PRACTICE **E**

1. Which of these values is written in scientific notation? _____

   a. 74   b. -14.7×10^{-24} c. -9.6×10^{14} d. 77×10^0

2. Which lettered parts in problem 3 below must have a negative power of 10 when written in scientific notation?

3. Convert these values to scientific notation:

   a. 6,280 =   b. 0.0093 =

   c. 0.741 =   d. −1,280,000 =

To summarize this lesson in your problem notebook:

- In your own words, list the shaded rules that were unfamiliar or that you found helpful.

- Then, write and recite your rules until you can write them from memory.

PRACTICE **F**

Check (✓) and do every *other* letter. If you miss one, do another letter for that set.

1. Write answers in fixed decimal notation for these operations:

   a. 924/10,000 =   b. 24.3×1000 =

   c. −0.024/10 =

2. Convert these values to scientific notation:

   a. 0.55×10^5 =   b. 0.0092×100 =

   c. -943×10^{-6} =   d. 0.00032×10^1 =

3. Convert these fixed decimal numbers to scientific notation:

   a. 7,700 =   b. 160,000,000 =

   c. 0.023 =   d. −0.00067 =

Lesson 1.4   Calculations Using Exponential Notation

> **PRETEST** If you can answer these two questions correctly, you may skip to Lesson 1.5. Otherwise, complete Lesson 1.4. Answers are at the end of this chapter. Do *not* use a calculator. Convert final answers to scientific notation.
>
> 1. $(2.0 \times 10^{-4}) (6.0 \times 10^{23}) =$
>
> 2. $\dfrac{10^{23}}{(100) (3.0 \times 10^{-8})} =$

Multiplying and Dividing Powers of 10

The following rules should be recited until they can be recalled from memory. Rules that apply to exponential terms include:

1. When you *multiply* exponentials, you *add* the exponents.

    Examples: $10^3 \times 10^2 = 10^5$ $10^{-5} \times 10^{-2} = 10^{-7}$ $10^{-3} \times 10^5 = 10^2$

2. When you *divide* exponentials, you *subtract* the exponents.

    Examples: $10^3/10^2 = 10^1$ $10^{-5}/10^2 = 10^{-7}$ $10^{-5}/10^{-2} = 10^{-3}$

    When subtracting, remember: Minus a minus is a plus.

    Example: $10^{6-(-3)} = 10^{6+3} = 10^9$

3. When you take the reciprocal of an exponential, change the exponent's sign.

    This rule is often remembered as:

    When you take an exponential term from the bottom to the top, change the exponent's sign.

    Example:

    $$\frac{1}{10^3} = 10^{-3}; \ 1/10^{-5} = 10^5$$

    Why does this work? By rule 2:

    $$\frac{1}{10^3} = \frac{10^0}{10^3} = 10^{0-3} = 10^{-3}$$

In later lessons, we will practice additional rules for exponential notation.

Finally, when simplifying a fraction that includes two or more exponential terms, it generally helps if you simplify the top and bottom exponential terms *separately* in the first step, then divide the terms in the next step.

Example:

$$\frac{10^{-2} \times 10^7}{10^2 \times 10^{-5}} = \frac{10^5}{10^{-3}} = 10^{5-(-3)} = 10^{5+3} = \mathbf{10^8}$$

► **TRY IT**

Q. Without using a calculator, write the simplified top, then the simplified bottom, and then divide:

$$\frac{10^{-3} \times 10^{-4}}{10^5 \times 10^{-8}} = \underline{\qquad} =$$

STOP

Answer:

$$\frac{10^{-3} \times 10^{-4}}{10^5 \times 10^{-8}} = \frac{\mathbf{10^{-7}}}{\mathbf{10^{-3}}} = 10^{-7-(-3)} = 10^{-7+3} = \mathbf{10^{-4}}$$

PRACTICE   **A**

Write these answers as 10 to a power. Work on this page if possible. Do *not* use a calculator.

1. $10^6 \times 10^2 =$

2. $10^{-5} \times 10^{-6} =$

3. $\dfrac{10^{-5}}{10^{-4}} =$

4. $\dfrac{10^{-3}}{10^5} =$

5. $\dfrac{1}{1/10^4} =$

6. $1/10^{23} =$

7. $\dfrac{10^3 \times 10^{-5}}{10^{-2} \times 10^{-4}} =$

8. $\dfrac{10^5 \times 10^{23}}{10^{-1} \times 10^{-6}} =$

9. $\dfrac{100 \times 10^{-2}}{1000 \times 10^6} =$

10. $\dfrac{10^{-3} \times 10^{23}}{10 \times 1000} =$

Multiplying and Dividing in Exponential Notation

To check calculator results, we need to be able to estimate answers (without using a calculator) for calculations that include exponential notation. In doing so, the following is the rule we use most often.

When multiplying and dividing in calculations that include exponential notation:

Handle the math of fixed decimals and exponential terms separately.

To do so:

- Do the math for the fixed decimal numbers (including significands) using number rules.
- Then, separately simplify the exponential terms using exponential rules.
- Finally, combine the two parts.

► TRY IT

Apply the rule to the following three problems.

Q1. Do not use a calculator: $(2 \times 10^3)(4 \times 10^{23}) =$

STOP

Answer:

For significands, use number rules: 2 multiplied by 4 is **8**

For exponentials, use exponential rules: $10^3 \times 10^{23} = 10^{3+23} = \mathbf{10^{26}}$

Then, combine the two parts: $(2 \times 10^3)(4 \times 10^{23}) = \mathbf{8 \times 10^{26}}$

Q2. Do the significand math on a calculator but the exponential math in your head for:

$$(2.4 \times 10^{-3})(3.5 \times 10^{23}) =$$

(We will review *how much* to round answers in Chapter 3. Until then, in your answers, round numbers and significands to *two* digits unless otherwise noted.)

STOP

Answer:

Handle significands and exponents separately.

Use a calculator for: $2.4 \times 3.5 = \mathbf{8.4}$

Do the exponentials in your head: $10^{-3} \times 10^{23} = \mathbf{10^{20}}$

Then, combine: $(2.4 \times 10^{-3})(3.5 \times 10^{23}) = \mathbf{8.4 \times 10^{20}}$

Q3. Do the significand math on a calculator but the exponential math without a calculator.

$$\frac{6.5 \times 10^{23}}{4.1 \times 10^{-8}} =$$

STOP

Answer:

$$\frac{6.5 \times 10^{23}}{4.1 \times 10^{-8}} = \frac{6.5}{4.1} \times \frac{10^{23}}{10^{-8}} = 1.585 \times \left[10^{23-(-8)}\right] = \mathbf{1.6 \times 10^{31}}$$

Here's one more rule that will help with problem solving:

> When dividing, if an exponential term on top does not have a significand, place a "1×" in front of the exponential so that the number–number division is clear.

► TRY IT

Q. Apply the rule to the following problem, then write the final answer in scientific notation. Do not use a calculator.

$$\frac{10^{-14}}{2.0 \times 10^{-8}} =$$

STOP

Answer:

$$\frac{10^{-14}}{2.0 \times 10^{-8}} = \frac{\mathbf{1} \times 10^{-14}}{2.0 \times 10^{-8}} = 0.50 \times 10^{-6} = \mathbf{5.0 \times 10^{-7}}$$

PRACTICE   B

In your own words, summarize any unfamiliar rules in this lesson, then apply them from memory to these problems. If more room is needed for careful work, solve in your notebook. Do the odd-numbered problems first, then the evens if you need more practice. Try these *first* without a calculator, then check your mental arithmetic with a calculator if needed. Write final answers in scientific notation, rounding significands to two digits.

1. $(2.0 \times 10^1)(6.0 \times 10^{23}) =$

2. $(4.0 \times 10^{-3})(1.5 \times 10^{15}) =$

3. $\dfrac{3.0 \times 10^{-21}}{-2.0 \times 10^3} =$

4. $\dfrac{6.0 \times 10^{-23}}{2.0 \times 10^{-4}} =$

5. $\dfrac{10^{-14}}{-5.0 \times 10^{-3}} =$

6. $\dfrac{10^{14}}{4.0 \times 10^{-4}} =$

7. Complete the two problems in the Pretest at the beginning of this lesson.

In your problem notebook, write a list of rules in Lesson 1.4 that were unfamiliar, need reinforcement, or were helpful to you. Write and recite your list until you can write all of the points from memory. Then, do the problems in Practice C.

PRACTICE   C

Start by doing every *other* letter. If you get those right, go to the next number. If not, do a few more of that number. Save a few parts for your next study session.

1. Try these *without* a calculator. Convert your final answers to scientific notation.

   a. $3 \times (6.0 \times 10^{23}) =$

   b. $1/2 \times (6.0 \times 10^{23}) =$

   c. $0.70 \times (6.0 \times 10^{23}) =$

   d. $10^3 \times (6.0 \times 10^{23}) =$

   e. $(-0.5 \times 10^{-2})(6.0 \times 10^{23}) =$

(continued)

f. $\dfrac{1}{10^{12}} =$

g. $1/(1/10^{-9}) =$

h. $\dfrac{2.0 \times 10^{18}}{6.0 \times 10^{23}} =$

i. $\dfrac{10^{-14}}{4.0 \times 10^{-5}} =$

2. Use a calculator for the fixed decimal math but not for the exponents. Write final answers in scientific notation.

a. $\dfrac{2.46 \times 10^{19}}{6.0 \times 10^{23}} =$

b. $\dfrac{10^{-14}}{0.0072} =$

3. Try these *without* a calculator. Write answers as a power of 10.

a. $\dfrac{10^{7} \times 10^{-2}}{10 \times 10^{-5}} =$

b. $\dfrac{10^{-23} \times 10^{-5}}{10^{-5} \times 100} =$

Lesson 1.5   Estimation and Exponential Calculations

> **PRETEST**   If you can solve *both* of the following problems correctly, skip this lesson. Convert your final answers to scientific notation. Check answers at the end of this chapter.
>
> 1. Solve this problem *without* a calculator:
>
> $$\dfrac{(10^{-9})(10^{15})}{(4 \times 10^{-4})(2 \times 10^{-2})} =$$
>
> 2. For this problem, use a calculator as needed:
>
> $$\dfrac{(3.15 \times 10^{3})(4.0 \times 10^{-24})}{(2.6 \times 10^{-2})(5.5 \times 10^{-5})} =$$

Choosing a Calculator

If you have not already done so, please read "Note to Students" on page xvii.

Complex Calculations

The prior lessons covered the fundamental rules for exponential notation. For longer calculations, the rules are the same. The challenges are keeping track of the numbers and using the calculator correctly. The steps below will help you to simplify complex calculations and quickly *check* your answers.

> **TRY IT**

Q. Let's try the following calculation two ways.

$$\frac{(7.4 \times 10^{-2})(6.02 \times 10^{23})}{(2.6 \times 10^{3})(5.5 \times 10^{-5})} =$$

Method 1: Do Numbers and Exponents Separately Work the calculation using the following steps:

1. **Do the numbers on the calculator.** Ignoring the exponentials, use the calculator to multiply all of the *significands* on top. Write the result. Then, multiply all the significands on the bottom, and write the result. Divide, write your answer, round to two digits for your final answer, then check below.

$$\frac{7.4 \times 6.02}{2.6 \times 5.5} = \frac{44.55}{14.3} = 3.1152 = \textbf{3.1}$$

2. **Then handle the exponents.** Starting from the original problem, look only at the powers of 10. Solve the exponential math using pencil and paper as needed, but *without* a calculator. Simplify the top, then the bottom, then divide, and write a single exponential term as your answer.

$$\frac{10^{-2} \times 10^{23}}{10^{3} \times 10^{-5}} = \frac{10^{21}}{10^{-2}} = 10^{21-(-2)} = \textbf{10}^{\textbf{23}}$$

3. **Now combine** the significand and exponential, and write the final answer.

Answer:

$$\textbf{3.1} \times \textbf{10}^{\textbf{23}}$$

Note that by handling the numbers and exponents separately, you did *not* need to enter the exponents into your calculator. To multiply and divide the powers of 10, you simply add and subtract whole numbers.

> **TRY IT**

Q. Method 1 above is a simple way to solve exponential notation without having to enter lots of numbers into the calculator. A second option is to enter *all* of the numbers and operations into the calculator. If you want to try that approach, try method 2.

Method 2: All on the Calculator Starting from the original Try It problem, enter *all* of the numbers and exponents into your calculator. Different calculators use different forms of data entry, but your calculator manual (usually available online) can help. Write your final answer in scientific notation. Round the significand to two digits.

On most calculators, you will need to use an $\boxed{E}$ or $\boxed{EE}$ or $\boxed{EXP}$ or $\boxed{\wedge}$ key, rather than the *multiplication* key, to enter a "10 to a power" term.

Answer:

Your calculator answer, rounded, should be the same as with method 1:

$$3.1 \times 10^{23}$$

Note how your calculator *displays* the *exponential* term in answers. The exponent may be set apart at the right, sometimes with an **E** in front.

Which way was easier: "Numbers on the calculator, exponents on paper" or "all on the calculator"? "Doing the exponents in your head" is often easier—and helps to keep your mental math sharp.

During calculations, try to use mental arithmetic to solve the exponential math, but always convert final answers that have exponential terms to scientific notation.

Answers in Fixed Notation or Scientific Notation?

In chemistry, as a very general rule, numeric values ranging from 0.01 to 9,999 are expressed as fixed decimals, whereas values outside that range are reported in scientific notation. When writing final answers to problems, a "rule of thumb" is: If a fixed decimal answer has more than two zeros at the beginning or end, convert it to scientific notation.

Checking Calculator Results

Whenever a complex calculation is solved on a calculator, to check your calculator use you *must* do the calculation a *second* time using different steps.

In these lessons, you will learn "mental arithmetic estimation" as a quick way to check that an answer "makes sense."

▶ TRY IT

Q. Let's start with the calculation that we used in the first Try It section of this lesson:

$$\frac{(7.4 \times 10^{-2})\,(6.02 \times 10^{23})}{(2.6 \times 10^3)\,(5.5 \times 10^{-5})} =$$

Apply the following steps to the numbers above:

1. *Estimate* **the numbers answer first.** Ignoring the exponentials and using a pencil, write a *rounded* whole-number substitute for each significand on top. Then multiply all of the top significands, and write the result. *Round* the bottom significands to whole numbers, multiply them, and write the result. Then write a *rounded estimate* of the answer when you divide the top and bottom numbers.

STOP

Answer:

Your rounding might be

$$\frac{7 \times 6}{3 \times 6} = \frac{7}{3} \approx 2 \text{ (The } \approx \text{ sign means "approximately equals.")}$$

Your "pencil and paper" estimate needs to be fast but does not need to be exact. With practice, your skill will improve in doing these estimates "in your head."

    2. **Simplify the exponents.** Use pencil and paper to simplify the top exponential terms, then the bottom, then divide.

$$\frac{10^{-2} \times 10^{23}}{10^3 \times 10^{-5}} = \frac{10^{21}}{10^{-2}} = 10^{21-(-2)} = \mathbf{10^{23}}$$

    3. **Combine** the two parts. Compare this estimate to the answer found in the earlier Try It section. Are they close?

STOP

Answer:

The estimate is $\mathbf{2 \times 10^{23}}$. The answer with the calculator was $\mathbf{3.1 \times 10^{23}}$. Allowing for rounding, the two results are close.

If your fast, rounded, pencil-and-paper answer is *close* to the answer where all or part was done on the calculator, it is probable that your more precise calculator answer is correct. If the two answers are far apart, check your work.

On timed tests, you may want to do the more precise answer with the help of the calculator first, and then go back at the end, if time is available, and use rounded numbers and mental arithmetic as a check. When doing a calculation the second time, try not to look back at the first answer until after you write the estimate. If you look back, by the power of suggestion you will often arrive at the first answer whether it is correct or not.

> For complex operations on a calculator, check the answer by estimation using rounded numbers and mental arithmetic.

PRACTICE

Do problems 1–4 without a calculator. Convert final answers to scientific notation. Round the significand in the answer to two digits.

1. $\dfrac{4 \times 10^3}{(2.00)(3.0 \times 10^7)} =$

2. $\dfrac{1}{(4.0 \times 10^9)(2.0 \times 10^3)} =$

(continued)

3. $\dfrac{(3 \times 10^{-3})(8.0 \times 10^{-5})}{(6.0 \times 10^{11})(2.0 \times 10^{-3})} =$

4. $\dfrac{(3 \times 10^{-3})(3.0 \times 10^{-2})}{(9.0 \times 10^{-6})(2.0 \times 10^{1})} =$

Complete problems 5–8 below in your notebook as follows:

- First, write an *estimate*. Use mental math to solve exponents and rounded significands.

- Then, calculate a more precise answer. You may

  - use the "significands on calculator, exponents on paper" method; or

  - plug the entire calculation into the calculator; or

  - experiment to see which approach is best for you.

Convert both the estimate and the final answer to *scientific notation*. Round the significand in the answer to two digits. Use the calculator that you will be allowed to use on quizzes and tests.

To start, complete the even-numbered problems. If you need more practice, do the odds.

5. $\dfrac{(3.62 \times 10^{4})(6.3 \times 10^{-10})}{(4.2 \times 10^{-4})(9.8 \times 10^{-5})} =$

6. $\dfrac{10^{-2}}{(750)(2.8 \times 10^{-15})} =$

7. $\dfrac{(1.6 \times 10^{-3})(4.49 \times 10^{-5})}{(2.1 \times 10^{3})(8.2 \times 10^{6})} =$

8. $\dfrac{1}{(4.9 \times 10^{-2})(7.2 \times 10^{-5})} =$

9. For additional practice, do the two Pretest problems at the beginning of this lesson.

SUMMARY

To prepare for a quiz that includes the topics in this chapter:

1. Be able to summarize in your own words any shaded facts, relationships, and rules that are unfamiliar.

2. In your head, by *quick recall*, you need to be able to add and subtract numbers from 1 through 20 and to know your times tables through 12. If your mental math is at all rusty, try different math flashcard apps (or paper flashcards) for a 10-minute workout for several days a week.

3. Be able to solve the problems in the Review Quiz and the chapter.

REVIEW QUIZ

Solve the problems that follow. Do not use a calculator.

1. 87 multiplied by 94 (Solve in your notebook.)

2. 2,601 divided by 9

3. Simplify. Answer in fixed decimal numbers that have two non-zero digits.

   a. $\dfrac{42 \times 6 \times 8}{3 \times 7} =$

   b. $\dfrac{7 \times 12}{4 \times 28} =$

4. Convert these to scientific notation:

   a. $-0.0068 =$

   b. $8{,}920 \times 10^{-1} =$

5. Answer in scientific notation:

   a. $10^{-2} \times (6.0 \times 10^{23}) =$

   b. $\dfrac{3.0 \times 10^{24}}{6.0 \times 10^{23}} =$

   c. $\dfrac{10^{10}}{2.0 \times 10^{-5}} =$

   d. $\dfrac{3}{(1000)(9.0 \times 10^{-8})} =$

6. Answer in scientific notation:

   $$\dfrac{10^{23}}{(2.5 \times 10^{10})(2.0 \times 10^{-6})} =$$

ANSWERS

To make answer pages easy to locate, use a sticky note.

Lesson 1.1

1. Repeated practice to perfection.
2. When processing information not previously memorized.
3. Distributed (spaced) practice.

Lesson 1.2

Practice B   1a. 6,975   1b. 5,248   1c. 1,862   2a. 42   2b. 89   2c. 56

Practice D   1a. 12   1b. 50   1c. 60   1d. 7
2a. 0.333   2b. 0.500   2c. 0.125   2d. 0.250

Practice E   1. 0.25   2. $\boxed{6.02} \times 10^{23}$

Lesson 1.3

Practice A   1. 10,000   2. 0.0001   3. 10,000,000   4. 0.00001   5. 1

Practice B   1. When dividing by 1000, move the decimal to the **left** by **three** places.

2a. 420   2b. 0.63 (Must have a zero in front.)   2c. -0.00746

Practice C   1. 3,000   2. 0.00055   3. 770,000   4. -0.0095

Practice D   1. 5.422×10^6 2. 6.7×10^{-7} 3. 2.4×10^1 4. -8.77×10^{-2} 5. 4.92×10^{-15} 6. -6.02×10^{23}

Practice E   1. c. -9.6×10^{14} (Significand *between 1 and 10* followed by *exponential* term.)

2. b and c   3a. 6.28×10^3 3b. 9.3×10^{-3} 3c. 7.41×10^{-1} 3d. -1.28×10^6

Practice F   1a. 0.0924   1b. 24,300   1c. -0.0024 2a. 5.5×10^4 2b. 9.2×10^{-1} 2c. -9.43×10^{-4}

2d. 3.2×10^{-3} 3a. 7.7×10^3 3b. 1.6×10^8 3c. 2.3×10^{-2} 3d. -6.7×10^{-4}

Lesson 1.4

Pretest   1. 1.2×10^{20} 2. 3.3×10^{28}

Practice A   1. 10^8 2. 10^{-11} 3. 10^{-1} 4. 10^{-8} 5. 10^4 6. 10^{-23} 7. 10^4 8. 10^{35}

9. $\dfrac{100 \times 10^{-2}}{1,000 \times 10^6} = \dfrac{10^2 \times 10^{-2}}{10^3 \times 10^6} = \dfrac{10^0}{10^9} = \mathbf{10^{-9}}$ 10. $\dfrac{10^{-3} \times 10^{23}}{10 \times 1,000} = \dfrac{10^{20}}{10^1 \times 10^3} = \dfrac{10^{20}}{10^4} = \mathbf{10^{16}}$

(For problems 9 and 10, you may use different steps, but you must arrive at the same answer.)

Practice B   1. 1.2×10^{25} 2. 6.0×10^{12} 3. -1.5×10^{-24} 4. 3.0×10^{-19} 5. -2.0×10^{-12} 6. 2.5×10^{17}

7. See Pretest answers.

Practice C   1a. 1.8×10^{24} 1b. 3.0×10^{23} 1c. 4.2×10^{23} 1d. 6.0×10^{26} 1e. -3.0×10^{21} 1f. 1.0×10^{-12}

1g. 1.0×10^{-9} 1h. $\dfrac{2.0 \times 10^{18}}{6.0 \times 10^{23}} = 0.33 \times 10^{-5} = \mathbf{3.3 \times 10^{-6}}$

1i. $\dfrac{1.0 \times 10^{-14}}{4.0 \times 10^{-5}} = 0.25 \times 10^{-9} = \mathbf{2.5 \times 10^{-10}}$ 2a. $\dfrac{2.46 \times 10^{19}}{6.0 \times 10^{23}} = 0.41 \times 10^{-4} = \mathbf{4.1 \times 10^{-5}}$

2b. $\dfrac{10^{-14}}{0.0072} = \dfrac{1.0 \times 10^{-14}}{7.2 \times 10^{-3}} = \dfrac{1.0}{7.2} \times \dfrac{10^{-14}}{10^{-3}} = 0.14 \times 10^{-11} = \mathbf{1.4 \times 10^{-12}}$

3a. $\dfrac{10^7 \times 10^{-2}}{10^1 \times 10^{-5}} = \dfrac{10^5}{10^{-4}} = \mathbf{10^9}$ 3b. $\dfrac{10^{-23} \times \cancel{10^{-5}}}{\cancel{10^{-5}} \times 10^2} = \mathbf{10^{-25}}$

Lesson 1.5

Pretest   1. 1.25×10^{11} or 1.3×10^{11} 2. 8.8×10^{-15}

Practice   You may do the arithmetic using different steps than shown below, but you must get the same answer.

1. $\dfrac{4 \times 10^3}{(2.00)(3.0 \times 10^7)} = \dfrac{4}{6} \times 10^{3-7} = \dfrac{2}{3} \times 10^{-4} = \mathbf{0.67 \times 10^{-4}} = \mathbf{6.7 \times 10^{-5}}$

2. $\dfrac{1}{(4.0 \times 10^9)(2.0 \times 10^3)} = \dfrac{1}{8 \times 10^{12}} = \dfrac{1}{8} \times 10^{-12} = \mathbf{0.125 \times 10^{-12}} = \mathbf{1.3 \times 10^{-13}}$

3. $\dfrac{(3 \times 10^{-3})(8.0 \times 10^{-5})}{(6.0 \times 10^{11})(2.0 \times 10^{-3})} = \dfrac{8}{4} \times \dfrac{10^{-3-5}}{10^{11-3}} = 2 \times \dfrac{10^{-8}}{10^{8}} = 2 \times 10^{-8-8} = \mathbf{2.0 \times 10^{-16}}$

4. $\dfrac{(3 \times 10^{-3})(3.0 \times 10^{-2})}{(9.0 \times 10^{-6})(2.0 \times 10^{1})} = \dfrac{9}{18} \times \dfrac{10^{-3-2}}{10^{-6+1}} = 0.50 \times \dfrac{\cancel{10^{-5}}}{\cancel{10^{-5}}} = 0.50 = \mathbf{5.0 \times 10^{-1}}$

5. Final in scientific notation: $0.55 \times 10^{3} = \mathbf{5.5 \times 10^{2}}$

6. Estimate: $\dfrac{1}{7 \times 3} = \dfrac{1}{20} = \mathbf{0.05}$; $\dfrac{10^{-2}}{(10^{2})(10^{-15})} = 10^{-2-(-13)} = \mathbf{10^{11}}$

   Estimate in scientific notation: $0.05 \times 10^{11} = \mathbf{5 \times 10^{9}}$

   Numbers on calculator: $\dfrac{1}{7.5 \times 2.8} = \mathbf{0.048}$

   Exponents: same as in estimate.

   Final in scientific notation: $\mathbf{0.048} \times 10^{11} = \mathbf{4.8 \times 10^{9}}$. Close to the estimate.

7. $\mathbf{4.2 \times 10^{-18}}$ 8. $\mathbf{2.8 \times 10^{5}}$ 9. See the Pretest answers.

Review Quiz

1. 8,178   2. 289   3a. 96   3b. 0.75   4a. -6.8×10^{-3} 4b. 8.92×10^{2}

5a. 6.0×10^{21} 5b. 5.0×10^{0}, or 5.0   5c. 5.0×10^{14} 5d. $0.33 \times 10^{5} = \mathbf{3.3 \times 10^{4}}$

6. $1/5 \times 10^{23-10+6} = 0.20 \times 10^{19} = \mathbf{2.0 \times 10^{18}}$ in scientific notation

2

The Metric System

Lesson 2.1   Metric Fundamentals

If you get a perfect score on the following Pretest, you may skip to Lesson 2.2. If not, complete Lesson 2.1.

> **PRETEST**   From memory, write answers to these, then check your answers at the end of the chapter.
>
> 1. Fill in these blanks with metric prefixes (not abbreviations).
>
>   a. 10^3 grams $\equiv$ 1 _____ gram
>
>   b. 10^{-3} second $\equiv$ 1 _____ second
>
> 2. Fill in the prefix abbreviations:
>
>   10 _____ m $\equiv$ 1 m $\equiv$ 1000 _____ m $\equiv$ 100 _____ m
>
> 3. Add fixed decimal numbers:
>
>   1 liter $\equiv$ _____ mL $\equiv$ _____ cm^3 $\equiv$ _____ dm^3
>
> 4. How many liters are in a kiloliter?
>
> 5. What is the mass of 15 milliliters of liquid water?
>
> 6. One liter of liquid water has what mass in grams?

The Importance of Units

The fastest and most effective way to solve calculations in chemistry is to focus on the **units** of measurements.

The physical universe can be described by the **fundamental quantities**, including **distance, mass,** and **time**. All measurement systems define **base units** to measure the fundamental quantities. In science, measurements and calculations are done using the **metric system**.

Distance The metric base unit for distance is the **meter**. A *meter* is about 10% longer than the *yard* of the English measurement system.

Just as a dollar can be divided into 100 *cent*s, a meter can be divided into 100 **centi**meters. A meter stick is usually numbered in centimeters.

A meter stick can also be divided into 10 equal **deci**meters. The smallest markings on a meter stick are its 1000 **milli**meters. These definitions mean that

> 1 meter ≡ **10 deci**meters ≡ **100 centi**meters ≡ **1000 milli**meters

The symbol ≡ is a form of an = sign that means "is *defined* as equal to" and "is *exactly* equal to."

For these definitions, other metric relationships can be written. For example, because 100 **centi**meters ≡ 1000 **milli**meters, we can divide both sides by 100 to get: 1 centimeter ≡ 10 millimeters. But rather than memorize all possible relationships, with the four-part equality in memory we can derive others as needed.

Long distances in the metric system are usually measured in **kilo**meters:

> 1000 meters ≡ **1 kilo**meter

One kilometer is approximately 0.62 miles.

Kilo-, deci-, centi-, and *milli-* are termed **metric prefixes**. As one way to define the prefixes, you will need to be able to write *rule* 1 from memory:

> 1. The "meter-stick" equalities are:
>
>   1 meter ≡ 10 **deci**meters ≡ 100 **centi**meters ≡ 1000 **milli**meters
>   *and* 1000 meters ≡ **1 kilo**meter

To help in remembering these definitions, visualize a meter stick. Recall what the marks and numbers on a meter stick mean. Use that image to help in writing the four-part equality above.

To remember the kilometer definition, visualize 1000 meter sticks in a row. That's a distance of 1 *kilo*meter: 1 kilometer ≡ 1000 meter sticks.

Rule 1 is especially important because of rule 2.

> 2. In the meter-stick equalities, you may substitute *any unit* for *meter*.

Rule 2 means that the *prefix* definitions for meters are true for *all* metric units. To use *kilo-*, *deci-*, *centi-*, or *milli-* with *any* metric units, you simply need to be able to recall and write the metric equalities in rule 1.

The two rules above allow us to write a wide range of equalities that we can use to solve science calculations, such as

$$1 \textit{ liter} \equiv 1000 \text{ milli}\textit{liters}$$
$$1 \textit{ gram} \equiv 100 \text{ centi}\textit{grams}$$
$$1 \text{ kilo}\textit{calorie} \equiv 10^3 \textit{ calories}$$

Abbreviations. *Meter* is abbreviated as **m.** In the metric system,

> 3. Prefix *abbreviations* include:
> kilo- = **k-**
> deci- = **d-**
> centi- = **c-**
> milli- = **m-**

Using these abbreviations, the meter-stick equalities can be abbreviated as

> 1 m ≡ 10 dm ≡ 100 cm ≡ 1000 mm   *and*   1000 m ≡ 1 km

One (and only one) prefix *abbreviation* can be written in front of any metric *base* unit abbreviation.

PRACTICE   **A**

Write rules 1, 2, and 3 until you can do so from memory, then complete these problems without looking back at the rules.

1. Add fixed decimal numbers to these blanks.

  a. 1 meter = _____ millimeters   b. 1 liter = _____ deciliters

(continued)

2. Add full metric prefixes to these blanks.

    a. 1000 grams = 1 _____ gram   b. 1000 _____ liters = 1 liter

3. Add exponential terms (powers of 10) to these blanks.

    a. 1 kilogram = _____ grams   b. _____ mm = 1 m = _____ cm

4. Add prefix *abbreviations* to these blanks.

    a. 10^2 __ m = 1 m = 10 __ m = 10^3 __ m   b. 10^3 m = 1 __ m

Volume Volume is the amount of three-dimensional space that a material or shape occupies. Volume is termed a **derived quantity** because it is derived from the fundamental quantity of distance. Any volume unit can be converted to a distance unit cubed.

A cube that is 1 centimeter *wide* by 1 cm *high* by 1 cm *long* has a volume of 1 **cubic centimeter** (1 cm^3). In biology and medicine, a cubic centimeter is often abbreviated as "**cc**," but cm^3 is the standard abbreviation in chemistry.

In chemistry, cubic centimeters are usually referred to as **milliliters**, abbreviated **mL**. One milliliter is defined as exactly 1 cubic centimeter. On the basis of this definition, because

- 1000 milli*meters* ≡ 1 *meter* and 1000 milli*anythings* ≡ 1 *anything*

- 1000 milli*liters* is *defined* as 1 **liter**, abbreviated **L**

The milliliter is a convenient measure for smaller volumes, while the liter (about 1.1 quarts) is preferred when measuring larger volumes.

One liter is the same as **1 cubic decimeter (1 dm^3)**. Note how these units are related:

- The volume of a cube that is

$$10 \text{ cm} \times 10 \text{ cm} \times 10 \text{ cm}$$

    has a volume of

$$1000 \text{ cm}^3 = 1000 \text{ mL}$$

- Because 10 cm ≡ 1 dm, the volume of this *same* cube can be calculated as

$$1 \text{ dm} \times 1 \text{ dm} \times 1 \text{ dm} \equiv 1 \text{ cubic } decimeter \equiv 1 \text{ dm}^3$$

These relationships mean that by definition, all of the following terms are *equal*:

$$1 \text{ L} \equiv 1000 \text{ mL} \equiv 1000 \text{ cm}^3 \equiv 1 \text{ dm}^3$$

What do you need to remember about volume? Just two more rules.

4. 1 milliliter (mL) ≡ 1 cm^3

5. 1 liter (L) ≡ 1000 mL ≡ 1000 cm^3 ≡ 1 dm^3

Mass **Mass** measures the amount of matter in an object. Mass and weight are not the same, but in chemistry, unless stated otherwise, we assume that mass is measured at the constant gravity of Earth's surface. In that case, mass and weight are directly proportional and can be measured with the same instruments.

The metric base unit for mass is the gram. One **gram (g)** was originally defined as the mass of *1 cubic centimeter* of *liquid water* at 4 degrees Celsius, the temperature at which water has its highest density. The modern definition for 1 gram is a bit more complex, but it is still very close to the historic definition, and in calculations involving liquid water we often use the historic definition if high precision is not required.

For most calculations involving *liquid* water at or near room temperature, the following *approximation* may be used.

6. For *liquid water*: $1 \text{ cm}^3 \text{ H}_2\text{O}(\ell) \equiv 1 \text{ mL } \text{H}_2\text{O}(\ell) \approx 1.00 \text{ gram } \text{H}_2\text{O}(\ell)$

The squiggly equal sign ($\approx$) means "approximately equals."

The substance H_2O is solid when it is ice, liquid when it is water, and gaseous when it is steam or vapor. The notation (*liquid*), abbreviated as (ℓ) after the chemical formula, means that this rule is true *only* if H_2O is in its *liquid* state.

Temperature Metric temperature scales are defined by the properties of water. Temperature in the metric system can be measured in **degrees Celsius** (°C).

0°C $\equiv$ the freezing point of water.

100°C $\equiv$ the boiling point of water if the gas above is at "1 atmosphere pressure."

Room temperature is generally between 20°C (which is 68°F) and 25°C (77°F).

Time The metric base unit for time is the **second** (abbreviated as a lowercase **s**).

Unit Abbreviations The following are metric base-unit abbreviations.

7. meter = **m**   gram = **g**   second = **s**

Abbreviations for metric units do not have a period at the end, and no distinction is made between singular and plural.

Note that an **m** by itself stands for meter, and an **m** *after* a prefix stands for meter, but an **m** in *front* of a unit abbreviation means *milli-*.

Examples: ms = millisecond   mm = millimeter   km = kilometer

PRACTICE **B**

Write rules 1–7 until you can do so from memory, then answer the first three Pretest problems for this lesson without looking at the rules.

SI Units

The modern metric system (*Le Système International d'Unités*) is based on what are termed the **SI units**. SI units choose one preferred metric unit as the standard for measuring each physical quantity. The SI unit for distance is the *meter*, for mass is the *kilogram*, and for time is the *second*.

In chemistry, when dealing with laboratory-scale quantities, measurements and calculations frequently use units that are metric but not SI. For example, chemists generally measure volume in liters or milliliters instead of cubic meters and mass in grams instead of kilograms. In Chapter 5, you will learn to convert between non-SI and the SI units required for some types of chemistry calculations.

Learning the Metric Fundamentals

We will use *equalities* to solve most chemistry calculations, and the metric definitions are the equalities that we use most often. A strategy that can help in problem solving is to start each homework assignment, quiz, or test by writing *recently* memorized rules at the top of your paper. By writing the rules at the beginning, you avoid having to remember them under time pressure later in the test.

A Note on Memorization   A goal of these lessons is to minimize what you must memorize. It is not possible, however, to eliminate memorization from science courses. When there are facts that you must memorize to solve problems, these lessons will tell you. This is one of those times. Memorize the metric basics in the table on the next page. You will need to recall them automatically from memory as part of most assignments in chemistry.

Memorization Tips   When you memorize, it helps to use as many *senses* as you can.

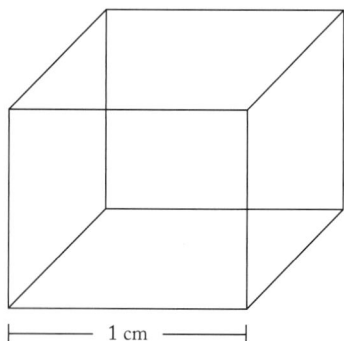

- Say the rules out loud, over and over, as you would learn lines for a play.
- Write the rules several times, in the same way and order each time.
- *Organize* the rules into patterns, rhymes, or mnemonics.
- *Number* the rules so you know which rule you forgot, and when to stop.
- *Picture* real objects:
  - Sketch a meter stick, then write metric rule 1 and compare it to your sketch.
  - For volume, mentally picture a 1 cm × 1 cm × 1 cm = **1 cm³** cube. Call it *1 milliliter*. Fill it with water to have a *mass* of 1.00 *gram*.

After repetition, you will recall new rules *automatically*. That's the goal.

Metric Basics

1. 1 *meter* $\equiv$ 10 decimeters
            $\equiv$ 100 centimeters
            $\equiv$ 1000 millimeters
   1000 meters $\equiv$ 1 kilometer

2. Any unit can be substituted for *meter* above.

3. Prefixes include: kilo- = **k-**; deci- = **d-**; centi- = **c-**; milli- = **m-**

4. 1 milliliter (mL) $\equiv$ 1 cm^3

5. 1 liter (L) $\equiv$ 1000 mL $\equiv$ 1000 cm^3 $\equiv$ 1 dm^3

6. 1 cm^3 H$_2$O(ℓ) $\equiv$ 1 mL H$_2$O(ℓ) $\approx$ 1.00 gram H$_2$O(ℓ)

7. Base unit abbreviations: meter = m; gram = g; second = s

PRACTICE   C

Study the metric basics table until you can write all parts of the table from memory, then answer the following problems in your notebook.

1. In your mind, picture a kilometer and a millimeter. Which is larger?

2. Which is larger, a kilogram or a milligram?

3. Name four units that can be used to measure volume in the metric system.

4. Answer Pretest problems 4–6 at the beginning of this lesson.

Lesson 2.2   Metric Prefixes

Additional Prefixes

For measurements of very large or very small quantities, prefixes larger than *kilo-* and smaller than *milli-* are often used. Ten frequently encountered prefixes, including the four from Lesson 2.1, are listed in the table at right. Note that

- Outside the range between 10^{-3} and 10^3, metric prefixes correspond to powers of 10 divisible by 3.

- When the full prefix name is written, the first letter is *not* normally capitalized.

- For prefixes above **k-**, the abbreviation *must* be *capitalized*.

- For the prefixes **k-** and below, the abbreviation *must* be lowercase.

Prefix	Abbreviation	Means
tera-	T-	$\times 10^{12}$
giga-	G-	$\times 10^9$
mega-	M-	$\times 10^6$
kilo-	k-	$\times 10^3$
deci-	d-	$\times 10^{-1}$
centi-	c-	$\times 10^{-2}$
milli-	m-	$\times 10^{-3}$
micro-	μ (mu) *or* u-	$\times 10^{-6}$
nano-	n-	$\times 10^{-9}$
pico-	p-	$\times 10^{-12}$

Using Prefixes

A metric prefix is interchangeable with the exponential term it represents.

1. An exponential term can be *substituted* for a prefix or prefix abbreviation on the basis of what the prefix means.

Examples:

$$7.0 \text{ } milliliters = 7.0 \times 10^{-3} \text{ liter}$$

$$5.6 \text{ kg} = 5.6 \times 10^3 \text{ g}$$

$$43 \text{ nanometers} = 43 \text{ nm} = 43 \times 10^{-9} \text{ meter}$$

2. A metric *prefix* can be substituted for its equivalent exponential term.

Examples:

$$3.5 \times 10^{-12} \text{ meter} = 3.5 \text{ } \textbf{pico}\text{meters} = 3.5 \text{ pm}$$

$$7.2 \times 10^6 \text{ grams} = 7.2 \text{ } \textbf{mega}\text{grams} = 7.2 \text{ Mg}$$

► TRY IT

(See "How to Use These Lessons," point 1, p. xvii.)

Q1. From memory, fill in these blanks with full prefixes (not the abbreviations).

    a. 10^3 grams = 1 _____ gram   b. 2×10^{-3} meter = 2 _____ meters

Q2. From memory, fill in these blanks with prefix *abbreviations*.

    a. 2.6×10^{-1} L = 2.6 __ L   b. 6×10^{-2} g = 6 __ g

Q3. Fill in these blanks with exponential terms (consult the table of prefixes above if needed).

    a. 1 gigawatt = $1 \times$ _____ watts   b. 9 μm = $9 \times$ _____ m

STOP

Answers:

1a. 10^3 grams = 1 **kilo**gram	1b. 2×10^{-3} meter = 2 **milli**meters
2a. 2.6×10^{-1} L = 2.6 **dL**	2b. 6×10^{-2} g = 6 **cg**
3a. 1 gigawatt = $1 \times \textbf{10}^\textbf{9}$ watts	3b. 9 μm = $9 \times \textbf{10}^{\textbf{-6}}$ m

---

From the prefix definitions, even if you are not yet familiar with the quantity that a unit is measuring, you can convert between a *prefix* and its equivalent exponential.

Learning the Prefixes

To solve calculations, you will need to recall each row in the table of 10 metric prefixes quickly and automatically. To help, look for patterns as a memory device. Note:

tera- = **T-** = $\times 10^{\text{twelve}}$

nano- (which connotes small) = **n-** = $\times 10^{-\text{nine}}$

Focusing on those two can help to "anchor" the prefixes near them in the table.

Then make a self-quiz: On a sheet of paper, draw a table 3 columns across and 11 rows down. In the top row, write

| Prefix | Abbreviation | Means |

Then fill in the table. Repeat writing the table until you can do so from memory, then try the problems below without looking at your table.

PRACTICE A

Use a sticky note to mark the answer page at the end of this chapter.

1. From memory, add exponential terms to these blanks.

    a. 7 microseconds = 7 × _____ second   b. 9 kg = 9 × _____ g

    c. 8 cm = 8 × _____ m   d. 1 ng = 1 × _____ g

2. From memory, add full metric prefixes to these blanks.

    a. 6×10^{-2} amps = 6 _____ amps   b. 45×10^{9} watts = 45 _____ watts

3. From memory, add prefix abbreviations to these blanks.

    a. 10^{12} g = 1 __ g   b. 10^{-12} s = 1 __ s

    c. 0.01 m = 1 __ m   d. 0.001 g = 1 __ g

    e. 6×10^{-9} m = 6 __ m   f. 5×10^{-1} L = 5 __ L

    g. 4×10^{-6} L = 4 __ L   h. 16×10^{6} Hz = 16 __ Hz

4. When writing prefix abbreviations *by hand*, write so that you can distinguish between (add a prefix abbreviation):

    5×10^{-3} g = 5 _____ g and 5×10^{6} g = 5 _____ g

5. For which prefix abbreviations is the first letter always capitalized?

6. Convert "0.30 gigameters/second" to a value in scientific notation without a prefix.

Converting between Prefix Formats

To solve calculations, we will often use conversion factors that are constructed from metric prefix definitions. For those definitions, we have learned two types of equalities:

- The meter-stick equalities are based on what *one unit* equals:

    1 meter ≡ 10 decimeters ≡ 100 centimeters ≡ 1000 millimeters

- The prefix definitions are based on what *one prefix* equals, such as *nano* $= \times 10^{-9}$

It is essential to be able to write *both* forms of metric definitions correctly, because work in science often uses both.

Example: When converting between milliliters and liters, you may see either

- $1 \text{ mL} = 10^{-3} \text{ L}$, based on what *milli-* means; *or*
- $1000 \text{ mL} = 1 \text{ L}$, which is an easy-to-visualize definition of 1 liter

Those two equalities are equivalent. The second is simply the first with the numbers on both sides multiplied by 1000. But the numbers in the equalities are different depending on whether the 1 is in front of the *prefix* or the *unit*. Which format should we use? How do we avoid errors?

In these lessons, we will generally use the 1-*prefix* equalities to solve problems. If you need to write or check prefix equalities in the "1 unit =" format, you can derive them from the 1-*prefix* definitions in the prefix definition table if needed.

Example: 1 gram = ? **micrograms**

From the prefix table, 1-micro-unit = 10^{-6} unit. Therefore,

$$1 \textbf{ microgram} = 10^{-6} \text{ gram}$$

To get a 1 in front of *gram*, we multiply both sides by $\textbf{10}^6$, so

$$10^6 \text{ micrograms} = 1 \text{ gram}$$

The steps above can be summarized as the *reciprocal* rule for prefixes:

Memorize the prefix equalities: 1 *prefix*- = $\textbf{10}^a$

If you need 1 *unit* =, write: **1** *prefix* units- = $\textbf{10}^a$ units, **1 unit** = $\textbf{10}^{-a}$ *prefix* units.

Another way to state the reciprocal rule for prefixes:

To change a prefix definition between the "1 **prefix** =" format and the "1 **unit** =" format, change the sign of the exponent.

▶ **TRY IT**

Q. Using those rules, provide the missing exponential terms below.

a. 1 nanogram = 1 × _____ gram, so 1 gram = 1 × _____ nanograms.

b. 1 dL = 1 × _____ liter, so 1 L = 1 × _____ dL

STOP

Answers:

a. 1 nanogram = 1 × $\textbf{10}^{-9}$ gram, so 1 gram = 1 × 10^9 nanograms.

b. 1 dL = 1 × 10^{-1} liter, so 1 L = 1 × 10^1 dL = 10 dL

To summarize:

- When using metric prefix definitions, be careful to note whether the **1** is in front of the *prefix* or the *unit*.

- To avoid confusing the signs of the exponential terms, commit the table of prefix definitions to memory. Then, if you need an equality with the "*1 unit* = 10^x prefix-unit" format, reverse the sign of the exponent in the table of definitions.

PRACTICE   B

Write the table of 10 metric prefixes until you can do so from memory; then answer these without consulting the table.

1. Fill in the blanks with exponential (10^X) terms.

   a. 1 terasecond = 1 × _____ seconds, so 1 second = 1 × _____ terasecond.

   b. 1 ng = 1 × _____ gram, so 1 g = 1 × _____ ng

2. Apply the reciprocal rule to add exponential terms to these *1-unit* equalities.

   a. 1 gram = _____ centigrams   b. 1 meter = _____ picometers

   c. 1 s = _____ ms   d. 1 s = _____ Ms

3. Add fixed decimal numbers to these blanks.

   a. 1000 cm^3 = _____ L   b. 100 cm^3 H$_2$O(ℓ) = _____ grams H$_2$O(ℓ)

4. Add exponential terms to these blanks. Watch where the 1 is!

   a. 1 micromole = _____ mole   b. 1 g = 1 × _____ Gg

   c. _____ ns = 1 s   d. 1 pL = _____ L

Lesson 2.3   Flashcards

In scientific fields, initial learning is in many respects like learning a new language. To start, we learn new words and their definitions (such as milli- = × 10^{-3}) then rules for their use (for example, can be put in front of a base unit).

With effort and practice, fundamental facts and procedures can be recalled automatically. This opens space in working memory to focus on the conceptual framework into which new knowledge fits. As additional practice gradually leads to fluency, you are able to use your new knowledge for higher-level work.

To learn the vocabulary of chemistry, in these lessons you will make two types of flashcards:

- "one-way" cards for questions that make sense in *one* direction, and
- "two-way" cards for facts that need to be recalled in *both* directions

If you have access to about thirty 3-in. × 5-in. index cards, you can get started now. Plan to buy about 100–200 additional index cards, lined or unlined. A variety of colors is helpful but not essential. With your first 30 cards, complete these steps:

1. On 12–15 cards (of the same color, if possible), cut a triangle off the top-right corner, making cards like this:

These cards will be used for questions that go in one direction. Keeping the notch at the *top right* will identify the front side.

2. Using the following table, cover the answers in the *right*-side column with an index card. For each question in the left column, verbally answer, then slide the cover sheet down to check your answer. Put a *check mark* beside questions that you answer accurately and without hesitation. When done, write the questions and answers without checks onto the notched cards. For these Chapter 2 cards, write a "2" at the bottom right corner of each card.

Front Side of Cards (with Notch at Top Right)	Back Side—Answers
To convert to scientific notation, move the decimal to _____	After the first number that is not a zero
If you make the significand larger, _____	Make the exponent smaller
42^0 = _____	Positive numbers to the zero power = 1
Simplify $1/(1/X)$ = _____	X
To divide exponentials, _____	Subtract the exponents
To bring an exponent from the bottom of a fraction to the top: _____	Change its sign
1 cc ≡ 1 _____ ≡ 1 _____	1 cc ≡ 1 cm^3 ≡ 1 ml
0.0018 in scientific notation = _____	1.8×10^{-3}
1 L ≡ _____ mL ≡ _____ dm^3	1 L ≡ 1000 mL ≡ 1 dm^3
To multiply exponential terms, _____	Add the exponents
Simplify: $1/10^x$ = _____	10^{-x}
74 in scientific notation = _____	7.4×10^1
The historical definition of 1 gram is: _____	The mass of 1 cm^3 of liquid water at 4°C
8×7 = _____	56
$42/6$ = _____	7

If there remain multiplication or division facts that you cannot reliably answer instantly, add them to your list of one-sided cards.

3. To make two-way cards, use the index cards as they are, *without* a notch.

For the following cards, first cover the *right* column, then put a check mark (✔) on the left if you can answer the left-column question quickly and correctly. Then cover the *left* column and check (✔) the right side if you can answer the right side *automatically*.

When done, if a row does not have *two* checks, make the flashcard.

Two-Way Cards (*without* a Notch)

10^3 g or 1000 g = 1 _____ g	1 kg = _____ g
Boiling temperature of water = _____	100 degrees Celsius (if 1 atm pressure) = _____
1 nanometer = 1 × _____ meter	1 _____ meter = 1 × 10^{-9} meters
Freezing temperature of water = _____	0 degrees Celsius = _____
4.7×10^{-3} = _____ (fixed decimal)	$0.0047 = 4.7 \times 10^?$

1 GHz = $10^?$ Hz	10^9 Hz = 1 _____ Hz	$2/3 \approx 0.?$	$0.666... \approx ?/?$
1 pL = $10^?$ L	10^{-12} L = 1 _____ L	$3/2 = ?.?$	$1.5 = ?/?$
$3/4 = 0.?$	$0.75 = ?/?$	$1 \ dm^3 = 1$ _____	1 L = 1 _____
$1/8 = 0.?$	$0.125 - 1/?$	$1/4 = 0.?$	$0.25 = 1/?$

More Two-Way Cards (*without* a Notch) for the Metric Prefix Definitions

kilo- = × $10^?$	× 10^3 = ? prefix	d- = × $10^?$	× 10^{-1} = ? abbrev.	micro- abbrev. = ?	μ- = ? prefix
nano- = × $10^?$	× 10^{-9} = ? prefix	m- = × $10^?$	× 10^{-3} = ? abbrev.	mega- abbrev. = ?	capital M- = ? prefix
giga- = × $10^?$	× 10^9 = ? prefix	T- = × $10^?$	× 10^{12} = ? abbrev.	kilo- abbrev. = ?	k- = ? prefix
milli- = × $10^?$	× 10^{-3} = ? prefix	k- = × $10^?$	× 10^3 = ? abbrev.	pico- abbrev. = ?	p- = ? prefix
deci- = × $10^?$	× 10^{-1} = ? prefix	n- = × $10^?$	× 10^{-9} = ? abbrev.	deci- abbrev. = ?	d- = ? prefix
tera- = × $10^?$	× 10^{12} = ? prefix	μ- = × $10^?$	× 10^{-6} = ? abbrev.	centi- abbrev. = ?	c- = ? prefix
pico- = × $10^?$	× 10^{-12} = ? prefix	G- = × $10^?$	× 10^9 = ? abbrev.	tera- abbrev. = ?	capital T- = ? prefix
mega- = × $10^?$	× 10^6 = ? prefix	M- = × $10^?$	× 10^6 = ? abbrev.	milli- abbrev. = ?	m- = ? prefix
micro- = × $10^?$	× 10^{-6} = ? prefix	p- = × $10^?$	× 10^{-12} = ? abbrev.	nano- abbrev. = ?	n- = ? prefix
centi- = × $10^?$	× 10^{-2} = ? prefix	c- = × $10^?$	× 10^{-2} = ? abbrev.	giga- abbrev. = ?	capital G- = ? prefix

Which cards you need will depend on your prior knowledge, but when in doubt, make the card. On fundamentals, you need quick, confident, accurate recall—every time.

4. **Practice** with one *type* of card at a time.

   - **For front-sided cards**, if you get a card right quickly, place it in the *got-it* stack. If you miss a card, recite the content. Close your eyes. Recite it again. And again. If needed, write it several times. Return that card to the bottom of the *do* stack. Practice until every card is in the *got-it* stack.

   - For two-sided cards, do the same steps in one direction, then the other.

5. Master the cards at least once, then apply them to the Practice on the topic of the new cards. Treat Practice as a practice test.

6. **For three days in a row**, repeat those steps. Repeat them again before working assigned problems and before your next test that includes this material.

7. Make cards for new topics *before* the lectures on a topic if possible. Studying fundamentals first will help in understanding the lecture.

8. Rubber band and carry new cards. Practice during "down times."

9. After a few chapters or topics, change card colors.

This system requires an initial investment of time, but in the long run it will save time and improve achievement. Add cards of your design and choosing as needed.

Flashcards, Charts, or Lists?

The best strategy for learning new information is to practice *multiple* strategies. Flashcards are good for learning simple rules and definitions. For more complex information, practice reciting mnemonics or phrases with meter or rhyme. Being able to write *charts, diagrams,* and *tables* is helpful in recalling information that falls into *patterns*.

For the metric system, writing the seven rules *and* the prefix chart *and* picturing the meter-stick relationships and running the flashcards will all help to fix these fundamentals in long-term memory.

PRACTICE

Run your set of flashcards until all cards are in the *got-it* stack. Then try these problems. Make additional cards if needed.

1. Fill in the following blanks with an exponential (10^x) term.

Format: 1 Prefix	**1 Base Unit**
1 micrometer = _____ meter	1 meter = _____ micrometers
1 gigawatt = _____ watts	1 watt = _____ gigawatt
1 nanoliter = _____ liter	_____ nanoliters = 1 liter

2. Add exponential terms to these blanks. Watch where the 1 is!

    a. 1 picosecond = _____ second   b. 1 megawatt = _____ watts

    c. 1 cg = _____ g   d. 1 mole = _____ millimoles

    e. 1 m = _____ nm   f. 1 μs = _____ s

3. Do these *without* a calculator. Write the fixed decimal equivalent.

    a. 1/5 = _____   b. 1/50 = _____

Lesson 2.4   Calculations with Units

Except as noted, try this lesson without a calculator.

Adding and Subtracting with Units

Most calculations in mathematics consist of numbers without units, but during calculations in science, it is essential to write the *unit* after numbers. Why?

- Scientific calculations are nearly always based on *measurements* of physical quantities. A measurement is a numeric value and its unit.
- Units indicate which steps to take to solve a problem.
- Units provide a check that you have done a calculation correctly.

When solving calculations, the math must take into account *both* the numbers and their units. To do so, apply the following four rules.

1. The *units* must be the *same* in quantities being *added and subtracted*, and those same units must be added to the answer.

> **TRY IT**

Q. Apply rule 1 to these two problems.

    1. 5 apples + 2 apples =

    2. 5 apples + 2 oranges =

STOP

Answers:

Answer 1 is *7 apples*, but you can't add apples and oranges.

By rule 1, you can add numbers that have the same units, but you *cannot* add numbers directly that do *not* have the same units.

▶ TRY IT

Q. Apply rule 1 to this problem:

$$\begin{array}{r} 14.0 \text{ grams} \\ -7.5 \text{ grams} \\ \hline \end{array}$$

Answer:

$$\begin{array}{r} 14.0 \text{ grams} \\ -7.5 \text{ grams} \\ \hline \mathbf{6.5 \text{ grams}} \end{array}$$

If all units are the same, you can add or subtract numbers, but you must add the common unit to the answer.

Multiplying and Dividing with Units

The rule for *multiplying* and *dividing* units is different, but logical.

> 2. When multiplying and dividing *units*, the units multiply and divide.

Example: cm $\times$ cm = $\mathbf{cm^2}$

Units obey the laws of algebra.

▶ TRY IT

Q. Simplify:

$$\frac{cm^5}{cm^2} =$$

Answer:

Either solve by cancellation:

$$\frac{cm \cdot cm \cdot cm \cdot \cancel{cm} \cdot \cancel{cm}}{\cancel{cm} \cdot \cancel{cm}} = \mathbf{cm^3}$$

or by the rules for exponents:

$$\frac{cm^5}{cm^2} = cm^{5-2} = \mathbf{cm^3}$$

Both methods arrive at the same answer (as they must).

> 3. When multiplying and dividing *group* numbers, exponentials, and units separately, solve the three parts separately, then recombine the terms.

━━━▶ TRY IT

Q. If a postage stamp has dimensions of 2.0 cm × 4.0 cm, the surface area of one side of the stamp =

Answer:

$$\text{Area of a rectangle} = l \times w = 2.0 \text{ cm} \times 4.0 \text{ cm}$$
$$= (2.0 \times 4.0) \times (\text{cm} \times \text{cm}) = \textbf{8.0 cm}^2$$
$$= 8.0 \text{ } square \text{ centimeters}$$

By rule 2, the units obey the rules for multiplication and division. By rule 3, unit math is done *separately* from number math.

Units follow the familiar laws of multiplication, division, and powers, including "like units cancel."

━━━▶ TRY IT

Q. Apply rule 3 to the following:

   a. $\dfrac{8.0 \text{ L}^6}{2.0 \text{ L}^2} =$ b. $\dfrac{9.0 \text{ m}^6}{3.0 \text{ m}^6} =$

Answer:

   a. $\dfrac{8.0 \text{ L}^6}{2.0 \text{ L}^2} = \dfrac{8.0}{2.0} \cdot \dfrac{\text{L}^6}{\text{L}^2} = \textbf{4.0 L}^4$ b. $\dfrac{9.0 \text{ } \cancel{\text{m}^6}}{3.0 \text{ } \cancel{\text{m}^6}} = \textbf{3.0}$ (with no unit)

In science, the *unit math* must be done in calculations. A *calculated* unit *must* be written in a calculated answer (except in rare cases, such as *part b* above, when all of the units cancel).

━━━▶ TRY IT

Q. Apply the rules for numbers, exponential terms, and units.

$$\dfrac{12 \times 10^{-3} \text{ m}^4}{3.0 \times 10^2 \text{ m}^2} =$$

Answer:

$$\dfrac{12 \times 10^{-3} \text{ m}^4}{3.0 \times 10^2 \text{ m}^2} = \dfrac{12}{3.0} \cdot \dfrac{10^{-3}}{10^2} \cdot \dfrac{\text{m}^4}{\text{m}^2} = \textbf{4.0} \times \textbf{10}^{-5} \textbf{ m}^2$$

In science calculations, you will often need a calculator for the number math, but both the *exponential* and *unit* math nearly always can (and should) be done *without* a calculator.

4. If *more than one* unit is being multiplied or divided, the math for *each unit* is done separately.

━━▶ TRY IT

Q. Use a calculator for the numbers, but not for the exponents and units.

$$4.8 \frac{g \bullet m}{s^2} \bullet 3.0 \text{ m} \bullet \frac{6.0 \text{ s}}{9.0 \times 10^{-4} \text{ m}^2} =$$

STOP

Answer:

Do the math for numbers, exponentials, and then *each* unit separately.

$$\frac{86.4}{9.0} \bullet \frac{1}{10^{-4}} \bullet \frac{g \bullet \cancel{m} \bullet \cancel{m} \bullet \cancel{s}}{s \bullet \cancel{s} \quad \cancel{m^2}} = 9.6 \times 10^4 \frac{g}{s}$$

This answer unit can also be written as $g \bullet s^{-1}$

PRACTICE

Do *not* use a calculator except as noted. After completing each problem, check your answer.
If you miss a problem, review the rules to find out why before continuing.

1. $16 \text{ cm} - 2 \text{ cm} =$

2. $12 \text{ cm} \bullet 2 \text{ cm} =$

3. $(m^4)(m) =$

4. $m^4/m =$

5. $\dfrac{cm^3}{cm^2} =$

6. $\dfrac{s^{-5}}{s^2} =$

7. $3.0 \text{ meters} \bullet 9.0 \text{ meters} =$

8. $3.0 \text{ g}/9.0 \text{ g} =$

9. $\dfrac{24 \text{ L}^5}{3.0 \text{ L}^{-4}} =$

10. $\dfrac{18 \times 10^{-3} \text{ g} \bullet m^5}{3.0 \times 10^1 \text{ m}^2} =$

11. *Without* a calculator, multiply:

   a. $2.0 \dfrac{g \bullet m}{s^2} \bullet \dfrac{3.0 \text{ m}}{4.0 \times 10^{-2}} \bullet 6.0 \times 10^2 \text{ s} =$

   b. $12 \times 10^{-2} \dfrac{L \bullet g}{s} \bullet 2.0 \text{ m} \bullet \dfrac{2.0 \text{ s}^3}{6.0 \times 10^{-5} \text{ L}^2} =$

12. A rectangular box has dimensions of $2.0 \text{ cm} \times 4.0 \text{ cm} \times 6.0 \text{ cm}$. Calculate its volume.

SUMMARY

To prepare for a quiz and/or test on the material in this chapter:

1. Be able to write the "seven metric basics" rules and the 10 rows of the metric prefix table from memory.

2. For several days, run your flashcards until you can quickly answer each card in both directions.

3. In your own words, be able to summarize any shaded rules in the chapter that are unfamiliar.

4. Be able to solve the problems in the Review Quiz and the chapter.

REVIEW QUIZ

Do not use a calculator. Once you start this quiz, do not look at rules written before the quiz.

1. Add exponential terms to these blanks.

   a. 4 centigrams = 4 × _____ gram   b. 3 pm = 3 × _____ m

   c. 5 dL = 5 × _____ L   d. 1 THz = 1 × _____ Hz

2. Add full metric *prefixes* to these blanks.

   a. 7×10^{-6} liter = 7 _____ liters

   b. 5×10^3 grams = 5 _____ grams

3. Add prefix *abbreviations* to these blanks.

   a. 10^6 g = 1 __ g   b. 10^{-3} g = 1 __ g

   c. 6×10^{-9} L = 6 __ L   d. 2×10^{-1} m = 2 __ m

4. Add exponential terms to these blanks. Watch where the 1 is!

   a. 1 micrometer = _____ meter   b. 1 watt = _____ gigawatt

   c. 1 MHz = _____ Hz   d. 1 mole = _____ nanomoles

   e. 1 m = _____ km   f. 1 ms = _____ s

5. Simplify:

   $$\frac{56 \times 10^{-3}\, m^4}{8 \times 10^2\, m^{-2}} =$$

6. Solve as if the question is *not* multiple choice, then circle your answer among the choices provided.

   $$5.0 \times 10^{-2}\, \frac{L^3 \cdot m}{s} \cdot 4.0\, m \cdot \frac{2.0\, s^3}{8.0 \times 10^{-5}\, L^2} =$$

   a. $1.0 \times 10^{-4}\, m^2 \cdot s^2 \cdot L$ b. $5.0 \times 10^{-7}\, m^2 \cdot s \cdot L$
   c. $5.0 \times 10^3\, m^2 \cdot s^2 \cdot L$ d. $1.0 \times 10^{-3}\, m \cdot s^2 \cdot L$
   e. $5.0 \times 10^{-3}\, m^2 \cdot s^2 \cdot L$

ANSWERS

Lesson 2.1

Pretest   1a. 10^3 grams ≡ 1 **kilo**gram   1b. 10^{-3} second ≡ 1 **milli**second   2. 10 **dm** ≡ 1 m ≡ 1000 **mm** ≡ 100 **cm**
3. 1 liter ≡ **1000** mL ≡ **1000** cm³ ≡ **1** dm³   4. 1000 liters   5. 15 grams   6. 1000 grams

Practice A   1a. 1000   1b. 10   2a. 1000 grams = 1 **kilo**gram   2b. 1000 **milli**liters = 1 liter
3a. 1 kilogram = **10^3** grams   3b. **10^3** mm = 1 m = **10^2** cm   4a. 10^2 **cm** = 1 m = 10 **dm** = 10^3 **mm**
4b. 10^3 m = 1 **km**

Practice B   See Pretest answers 1–3 above.

Practice C   1. A kilometer   2. A kilogram   3. Possible answers include cubic centimeters, milliliters, liters, cubic decimeters, cubic meters, and any metric distance unit, cubed.

4. See Pretest answers 4, 5, and 6 above.

Lesson 2.2

Practice A   1a. 7 microseconds $= 7 \times 10^{-6}$ seconds   1b. 9 kg $= 9 \times 10^{3}$ g   1c. 8 cm $= 8 \times 10^{-2}$ m

1d. 1 ng $= 1 \times 10^{-9}$ g   2a. 6×10^{-2} amps $= 6$ **centi**amps   2b. 45×10^{9} watts $= 45$ **giga**watts

3a. 10^{12} g $= 1$ **Tg**   3b. 10^{-12} s $= 1$ **ps**   3c. 0.01 m $= 10^{-2}$ m $= 1$ **cm**   3d. 0.001 g $= 10^{-3}$ g $= 1$ **mg**

3e. 6×10^{-9} m $= 6$ **nm**   3f. 5×10^{-1} L $= 5$ **dL**   3g. 4×10^{-6} L $= 4$ **μL**

3h. 16×10^{6} Hz $= 16$ **MHz**   4. 5 **mg** and 5 **Mg**   5. **M-, G-, and T-**   6. 3.0×10^{8} meters/second

Practice B   1a. 1 terasecond $= 1 \times 10^{12}$ seconds, so 1 second $= 1 \times 10^{-12}$ terasecond.   1b. 1 ng $= 1 \times 10^{-9}$ gram, so 1 g $= 1 \times 10^{9}$ ng.

2a. 1 gram $= 10^{2}$ centigrams (For "1 *unit* $=$," reverse sign of prefix definition.)

2b. 1 meter $= 10^{12}$ picometers   2c. 1 s $= 10^{3}$ ms   2d. 1 s $= 1 \times 10^{-6}$ Ms   3a. 1000 $cm^{3} = 1$ L

3b. 100 cm^{3} $H_2O(\ell) = 100$ grams $H_2O(\ell)$ 4a. 1 micromole $= 10^{-6}$ moles   4b. 1 g $= 1 \times 10^{-9}$ Gg

4c. 10^{9} ns $= 1$ s   4d. 1 pL $= 10^{-12}$ L

Lesson 2.3

Practice

1. 1 micrometer $= 10^{-6}$ meter	1 meter $= 10^{6}$ micrometers
1 gigawatt $= 10^{9}$ watts	1 watt $= 10^{-9}$ gigawatt
1 nanoliter $= 10^{-9}$ liter	10^{9} nanoliters $= 1$ liter

2a. 1 picosecond $= 10^{-12}$ second   2b. 1 megawatt $= 10^{6}$ watts   2c. 1 cg $= 10^{-2}$ g

2d. 1 mole $= 10^{3}$ millimoles   2e. 1 m $= 10^{9}$ nm   2f. 1 μs $= 10^{-6}$ s   3a. $1/5 = 0.20$

3b. $1/50 = 0.020$

Lesson 2.4

Practice   Both the *number* and the *unit* must be written and correct.

1. **14 cm**   2. **24 cm^{2}**   3. $m^{(4+1)} = m^{5}$ 4. $m^{(4-1)} = m^{3}$ 5. **cm**   6. s^{-7} 7. **27 meters^{2}**

8. **0.33** (*no unit*)   9. **8.0 L^{9}**   10. **6.0×10^{-4} g • m^{3}**   11a. 9.0×10^{4} $\underline{\mathbf{g \bullet m^{2}}}$ (Answer unit could also be written as: g • m^{2} • s^{-1}) $\mathbf{s}$

11b. $\mathbf{8.0 \times 10^{3}}$ $\underline{\mathbf{g \bullet m \bullet s^{2}}}$ (Answer unit could also be written as: g • m • s^{2} • L^{-1}) $\mathbf{L}$

12. $V_{\text{rectangular solid}} =$ length *multiplied by* width *multiplied by* height $= \mathbf{48\ cm^{3}}$

Review Quiz

1a. 4 centigrams $= 4 \times 10^{-2}$ gram   1b. 3 pm $= 3 \times 10^{-12}$ m   1c. 5 dL $= 5 \times 10^{-1}$ L

1d. 1 THz $= 1 \times 10^{12}$ Hz   2a. 7×10^{-6} liter $= 7$ **micro**liters   2b. 5×10^{3} grams $= 5$ **kilo**grams

3a. 10^{6} g $= 1$ **Mg**   3b. 10^{-3} g $= 1$ **mg**   3c. 6×10^{-9} L $= 6$ **nL**   3d. 2×10^{-1} m $= 2$ **dm**

4a. 1 micrometer $= 10^{-6}$ meter   4b. 1 watt $= 10^{-9}$ gigawatt   4c. 1 MHz $= 10^{6}$ Hz

4d. 1 mole $= 10^{9}$ nanomoles   4e. 1 m $= 10^{-3}$ km   4f. 1 ms $= 10^{-3}$ s   5. 7×10^{-5} m^{6} 6. **c**

3

Atoms—and Significant Figures

Lesson 3.1   The Atoms (Part 1)

Atoms are the building blocks of matter. In Earth's crust are 92 different kinds of atoms, from hydrogen, the lightest atom, to uranium, the heaviest. Additional (heavier) atoms can be formed in high-energy nuclear reactions.

The periodic table organizes the atoms in a way that facilitates prediction of their properties. If you know where to *look* for an atom in the table, you can quickly find additional data that a detailed periodic table provides. When you can automatically associate the name, symbol, and periodic table position for each of the ~40 atoms that are most frequently encountered in first-year chemistry, it opens space for working memory to note concepts and move them into long-term memory.

Your brain gains fluency with new vocabulary gradually, with repeated effort. To begin to learn the language of chemistry, your assignment is:

- For the first 20 atoms, be able to fill in an empty chart like the one below with the names and symbols entered in their correct locations.

- Note how the atoms are numbered going across each row. Each of these whole numbers represents the **atomic number** of the respective atom.

- Space your practice to learn the first 12 atoms by the end of Chapter 3 and all 20 atoms shown by the end of Chapter 4.

PERIODIC TABLE

1A	2A		3A	4A	5A	6A	7A	8A
1 **H** Hydrogen								2 **He** Helium
3 **Li** Lithium	4 **Be** Beryllium		5 **B** Boron	6 **C** Carbon	7 **N** Nitrogen	8 **O** Oxygen	9 **F** Fluorine	10 **Ne** Neon
11 **Na** Sodium	12 **Mg** Magnesium		13 **Al** Aluminum	14 **Si** Silicon	15 **P** Phosphorus	16 **S** Sulfur	17 **Cl** Chlorine	18 **Ar** Argon
19 **K** Potassium	20 **Ca** Calcium							

Lesson 3.2   Uncertainty and Significant Figures

Rounding Numbers

Most chemistry calculations require *rounding* of numbers to obtain a final answer. The rules for rounding in science follow (but may be more detailed than) those you have learned in mathematics. In chemistry, use these terms and rules.

1. **Place value.** Recall that in the number 12.345, the 1 is said to be in the *tens place*, the 2 in the *ones* place, the 3 in the *tenths* place (one place to the right of the decimal), the 4 in the *hundredths* place, and the 5 in the *thousandths* place.

   In 12.345, the *highest* place with a digit is the tens place and the *lowest* is the thousandths place.

2. **Rounding: Up or Down?** When rounding, if the number *beyond* the place you are rounding to is

   a. *less* than 5: drop it (round *down*).

      Example: 1.3̲42 rounded to *tenths* = 1.3

   b. *greater* than 5: round *up*.

      Example: 1.7̲38 rounded to the <u>underlined</u> place = 1.74

   c. a **5** followed by *any non-zero* digits: round *up*.

      Example: 1.02̲502 = 1.03

3. **Look only *one* place past** the place you are rounding to.

   Example: Rounding 9.7̲49 to tenths = 9.**7**

   When rounding the 7, look *only* at the 4. The 4 rounds down.

If rounding from the tenths place is doubtful, the hundredths place is used as a basis for rounding to tenths, but numbers past the hundredths place have no reliability and are not used as a basis for rounding.

4. **Rounding a lone 5.** A "lone 5" is a 5 to be rounded off that either does not have "following" digits (as in 7.35) *or* is a 5 followed by one or more zeros (19.6500).

To round a lone 5, some instructors prefer the simple "round 5 up" rule. Others prefer a slightly more precise "engineer's rule," which rounds 5 up half the time and down half the time. Rounding a lone 5 is not a case that occurs often in calculations, but when it does, in these lessons we will always round 5 up.

   Examples: 7.35 rounds to 7.4; 19.6500 rounds to 19.7

You should use the rounding rule preferred by the instructor in *your* course.

PRACTICE   A

Round to the underlined place. Check answers at the end of this chapter.

1. 0.00212

2. 0.0994

3. 20.0561

4. 23.25

5. 0.1950

Uncertainty

Nearly all measurements have **uncertainty.** In science, we need to express

- how much uncertainty exists in measurements, and
- the uncertainty in calculations based on measurements.

One way to indicate how much uncertainty is present in a measurement is "plus-minus notation."

   Example: If a balance reads mass to the thousandths place, but under the conditions of the experiment the uncertainty in the measurement is ±0.02 gram, we can write

$$12.437 \text{ g} \pm 0.02 \text{ g}$$

This use of **plus-minus notation** is a precise way to record estimated uncertainty.

In calculations, however, we often need to multiply and divide measurements, and the math to include the plus-minus uncertainty can be time-consuming. A less complicated method that conveys an *approximate* uncertainty is **significant figures**, abbreviated in these lessons as *s.f.* Because calculations using significant figures rather than plus-minus notation can be done more quickly, in most calculations it will be the method of choice to track uncertainty.

> In chemistry problems, unless otherwise noted, you should assume all measured values appear in significant figures notation, and all calculations must follow the rules for significant figures.

Significant Figures: Fundamentals

Use the rules that follow when recording measurements and rounding calculations.

1. **When writing a measurement:** Write all the digits you are sure of, plus the *first* digit that you must *estimate* in the measurement: the first **doubtful digit** (the first **uncertain digit**). Then *stop.*

When writing a measurement using significant figures, the *last* digit is the first *doubtful* digit. The *significant* digits *include* the doubtful digit.

▶ **TRY IT**

(See "How to Use These Lessons," point 1, p. xvii.)

Q. Find a 3-in. × 5-in. index card and a ruler with a *centimeter* scale. Measure and record the diagonal distance between two corners of the card in centimeters.

Index card diagonal distance = _____ cm

(STOP)

Answer:

Index cards are likely not cut to exact and consistent dimensions, but on our card, the diagonal distance was definitely between 14.6 and 14.7 cm, but closer to 14.6 cm. Estimating the additional digit, you might write

Index card diagonal distance = **14.63** cm

The 14.6 are certain digits. The 3 has doubt. Measurements in significant figures include the *first doubtful* digit, then *stop.*

2. **Converting plus-minus to significant figures notation.** A fundamental rule for significant figures notation is:

Report measurements *rounded* to the *highest place* with doubt.

Example: In the earlier example of the balance reading, the measured value using ± uncertainty was

12.437 g ± 0.02 g

The ± shows the highest *place* with doubt is the hundredths place. To convert this value to significant figures, round the number back to the hundredths place, dropping the ±. The 7 rounds up. In *s.f.*, this measurement is written as

12.44 g

Example: If a recorded value was

538.6 mL ± 0.4 mL

the highest *place* with doubt, as indicated by the ±, is the tenths place. Rounding to the tenths place—as there is no number after the 6 as a basis for rounding—leave the 6 and drop the ±. In *s.f.* the value is written as

538.6 mL

We record measurements in significant figures because in calculations, the arithmetic for numbers such as 12.44 and 538.6 follows familiar rules.

PRACTICE B

Convert these from *plus-minus* notation to significant figures (*s.f.*) notation.

1. 65.316 mL ±0.05 mL   2. 5.2 cm ±0.1 cm

3. 1.8642 km ±0.2 km   4. 16.8°C ±1°C

Lesson 3.3   Calculating with Significant Figures

For significant figures, there are *two* sets of calculation rules: one for addition and subtraction, the other for multiplication and division (including powers and roots). This means in any operation with numbers, a question you must ask before writing the final answer is: Was this addition or subtraction or was it multiplication or division?

Addition and Subtraction

To estimate the uncertainty in a calculated answer, a fundamental rule is:

> In addition and subtraction, an answer has doubt in the same *place* as the *highest* place with doubt in the numbers being added or subtracted. Round the answer to that *place*.

This rule is the same whether you are using "pencil-and-paper math" or a calculator, but the mechanics of applying it differ just a bit.

1. **Adding and subtracting with a pencil.** Let's learn the steps with an example.

   Example: Add 23.1 + 16.01 + 1.008 without a calculator.

   First, write the numbers in columns, lining up the decimal points in the same column, then add or subtract as you normally would.

   Next, search the numbers for the doubtful digit in the *highest place*. The answer's *doubtful* digit must be in that *place*. *Round* the answer to that *place*.

   $$
   \begin{array}{r}
   23.1 \\
   16.01 \\
   \underline{1.008} \\
   40.118 = \textbf{40.1}
   \end{array}
   $$

   This answer must be rounded to **40.1** because the tenths place is the highest *place* with doubt among the numbers added.

The logic is: If you add or subtract a number with doubt in the tenths place and a number with doubt in the hundredths place, the answer will have doubt in the higher place: the tenths place. Numbers in the answer after the doubtful place would have no reliability.

2. **Adding and subtracting with a calculator.** Follow these steps:

   a. Add or subtract on the calculator and write the calculator answer.

   b. Among the numbers added and subtracted, find the *highest place* that has doubt. In one of the numbers that has doubt in that place, underline the doubtful digit.

   c. Round your calculator answer to the underlined *place.*

► TRY IT

Q. Using a calculator, add and subtract:

$$43 + 1.00 - 2.008 =$$

Answer:

$$4\underline{3} + 1.00 - 2.008 = 4\underline{1}.992 = \textbf{42} \text{ in significant figures}$$

Among the numbers being added and subtracted, the *highest* doubt is in the ones place. In the answer, that place must have doubt, so you must round your answer to that *place.*

In your notebook, summarize the steps above in your own words. Practice recalling your summary. Then complete the practice below.

PRACTICE   A

1. Add these numbers *without* a calculator. Round your final answer to the proper number of *s.f.*

$$\begin{array}{r} 23.1 \\ 23.1 \\ + \ 16.01 \\ \hline \end{array}$$

2. Use a calculator. Round your final answer to the proper number of significant figures.

   a. $2.016 + 32.18 + 64.5 =$ b. $16.00 - 4.034 - 1.008 =$

3. Write the atom symbol for:

   a. Helium   b. Beryllium   c. Boron

Counting Significant Figures

When multiplying and dividing, we will need to *count* the number of significant figures in a numeric value.

1. **In fixed decimal notation.** When counting *s.f.*, count the sure digits *plus* the doubtful digit. The doubtful digit is significant.

   This rule means that for numbers that do *not* include zeros, the *count* of *s.f.* is simply the number of digits shown.

   Examples: 123 meters has **3** *s.f.*   14.27 grams has **4** *s.f.*

2. **In exponential notation.** In exponential notation, the significand contains the significant figures. In the special case of *scientific* notation, all digits shown in the significand are significant. (We call it the significand because it contains the significant figures.) This means:

   a. **When *rounding* a value in exponential notation, you change only the significand (*not* the sign *or* the exponential term).

   b. **To *count* the number of *s.f.* for a value in exponential notation, look *only* at the significand.** The exponential term does not affect the number of significant figures.

   Example: 2.99×10^8 meters/second has **3** *s.f.*

   c. **When converting** among fixed decimal, exponential, and scientific notation, do not change the *number* of *s.f.* in the fixed decimal terms (including the significands).

► TRY IT

Q. Convert 424.7×10^{-11} to scientific notation.

Answer:

$$4.247 \times 10^{-9}$$

The original has **4** *s.f.* Moving the decimal must keep the same number of *s.f.*

PRACTICE   B

1. Count the number of *s.f.* in:

   a. 1,261   b. 4.7×10^{-3} c. 6.02×10^{23}

2. Convert to scientific notation:

   a. 41.7×10^{-15} b. $1,827 \times 10^{-1}$

3. Round to the underlined place, then convert to scientific notation:

   a. $2.\underline{6}48 \times 10^{-3}$ b. $5\underline{5}.5 \times 10^6$

Multiplying and Dividing

In chemistry calculations, multiplication and division (which includes in this case powers and roots) are the operations that we perform most often, so the following are the steps that we use *most often*.

1. First, multiply or divide as you normally would.

2. Then, *count* the *number* of *s.f.* in *each* of the numbers you are multiplying or dividing.

3. Your answer can have *no more s.f.* than that of the measurement with the *least* number of *s.f.* that you multiply or divide by. *Round* the answer to that *number* of *s.f.*

   Example: 3.1865 cm $\times$ **8.8** cm = 28.041 = **28** cm^2 (Must *round* to **2** *s.f.*)

           5 *s.f.*   2 *s.f.*   2 *s.f.*

We can summarize with this rule:

> If you **multiply** and/or **divide** a 10-*s.f.* number and a 9-*s.f.* number and a 2-*s.f.* number, you must round your answer to 2 *s.f.*

PRACTICE **C**

Multiply and divide using a calculator. Write the first six digits of the calculator result, then write the final answer with units and the proper number of *s.f.*

1. 3.42 cm $\times$ 2.3 cm^2 =   2. 74.3 L^2 ÷ 12.4 L =

Calculations with Steps or Parts

The rules for *s.f.* are applied at the *end* of a calculation. If a calculation has several steps, you should carry at least one extra *s.f.* until the final step, and round to the proper number of *s.f.* at the final step.

In problems that have several separate parts, and earlier answers may be used for later parts, many instructors prefer that you include one extra *s.f.* in each part until the end of a calculation, then round to the proper number of *s.f.* at the *final* step.

These rules minimize changes in the final doubtful digit due to rounding in the middle steps. Small variations in the value of a doubtful digit are generally acceptable in answers, but in this text we will use these "carry an extra digit until the last step" rules to limit the impact of "middle-step rounding."

Calculations Summary

When using significant figures, a precise statement of the rules is less important than being able to apply them, but thinking about the rules will help to move them into memory.

In your notebook write a summary of the steps and rules in this lesson in whatever form you think will make them easiest to remember. Try to make your rules into useful flashcards. Use your summary in the practice below and the next lesson.

PRACTICE **D**

Use a calculator as needed.

1. Solve this "two-part" question using the rules for significant figures.

    a. $9.76573 \times 1.3 =$ part a answer =   b. (Part a answer)/2.5 =

2. Multiply, then write your answer with proper significant figures.

    $3.5 \text{ m} \times (2.762 \times 10^2 \text{ m}) =$

Lesson 3.4   Zeros and Exact Numbers

For significant figures, zeros and exact numbers have special rules.

Zeros

When using significant figures to express uncertainty, the non-zero digits (1–9) are *always* significant, but zeros may or may not be significant.

Zeros are a special case because a zero has two uses in our number system: one as part of a numeric value (as in 29, 30, 31), and the other as a way to indicate the position of a decimal point (as in 0.0025).

These two uses mean there are six rules for zeros:

1. *Leading* zeros (zeros in *front* of *all* other digits) are *never* significant.

    Example: 0.0006 has 1 *s.f.* (Zeros in front never count.)

2. Zeros embedded *between* other digits *are* always significant.

    Example: 300.07 has 5 *s.f.* (Zeros sandwiched by *s.f.* count.)

3. Zeros *after* all other digits and the decimal point *are* significant.

    Example: 565.0 has 4 *s.f.* (You would not need to include that zero if it were not significant.)

4. Zeros *after* all other digits but *before* the decimal point are assumed to be *not* significant.

    Example: 300 is assumed to have 1 *s.f.*, meaning "give or take at least 100."

When a number is written as 300 or 250 or 90, it is not *clear* whether the zeros are significant. Many science textbooks address this problem by using this rule:

- "500 meters" means 1 *s.f.*, but
- "500. meters," with an *unneeded decimal point* added after a zero, means 3 *s.f.*

These lessons will use that convention as well. Our rule will be:

5. The zeros immediately to the left of an unneeded decimal point are significant.

But the best way to avoid this ambiguity is to write numbers in *scientific* notation.

Example: 4×10^2 has 1 *s.f.*; 4.00×10^2 has 3 *s.f.*

In general exponential notation, the significand contains the significant figures, and how zeros are treated follows the rules above.

In *scientific* notation, however, significands are always a single number between 1 and 9 followed by a decimal. This means that the exponent, and not the significand, indicates the position of the decimal point in the value. As a result:

6. In *scientific* notation, all of the digits in the significand are significant.

When zeros are simply "indicating the place for the decimal," they are *not* significant as measurements. When the zero represents "a number between a 9 and a 1 in a measurement," it is significant.

PRACTICE   A

For each of the six rules above, in your notebook write a version you find easy to remember. Then write a version of your rule that works on a one-way flashcard. Make the flashcards, run them to perfection, then try these questions.

Write the number of *s.f.* in these:

1. 0.0075   2. 600.3   3. 178.40   4. 4,640.   5. 800

Exact Numbers

Exact numbers do not add uncertainty to calculations. For this reason, measurements with *no* uncertainty are considered to have an *infinite* number of *s.f.*, and *exact* numbers are *ignored* when deciding the *s.f.* in a calculated answer.

Example: 1.008 × (an exact 5) = 5.040 (4 *s.f.*)

In chemistry, we use this rule in situations including the following.

1. Numbers in *definitions* are exact.

    Example: "1 km = 1000 meters" is an exact definition of *kilo-* and not a measurement with uncertainty. Both the 1 and the 1000 are exact. Multiplying or dividing by those *exact* numbers will not limit or affect the number of *s.f.* in your answer.

2. The number **1** in nearly all cases is *exact*.

    Example: The conversion "**1** km = 0.62 miles" is a legitimate approximation, but it is not a *definition* (≡) and is not *exactly* correct. The **1** is therefore assumed to be exact, but the 0.62 has uncertainty and has 2 *s.f.*

3. Whole numbers (such as 2 or 6), *if* they are a count of exact quantities (such as 2 people or 6 molecules), are also exact numbers with infinite *s.f.*

4. *Coefficients* and *subscripts* in chemical formulas and equations are exact.

    Example: $2\,H_2 + 1\,O_2 \rightarrow 2\,H_2O$ All of those *numbers* are exact.

You will be reminded about these exact-number cases as we encounter them. For now, simply remember that exact numbers

- have infinite *s.f.*, and
- do not limit the *s.f.* in an answer.

Flashcards

Below, cover the answers on the right, then check whether you can answer each left-side question with confidence. When done, add cards to your cards for this chapter as needed.

Front Side (with Notch at Top Right)	Back Side—Answers
Writing measurements in *s.f.*, stop where?	At the first doubtful digit
Counting the number of *s.f.*, which digits count?	All the sure, plus the doubtful digit
Adding and subtracting, round to where?	The *column* with doubt in highest place
Multiplying and dividing, rounding rule = _____?	Least number of *s.f.* in calculation = number of *s.f.* allowed
In counting *s.f.*, zeros in front _____	Never count
Sandwiched zeros _____	Always count
Zeros after numbers and the decimal _____	Always count
Zeros after numbers but before the decimal _____	Probably don't count
Zeros followed by unneeded decimal ____	Always count
Zeros in scientific notation _____	Always count
Exact numbers have _____ *s.f.*	Infinite *s.f.*

PRACTICE B

Run your flashcards, then complete these. On multipart problems, save one part for your next practice session.

1. Write the number of *s.f.* in

    a. 0.00200 b. 2.06×10^{-9} c. 0.060×10^3

    d. 0.02090×10^5 e. 3 (exact) f. 900. g. 1,320

2. Convert these to *s.f.* notation.

    a. 2.0646 m ± 0.050 m b. 5.04 nm ± 0.1 nm

3. Round to the place indicated.

    a. 0.04070 g (2 *s.f.*) b. 6.255 cm (tenths place)

4. Using a calculator, add then round to the proper *s.f.*: 203.0 + 16.03 + 1.008 =

5. Use a calculator. Write final answers using proper *s.f.* and proper units.

    a. 4.4 meters $\times$ 8.312 meters^2 = b. 2.03 cm^2/1.2 cm =

6. Without looking at a periodic table, write the symbols for the first 12 atoms in order.

SUMMARY

For this chapter, practice drawing the form of the first three rows of the periodic table and then filling in the first 12 atoms.

Note the form of the graphic near the bottom of page 53 to help in remembering the method and concept.

For any of the steps or rules for significant figures that you cannot recall automatically, make a flashcard that works for you. Run the cards, then practice the problems, and you will quickly be able to apply the rules automatically. That's the goal.

REVIEW QUIZ

Before your quiz on this material, but after your flashcard review, complete these.

1. Write the number of *s.f.* in the following.

   a. 107.42   b. 10.04   c. 13.40

   d. 0.00640   e. 0.043×10^{-4} f. 1,590.0

   g. 320×10^{9} h. 14 (exact)   i. 2,500

2. Round to the place indicated.

   a. 5.15 cm (tenths place)   b. 31.84 meters (3 *s.f.*)

   c. 0.819 mL (hundredths place)   d. 0.06349 cm^2 (2 *s.f.*)

3. Calculate with or without a calculator, then write *final* answers with proper *s.f.*

   a.   17.65   b. $39.1 + 124.0 + 14.05 =$
        $-\ 9.7$

4. Use a calculator. Write final answers using proper *s.f.* and proper units.

   a. 13.8612 cm $\times 2.02$ cm^2 =   b. 0.5223 cm^3/0.040 cm =

5. Use a calculator. Answer in scientific notation with proper *s.f.*

   a. $(2.25 \times 10^{-2})(6.0 \times 10^{23}) =$

   b. $(6.022 \times 10^{23}) / (1.50 \times 10^{-2}) =$

6. Convert these to *s.f.* notation.

   a. 12.675 g $\pm\ 0.2$ g   b. $24.81°$C $\pm\ 1.0°$C

7. Answer in scientific notation with proper units and proper *s.f.* Use a calculator as needed.

$$5.60 \times 10^{-2}\ \frac{\text{L}^2 \cdot \text{g}}{\text{s}} \cdot 0.090\ \text{s}^{-3} \cdot \frac{4.00\ \text{s}^4}{6.02 \times 10^{-5}\ \text{L}^3} \cdot (\text{an exact 2}) =$$

8. Without a calculator, add, then select the correct answer in proper significant figures.

$$1.008$$
$$238.00$$
$$\underline{+\ 16.00}$$

   a. 255.00   b. 255.0   c. 255.008   d. 255.1   e. 255.01

9. Without viewing a periodic table, fill in the names and symbols for the first 12 atoms in the portion of the table below.

ANSWERS

Lesson 3.2

Practice A   1. 0.002<u>1</u>2 rounds to **0.0021**

2. 0.0<u>9</u>94 rounds to **0.10**

3. 20.0<u>5</u>61 rounds to **20.06**

4. 23.<u>2</u>5 rounds to **23.3** by the "round lone 5 up" rule

5. 0.1<u>9</u>50 rounds to **0.20**

Practice B   Your answers must match these exactly.

1a. 65.316 mL ±0.05 mL   **65.32 mL**   The highest place with doubt is the hundredths.

1b. 5.2 cm ±0.1 cm   **5.2 cm**   Highest doubt is in the tenths place. Round to tenths.

1c. 1.8642 km ±0.2 km   **1.9 km**   Highest doubt is in the tenths place. Round to tenths.

1d. 16.8°C ±1°C   **17°C**   Highest doubt is in the ones place.

Lesson 3.3

Practice A   1.   23.1
             23.1
          $\underline{+\ 16.01}$
             62.<u>2</u>1   Must round to **62.2**

2a. 2.016 + 32.18 + 64.<u>5</u> = 98.<u>6</u>96   Must round to **98.7**

2b. 16.0<u>0</u> − 4.034 − 1.008 = 10.9<u>5</u>8   Must round to **10.96**

3a. He   3b. Be   3c. B

62   Chapter 3 | Atoms—and Significant Figures

Practice B   1a. 4   1b. 2   1c. 3

2a. 4.17×10^{-14} 2b. 1.827×10^2

3a. $2.\underline{6}48 \times 10^{-3}$ rounds to **2.6×10^{-3}**.   Look one place past the place to which you are rounding.

3b. $55.\underline{5} \times 10^6$ rounds to $56 \times 10^6 =$ **5.6×10^7** in scientific notation.

Practice C   1a. 7.9 cm³ (2 *s.f.*)   1b. 5.99 L (3 *s.f.*)

Practice D   1a. **12.7**   If this answer were not used in part b, the proper answer would be 13 (2 *s.f.*), but because we need the answer in part b, it is preferred to carry an extra *s.f.*

1b. $12.7/2.5 =$ **5.1**   2. 9.7×10^2 m²

Lesson 3.4

Practice A   1. 0.0075 has **2** *s.f.*   Zeros in front never count.

2. 600.3 has **4** *s.f.*   Sandwiched zeros count.

3. 178.40 has **5** *s.f.*   Zeros after the decimal and after all the numbers count.

4. 4,640. has **4** *s.f.*   Zeros after all the numbers but before an unneeded decimal point count.

5. 800 has **1** *s.f.*   Zeros after all numbers but before the decimal place usually don't count.

Practice B   1a. 0.00200 has **3** *s.f.*   Zeros in front never count. Zeros after the decimal and after all the numbers count.

1b. 2.06×10^{-9} has **3** *s.f.*   The significand in front contains and determines the *s.f.*

1c. 0.060×10^3 has **2** *s.f.*   Significand holds the *s.f.* Leading zeros never count.

1d. 0.02090×10^5 has **4** *s.f.*   Leading zeros never count, the rest here do.

1e. 3 (exact)   **Infinite** *s.f.*   Exact numbers have no uncertainty and infinite *s.f.*

1f. 900. has **3** *s.f.*   The decimal point is one way to say that the zeros in front of it count.

1g. 1,320 has **3** *s.f.*   Zeros after all numbers but before the decimal place usually don't count.

2a. 2.0646 m±0.050 m   **2.06 m**   Highest doubt in hundredths. Round to that place.

2b. 5.04 nm ±0.1 nm   **5.0 nm**   Highest doubt in tenths. Round to that place.

3a. 0.04070 g (2 *s.f.*)   **0.041 g**   Zeros in front never count.

3b. 6.255 cm (tenths place)   **6.3 cm**   Rounding a 5 *followed* by other digits, always round up.

4. $203.\underline{0} + 16.03 + 1.008 = 220.038 =$ **220.0**   Must round to tenths (highest *place* with doubt)

5a. 4.4 meters $\times$ 8.312 meters² $= 36.5728 =$ **37 meter³** (**2** *s.f.*)   5 plus following digits, always round up.

5b. 2.03 cm²/1.2 cm $= 1.69166 =$ **1.7 cm** (2 *s.f.*)

6. H, He, Li, Be, B, C, N, O, F, Ne, Na, Mg

Review Quiz

1a. 107.42 **5** *s.f.*   1b. 10.04 **4** *s.f.*   1c. 13.40 **4** *s.f.*   1d. 0.00640 **3** *s.f.*

1e. 0.043×10^{-4} **2** *s.f.*   1f. 1,590.0 **5** *s.f.*   1g. 320×10^9 **2** *s.f.*

1h. 14 (exact) **Infinite** *s.f.*   1i. 2,500   **2** *s.f.*

2a. **5.2 cm**   Round lone 5 up.   2b. **31.8 meters**   Third digit is last digit: rounding off a 4, round down.

2c. **0.82 mL**   9 rounds up.   2d. **0.063 cm²**   Round to 2 *s.f.* based only on following digit.

3a. **8.0** Round 5 up   3b. $39.\underline{1} + 124.\underline{0} + 14.05 = 177.15 =$ **177.2**

4a. 28.0 cm³ (**3** *s.f.*)   4b. 13 cm² (**2** *s.f.*)

5a. 1.4×10^{22} in scientific notation (**2** *s.f.*)   5b. 4.01×10^{25} (**3** *s.f.*)

6a. 12.675 g ±0.2 g   **12.7 g**   6b. 24.81°C ±1.0°C   **25°C**

7. 6.7×10^2 g/L The 0.090 limits the answer to 2 *s.f.*

8. **e. 255.01**

9. See a periodic table.

4

Conversion Factors

Lesson 4.1   Conversion Factor Basics

Conversion factors can be used to make precise predictions about a wide range of scientific relationships and processes. A conversion factor is a *ratio* (a *fraction*) made from two measured quantities that are *always equal* (such as 1 m ≡ 100 cm) or are *equivalent* in a given situation. A conversion factor is a fraction that equals *one*.

Conversion factors have a value of unity (1) because when a fraction has a **numerator** (top) and **denominator** (bottom) that are equal or equivalent, its value is *one*.

Example:

$$\frac{7}{7} = 1$$

Or, because 1 milliliter $= 10^{-3}$ liters,

$$\frac{10^{-3}\,L}{1\,mL} = \mathbf{1} \quad \text{and} \quad \frac{1\,mL}{10^{-3}\,L} = \mathbf{1}$$

These last two fractions are typical conversion factors. Any fraction that equals *one* right-side up will also equal *one* upside down. Any conversion factor can be inverted (flipped over) for use if necessary, and it will still equal one.

Example: When converting between liters and milliliters, all of these are legal conversion factors:

$$\frac{1 \text{ mL}}{10^{-3} \text{ L}} \qquad \frac{1000 \text{ mL}}{1 \text{ L}} \qquad \frac{10^3 \text{ mL}}{1 \text{ L}} \qquad \frac{3{,}000 \text{ mL}}{3 \text{ L}}$$

Upside down, each fraction is also a legal conversion factor because its numerator and denominator are equal.

In solving calculations, the conversions that are preferred (if available) are those that are made from *definitions*, such as "milli- $= 10^{-3}$." But all of the first three conversion-factor forms given above are familiar ways to define *milli*units, and any of the first three forms may be encountered in calculations shown in textbooks.

If the terms in a *series* are equal, any *two* of the terms can be used as a conversion factor.

Example: Because 1 meter $\equiv$ 10 decimeters $\equiv$ 100 centimeters $\equiv$ 1000 millimeters, each of the following (and others) is a legitimate conversion factor:

$$\frac{1000 \text{ mm}}{1 \text{ m}} \qquad \frac{1 \text{ mm}}{10^{-3} \text{ m}} \qquad \frac{10^2 \text{ cm}}{1 \text{ m}} \qquad \frac{100 \text{ cm}}{10 \text{ dm}} \qquad \frac{10 \text{ cm}}{1 \text{ dm}}$$

Each conversion factor represents a way to write the *ratio* between two terms in the series of equal terms.

TRY IT

(See "How to Use These Lessons," point 1, p. xvii.)

Let's work an example of conversion-factor math. Show your work on this page, then check your answer below.

Q. Multiply the following two terms:

$$7.5 \text{ km} \cdot \frac{10^3 \text{ m}}{1 \text{ km}} =$$

Answer:

$$7.5 \text{ k\!m} \cdot \frac{10^3 \text{ m}}{1 \text{ k\!m}} = \frac{(7.5 \cdot 10^3)}{1} \text{ m} = \mathbf{7.5 \times 10^3 \, m}$$

When these terms are multiplied, the "like units" on the top and bottom cancel, leaving meters as the unit on top.

Because the conversion factor multiplies the *given* quantity by *one*, the result is the same given *amount*, but measured in a different unit. Multiplying by the conversion is a form of proportional reasoning. To the *given* amount you apply the proportion: there are 1000 meters *per* 1 kilometer.

The answer means that 7,500 meters is the same as the given 7.5 kilometers.

The conversion above answers a question posed in many science problems: From the measurement units we are given, how can we obtain the units we want?

Conversion Factors

- Conversion factors are made from two measured quantities that are defined as equal *or* are equivalent in the problem.

- Conversion factors have a value of *one* because the top and bottom terms are equal or equivalent.

- Any *equality* can be made into a conversion (a *fraction* or *ratio*) equal to *one*.

PRACTICE

Try every other lettered problem. Check your answers frequently. If you miss one in a section, try a few more. For time abbreviations, use min (minute), hr (hour), and yr (year).

1. Multiply these terms. Cancel units that cancel, then group the numbers and do the math. Write the answer in scientific notation, and include the unit.

    a. $225 \text{ cg} \cdot \dfrac{10^{-2} \text{ g}}{1 \text{ cg}} \cdot \dfrac{1 \text{ kg}}{10^3 \text{ g}} =$

    b. $1.5 \text{ hr} \cdot \dfrac{60 \text{ min}}{1 \text{ hr}} \cdot \dfrac{60 \text{ s}}{1 \text{ min}} =$

2. To be legal, the top and bottom of conversion factors must be equal. Under each conversion, label these as *legal* or *illegal*.

    a. $\dfrac{1000 \text{ mL}}{1 \text{ L}}$ b. $\dfrac{1000 \text{ L}}{1 \text{ mL}}$ c. $\dfrac{1.00 \text{ g H}_2\text{O}}{1 \text{ mL H}_2\text{O}}$ d. $\dfrac{10^{-2} \text{ volt}}{1 \text{ centivolt}}$

    e. $\dfrac{1 \text{ mL}}{1 \text{ cm}^3}$ f. $\dfrac{10^3 \text{ cm}^3}{1 \text{ L}}$ g. $\dfrac{10^3 \text{ km}}{1 \text{ m}}$ h. $\dfrac{1 \text{ kilocalorie}}{10^3 \text{ calories}}$

3. Place a 1 in front of the unit with a prefix, then complete the conversion factor.

    a. $\dfrac{\text{g}}{\text{kg}}$ b. $\dfrac{\text{s}}{\text{ns}}$ c. $\dfrac{\text{picometer}}{\text{meter}}$ d. $\dfrac{\text{centigram}}{\text{gram}}$

4. Use a calculator. Multiply the terms and write the answer with its unit.

    a. $\dfrac{95 \text{ km}}{\text{hr}} \cdot \dfrac{0.621 \text{ mi.}}{1 \text{ km}} =$ b. $\dfrac{27 \text{ m}}{\text{s}} \cdot \dfrac{60 \text{ s}}{1 \text{ min}} \cdot \dfrac{1 \text{ km}}{10^3 \text{ m}} =$

5. Multiply. Do the math without a calculator, and write the answer with its unit.

    $2.5 \text{ m} \cdot \dfrac{1 \text{ cm}}{10^{-2} \text{ m}} =$

Lesson 4.2   Single-Step Conversions

In the previous lesson, conversion factors were supplied. In this lesson, you will learn to make your own conversion factors to solve problems. Let's learn the method with a simple example.

────────────────────────────── **TRY IT** ──────────────────────────────

Q. How many years is 925 days?

In your notebook, write an answer to each step below.

Steps for Solving with Conversion Factors

1. Begin by writing a question mark (?) and then the unit you are looking for in the problem: the answer unit that is WANTED in the problem.

2. Next, write an equal (=) sign. It means, "Okay, that part of the problem is done. From here on, leave the WANTED unit alone." You don't cancel the WANTED unit, and you don't multiply by it.

3. After the = sign, write the number and unit you are *given* (the known quantity).

At this point, in your notebook should be

? years = 925 days

4. Next, write a • or × and then a line _____ for a conversion factor to multiply by.

5. A key step: Write the *unit* of the *given* quantity in the denominator (on the bottom) of the conversion factor. Leave room for a number in front.

   Do *not* put the given *number* in the conversion factor—just the given *unit*.

$$? \text{ years} = 925 \text{ days} \cdot \frac{\qquad}{\textbf{days}}$$

This step puts the *given* unit where it must be to cancel. It also tells you one unit that the next conversion must include.

6. Next, write the answer unit on the top of the conversion factor.

$$? \text{ years} = 925 \text{ days} \cdot \frac{\textbf{year}}{\text{days}}$$

7. Add *numbers* that make the numerator and denominator of the conversion factor *equal*. In a legal conversion factor, the top and bottom quantities must be equal or equivalent.

8. Cancel the *units* that you set up to cancel.

9. *If* the unit on the right side after cancellation is the answer unit, stop adding conversions. Write an = sign. Multiply the *given* quantity by the conversion factor. Write the number and the uncanceled unit. Done!

Finish the above steps, then check your answer below.

STOP

Answer:

$$? \text{ years } = 925 \,\cancel{\text{days}} \cdot \frac{\mathbf{1} \text{ year}}{\mathbf{365} \,\cancel{\text{days}}} = \frac{925 \text{ years}}{365} = \mathbf{2.53 \text{ years}}$$

Significant Figures: **1** is *exact*, 925 has 3 *s.f.*, 365 is not exact (1 year = 365.24 days is more precise) so 365 has 3 *s.f.* Round the answer to 3 *s.f.*

You may need to look back at the above steps, but you should not need to memorize them. By doing the following problems, you will quickly gain an intuitive sense of what to do to solve problems.

To summarize:

- When the units are set up to cancel correctly, the *given* numbers and units, multiplied by conversions, will result in the WANTED numbers and units.

- *Units* tell you where to write the *numbers* to solve a calculation correctly.

PRACTICE

If these are easy, do the last lettered part of each numbered problem. If you miss one, do another part of that problem.

Round answers to the correct number of significant figures. After each numbered problem, check your answers.

1. Add numbers to make legal conversions, cancel units that cancel, multiply, and write your answer.

    a. $? \text{ days} = 96 \text{ hr} \cdot \dfrac{\text{days}}{24 \text{ hr}} =$

    b. $? \text{ ml} = 3.50 \text{ L} \cdot \dfrac{1 \text{ mL}}{\text{L}} =$

2. To start these, put the *unit* of the *given* quantity where it will cancel. Then finish the conversion factor, do the math, and write your answer with its unit.

    a. $? \text{ s} = 0.25 \text{ min} \cdot \dfrac{\text{s}}{1} =$

    b. $? \text{ kg} = 250 \text{ g} \cdot \dfrac{\text{kg}}{10^3} =$

    c. $? \text{ days} = 2.73 \text{ yr} \cdot \dfrac{365}{} =$

    d. $? \text{ yr} = 200. \text{ days} \cdot \dfrac{1}{} =$

(continued)

3. It is helpful to keep in mind this "single-unit starting template."

> When solving for single units, begin with
>
> ? unit WANTED = number and unit *given* • _____
>
> $\overline{\qquad\qquad}$ **unit *given***

The template emphasizes that your first conversion factor puts the given *unit* (but *not* the given number) where it will cancel. Apply the template rule to these.

a. ? months = 5.0 yr • _____ =

b. ? L = 350 mL •

4. Use the starting template to find how many hours equal 390 minutes.

?

5. ? mg = 0.85 kg • $\dfrac{\text{g}}{\text{kg}}$ • $\dfrac{\qquad}{\text{g}}$ =

Lesson 4.3   Multistep Conversions

In problem 5 at the end of the previous lesson, we did not know a direct conversion from kilograms to milligrams. But we knew a conversion from kilograms to grams and another from grams to milligrams.

In most problems, you will not know a single conversion from the *given* to the WANTED unit, but there will be known conversions that you can *chain together* to solve the calculation.

▸ TRY IT

Q. Complete this two-step conversion as done in problem 5 from Lesson 4.2. Answer in scientific notation.

? ms = 0.25 min

STOP

Answer:

$$? \text{ ms} = 0.25 \text{ min} \cdot \frac{60 \text{ s}}{1 \text{ min}} \cdot \frac{1 \text{ ms}}{10^{-3} \text{ s}} = 15 \times 10^3 \text{ ms} = \mathbf{1.5 \times 10^4 \text{ ms}}$$

The 0.25 has 2 *s.f.*, and both conversions are exact definitions that do not affect the significant figures in the answer, so the answer must be rounded to 2 *s.f.*

Mathematically, multiplying by conversions can be viewed as applying successive proportions to the *given* amount. Because each conversion has a value of 1, the unit of measure changes, but the quantity that was *given* remains equivalent.

Apply the following rules to arrange conversions.

Using Multiple Conversions to Solve for Single Units

- If a unit remaining on the right side after you cancel units is *not* the answer unit, write it in the next conversion factor where it will cancel.
- Finish the next conversion by writing a known conversion, one that either *includes* the answer unit or gets you *closer* to the answer unit.
- In writing a conversion, first set up a *unit* to cancel, then finish by adding the second unit and *numbers* in front of both units that make the conversion legal.

PRACTICE

The first two problems occur in pairs. If part a is easy, go to part a of the next question. If you need help with part a, do part b for more practice. Answers must *always* include an answer *unit* and proper number of *s.f.*

1a. ? Gg = 760 mg • _____ g • _____ =

1b. ? cg = 4.2 kg • _____ • _____ =
$$g$$

2a. ? yr = 2.63×10^4 hr • _____ • _____ =

2b. ? s = 1.00 day • _____ hr • _____ • _____ =

3. ? µg $H_2O(\ell)$ = 1.5 cm³ $H_2O(\ell)$ _____ g $H_2O(\ell)$ • _____ =

4. In your problem notebook, draw the chart that is supplied with problem 9 of the Review Quiz at the end of Chapter 3. Practice until, given a blank chart, you can write the names and symbols for the atoms with atomic numbers **1–18** in their proper locations.

Lesson 4.4   Conversions between the English and Metric Systems

Most conversions we have used so far have had an exact number **1** on either the top or the bottom. But a 1 is not required in a legal conversion. Both "1 kilometer = 1000 meters" and "3 kilometers = 3,000 meters" are true equalities, and both equalities could be used to make legal conversion factors. In most cases, however, conversions with a 1 are preferred.

Why? We want conversions to be *familiar* so that we can write them automatically and quickly check that they are correct. Definitions are usually based on *one* component, such as "1 kilometer = 10^3 meters." Because they are the most familiar equalities, definitions are preferred in conversions.

But some conversions may be familiar even if they do not include a **1**. For example, many cans of soft drinks are labeled "12.0 fluid ounces (355 mL)." This supplies an equality for English-to-metric volume units: 12.0 fluid ounces = 355 mL. That is a legal conversion, and, because its numbers and units are seen often, it is easy to remember and check.

Bridge Conversions

Scientific calculations often involve a conversion that provides a *bridge* between one unit system, quantity, or substance and another. Conversions between the metric and English systems provide practice in the *bridge-conversion methods* that we will use to solve chemical reaction calculations.

An example of a *bridge conversion* between distance units of the metric and English systems is

$$2.54 \text{ centimeters} \equiv 1 \text{ inch}$$

In countries that use English units, this is now the exact definition of an inch.

Any metric–English distance equality can be used to convert between distance measurements in the two systems. Another metric–English distance conversion that is frequently used (but is not exact) is 1 kilometer = 0.621 mile.

In problems that require conversion from one system or substance to another, our strategy will be to "head for the bridge"—to begin by converting to one of the two units in the bridge conversion. When a problem needs a bridge conversion between systems, use these steps.

1. First, convert the *given* unit to the unit in the *bridge conversion* that is in the *same system* as the *given* unit.

2. Next, multiply by the bridge conversion. In the bridge conversion, place the *given* system on the bottom and the WANTED system on top.

3. Multiply by other conversions in the WANTED system to get the answer *unit* WANTED.

Add these English distance-unit definitions to your list of memorized conversions:

12 inches ≡ 1 foot   3 feet ≡ 1 yard   5,280 feet ≡ 1 mile

(English system abbreviations: inch = in., foot = ft., yard = yd., and mile = mi.)

Also commit to memory this metric-to-English bridge conversion for distance:

2.54 cm ≡ 1 inch

▶ **TRY IT**

Q. Apply the steps and conversions above to solve this problem:

STOP

? feet = 1.00 meter

Answer:

1. Because the WANTED unit is English and the *given* unit is metric, a metric-to-English bridge is needed. Head for the bridge.

   Because the *given* unit (meters) is in the metric system, convert to the *metric unit* used in the bridge conversion: *centimeters.*

   $$? \text{ ft.} = 1.00 \text{ m} \cdot \frac{1 \text{ cm}}{10^{-2} \text{ m}} \cdot \frac{}{\text{cm}}$$

   Note the start of the next conversion. Because centimeters is not the wanted answer unit, centimeters *must* be put in the next conversion where it will cancel. If you *start* the "next unit to cancel" conversion automatically after finishing the prior conversion, it helps to arrange and choose the next conversion.

   Your "m-to-cm" conversion could also be 100 cm/1 m. In all conversion calculations, a conversion that is equivalent may be used and will not change the answer.

   Adjust and complete the steps if needed.

STOP

2. Complete the bridge that converts to the *system* of the answer: English units.

   $$? \text{ ft.} = 100 \text{ m} \cdot \frac{1 \text{ cm}}{10^{-2} \text{ m}} \cdot \frac{1 \text{ in.}}{2.54 \text{ cm}} \cdot \frac{}{\text{in.}}$$

STOP

3. Get rid of the unit you've got. Get to the unit you WANT.

   $$? \text{ ft.} = 1.00 \text{ m} \cdot \frac{1 \text{ cm}}{10^{-2} \text{ m}} \cdot \frac{1 \text{ in.}}{2.54 \text{ cm}} \cdot \frac{1 \text{ ft.}}{12 \text{ in.}} = 3.28 \text{ ft.}$$

   For *s.f.*, 1.00 has 3 *s.f.*, and all of the other numbers are part of exact definitions, so the answer is rounded to 3 *s.f.*

   The answer tells us that 1.00 meter (the *given* quantity) is equal to 3.28 feet.

Some science problems take 10 or more conversions to solve. But if you know that a bridge conversion is needed, "heading for the bridge" breaks the problem into pieces that will simplify your navigation to the answer.

PRACTICE

Use the inch-to-centimeter bridge conversion above. Start by doing every other problem. Do more if you need more practice. Be sure to do problem 7.

1. ? cm = 12.0 in. • _____ =

2. ? in. = 1.00 m • _____ • _____

(continued)

3. For the problem: ? in. = 760. mm

    a. To what unit do you aim to convert the *given* in the initial conversions? Why?

    b. Solve: ? in. = 760. mm

4. Using 1 in. = 2.54 cm,

    Solve: ? km = 1.00 mi.

5. To bridge between metric and English units, use 1 kg = 2.2 lb. (lb. is the abbreviation for "pounds").

    ? g = 7.7 lb.

6. Use the "soda can" volume conversion (12.0 fl. oz. = 355 mL).

    ? fl. oz. = 2.00 L

7. In your problem notebook, draw the chart below. Practice until you can write in the chart the names and symbols for the atoms with atomic numbers **1–20** in their proper locations.

PERIODIC TABLE

Lesson 4.5   Ratio-Unit Conversions

The order in which numbers are multiplied does not affect the result. For example, 1 × 2 × 3 has the same answer as 3 × 2 × 1. The same is true when multiplying symbols or units. Our rules for arranging conversions provide a sequence that is easy to set up and understand, but mathematically the order of multiplication does not affect the answer.

The following problem is an example of how units can cancel in separated as well as adjacent conversions.

TRY IT

Q. Cancel units that cancel, then multiply. Write the answer number and unit.

$$\frac{12 \ m}{s} \cdot \frac{60 \ s}{1 \ min} \cdot \frac{60 \ min}{1 \ hr} \cdot \frac{1 \ km}{1000 \ m} \cdot \frac{0.62 \ mi.}{1 \ km} =$$

Answer:

$$\frac{12 \ \cancel{m}}{s} \cdot \frac{60 \ s}{1 \ \cancel{min}} \cdot \frac{60 \ \cancel{min}}{1 \ hr} \cdot \frac{1 \ \cancel{km}}{1000 \ \cancel{m}} \cdot \frac{0.62 \ mi.}{1 \ \cancel{km}} = \frac{\textbf{27 mi.}}{\textbf{hr}}$$

This answer means that a speed of 12 meters/second is the *same* as 27 miles/hour.

Ratio Units in the Answer

In these lessons, we will use the term *single unit* to describe a unit that has one base unit in the numerator but no denominator (which means the denominator is 1). The base units may have prefixes or powers. Meters, kilograms, minutes, milliliters, and cubic centimeters are all single units.

We will use *ratio unit* to describe a fraction that has *one* kind of base unit in the numerator and *one* different kind of base unit in the denominator. If a problem asks you to find

$$\text{meters per second} \quad or \quad \text{meters/second} \quad or \quad \frac{\text{meters}}{\text{second}} \quad or \quad \text{m} \cdot \text{s}^{-1}$$

all of those terms are equivalent, and the problem is asking for a ratio unit. During *conversion* calculations, ratio units should be written in the *fraction* form (with a top and bottom) rather than the form using units with negative exponents.

In Chapter 11, we will address in detail the different characteristics of single units and ratio units. For now, the distinctions above will allow us to solve problems.

Converting the Denominator

When solving for single units, we have used a starting template that includes canceling a *given* single unit.

> When solving for a single unit, begin with:
>
> ? unit WANTED = # and *unit given* • $\dfrac{\rule{3cm}{0.4pt}}{\textit{unit given}}$

When solving for ratio units, we may need to cancel a denominator (bottom) unit to start a problem. To do so, we will loosen our *starting* rule to say the following.

Solving with Conversion Factors

1. If a unit to the right of the equal (=) sign, in or after the *given*, on the top or the bottom:
   • *Matches* a unit in the answer unit, in both what it is and where it is, (circle) that unit on the right side and do not convert it further.
   • Is *not* what you WANT, put it where it will cancel, and convert until it matches what you WANT.

2. After canceling units, if the unit or units to the right of the equal sign *match* the WANTED answer unit, no more conversions are needed. Write an = sign, do the math, and write the answer.

➤ **TRY IT**

Q. Use the rule above to solve this problem.

$$\frac{?\ cm}{min} = 0.50\ \frac{cm}{s} \cdot \underline{\hspace{2cm}} =$$

STOP

Answer:

$$\frac{?\ cm}{min} = 0.50\ \frac{\boxed{cm}}{s} \cdot \frac{60\ s}{1\ \boxed{min}} = \frac{30.\ cm}{min}$$

Start by comparing the WANTED units to the *given* units.

Because you WANT cm on top, and are *given* cm on top, circle (cm) to say, "The top is done. Leave the top alone."

On the bottom, you have seconds, but you WANT minutes. Put seconds where it will cancel. Convert to minutes on the bottom.

When the units on the right of the equal sign match the units you WANT in the answer unit, stop conversions and do the math.

PRACTICE   A

Do problem 2. Then do problem 1 if you need more practice.

1. (Do not use a calculator.) $\dfrac{?\ g}{dL} = 355\ \dfrac{g}{L} \cdot \underline{\hspace{3cm}} =$

2. $\dfrac{?\ m}{s} = 4.2 \times 10^5\ \dfrac{m}{hr} \cdot \underline{\hspace{2.5cm}} \cdot \underline{\hspace{2.5cm}} =$

Converting Both Top and Bottom Units

Many problems require converting *both* numerator and denominator units. In the following problem, an order to convert both units is specified.

> ▶ **TRY IT**

Q. Write what must be placed in the blanks to make legal conversions, cancel units, do the math, and write the answer.

$$? \; \frac{m}{s} = 740 \; \frac{cm}{min} \cdot \frac{\underline{\qquad}}{cm} \cdot \frac{min}{\underline{\qquad}} =$$

🛑 **Answer:**

$$? \; \frac{m}{s} = 740 \; \frac{cm}{min} \cdot \frac{10^{-2} \, \text{ⓜ}}{1 \; cm} \cdot \frac{1 \; min}{60 \; \text{Ⓢ}} = 0.12 \; \frac{m}{s}$$

In the *given* on the right, cm is *not* the unit WANTED on top, so put it where it will cancel, and convert to the unit you want on top.

Next, because minutes are on the bottom on the right, but seconds are WANTED, put minutes where it will cancel. Convert to the seconds WANTED.

When chaining conversions, which unit you convert first—the top or bottom unit—makes no difference. The order in which you multiply factors does not change the answer.

> ▶ **TRY IT**

In the following problem, no order for the conversions is specified.

Q. Add legal conversions in any order, solve, then check your answer below. Try without a calculator.

$$? \; \frac{cg}{L} = 0.550 \times 10^{-2} \; \frac{g}{mL}$$

🛑 **Answer:**

Your conversions may be in a different order.

$$? \; \frac{cg}{L} = 0.550 \times 10^{-2} \; \frac{g}{mL} \cdot \frac{1 \, \text{ⓒⓖ}}{10^{-2} \, g} \cdot \frac{1 \; mL}{10^{-3} \, \text{Ⓛ}} = 5.50 \times 10^{2} \; \frac{cg}{L}$$

PRACTICE   B

Do every other part, and do more if you need more practice.

1. In these, an order of conversion is specified. Write what must be placed in the blanks to make legal conversions, then solve.

   a. $? \dfrac{\text{mi.}}{\text{hr}} = \dfrac{80.7 \text{ ft.}}{\text{s}} \cdot \dfrac{\text{mi.}}{\underline{\hspace{2em}}} \cdot \dfrac{\underline{\hspace{2em}}}{\text{min}} \cdot \underline{\hspace{2em}} =$

   b. $? \dfrac{\text{m}}{\text{s}} = \dfrac{250. \text{ ft.}}{\text{min}} \cdot \dfrac{\text{min}}{\underline{\hspace{2em}}} \cdot \dfrac{\underline{\hspace{2em}}}{\underline{\hspace{2em}}} \cdot \dfrac{\underline{\hspace{2em}}}{1 \text{ in.}} \cdot \underline{\hspace{2em}} =$

2. Add conversions in any order and solve.

   a. $? \dfrac{\text{km}}{\text{hr}} = \dfrac{1.17 \times 10^4 \text{ mm}}{\text{s}}$

   b. Solve part b without a calculator. Answer in scientific notation.

   $? \dfrac{\text{ng}}{\text{mL}} = \dfrac{47 \times 10^2 \text{ mg}}{\text{dm}^3}$

   c. $? \dfrac{\text{ft.}}{\text{s}} = \dfrac{95 \text{ m}}{\text{min}}$

Flashcards

The best flashcards are those you design yourself as you work through a chapter. Thinking about what should be on a card helps to move new knowledge into memory.

Design two-sided cards when possible: Two-sided cards prompt additional recall. Some flashcard possibilities for this chapter might include the following.

One-Sided Cards

Front Side (with Notch at Top Right)	Back Side—Answers
Conversion factors are made from _____	Two amounts equal or, in a problem, equivalent
Conversion factors have a value of _____	1
1 mile = ? feet	5,280 feet

Two-Sided Card

Stop adding conversions when?	When unit WANTED = unit remaining on right

More Two-Sided Cards

1 foot = ? in.	12 in. = 1 ?	1 in. = ? cm	2.54 cm = ?
1 yard = ? feet	3 feet = ? yards	1 kg = ? lb.	2.20 lb. = 1 ?

SUMMARY

Previous chapters have been primarily about *factual* knowledge: learning new definitions and vocabulary. In this chapter, we applied those facts using *procedures*. Procedural knowledge can be "automated" by applying a stepwise procedure repeatedly to a variety of problems. Being able to recall a procedure automatically opens slots in working memory to see the "bigger picture": patterns and concepts.

To be "quiz ready" for this chapter, from memory you should be able to:

- Write the "single-unit starting template."
- Answers your flashcards quickly.
- Convert measurement units using definitions from the metric and English systems.
- Fill in a blank periodic table with the names and symbols of the elements with atomic numbers 1–20.

When solving with conversions, the following "organizing concepts" are helpful to remember.

1. Conversion factors are fractions made from two entities that are equal or equivalent. Conversions can be described as ratios or proportions or fractions that have a value of one.
2. Chain conversions so that units cancel to get rid of the unit you've got and get to the unit you WANT.
3. When the unit on the right is the unit of the answer on the left, stop use of conversion factors. Do the number math. Write the answer with its unit.

REVIEW QUIZ

1. Label these as legal or illegal conversion factors:

    a. $\dfrac{1\ \text{L}}{10^3\ \text{cm}^3}$ b. $\dfrac{1\ \text{dm}^3}{1\ \text{L}}$ c. $\dfrac{100\ \text{g}}{1\ \text{cg}}$ d. $\dfrac{1\ \text{Mg}}{10^{-6}\ \text{g}}$

Solve problems 2–5 in your notebook.

2. Without a calculator, divide: $6\overline{)204}$

3. ? hours $= 1.53 \times 10^4$ seconds

4. ? inches $= 2.75$ meters

5. ? feet $= 2.00$ km

6. Try without a calculator. Convert your final answer to scientific notation.

$$\frac{?\ \text{m}}{\text{s}} = 180 \times 10^5\ \frac{\text{km}}{\text{min}}$$

7. Using 1 kg = 2.20 lb., solve:

$$?\ \text{mg} = 4.0 \times 10^{-2}\ \text{lb.}$$

8. $\dfrac{?\ \text{kg}}{\text{mL}} = \dfrac{2.4 \times 10^5\ \mu\text{g}}{\text{dm}^3}$

ANSWERS

Lesson 4.1

Practice

1a. $225 \text{ cg} \cdot \dfrac{10^{-2}\text{ g}}{1\text{ cg}} \cdot \dfrac{1\text{ kg}}{10^{3}\text{ g}} = \dfrac{225 \times 10^{-2} \times 1}{1 \times 10^{3}}\text{ kg} = \textbf{2.25} \times \textbf{10}^{-3}\textbf{ kg}$

1b. $1.5 \text{ hr} \cdot \dfrac{60\text{ min}}{1\text{ hr}} \cdot \dfrac{60\text{ s}}{1\text{ min}} = \dfrac{1.5 \times 60 \times 60}{1}\text{ s} = \textbf{5,400 s } \textit{or}\textbf{ 5.4} \times \textbf{10}^{3}\textbf{ s}$

Recall that **s** is the abbreviation for seconds. This answer means that 1.5 hours is equal to 5,400 seconds.

2a. $\dfrac{1000\text{ mL}}{1\text{ L}}$ **Legal**
2b. $\dfrac{1000\text{ L}}{1\text{ mL}}$ **Illegal**
2c. $\dfrac{1.00\text{ g }H_2O}{1\text{ mL }H_2O}$ **Legal if liquid water**
2d. $\dfrac{10^{-2}\text{ volt}}{1\text{ centivolt}}$ **Legal**
2e. $\dfrac{1\text{ mL}}{1\text{ cm}^3}$ **Legal**

2f. $\dfrac{10^{3}\text{ cm}^3}{1\text{ L}}$ **Legal**
2g. $\dfrac{10^{3}\text{ km}}{1\text{ m}}$ **Illegal**
2h. $\dfrac{1\text{ kilocalorie}}{10^{3}\text{ calories}}$ **Legal**

3a. $\dfrac{\textbf{10}^{3}\textbf{ g}}{\textbf{1 kg}}$
3b. $\dfrac{\textbf{10}^{-9}\textbf{ s}}{\textbf{1 ns}}$
3c. $\dfrac{\textbf{1 picometer}}{\textbf{10}^{-12}\textbf{ meter}}$
3d. $\dfrac{\textbf{1 centigram}}{\textbf{10}^{-2}\textbf{ gram}}$

4a. $\dfrac{95 \text{ km}}{\text{hr}} \cdot \dfrac{0.621\text{ mi}}{1\text{ km}} = \dfrac{95 \cdot 0.621}{1}\dfrac{\text{mi.}}{\text{hr}} = \textbf{59}\dfrac{\textbf{mi.}}{\textbf{hr}}$
4b. $\dfrac{27\text{ m}}{\text{s}} \cdot \dfrac{60\text{ s}}{1\text{ min}} \cdot \dfrac{1\text{ km}}{10^{3}\text{ m}} = \dfrac{27 \cdot 60}{10^{3}}\dfrac{\text{km}}{\text{min}} = \textbf{1.6}\dfrac{\textbf{km}}{\textbf{min}}$

5. $2.5 \text{ m} \cdot \dfrac{1\text{ cm}}{10^{-2}\text{ m}} = 2.5 \times 10^{2}\text{ cm} = \textbf{250 cm}$

Lesson 4.2

Practice

For visibility, not all cancellations are shown, but cancellations should be marked on your paper. For your final answer to be correct, it must include its unit and the same number of significant figures as in these answers.

In all calculations, your conversions may be in different formats, such as 1 m = 100 cm *or* 1 cm = 10^{-2} m, as long as the conversion's top and bottom are equal and your answer is the same as shown here.

1a. $? \text{ days} = 96 \text{ hr} \cdot \dfrac{\textbf{1 day}}{24\text{ hr}} = \dfrac{96}{24}\text{ days} = \textbf{4.0 days}$

1b. $? \text{ mL} = 3.50 \text{ L} \cdot \dfrac{\textbf{1 mL}}{\textbf{10}^{-3}\textbf{ L}} = 3.50 \cdot 10^{3}\text{ mL} = \textbf{3.50} \times \textbf{10}^{3}\textbf{ mL}$

(*S.F.*: 3.50 has 3 *s.f.*, prefix definitions are exact with infinite *s.f.*, answer is rounded to 3 *s.f.*)

2a. $? \text{ s} = 0.25 \text{ min} \cdot \dfrac{\textbf{60 s}}{1\text{ min}} = (0.25 \cdot 60)\text{ s} = \textbf{15 s}$

(*S.F.*: 0.25 has 2 *s.f.*, 1 min = 60 s is a definition with infinite *s.f.*, answer is rounded to 2 *s.f.*)

2b. $? \text{ kg} = 250 \text{ g} \cdot \dfrac{\textbf{1 kg}}{10^{3}\text{ g}} = \dfrac{250}{10^{3}}\text{ kg} = \textbf{0.25 kg}$

2c. $? \text{ days} = 2.73 \text{ yr} \cdot \dfrac{365\text{ \textbf{days}}}{\textbf{1 yr}} = (2.73 \cdot 365)\text{ days} = \textbf{996 days}$

2d. $? \text{ yr} = 200.\text{ days} \cdot \dfrac{\textbf{1 yr}}{\textbf{365 days}} = \dfrac{200}{365}\text{ yr} = \textbf{0.548 yr}$

3a. $? \text{ months} = 5.0 \text{ yr} \cdot \dfrac{\textbf{12 months}}{\textbf{1 yr}} = \textbf{60. months}$

(*S.F.*: 5.0 has 2 *s.f.*, 12 months = 1 yr is a definition with infinite *s.f.*, round to 2 *s.f.*, the 60. decimal means 2 *s.f.*)

3b. $? \text{ L} = 350 \text{ mL} \cdot \dfrac{\textbf{10}^{-3}\textbf{L}}{\textbf{1 mL}} = 350 \times 10^{-3}\,\text{L} = \textbf{0.35 L}$

(*S.F.*: 350 has 2 *s.f.*, prefix definitions are *exact* with infinite *s.f.*, round to 2 *s.f.*)

4. $\textbf{? hr} = \textbf{390 min} \cdot \dfrac{\textbf{1 hr}}{\textbf{60 min}} = \textbf{6.5 hr}$

5. $? \text{ mg} = 0.85 \text{ kg} \cdot \dfrac{\textbf{10}^3\textbf{g}}{\textbf{1 kg}} \cdot \dfrac{\textbf{1 mg}}{\textbf{10}^{-3}\textbf{g}} = 0.85 \times 10^6\,\text{mg} = \textbf{8.5} \times \textbf{10}^5\,\textbf{mg}$

Lesson 4.3

Practice

1a. $? \text{ Gg} = 760 \text{ mg} \cdot \dfrac{10^{-3}\text{g}}{1 \text{ mg}} \cdot \dfrac{1 \text{ Gg}}{10^9\text{g}} = 760 \times 10^{-12}\,\text{Gg} = \textbf{7.6} \times \textbf{10}^{-10}\,\textbf{Gg}$

1b. $? \text{ cg} = 4.2 \text{ kg} \cdot \dfrac{10^3\text{g}}{1 \text{ kg}} \cdot \dfrac{1 \text{ cg}}{10^{-2}\text{g}} = \textbf{4.2} \times \textbf{10}^5\,\textbf{cg}$

2a. $? \text{ yr} = 2.63 \times 10^4 \text{ hr} \cdot \dfrac{1 \text{ day}}{24 \textbf{ hr}} \cdot \dfrac{\textbf{1 yr}}{365 \text{ days}} = \dfrac{2.63 \times 10^4}{24 \cdot 365}\,\text{yr} = \textbf{3.00 yr}$

2b. $? \text{ s} = 1.00 \text{ day} \cdot \dfrac{24 \textbf{ hr}}{1 \text{ day}} \cdot \dfrac{60 \textbf{ min}}{1 \textbf{ hr}} \cdot \dfrac{60 \text{ s}}{1 \textbf{ min}} = \textbf{8.64} \times \textbf{10}^4\,\textbf{s}$

3. $? \,\mu\text{g H}_2\text{O}(\ell) = 1.5 \text{ cm}^3\,\text{H}_2\text{O}(\ell) \cdot \dfrac{1.00 \text{ g H}_2\text{O}(\ell)}{1 \text{ cm}^3\,\text{H}_2\text{O}(\ell)} \cdot \dfrac{1 \,\mu\text{g}}{10^{-6}\text{g}} = \textbf{1.5} \times \textbf{10}^6\,\boldsymbol{\mu}\textbf{g H}_2\textbf{O}(\boldsymbol{\ell})$

Lesson 4.4

Practice

The definition 1 cm = 10 mm may be used for *mm*-to-*cm* conversions. Doing so will change the number of conversions but not the answer.

To be correct, answers must be written that include correct units and the same number of significant figures as in these answers.

1. $? \text{ cm} = 12.0 \text{ in.} \cdot \dfrac{\textbf{2.54 cm}}{\textbf{1 in.}} = 12.0 \cdot 2.54 \text{ cm} = \textbf{30.5 cm}$ (Check how many centimeters are on a 12-inch ruler.)

2. $? \text{ in.} = 1.00 \text{ m} \cdot \dfrac{1 \text{ cm}}{10^{-2}\text{ m}} \cdot \dfrac{\textbf{1 in.}}{\textbf{2.54 cm}} = \dfrac{1}{2.54} \times 10^2\,\text{in.} = 0.394 \times 10^2\,\text{in.} = \textbf{39.4 in.}$

3a. Aim to convert the *given* unit (mm) to the unit in the *bridge conversion* that is in the same system (English or metric) as the *given*. The bridge unit **cm** is in the same measurement *system* as mm.

3b. $? \text{ in.} = 760. \text{ mm} \cdot \dfrac{10^{-3}\text{m}}{1 \text{ mm}} \cdot \dfrac{1 \text{ cm}}{10^{-2}\text{m}} \cdot \dfrac{\textbf{1 in.}}{\textbf{2.54 cm}} = 760 \times 10^{-1}\,\text{in.} = \textbf{29.9 in.}$

(*S.F.*: 760., with the *decimal* after the zero, means 3 *s.f.* Metric definitions and the number 1 have infinite *s.f.* The answer must be rounded to 3 *s.f.*)

4. $? \text{ km} = 1.00 \text{ } mi. \cdot \dfrac{5{,}280 \text{ ft.}}{1 \text{ } mi} \cdot \dfrac{12 \text{ in}}{1 \text{ ft}} \cdot \dfrac{\textbf{2.54 cm}}{\textbf{1 in.}} \cdot \dfrac{10^{-2}\text{m}}{1 \text{ cm}} \cdot \dfrac{1 \text{ km}}{10^3\text{m}} = \textbf{1.61 km}$

5. $? \text{ g} = 7.7 \text{ lb.} \cdot \dfrac{1 \text{ kg}}{2.2 \text{ lb.}} \cdot \dfrac{10^3 \text{ g}}{1 \text{ kg}} = \mathbf{3.5 \times 10^3 \text{ g}}$

   (*S.F.*: 7.7 and 2.2 have 2 *s.f.* The number 1 and all metric definitions have infinite *s.f.* Round the answer to 2 *s.f.*)

6. $? \text{ fl. oz.} = 2.00 \text{ L} \cdot \dfrac{1 \text{ mL}}{10^{-3} \text{ L}} \cdot \dfrac{12.0 \text{ fl. oz.}}{355 \text{ mL}} = \mathbf{67.6 \text{ fl. oz.}}$ (Check this answer on any 2-liter soda bottle.)

Lesson 4.5

Practice A   1. $? \dfrac{\text{g}}{\text{dL}} = 355 \dfrac{\text{ⓖ}}{\text{L}} \cdot \dfrac{10^{-1} \text{ L}}{1 \text{ ⓓⓁ}} = \mathbf{35.5 \dfrac{\text{g}}{\text{dL}}}$

2. $? \dfrac{\text{m}}{\text{s}} = 4.2 \times 10^5 \dfrac{\text{ⓜ}}{\text{hr}} \cdot \dfrac{1 \text{ hr}}{60 \text{ min}} \cdot \dfrac{1 \text{ min}}{60 \text{ⓢ}} = \mathbf{1.2 \times 10^2 \dfrac{\text{m}}{\text{s}}}$

Practice B   1a. $? \dfrac{\text{mi.}}{\text{hr}} = 80.7 \dfrac{\text{ft.}}{\text{s}} \cdot \dfrac{1 \text{ ⓜⓘ}}{5{,}280 \text{ ft.}} \cdot \dfrac{60 \text{ s}}{1 \text{ min}} \cdot \dfrac{60 \text{ min}}{1 \text{ ⓗⓡ}} = \mathbf{55.0 \dfrac{\text{mi.}}{\text{hr}}}$ (3 *s.f.*)

1b. $? \dfrac{\text{m}}{\text{s}} = 250. \dfrac{\text{ft.}}{\text{min}} \cdot \dfrac{1 \text{ min}}{60 \text{ⓢ}} \cdot \dfrac{12 \text{ in.}}{1 \text{ ft.}} \cdot \dfrac{2.54 \text{ cm}}{1 \text{ in.}} \cdot \dfrac{10^{-2} \text{ⓜ}}{1 \text{ cm}} = \mathbf{1.27 \dfrac{\text{m}}{\text{s}}}$

   (*S.F.*: 250. has 3 *s.f.* because of the decimal. All other conversions are definitions. The answer is rounded to 3 *s.f.*)

2a. $? \dfrac{\text{km}}{\text{hr}} = 1.17 \times 10^4 \dfrac{\text{mm}}{\text{s}} \cdot \dfrac{10^{-3} \text{ m}}{1 \text{ mm}} \cdot \dfrac{1 \text{ ⓚⓜ}}{10^3 \text{ m}} \cdot \dfrac{60 \text{ s}}{1 \text{ min}} \cdot \dfrac{60 \text{ min}}{1 \text{ ⓗⓡ}} = \mathbf{42.1 \dfrac{\text{km}}{\text{hr}}}$

2b. $? \dfrac{\text{ng}}{\text{mL}} = 47 \times 10^2 \dfrac{\text{mg}}{\text{dm}^3} \cdot \dfrac{1 \text{ dm}^3}{1 \text{ L}} \cdot \dfrac{10^{-3} \text{ L}}{1 \text{ ⓜⓁ}} \cdot \dfrac{10^{-3} \text{ g}}{1 \text{ mg}} \cdot \dfrac{1 \text{ ⓝⓖ}}{10^{-9} \text{ g}} = \mathbf{4.7 \times 10^6 \dfrac{\text{ng}}{\text{mL}}}$

2c. Hint: A metric-to-English bridge conversion for distance units is needed. Head for the bridge: first convert the given metric distance unit to the metric distance unit used in your known bridge conversion.

$? \dfrac{\text{ft.}}{\text{s}} = 95 \dfrac{\text{m}}{\text{min}} \cdot \dfrac{1 \text{ min}}{60 \text{ⓢ}} \cdot \dfrac{1 \text{ cm}}{10^{-2} \text{ m}} \cdot \dfrac{1 \text{ in.}}{2.54 \text{ cm}} \cdot \dfrac{1 \text{ ⓕⓣ}}{12 \text{ in.}} = \mathbf{5.2 \dfrac{\text{ft.}}{\text{s}}}$

Review Quiz

1a. Legal   1b. Legal   1c. Illegal   1d. Illegal

2. 34

3. 4.25 hours

4. 108 inches

5. 6,560 feet

6. 3.0×10^8 m/s

7. 1.8×10^4 mg

8. 2.4×10^{-7} kg/mL (L = dm^3)

5

Word Problems

Lesson 5.1	**Answer Units**

In this chapter, you will learn a system to organize your DATA before you solve word problems. Most students report that by using this structured approach, they have a better understanding of the steps to take to solve science calculations. Our fundamental rule will be:

To solve word problems, get rid of the words.

By distilling the words of a problem into numbers, units, and labels, you can solve most of the initial word problems in chemistry by chaining conversions. We will divide solving into three parts:

WANTED

DATA

SOLVE

Types of Units

Measurement units can be divided into three types:

- **Single units** such as meters, cm^3, kilograms, and hours
- **Ratio units** such as meters/second and kg/dm^3
- **Complex units** (which are all other units) such as 1/seconds or $(kg \cdot m^2)/s^2$

Rules to solve for single units will be covered in this chapter. Problems that ask you to solve for both single and ratio units are covered in Chapters 10 and 11. Complex units will be addressed in Chapter 16.

Solving Calculations: Step 1

To solve a problem, begin by identifying the goal. Chemistry calculations nearly always ask, "In this situation *or* at the end of this process, *how many* of these answer *units* will be present?" The answer will therefore include two (or three) parts: a *number* and a *measurement unit*. (The unit may also include a *label* describing what is being measured).

Solving for *how many* will require conversion or mathematical equations, but the measurement unit of the answer will be evident from a careful reading of the problem. So, we start there. Step 1 will be:

> The *first* time you read a word problem, ask *only*: "What will be the *unit* of the answer?" Then *write*
>
> WANTED: **?** (answer unit) =

The "?" represents a not-yet-known *numeric value*—the *how many* you will use conversions or equations to find.

TRY IT

(See "How to Use These Lessons," point 1, p. xvii.)

Q. At an average speed of 25 miles/hour, how many hours will it take to travel 450 miles?

Begin by writing:

STOP

Answer:

WANTED: **? hours** =

Writing the *answer* unit first will

- Help you to choose the correct *given* to start your conversions
- Prompt you to write DATA conversions you will need
- Tell you when to stop conversions and do the math

WANTED **Single Units**

We will call a WANTED unit a *single unit* if it has *one kind* of measurement base unit and does not include the word *per,* or a slash mark (/), or their equivalents.

> **Example:** If a problem asks you to find miles, kg, cm^3, or dollars, it is asking for a single unit. Each of those measurement units contains *one* base unit and no *per* or slash mark. Exponents and metric prefixes do not affect the *count* of the *kinds* of base units in an overall unit.

> Note that the units grams per liter, miles/hour, and 1/seconds are *not* single units.

To write the WANTED unit, use the following steps.

1. If a calculation WANTS a single unit, begin by writing

   WANTED: ? (single unit) =

   > **Example:** If the answer unit is cubic centimeters, write:

   > WANTED: **? cm^3 =**

   The WANTED unit contains *one kind* of unit and is therefore a single unit. When calculations to solve a problem are written, use of *abbreviations* for familiar units is preferred.

2. If a problem asks for a "unit *per more than one* of *one* other unit," it WANTS the first unit: a *single* unit.

   > **Example:** If a problem asks for "grams per 100 milliliters," the WANTED unit is grams. Write

   > WANTED: **? g =**

   A ratio unit must be a unit *per one* of one other unit. Because this is *per 100,* a single unit is WANTED.

3. A *label* can be added after a unit to identify what the unit is measuring.

   > **Example:** If a problem asks for the grams of water produced by a reaction, begin by writing

   > WANTED: **? g H$_2$O =**

   The label is an inseparable part of the unit. When problems involve either *two* different entities (objects or processes), a label will be *required* after every unit. But if a problem is about only *one* entity, what is being measured is clear, and the identifying label is often omitted from conversions.

4. A *count* is a measurement. In a count, a word such as *individual, pair,* or *dozen* that is used to count is the *unit.*

   > **Example:** "How many dozen eggs were needed?" Write

   > WANTED: **? dozen eggs =**

   > **Example**: "How many students attended the picnic?" Write

   > WANTED: **? students =**

   The question means "How many individual students?" In a count, however, if the unit is *individual,* it is usually omitted as understood.

Applying these steps will become intuitive with practice.

PRACTICE

In your notebook, for each problem below, write *only* what you will write at the *first* step when you solve a word problem.

1. At an average speed of 25 miles per hour, how many minutes will it take to travel 15 miles?

2. Copper (symbol Cu) has a density of 8.94 g/cm³. What will be the volume in milliliters of 13.5 grams of copper?

3. Paper is sold as "reams," and each ream contains 500. sheets. If a ream weighs 6.00 pounds (lb.) and you need 7,500 sheets of paper, how many kilograms of paper do you need? (Use 0.454 kg/lb.)

4. A U.S. quarter is a coin with a mass of 5.67 grams. How many quarters are in 2.00 lb. of these coins? (Use 2.20 lb. • kg⁻¹.)

Lesson 5.2   Mining the DATA

In studying science, a primary goal is to be able to predict, with as much precision as possible, the outcome of an action or process. Those predictions are most often based on *relationships* that science has discovered between quantities. Relationships unique to a problem are described in the problem DATA. In this lesson, you will learn to translate a word problem into DATA that can be used to solve calculations using conversions.

Rules for Listing DATA

1. On your paper, after writing

   WANTED: ? (answer unit) =

   skip a line and write

   DATA:

2. Read the problem a second time.

   • Each time you encounter a *number*, stop. Write the number in the DATA section.

   • After the number, write its unit.

   • Decide if that number and unit is *equal* or *equivalent* to another number and unit.

3. To distinguish numbers and units that are parts of equalities from those that are not, we need to learn the many ways in which DATA quantities that are *equal* or *equivalent* can be expressed in words and symbols. Use the following rules.

   a. If a number and unit are designated as *equal* to another number and unit, *write* the equality in your DATA.

      Example: If a problem states "Use 0.621 km = 1 mile," write

      DATA: 0.621 km = 1 mi.   *or*   1 mi. = 0.621 km

      In DATA, any equality may be written in the reverse order.

   b. Watch for *per*. The word *per* means two quantities are equal, equivalent, or proportional. In DATA, write *per* as an equal sign (=).

Example: A problem states "The apples were priced at $8 *per* 3 kilograms." In your DATA, write

DATA: 8 dollars = 3 kg   *or* $8 = 3 kg

The stated relationship is: $8 worth of apples is *equal* to 3 kg of apples. Some abbreviations for *money* units (such as $) are written in front of the number rather than after.

c. Treat a slash mark (/) as *per*. A slash mark between two units is equivalent to a *per* between two units.

   Example: If a problem states "The tomatoes were priced at 7 dollars/ 2 kilograms," in your DATA write

   DATA: $7 = 2 kg

d. Reading for DATA, if *no number* is between the *per* or slash mark and the *unit* that follows, write "*per* unit" or "/unit" as " **= 1** unit."

   Example: If the problem states "A car is traveling 25 km/hour," write

   DATA: 25 km = **1** hour

The unit means the car is traveling 25 km per 1 hour.

▭▬▬▶ **TRY IT**

Q. Write what you would list as DATA after reading these two phrases.

   a. The candies were sold at 3 per 50 cents.

   b. The density is 10.5 g/mL.

🛑

Answers:

   a. 3 candies = 50 cents   b. 10.5 g = **1** mL

Listing Single-Unit DATA

Rules 3a–d above for writing DATA equalities apply whether conversion calculations WANT single or ratio units. But if a *single* unit (*not* a ratio) is WANTED, usually the DATA is in a special format that simplifies problem solving.

If a *single* unit is WANTED, write the DATA with the help of these additional rules:

4. Single-unit DATA is a "known number and unit" that is *not* paired in the DATA with another "known number and unit" by rules 3a–d for identifying equalities. Single-unit DATA is written on a *line by itself* in the DATA section.

5. If the WANTED unit is a single unit, *at least one* number and unit *must* be "single-unit DATA." *Watch* for the single-unit DATA as you read the problem.

6. If a single unit is WANTED, the question is often in this format:

   "*How many* of this single unit WANTED are equivalent to (known number more than one) of this different single unit?"

Write the "(more than one) of this different single unit" on a *line by itself* in the DATA section.

> **Example:** If a problem states, "What would be the mass in grams of 125 mL of gasoline?" then the problem matches the question format above. Grams is the single unit WANTED, and "125 mL gasoline" is the number and single-unit DATA. On your paper, write
>
> WANTED: ? g gasoline =
>
> DATA: (125 mL gasoline)

7. If a problem WANTS a single-unit amount, most often *one* measurement will be single-unit DATA and the rest of the DATA will be equalities.

8. When a single unit is WANTED, it is a good practice to ⓒircle the single-unit amount in the DATA, because it will be the *given* number and unit that is used to start your conversions.

▭━━▶ TRY IT

In your notebook, write the WANTED and DATA steps for this problem.

Q. If a runner completes a marathon (26.2 miles) at an average speed of 10.72 mi./hr, what is her time for the race in minutes?

🛑

Answer:

The question can be translated as: "For this race, how many minutes are equivalent to 26.2 miles?" Your paper should look like this.

WANTED: ? min =

DATA: (26.2 mi.)

  10.72 mi. = 1 hr

If a single unit is WANTED, usually the DATA will be *one* item of single-unit DATA and the rest of the DATA will be equalities.

In Chapter 9, we will discuss exceptions to the single-unit steps above, but in all problems until then, rules 4–8 above will apply.

PRACTICE   A

In your notebook, write the WANTED and DATA sections for problems 1 and 2 in the Practice for Lesson 5.1.

Additional Rules for DATA Equalities

9. The following are additional rules for recognizing *equalities* that apply when *both* single and ratio units are WANTED.

   a. Treat units in the format "unit X • unit Y^{-1}" as "X unit *per one* Y unit."

Example: "7.5 g • L^{-1}" means 7.5 g *per one* L. Write:

    DATA: 7.5 g = **1** L

b. Write an equality if the same *entity* is measured using two different units.

    Example: If a problem states, "0.920 grams of a sample of gas has a volume of 448 mL," write:

        DATA: 0.920 g gas = 448 mL gas   (Here, the label *gas* is optional.)

    Example: If a problem states "A sample has a mass of 0.50 kg (1.10 lb.)," write:

        DATA: 0.50 kg = 1.10 lb.

    In each of these two examples, the *same* physical entity is being measured in two different units.

c. Write an equality if one *process* is measured by two different units.

    Example: If a problem states, "Burning 0.25 grams of candle wax releases 1,700 calories of energy," write

        DATA: 0.25 g wax burned = 1,700 calories of energy released

    Both sides measure what happened as this candle burned.

d. If *two* different entities are measured in a problem, writing a *label* after each unit (such as "wax burned" and "energy released" above) is *required* to identify which entity the numbers and units are measuring. The *format* for equalities in DATA is always

    DATA: # unit label = # unit label   (where # is a stated number)
    but in problems about *one* entity, the label is often omitted as understood.

e. Words such as *individual*, *pair*, and *dozen* are *units* used to count objects. If the unit is *individual*, it is often omitted as understood.

f. Watch for words such as *each* and *every* that mean *per one*. One is a number, and you want *all* numbers and their attached units and labels in your DATA.

    Example: If you read "Every student was given 2 pairs of gloves," write

        DATA: 1 student = 2 pairs gloves

    The counting unit *individual* in front of the label *student* is understood.

g. Watch for words that can be *translated* to mean *per*. Treat them as *per*.

    Example: "There are 36 eggs in 3 cartons" means there are 36 eggs *per* 3 cartons. Write

        DATA: 36 eggs = 3 cartons

h. If your DATA seems to show *more than one* number and unit that are not part of a DATA equality, check again for words that can mean a *per* relationship. In conversion problems, usually, only one DATA term is not part of an equality.

10. Make sure *all* of the *numbers* in the problem are written in your DATA.

▬▬▬▬▬▬ ✏▬▬▬▶ **TRY IT**

Fill in the WANTED and DATA sections for this problem.

Q. Exactly 200 "dime" coins have a mass of 0.998 lb. What is the mass of exactly 15 dimes in grams? (Use 1 kg = 2.20 lb.)

WANTED:

DATA:

🛑

Answer:

WANTED: ? g dimes =   (See Lesson 5.1.)

DATA: 200 dimes (exact) = 0.998 lb.   (This lesson, steps 9b, 9e.)

            (15 dimes (exact))   (This lesson, steps 4–8.)

            1 kg = 2.2 lb.   (This lesson, step 3a.)

To summarize, when solving word problems:

> In the DATA section,
>
> - Write *every number* in the problem, followed by its *unit*.
> - See if that quantity is *paired* with another number and unit to form an equality.
> - If a *single* unit is WANTED, *one* item of DATA will probably be a single unit, but the rest will be equalities. Circle the single-unit DATA.
> - In problems involving more than one entity, add a *label* after the unit that describes the entity being measured.

Being able to recite the steps is not necessary. To use the steps as a tool, you simply need to be able to apply them, a skill that becomes intuitive with practice.

PRACTICE   B

In your notebook, list *only* the WANTED and DATA sections for problems 3 and 4 in the Practice for Lesson 5.1.

Lesson 5.3   Solving for Single Units

By the law of **dimensional homogeneity**, if a single-unit amount is WANTED in a calculation, at least one single-unit amount must be supplied in the DATA. It will simplify conversion calculations if we use that supplied single-unit amount as our *given* value: the starting point for chaining conversions.

To SOLVE

Below the WANTED and DATA sections, *start* by writing:

> SOLVE:
>
> ? unit and label WANTED = (circled) # unit and label *given*

Example: Solve: ? m = 0.050 km

For the SOLVE term starting with ?, write the unit and label you wrote after WANTED: above your DATA section. If a single unit is WANTED, the *given* measurement written after the = sign will be the (circled) single-unit DATA.

At the SOLVE step, the rule is:

> If you WANT a single unit, *start* with a single unit as your *given*.

Finish by converting to the WANTED unit. As conversions, use the equalities in the DATA (plus other fundamental equalities, such as metric definitions, if needed).

▭▭▭▭▶ **TRY IT**

Q. If a car's average speed is 1,470 meters per minute, how many meters does it travel in 5.00 seconds? (Fill in the three parts below.)

WANTED:

DATA:

SOLVE:

   ?

STOP

Answer:

WANTED: ? m =   (Means "how many meters?")

DATA: 1,470 m = 1 min

    (5.00 s)

SOLVE:

$$? \text{ m} = 5.00 \text{ s} \cdot \frac{1 \text{ min}}{60 \text{ s}} \cdot \frac{1{,}470 \text{ m}}{1 \text{ min}} = 122.50 \text{ m} = \textbf{123 m} \ (3 \ s.f.)$$

Note that you need "1 min = 60 s" to SOLVE, but it is not listed in the DATA. Familiar conversions will often be needed to solve, but listing them in the DATA is not required.

At the SOLVE step, in these lessons, the unit cancellation is generally not shown so that the conversion units are clear, but lightly marking cancellations in text examples and on your paper is a good idea. Lightly mark the cancellations at the SOLVE step above.

For significant figures: 1 is always exact, 60 is part of an exact definition, both 5.00 and 1,470 have 3 *s.f.*, so the answer must be rounded to 3 *s.f.* (the lone 5 rounds up).

You can solve simple problems without listing WANTED, DATA, and SOLVE, but this three-part method works for *all* problems. It works especially well for the more complex problems that you will soon encounter. By using the same three steps for every problem, you will know what steps to take to solve *all* problems. That's the goal.

PRACTICE

If you get stuck, read a *part* of the answer, adjust your work, and try again.

1. Solve on this page. Knowing that 2.20 lb. = 1 kg, what is the mass in grams of 12 lb.?

   WANTED:

   DATA:

   SOLVE:

   ?

Solve problems 2 and 3 by writing the WANTED, DATA, and SOLVE sections in your notebook.

2. One mile is 5,280 feet, and there are 1.61 km/mi. How many feet are in 0.500 km?

3. A bottle of drinking water is labeled "12 fl. oz. (355 mL)." What is the mass in centigrams of 0.55 fluid ounces of the H_2O? (Assume a water density of 1.00 g/mL.)

Lesson 5.4   Finding the *Given* for Single Units

When solving for single units, the *given* quantity is not always clear.

> **Example:** A student needs special postage stamps. The stamps are sold six per sheet, each stamp booklet has three sheets, 420 stamps are needed, and the cost is $43.20 per five booklets. What is the cost of the stamps?
>
> Among all those numbers, which is the *given* needed at the start of your chained conversions?

For a single-unit answer, finding the *given* is often a process of elimination. If all the numbers and units are paired into equalities except for one, that one is your *given*.

──────── **TRY IT**

Q. In your notebook, for the stamps problem above, write the WANTED and DATA sections (don't SOLVE yet). Then check your work below.

STOP

Answer:

Your paper should look like this.

WANTED: ? $ =   *or*   ? dollars =

DATA: 1 sheet = 6 stamps

3 sheets = 1 booklet

(420 stamps)

$43.20 = 5 booklets

Because you are looking for a *single* unit, dollars, your DATA has one number and unit that did not pair up in an equality: 420 stamps. That is your *given*.

──────── **TRY IT**

Q. In your notebook, add the SOLVE step to the stamps problem. Assume all of these numbers are exact.

STOP

Answer:

SOLVE: If you *WANT* a single unit, *start* with a single unit. Begin with

STOP

$$? \$ = 420 \text{ stamps}$$

$$? \$ = 420 \text{ stamps} \cdot \frac{1 \text{ sheet}}{6 \text{ stamps}} \cdot \frac{1 \text{ booklet}}{3 \text{ sheets}} \cdot \frac{\$43.20}{5 \text{ booklets}} = \mathbf{\$201.60}$$

Mathematically, the order in which you multiply factors does not matter, but starting with the "single-unit *given*" helps in arranging your conversions "right-side up."

PRACTICE   A

The conversion factor methods learned in chemistry can be useful when solving a variety of practical problems. In your notebook, use the WANTED, DATA, and SOLVE steps to solve these.

1. If a strawberry smoothie in Munich costs 4.50 euros and the exchange rate is 0.935 euros for each U.S. dollar (USD), what would be the cost of two smoothies in USD?

2. If a baseball pitcher throws a fastball at 95 mi./hr and the distance from the pitcher's mound to home plate is 60.5 feet, how many seconds does it take for the ball to reach the plate?

3. Without consulting a periodic table, write the symbols, in order from the top down, for the *top four atoms* in the *first column* of the periodic table (columns go up and down).

Some Chemistry Practice

For problems involving chemical substances, an essential step is to write the *chemical formula* of the substance being measured *after* every unit in which it is being measured.

> For measurements of chemical substances, write *number, unit, chemical formula.*

Examples:

In WANTED units, write: WANTED: ? g **NaCl** =

Write *single-unit* DATA as: 9.00 g **H₂O**

Write DATA *equalities* in the format: 4.00 g **He** = 1 mol **He**

The problems in Practice B supply the *chemistry* DATA needed for conversions. In upcoming chapters, you will learn how to write these conversions automatically even when the problem does not supply them. That small amount of additional information will be all you need to solve most initial chemistry calculations.

PRACTICE B

Place a check mark (✓) next to two of these problems and solve them in your notebook now. Do the remaining problem in your next study session. Use the WANTED, DATA, SOLVE method. If you get stuck, peek at the answer and try again.

The *mole* referred to in these problems is a counting unit, such as a dozen (only larger). For now, simply treat *mole* (abbreviated *mol*) like g or mL, as a *unit* of measure.

1. Gold (symbol Au) has a density of 19.3 g/mL and a molar mass of 197.0 grams per mole. What will be the volume in milliliters of 4.50 moles of gold?

2. Water has a molar mass of 18.0 grams H_2O per mole of H_2O. How many moles of H_2O are in 9.00 milligrams of H_2O? (Try the math at the end without a calculator.)

3. 1 mole of helium gas (formula He) has a volume of 22.4 liters at standard conditions. The molar mass of He is 4.00 grams He per mole He. At standard conditions, what is the volume in liters of 8.00 grams of He gas? (Try the math at the end without a calculator.)

SUMMARY

By the start of Chapter 5, you should be able to fill in a blank periodic table with the names and symbols for the first 20 elements. Practice as needed.

To summarize the steps for solving word problems:

1. *Get rid of the words.* Distill the words into numbers with units and labels.

2. Organize your work in three sections: WANTED, DATA, and SOLVE.

    WANTED: Reading the first time, ask, "What will be the unit of the answer?" Write

    WANTED: ? (answer unit) =

If a problem concerns either chemical substances *or* two entities, add a *label* after every unit that identifies what the WANTED unit is measuring.

DATA: Read the problem again. In the DATA section, write every number and attach its unit and label. Then, if a number and unit are *paired* with another number and unit as equal, equivalent, or proportional, write them as

   DATA: # unit label = # unit label

If a problem WANTS a single unit, most often *one* measurement will be "single-unit DATA" and the rest will be equalities. Circle the *single* unit in the DATA.

SOLVE: If you WANT a single unit, *start* with a single unit as your *given*.

Write

   SOLVE: ? unit and label WANTED = circled # unit and label *given*

3. Using the DATA and other equalities, *convert* to the unit WANTED.

REVIEW QUIZ

Complete these to prepare for a quiz or test on this chapter. Earlier problems in this chapter that you did not previously complete can be used for preparation as well.

1. In 2009, Jamaican sprinter Usain Bolt set a new world record in the 100.-meter dash with an average speed of 37.58 km/hr. What was his time in seconds?

2. A gallon (gal.) contains 4 quarts (qt.) and there are 3.78 L/gal. On a 0.500-quart container, what will be the listed volume in milliliters?

3. Aluminum has a density of 2.70 g/mL and a molar mass of 27.0 grams per mole. What is the volume in milliliters of 3.5 moles of Al? (Try the math at the end without a calculator.)

4. Recently, the cost of gasoline in Paris, France, averaged 1.48 euros per liter, and the exchange rate for currency was 1 U.S. dollar (USD) = 0.907 euro. At these rates, what is the price in USD of 10.0 gallons of Parisian gas? (Use 3.78 L/gal.)

5. Each mole of iron (Fe) has a mass of 55.8 g. The density of iron is 7.87 g/mL, and there are 6.02×10^{23} atoms per mole of Fe. What would be the volume in milliliters of 3.01×10^{23} Fe atoms? (First calculate, then circle the right answer.)

   a. 14.2 mL Fe   b. 2.22 mL Fe   c. 216 mL Fe

   d. 0.0705 mL Fe   e. 3.55 mL Fe

ANSWERS

Lesson 5.1

Practice   1.   WANTED: ? min =   (A single unit is WANTED.)

2.   WANTED: ? mL Cu =   (Include chemical symbols if known.)

3.   WANTED: ? kg paper =   (The label *paper* is recommended.)

4.   WANTED: ? quarters =   (See step 4 for counts.)

Lesson 5.2

Practice A   For an explanation of the WANTED unit, see the answers to the Practice for Lesson 5.1.

1.   WANTED: ? min =

DATA: 25 mi. = 1 hr   (Step 3b.)

   15 mi.   (Single-unit DATA, steps 4–8.)

2.   WANTED: ? mL Cu =

DATA: 8.94 g Cu = 1 cm^3 Cu   (Steps 3c, 3d.)

   13.5 g Cu   (Steps 4–8, add label for chemical if known.)

Practice B   3.   WANTED: ? kg paper =

DATA: 7,500 sheets paper   (Single-unit DATA, steps 4–8.)

   1 ream = 500 sheets paper   (Steps 9f, 9g.)

   1 ream = 6.00 lb. paper   (Steps 9f, 9g.)

   0.454 kg = 1 lb.   (Step 3c.)

4.   WANTED: ? quarters =   (A count of objects is a unit and label.)

DATA: 1 quarter = 5.67 g   (Steps 9f, 9g.)

   2.00 lb. quarters   (Single-unit DATA, steps 4–8.)

   2.20 lb. = 1 kg   (Step 9a.)

Lesson 5.3

Practice   1.   WANTED: ? g =

DATA: 2.20 lb. = 1 kg

   12 lb.

SOLVE: $? \text{ g} = 12 \text{ lb.} \cdot \dfrac{1 \text{ kg}}{2.20 \text{ lb.}} \cdot \dfrac{10^3 \text{ g}}{1 \text{ kg}} = \dfrac{12 \cdot 10^3 \text{ g}}{2.2} = \mathbf{5.5 \times 10^3 \text{ g}}$

2.   WANTED: ? ft. =

DATA: 1 mi. = 5,280 ft.

   1.61 km = 1 mi.

   0.500 km

SOLVE: $? \text{ ft.} = 0.500 \text{ km} \cdot \dfrac{1 \text{ mi.}}{1.61 \text{ km}} \cdot \dfrac{5,280 \text{ ft.}}{1 \text{ mi.}} = \dfrac{0.500 \cdot 5,280 \text{ ft.}}{1.61} = \mathbf{1,640 \text{ ft.}}$

3. WANTED: $? \, cg =$

   DATA: 12 fl. oz. = 355 mL

   $\boxed{\text{0.55 fl. oz.}}$

   1.00 g $H_2O(\ell)$ = 1 mL $H_2O(\ell)$

   SOLVE: $? \, cg = 0.55 \text{ fl. oz.} \cdot \dfrac{355 \text{ mL}}{12 \text{ fl. oz.}} \cdot \dfrac{1.00 \text{ g } H_2O(\ell)}{1 \text{ mL } H_2O(\ell)} \cdot \dfrac{1 \text{ cg}}{10^{-2} \text{ g}} = \textbf{1,600 cg}$ (2 *s.f.*)

Lesson 5.4

Practice A   1. WANTED: $? \, USD =$

   DATA: 1 smoothie = 4.50 euro

   0.935 euro = 1 USD

   $\boxed{\text{2 smoothies}}$ (exact)

   SOLVE: $? \, USD = 2 \text{ smoothies} \cdot \dfrac{4.50 \text{ euro}}{1 \text{ smoothie}} \cdot \dfrac{1 \text{ USD}}{0.935 \text{ euro}} = \textbf{9.63 USD}$

2. WANTED: $? \, s =$

   DATA: 95 mi. = 1 hr

   $\boxed{\text{60.5 ft.}}$

   5,280 feet = 1 mile   (but standard definitions are not *required* in the DATA)

   SOLVE: $? \, s = 60.5 \text{ ft.} \cdot \dfrac{1 \text{ mi.}}{5{,}280 \text{ ft.}} \cdot \dfrac{1 \text{ hr}}{95 \text{ mi.}} \cdot \dfrac{60 \text{ min}}{1 \text{ hr}} \cdot \dfrac{60 \text{ s}}{1 \text{ min}} = \textbf{0.43 s}$

3. H, Li, Na, K

Practice B   1. WANTED: $? \, mL \, Au =$

   DATA: 19.3 g Au = 1 mL Au

   197.0 g Au = 1 mol Au

   $\boxed{\text{4.50 mol Au}}$

   SOLVE: $? \, mL \, Au = 4.50 \text{ mol Au} \cdot \dfrac{197.0 \text{ g Au}}{1 \text{ mol Au}} \cdot \dfrac{1 \text{ mL Au}}{19.3 \text{ g Au}} = \textbf{45.9 mL Au}$

2. WANTED: $? \, mol \, H_2O =$

   DATA: 18.0 g H_2O = 1 mol H_2O

   $\boxed{\text{9.00 mg } H_2O}$

   SOLVE:

   $? \, mol \, H_2O = {}^1\cancel{9.00} \text{ mg } H_2O \cdot \dfrac{10^{-3} \text{ g}}{1 \text{ mg}} \cdot \dfrac{1 \text{ mol } H_2O}{{}^2\cancel{18.0} \text{ g } H_2O} = \textbf{0.500} \times \textbf{10}^{-3} \textbf{ mol } H_2O = \textbf{5.00} \times \textbf{10}^{-4} \textbf{ mol } H_2O$

3. WANTED: $? \text{ L He} =$

   DATA: $1 \text{ mol He} = 22.4 \text{ L He}$

   $4.00 \text{ g He} = 1 \text{ mol He}$

   $\boxed{8.00 \text{ g He}}$

   SOLVE: $? \text{ L He} = {}^2\cancel{8.00} \text{ g He} \cdot \dfrac{1 \text{ mol He}}{\cancel{4.00} \text{ g He}} \cdot \dfrac{22.4 \text{ L He}}{1 \text{ mol He}} = \textbf{44.8 L He}$

Review Quiz

Partial solutions are included below, but your work must include WANTED, DATA, and SOLVE.

1. SOLVE: $? \text{ s} = 100. \text{ m} \cdot \dfrac{1 \text{ km}}{1000 \text{ m}} \cdot \dfrac{1 \text{ hr}}{37.58 \text{ km}} \cdot \dfrac{60 \text{ min}}{1 \text{ hr}} \cdot \dfrac{60 \text{ s}}{1 \text{ min}} = \textbf{9.58 s}$

2. SOLVE: $? \text{ mL} = 0.500 \text{ qt.} \cdot \dfrac{1 \text{ gal.}}{4 \text{ qt.}} \cdot \dfrac{3.78 \text{ L}}{1 \text{ gal.}} \cdot \dfrac{1 \text{ mL}}{10^{-3} \text{ L}} = \textbf{473 mL}$

3. SOLVE: $? \text{ mL Al} = 3.5 \text{ mol Al} \cdot \dfrac{{}^{10}\cancel{27.0} \text{ g Al}}{1 \text{ mol Al}} \cdot \dfrac{1 \text{ mL Al}}{\cancel{2.7} \text{ g Al}} = \textbf{35 mL Al}$

4. SOLVE: $? \text{ USD} = 10.0 \text{ gal.} \cdot \dfrac{3.78 \text{ L}}{1 \text{ gal.}} \cdot \dfrac{1.48 \text{ euro}}{1 \text{ L}} \cdot \dfrac{1 \text{ USD}}{0.907 \text{ euro}} = \textbf{\$61.70}$

   (Gasoline in Europe tends to be more expensive than in the United States, due in part to a taxation policy encouraging alternatives to burning fossil fuels.)

5. e. **3.55 mL Fe**

   $? \text{ mL Fe} = {}^1\cancel{3.01 \times 10^{23}} \text{ atoms Fe} \cdot \dfrac{1 \text{ mol Fe}}{{}^2\cancel{6.02 \times 10^{23}} \text{ atoms Fe}} \cdot \dfrac{55.8 \text{ g Fe}}{1 \text{ mol Fe}} \cdot \dfrac{1 \text{ mL Fe}}{7.87 \text{ g Fe}} = 3.55 \text{ mL Fe}$

6

Atoms, Ions, and Periodicity

Lesson 6.1   The Atoms (Part 2)

To continue to learn the most frequently encountered atoms:

- For the first 20 atoms **plus** those in the first two and last two *columns* in the periodic table shown below, given either the atom *symbol* or *name*, be able to write the other.

- Be able to fill in a blank table with the symbols of these atoms in their correct positions.

- For the first and last two columns, be able to write the symbols in order from the top down.

PERIODIC TABLE

1A	2A		3A	4A	5A	6A	7A	8A
1 **H** Hydrogen								2 **He** Helium
3 **Li** Lithium	4 **Be** Beryllium		5 **B** Boron	6 **C** Carbon	7 **N** Nitrogen	8 **O** Oxygen	9 **F** Fluorine	10 **Ne** Neon
11 **Na** Sodium	12 **Mg** Magnesium		13 **Al** Aluminum	14 **Si** Silicon	15 **P** Phosphorus	16 **S** Sulfur	17 **Cl** Chlorine	18 **Ar** Argon
19 **K** Potassium	20 **Ca** Calcium						**Br** Bromine	**Kr** Krypton
Rb Rubidium	**Sr** Strontium						**I** Iodine	**Xe** Xenon
Cs Cesium	**Ba** Barium						**At** Astatine	**Rn** Radon
Fr Francium	**Ra** Radium							

Lesson 6.2 — Atoms: Terms and Definitions

The precise definition for some of the fundamental particles in chemistry is a matter of occasional debate, but the following simplified descriptions will provide us with a starting point for discussion of atoms.

1. **Matter.** Chemistry is concerned primarily with the measurement and description of the properties of matter and energy. Matter is anything that has mass and volume. In planetary environments, nearly all matter is composed of extremely small particles called atoms.

2. **Electric charge.** Some particles have a property known as electric charge. There are two types of charges: positive and negative. Particles with the same charge repel. Particles with opposite charges attract.

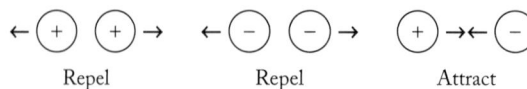

Repel   Repel   Attract

3. **Atoms.** Ninety-two (92) different kinds of atoms are found in Earth's crust. More than 20 additional atoms have been synthesized by scientists in nuclear reactions. All of the millions of different substances found in nature on Earth are assembled from these atoms. The different substances result from the different ways atoms can bond and be arranged in three-dimensional space.

An alphabetical list of the atoms is provided on page A-1 at the back of this book. Each atom has a name and a one- or two-letter **symbol**. The first letter of the symbol is always uppercase. The second letter, if any, is always lowercase.

4. **Atomic structure.** Atoms can be described as combinations of three **subatomic particles**: protons, neutrons, and electrons. An atom is a particle with a single nucleus that contains protons and (nearly always) neutrons. Associated electrons surround the nucleus.

   a. **Protons** (abbreviated $\mathbf{p^+}$)

   - Each proton has a **1+** electrical *charge* (one unit of positive charge).
   - Protons have a *mass* of about 1.0 atomic mass unit (amu).
   - Protons are found in the center of the atom, called the **nucleus** (plural **nuclei**).
   - The number of protons in an atom is defined as the **atomic number** of the atom.
   - The number of protons determines the **name** (and thus the symbol) of the atom.
   - The number of protons in an atom is never changed by chemical reactions.

   b. **Neutrons** (abbreviated $\mathbf{n^0}$)

   - A neutron has an electrical charge of zero.
   - A neutron has about the same mass as a proton: 1.0 amu.
   - Neutrons are located in the nucleus of an atom, along with the protons.
   - Neutrons are thought to act as the glue of the nucleus: particles that play a role in keeping the repelling protons from flying apart.
   - Neutrons, like protons, are never gained or lost in chemical reactions.
   - The neutrons in most cases have very little influence on the chemical behavior of an atom.

   c. **Electrons** (abbreviated $\mathbf{e^-}$)

   - Each electron has a **1−** electrical charge: equal in magnitude but opposite the proton's charge.
   - Electrons have very little mass, each weighing less than 0.001 amu.
   - Electrons are found outside the nucleus of an atom, in regions of space called **orbitals**.
   - Nearly all of the volume of an atom is due to the space occupied by the electrons around the nucleus.
   - Electrons are the only subatomic particles that can be gained or lost during chemical reactions.

5. **Neutral atoms.** If an atom has an *equal* number of protons and electrons, the balance between positive and negative charges gives the atom a *net* charge of zero. The charges are said to "cancel" to produce an overall **electrically neutral** atom.

Flashcards

The flashcards and problems below identify facts and vocabulary that you need in memory at this point in the course. For answers not firmly in memory, make and run the flashcards, then try the problems in Practice A.

One-Way Cards (with Notch at Top Right)	Back Side—Answers
The properties of matter include _____	Mass and volume
On planets, nearly all matter is made of _____	Atoms
Electric behavior: Like charges _____	Repel
Electric behavior: Unlike charges _____	Attract
Subatomic particle that decides atom name = _____	Proton
Subatomic particles with mass of 1.0 amu = _____	Protons and neutrons
amu is an abbreviation for _____	Atomic mass unit
Subatomic particle with lowest mass = _____	Electron
Subatomic particles with a charge = _____	Protons and electrons
Location of nearly all mass in an atom = _____	Nucleus
Protons minus electrons = _____	Net charge on a particle
Subatomic particles not changed by chemical reactions = _____	Protons and neutrons
Zero charge on an atom means = _____	# of protons = # of electrons
Symbols for the three subatomic particles	p^+, n^0, e^-

Two-Way Cards (*without* Notch)

Number of protons = _____ number	Atomic number = number of _____

PRACTICE **A**

1. From memory, write the symbols for these atoms.

    a. Carbon   b. Oxygen   c. Chlorine   d. Calcium

2. From memory, name the atoms represented by these symbols.

    a. K   b. Na   c. F

3. Which subatomic particles have minimal impact on chemical behavior?

4. Fill in the blanks for these subatomic particles.

Particle	Symbol	Charge	Mass	Location
Proton				
Neutron				
Electron				

5. In your notebook, draw a diagram of the "boxes" in the first three rows (rows go across) of the periodic table, then fill in the atom symbols and atomic numbers for the first 18 elements.

6. Use your problem 5 answer to fill in the blanks below, assuming each atom is electrically neutral.

Atom	Symbol	Protons	Electrons	Atomic Number
Sodium				
	N			
		6		
			15	
				9

6. **Ions.** Any particle (atom or group of bonded atoms) that does not have an equal number of protons and electrons is termed an **ion**, which is a particle with a non-zero electrical charge.

- Neutral particles that *lose electrons* become **positive ions**. A positive ion is termed a **cation** (pronounced KAT-eye-un).

- Neutral particles that *gain electrons* become **negative ions**. A negative ion is termed an **anion** (pronounced ANN-eye-un).

The number after the sign on the ion charge equals the *difference* between the number of protons and electrons in the particle. If a particle has more protons than electrons, it is positive. If it has more electrons than protons, it is negative.

Net charge on an ion = number of p^+ *minus* number of e^-

An ion is not the same as a *neutral* particle with the same atomic nucleus (or multiple nuclei). The ion has a different number of electrons and different chemical behavior.

The symbol or formula for an ion displays the value of the net charge as a superscript to the right of the symbol.

Example: All atoms with a nucleus containing 16 protons are examples of **sulfur** (symbol S).

An electrically neutral atom of sulfur has 16 protons and 16 electrons. The charges balance to give a net charge of zero. The symbol for a neutral sulfur atom is written as S, but S^0 may also be written to emphasize that the sulfur atom has a neutral charge.

Some substances contain an ion of sulfur that has 16 protons and 18 electrons. The 16 protons cancel the charge of 16 electrons, leaving two uncancelled electrons and an overall charge of **2−**. The symbol for this particle is S^{2-}.

Example: All atoms with 19 protons are named **potassium** (symbol K). In its elemental state, potassium is a soft metal that reacts with many substances, including water. In reactions, a neutral potassium atom loses one electron to form an *ion* with 18 electrons, which balance the charge of 18 protons. This leaves one positive charge uncancelled, so the ion has a net charge of **1+** and its symbol is K^+. If no number is shown in front of a charge, a 1 is understood.

➤ TRY IT

(See "How to Use These Lessons," point 1, p. xvii.)

Q. All atoms with 88 protons are named **radium** (symbol Ra). Because a radium nucleus must have 88 protons, Ra^{2+} ion must have how many electrons?

STOP

Answer:

86 electrons. The 2+ charge means the ion has two more protons than electrons.

Flashcards

Practice these to initial mastery, then complete Practice B.

One-Way Cards (with Notch at Top Right)	Back Side—Answers
Negative ions have more _____ than _____	More electrons than protons
If an atom is electrically neutral, what are the subatomic particles that are equal in number?	# protons = # electrons

Two-Way Cards (*without* Notch)	
Ion (definition) = _____	An atom or atom group with an electric charge = _____
Protons minus electrons = _____	Net charge on a particle = _____
Classification for all positive ions = _____	Cations = _____
Classification for all negative ions = _____	Anions = _____

PRACTICE   B

For the problems below, use the alphabetical list of atoms or the periodic table at the back of the book.

1. During chemical reactions, atoms do *not* gain or lose _____, which have a positive charge. Atoms can take on a charge to become *ions* by gaining or losing _____, which have a _____ charge.

2. Calcium has atomic number 20.

    a. A neutral Ca atom has how many protons?   How many electrons?

    b. How many protons are found in a Ca^{2+} ion?   How many electrons?

3. In terms of subatomic particles, an atom that is a cation will always have more _____ than _____.

4. For these symbols, write the atom names from memory.

    a. S =   b. Si =   c. P =

5. For these atom names, write their symbols from memory.

    a. Bromine =   b. Boron =   c. Barium =

6. For the particles below, fill in the blanks.

Symbol	Protons	Electrons
O		
O^{2-}		
Mg^{2+}		
	13	10
	79	79

Symbol	Protons	Electrons
	1	0
	35	36
	34	36
I^-		
Cl^-		

7. Write the ion symbols for each of the *cations* in the tables in problem 6.

Lesson 6.3   The Nucleus

Nuclear Structure

The Nucleus At the center of an atom is the nucleus. The nucleus is very small, with a diameter that is roughly 100,000 times smaller than the effective diameter of most atoms, yet the nucleus contains all of the atom's protons, neutrons, positive charge, and nearly all of its mass.

Because the nucleus contains close to all of the atom's mass in a tiny volume, it is extremely dense. Outside of the nucleus, nearly all of the volume of an atom is occupied by its electrons. Because electrons have low mass but occupy a large volume compared to the nucleus, in terms of mass an atom is mostly empty space. An electron, however, has a charge that is equal in magnitude (though opposite) to that of the much more massive proton.

Types of Nuclei Only certain combinations of protons and neutrons form a nucleus that is stable. Nuclei can be divided into three types.

- **Stable** nuclei have not changed on Earth in billions of years.
- **Radioactive** nuclei are *somewhat* stable. Once formed, they exist for a time (from a few seconds to several billion years), but they change by emitting or capturing subatomic particles (a process called **decay**) at a constant, characteristic rate.
- **Unstable** combinations of protons and neutrons, if formed in a nuclear reaction, will decay within a few seconds.

Atoms that have 1 to 92 protons are all found in Earth's crust. All atoms with 43, 61, and 83–92 protons are radioactive. All atoms with 93 or more protons exist on Earth only when they are created in nuclear reactors, and all are radioactive.

Radioactive atoms compose a very small percentage of the matter found on Earth. More than 99.99% of Earth's atoms have stable nuclei that have not changed since atoms came together to form Earth billions of years ago.

Terminology Each possible combination of protons and neutrons is called a **nuclide**. A group of nuclides that have the same number of protons (so they are all the same kind of atom) but different numbers of neutrons are called the **isotopes** of the atom.

Nuclide Symbols Each nuclide can be assigned a **mass number** that is the *sum* of its *number* of protons and neutrons.

$$\textbf{Mass number} \text{ of a nucleus} = \textbf{protons} + \textbf{neutrons} = \textbf{p}^+ + \textbf{n}^0$$

Example: A nucleus with 2 p^+ and 2 n^0 is helium with a mass number of **4**.

A nuclide can be identified in three ways: by its number of protons and neutrons, by its **nuclide symbol** (also termed its **isotope symbol**), and by its **nuclide name**.

A nuclide *symbol* has two required parts: the *atom symbol* and the *mass number*. The mass number is written as a superscript in front of the atom symbol. The format for a nuclide *name* is "atom name *hyphen* mass number."

Examples: Chlorine has two stable isotopes that can be represented as:

- 17 protons + 18 neutrons, or ^{35}Cl, named chlorine-35 (pronounced "chlorine thirty-five"); and
- 17 protons + 20 neutrons, or ^{37}Cl, named chlorine-37.

Knowing one representation for the composition of a nucleus, you need to be able to write the other two.

TRY IT

Use the periodic table at the back of this book as needed.

Q1. A nucleus with 6 protons and 8 neutrons has what nuclide name and symbol?

Answer:

Atoms with 6 protons are always named carbon, symbol **C**. The mass number of this nuclide is 6 protons + 8 neutrons = 14. This isotope of carbon, which is measured in *radiocarbon dating*, is named **carbon-14** and its symbol is written ^{14}C.

Q2. Fill in the blanks: How many p^+ and n^0 are in ^{20}Ne? p^+ = _____ and n^0 = _____

Answer:

Atoms named neon must contain 10 protons. The mass number 20 is the total number of protons *plus* neutrons, so neon-20 contains **10** protons and **10** neutrons.

Additional ways to represent nuclides will be discussed in Chapter 24.

Flashcards

One-Way Cards (with Notch)	Back Side—Answers
Location of nearly all mass in an atom = _____	The nucleus
Define a nuclide	A certain combination of protons and neutrons
Different nuclides with the same chemical behavior are called _____	Isotopes
How are isotopes the same?	Same number of protons
How do isotopes differ?	Different numbers of neutrons
% of nuclei unchanged since Earth formed = _____	More than 99.99%
Name for the isotope ^{131}I = _____	Iodine-131

Two-Way Cards (*without* Notch)	
# protons + # neutrons = _____ number	Mass number (definition) = _____
1 proton and 0 neutrons = ? nuclide symbol	^1H contains what particles?
1 proton and 2 neutrons = ? nuclide symbol	^3H contains what particles?

PRACTICE   A

Consult the table of atoms or periodic table at the back of this book to fill in the blanks below.

1.

Protons	Neutrons	Atomic Number	Mass Number	Nuclide Symbol	Nuclide Name
	6	6			
7	7				
					Iodine-131
				^{235}U	
		2	4		

2. Write the name of the atom in problem 1 for which all isotopes are radioactive.

The Mass of Nuclides

The mass of a single nuclide is usually measured in **atomic mass units**, abbreviated **amu,** a unit also called the dalton (Da).

Protons and neutrons have close to the same mass, and each is much heavier than an electron. The mass of both the proton and the neutron is 1.0 amu, but the mass of an electron is less than 0.001 amu.

On the basis of those masses, you might expect that the mass of a ^{35}Cl atom would be just over 35.0 amu, because it is composed of 17 protons, 18 neutrons, and some very light electrons. In fact, for atoms of ^{35}Cl, the actual mass is 34.97 amu, slightly *lighter* than the combined mass of its protons, neutrons, and electrons.

Why do the masses of the three subatomic particles *not* add exactly to the mass of the atom? When protons and neutrons combine to form nuclei, a small amount of mass is either converted to or created from energy. This change is the relationship postulated by Albert Einstein:

Energy gained or lost = mass lost or gained times the speed of light squared

In equation form, this is written $E = mc^2$.

In nuclear reactions, if a small amount of mass is lost, a very large amount of energy is created. In forming nuclei, however, because the gain or loss in *mass* is relatively small, the mass of a nuclide or atom in atomic mass units will *approximately* (but not exactly) *equal* its mass number.

The sum of the protons and neutrons of a nuclide *approximately* equals its mass in *amu.*

Atomic Mass

Some atoms have one stable nuclide; other atoms have as many as 10 stable isotopes.

Though one kind of atom may have several stable isotopes, for most atoms (those not formed by radioactive decay), in all samples of that atom found on Earth, the *percentage* of each isotope will be the same. Most atoms will therefore have the same *average mass* in any sample found on Earth.

This average mass of an atom, called its **atomic mass**, can be calculated from the **weighted average** of the mass of its isotopes.

> **Example:** In Earth's crust, atoms of chlorine have two frequently occurring isotopes. In all samples, close to 75% are ^{35}Cl with a mass of 35.0 amu, and close to 25% are ^{37}Cl with a mass of 37.0 amu.
>
> The average mass of a chlorine atom is the *weighted* average of its isotopes. Given this close to 3-to-1 isotopic ratio, this average can be approximated by
>
> $$\frac{(35.0 + 35.0 + 35.0 + 37.0)}{4} \text{ amu} = \textbf{35.5 amu}$$
>
> No single atom of chlorine will have this average mass, but in visible amounts of substances containing chlorine, all chlorine atoms can be assumed to have this *average* mass.

Use of this average mass (the atomic mass) will simplify chemistry calculations involving mass.

Isotopes and Chemistry

The rules and the reactions for "standard chemistry" are very different from those of *nuclear* chemistry. For example:

*   Chemical reactions can release substantial amounts of energy, such as in the burning of fuels or in conventional explosives, but nuclear reactions can release *much* larger amounts of energy, as seen in stars or nuclear weapons.

*   In chemical reactions, atoms can neither be created nor destroyed. Nuclear reactions often change one atom into another.

Because the rules are very different, a clear distinction must be made between *chemistry* and *nuclear chemistry*. By convention, the rules that are cited as part of "chemistry" refer to processes that do *not* involve changes in nuclei (unless *nuclear* chemistry is specified). Processes that change the composition of a nucleus are termed *nuclear* reactions, which by definition are not chemical reactions.

The good news is that all isotopes of an atom nearly always have the same *chemical* behavior, and nearly all atoms have the same average mass in visible amounts of substances. Therefore, during nearly all *chemical* reactions and processes, we can ignore the fact that atoms have isotopes. We will return to the differences among isotopes when we consider nuclear reactions such as radioactive decay.

A Flashcard Review System

At this point, you may have a sizeable stack of flashcards, and soon you may add more. Let's adopt a system to organize the cards.

Separate your existing flashcards into three stacks:

1. *Daily*: Those you need to practice until 100% correct on at least three separate days.

2. *Test prep*: Those you have practiced for three or more days. Run these again before your next quiz or examination on this material.

3. *Final exam review*: Those you have retired until the final.

For each stack, label a card with the name of the stack and add it to the top of the stack.

Place a rubber band around each stack. You may want to carry the *daily* pack with you for practice during downtime.

Flashcards

For this lesson, be sure these are moved into memory.

One-Way Cards (with Notch)	Back Side—Answers
Weighted average of mass of isotopes is termed _____	Atomic mass of an atom
Approximate mass of one carbon-14 nucleus = _____	14.0 amu
Approximate mass of one electron = _____	Less than 0.001 amu
Approximate mass of 2 protons + 2 neutrons + 2 electrons = _____	4.0 amu (electron mass is negligible)

PRACTICE B

1. Fill in the blanks.

Protons	Neutrons	Electrons	Atomic Number	Mass Number	Nuclide Symbol	Ion Symbol
	144	88	90			
	148					Pu^{2+}
		78			^{206}Pb	
	0					H^+
					^{3}H	H^-
		36			^{90}Sr	
11		10		23		
15	16	18				

2. Which atom in problem 1 is not found in Earth's crust?

Lesson 6.4   Elements and Compounds

Substance Terminology

The definitions below are general and simplified, but they will give us a starting point for discussing how atoms may combine to yield different substances.

1. A chemical **substance** is matter that contains *one* kind of chemical particle: All of its electrically neutral units are composed of atoms, chemically *bonded* in the same manner and geometry. There are two types of substances: *elements* and *compounds*. **Chemical formulas** can be written to represent a substance.

2. Except at extremes of temperature and/or pressure, substances will be found in one of three **phases** (also called **states**): solid, liquid, or gas. When a substance *changes* phase, such as in melting or boiling, it remains the same substance: its composition does not change.

3. Substances have *characteristic* properties. One characteristic is that under a given atmospheric pressure, the temperature at which a substance melts (changes from a solid to a liquid) or boils (becomes a gas inside its liquid state) will be the same no matter how the substance is created. Because color and density as solid or liquid are characteristic for a substance, they are some of the other properties that can help in identifying the substance.

4. If the smallest particles of a substance that are stable independent units are neutral particles with two or more atoms, the particles are called **molecules**. If a substance consists of charged particles, the particles are called ions, and the smallest electrically neutral combination of ions is called a **formula unit**.

5. **Elements** are composed of electrically *neutral* particles that contain only one *kind* of atom. Each atom has an **elemental state**: the substance formula and phase (solid, liquid, or gas) that is the most stable form of the element at room temperature and pressure.

   Eleven atoms are gases in their elemental state: hydrogen, nitrogen, oxygen, fluorine, chlorine, and the noble gases. Two are liquids: bromine and mercury. The rest are solids.

   The **monatomic elements** are individual atoms in their elemental state. Formulas for monatomic elements are written as a single atom.

   > **Example:** Particles of the *noble gases* are single atoms. The formulas for these elements include He for helium and Ne for neon.

6. **Chemical bonds** are forces that hold particles together. Some elements are **polyatomic**, consisting of two or more atoms of the same kind that are chemically bonded by shared electrons to form a larger unit. Polyatomic elements include the **diatomic elements** that are composed of two atoms bonded together (*di-* means "two").

   > **Example:** The most stable elemental forms of oxygen, nitrogen, and chlorine are all diatomic. Their chemical formulas are O_2, N_2, and Cl_2. The most stable elemental formula for sulfur has eight atoms bonded together, written as S_8.

   In chemical formulas, a **subscript** is a number written after a symbol and represents the number of that kind of atom or ion that is bonded within a particle.

7. **Metal atoms** compose more than 70% of the different kinds of atoms found in Earth's crust. In their elemental state, all metals except mercury (Hg) are solids. All metals conduct electricity. All solid metals are shiny if polished, and they are **malleable**: They can be hammered to form shapes. Metals have a structure that is more complex than the monatomic or polyatomic elements; however, a metal in its elemental state is represented by a formula with a single atom, such as Na for sodium and Ag for silver.

8. A **compound** consists of two or more *different kinds* of atoms that are electrically neutral overall but are held together by chemical bonds. While there are fewer than 100 elements in nature, there are millions of known compounds. In a given compound, the ratio of the atoms is always the same and is shown by their formulas. H_2O, NaCl, and H_2SO_4 represent compounds because they contain two or more different kinds of atoms. Compounds can be classified as either *covalent* or *ionic*, depending on the nature of their bonds.

9. The basic particles for **molecular compounds** (also known as **covalent compounds**) are **molecules**. Molecules are held together by **covalent bonds** in which electrons are *shared* between two neighboring atoms. Covalent bonds can be single bonds (involving two shared electrons), double bonds (four shared electrons), or triple bonds (six shared electrons), and the bonds hold atoms at predictable average angles within a molecule.

10. **Molecular formulas** use atomic symbols and subscripts to represent the number and kind of atoms covalently bonded to form a single molecule.

   • A water molecule consists of two hydrogen atoms and one oxygen atom, represented by the molecular formula H_2O. In chemical formulas, if no subscript is written after a symbol, a subscript of **1** is understood.

   • Carbon dioxide molecules consist of two oxygen atoms and one carbon atom. The molecular formula is CO_2.

Flashcards

Before trying the problems below, make and practice these.

Two-Way Cards (*without* Notch)

Substance (definition) = _____	Electrically neutral matter in which all particles have the same formula is termed a _____
Elements (definition) = _____	Stable neutral substances with one kind of atom are termed _____
Compounds (definition) = _____	Neutral substances with more than one kind of atom are termed _____
Bonds (definition) = _____	Forces holding atoms together are termed _____
Covalent bonds (definition) = _____	Shared electrons holding atoms together are termed _____
Molecule (definition) = _____	Neutral particle with all covalent bonds is termed a _____

PRACTICE **A**

Consult the atoms table at the back of the book as needed.

1. Which of the following formulas represent elements?

    a. Ne   b. H_2O c. NaCl

    d. S_8 e. $C_6H_{12}O_6$

2. Which of these substances contain chemical bonds?

    a. H_2 b. CO_2 c. NH_3 d. He

3. In problems 1 and 2, which formula(s) represent

    a. Diatomic elements?   b. Monatomic elements?   c. Four atoms?

11. **Structural formulas** can be used to represent chemical particles held together by *covalent* bonds. These formulas show each of the atoms present along with information about their positions within the particle.

    Example: Water, H_2O, is sometimes written as $H-O-H$ to show that the oxygen atom is between the two hydrogens and that the molecule has two single bonds. The structural formula below conveys additional information, showing that water has a *bent* shape.

$$\begin{array}{c} O-H \\ \diagup \\ H \end{array}$$

    Example: The structural formula for carbon dioxide, CO_2, is

$$O=C=O$$

    Carbon dioxide has two double bonds and a linear shape with the carbon atom in the middle.

    We generally write structural formulas if knowing the shape of the molecule is important, but we write the more compact molecular formulas when it is not.

12. **Ionic compounds** are substances consisting of an array of positive and negative **ions** (particles with a net electrical charge). Ions can be *monatomic* (single atoms) or *polyatomic* particles that have an unequal number of protons and electrons. An **ionic bond** is the electrostatic attraction between oppositely charged ions.

13. **Ionic compound formulas** represent the ratio and kind of ions present in an ionic compound. A **formula unit** is defined as the smallest combination of ions for which the sum of the electrical charges is zero. The ions in an ionic compound are always present in a ratio that guarantees overall electric neutrality. Formulas for ionic compounds show the ion ratios in a single neutral formula unit.

Example: Table salt consists of a 1:1 ratio of positively charged sodium ions (Na^+) and negatively charged chloride ions (Cl^-). The formal name of table salt is sodium chloride, and its formula is written **NaCl**. The formula unit NaCl contains two ions.

Parentheses are used if a formula unit contains more than one of a polyatomic ion.

Example: Calcium phosphate is an ionic compound composed of three monatomic Ca^{2+} ions for every two polyatomic $PO_4{}^{3-}$ ions. The ionic compound formula is **$Ca_3(PO_4)_2$**. One formula unit contains five ions and 13 atoms.

14. **Empirical formulas** show the lowest whole-number ratio of the atoms in a formula.

    Example: The formula for glucose (a biologically important sugar) is most often written as $C_6H_{12}O_6$ on the basis of the structure of its neutral molecule. The empirical formula for glucose is CH_2O, reporting the lowest whole-number ratio of the atoms.

    Example: For the ionic compound ammonium nitrate, the formula is written as NH_4NO_3 to indicate its ions, but the empirical formula is $N_2H_4O_3$.

15. **Standard formulas** will be the term used in this text to refer to *molecular* and *ionic compound* formulas: those that use subscripts to represent the number of atoms in the simplest *neutral* particle. When a reference to a *formula* is made, it will mean a *standard* formula unless otherwise noted.

16. **When you *write* formulas** by hand, be careful to print in a way that distinguishes between uppercase and lowercase letter combinations such as CS and Cs, Co and CO, No and NO.

    Examples

    - $Co(OH)_2$ has one cobalt atom, two oxygen atoms, and two hydrogen atoms.
    - CH_3COOH has two carbon atoms, four hydrogen atoms, and two oxygen atoms.

To summarize, although molecules of covalent substances and formula units of ionic compounds have different types of bonds, all substance formulas refer to a single, overall electrically neutral unit of the substance.

Physical and Chemical Changes

17. A **mixture** is a combination of two or more substances. One type of mixture is an *aqueous solution* in which a substance is dissolved in water.

18. **Physical changes.** Melting, freezing (solidifying), boiling, and vapor condensing are examples of a physical change: a change in which substances do not change their identity. A physical change is not considered to be a chemical reaction.

19. **Chemical reactions.** A *chemical reaction* (also called a *chemical change*) cannot create or destroy atoms, or change an atom from one kind to another, or make any changes to the nuclei of the atoms involved in the reaction. But during a chemical reaction, atoms may gain or lose electrons, or they may change the way their electrons are shared or arranged in space. A chemical reaction results in one or more new substances being formed.

Flashcards

One-Way Cards (with Notch)	Back Side—Answers
A physical change does *not* change _____	Substance formulas and bond positions
A chemical change does *not* change _____	The nuclei of the reacting atoms

Two-Way Cards (*without* Notch)	
Standard formula (definition) = _____	Formula showing number of atoms inside a neutral molecule or ionic compound is termed a _____
Structural formula (definition) = _____	Formula indicating atom positions in a particle is termed a _____
A mixture (definition)	A combination of substances is termed a _____
Three phases or states (definition)	Solid, liquid, gas are three _____

PRACTICE **B**

1. For the four particles Na^+, NO_3^-, O_2, and H_2O, write the formula for the particle that

    a. Is an anion   b. Has 11 protons

    c. Is an element   d. Is a compound

    e. Is diatomic   f. Is monatomic

2. Draw the structural formula for a water molecule.

3. Classify these as physical changes, chemical reactions, or nuclear reactions.

    a. Burning octane (C_8H_{18}) to form carbon dioxide and water

    b. Converting hydrogen to helium in a star

    c. Melting ice

Lesson 6.5 The Periodic Table

Learning the behavior of more than 100 different atoms would be a formidable task. Fortunately, the atoms can be organized into families. The chemical behavior of one atom in a family helps to predict the behavior of other atoms in the family. The grouping of atoms into families results in the **periodic table**. A periodic table is included at the back of this book.

To build the table, the atoms are arranged in **rows** (also called **periods**) in order of the number of protons in each atom. This order usually, but not always, matches the order of the increasing atomic mass of the atoms. At certain points, the chemical properties of the atoms begin to repeat, somewhat like the octaves on a musical scale.

In the periodic table, under most graphic designs, when a noble gas atom is reached, it marks the end of a row. The next atom, with one more proton, starts a new row of the table.

The main group atoms are those found in the *tall* column blocks on both sides of the table. They are termed either groups 1, 2, and 13 to 18 or groups 1A, 2A, and 3A–8A, depending on the version of the periodic table that you are using. **The transition metals** are in the "middle dip" block of the periodic table. There are 10 columns of transition metals. **The inner transition metals** include the 14 **lanthanides** (or **rare earth metals**) that begin with lanthanum in the sixth row and the 14 **actinides** that begin with actinium in the seventh row. These atoms are usually listed in a block below the rest of the periodic table in order to display a table that fits easily on a chart or page.

Families in the Table

The standard periodic table design places the atoms into **columns** (called **families** or **groups**). Within each column, the atoms tend to have similar chemical behavior.

The noble gases (He, Ne, Ar, Kr, Xe, Rn) are in **group 18**, or **8A**, of the periodic table at the far right. As elements they are monatomic (composed of single atoms), and in their elemental state at room temperature and pressure, all are gases. These atoms are termed "noble" because they are "content" with their status: These atoms rarely bond with other atoms or with each other.

The alkali metals are those below hydrogen in **group 1**, or **1A**, at the far left of the table. As elements, all are soft, shiny metals that tend to react with many substances, including water. In reactions, an alkali metal atom tends to *lose* one electron to become a **1+** *ion*. Once an alkali metal atom forms a 1+ ion, it becomes quite stable, and most reactions do not change its 1+ charge.

The alkaline earth metals are in **group 2**, or **2A**: The second "tall column." In reactions, each alkaline earth metal atom tends to *lose* two electrons to become a **2+** *ion*, and once the ion is formed, most reactions do not change its 2+ charge.

The halogens are in **group 17**, or **7A**, just to the left of the noble gas column. As neutral elements, halogen atoms are stable only when they are found in the diatomic molecules F_2, Cl_2, Br_2, I_2, and At_2. Like the alkali metals, the halogens are very reactive. In many reactions, a neutral halogen atom tends to *gain* one electron to become a **halide ion** with a **1−** charge. Halogen atoms can also covalently bond (share electrons) to form compounds and ions.

Other families are often named by the top atom in their column, such as the **carbon family**, the **nitrogen family**, and the **oxygen family**. For example, silicon is said to be "in the carbon family."

Hydrogen is often placed in column 1 of the table, and the reactions of hydrogen are often like those of the alkali metals; however, other hydrogen reactions are like those of the halogens. Hydrogen is probably best portrayed as a unique family of one that has characteristics of both alkali metals and halogens.

Predicting Behavior

For the following diagram of the periodic table groups, given the group numbers, be able to fill in the family names and likely ion charges from memory.

Group	1A	2A	3B → 2B	3A	4A	5A	6A	7A	8A
	1	2	3–12	13	14	15	16	17	18
Family Name	Alkali metals (except H)	Alkaline earth metals	Transition metals			Nitrogen family	Oxygen family	Halogens	Noble gases
Charge if Monatomic Ion	1+	2+	Positive, but varies			3−	2−	1−	None

Using this table, chemical behavior and (for many atoms) ion charges can be predicted.

> **Example:** Cesium (Cs) is in column 1 of the periodic table. On the basis of this placement, we can predict that it will
> - behave like other alkali metal atoms, and
> - exist as a Cs^+ ion in compounds.

PRACTICE A

Use a copy of the periodic table to answer these questions.

1. Describe the location in the periodic table of the
   a. Noble gases
   b. Alkali metals
   c. Halogens
   d. Transition metals

2. Add a charge to these symbols to show the ion that a single atom of each tends to form.
   a. Br   b. Ra   c. Cs   d. O

Metalloids, Nonmetals, and Metals

As elements, atoms can be divided into metalloids (also called semimetals), nonmetals, and metals.

Metalloids Many periodic tables include a thick line, like a staircase, shown in the partial periodic table below. This line separates the metal and nonmetal atoms. The six atoms in boldface bordering the thick line are the **metalloids** that have chemical behaviors in between those of metal and nonmetal atoms.

					(H)	He
B	C	N	O	F	Ne	
	Si	P	S	Cl	Ar	
	Ge	As	Se	Br	Kr	
		Sb	Te	I	Xe	
			(Po)	At	Rn	

For exams, if you are not allowed to use a periodic table that has the staircase and identifies the metalloids, you should memorize the location of the staircase and the six metalloids. If you memorize how the staircase looks at boron (B), the rest of the staircase is easy.

Some textbooks include polonium (Po) as a seventh metalloid, others do not. The halogen astatine (At) is usually, but not always, considered to be a nonmetal rather than a metalloid. In these lessons, we will consider all of the halogens to be nonmetals.

The metalloids tend to be *semiconductors* of electricity, a property that makes metalloids a central component in integrated circuits ("computer chips").

Nonmetals Shown below are the 18 nonmetals. The nonmetals must be *memorized*: H, C, N, O, P, S, Se, plus the five halogens and six noble gases. Note the shape of their positions: The nonmetals are all to the right of the staircase and to the right of the metalloids. All atoms in the last two columns are nonmetals.

			(H)	He
C	N	O	F	Ne
	P	S	Cl	Ar
		Se	Br	Kr
			I	Xe
			At	Rn

Note also that hydrogen, although it is often shown in column 1, is considered to be a *nonmetal*. Hydrogen has unique properties, but it most often behaves as a nonmetal.

Metals The metals are *all* of the elements (except hydrogen) to the left of the staircase and the six metalloids. The metals include the inner transition elements usually listed below the rest of the chart. Of the 100-plus elements, more than 75% are metals. To learn the elements that are metals, memorize the six metalloids and 18 nonmetals. All of the remaining elements are metals.

Flashcards

One-Way Cards (with Notch)	Back Side—Answers
Family that rarely bonds to other atoms or each other = _____	Noble gases
Lightest nonmetal atom = _____	Hydrogen (H)
Lightest metalloid atom = _____	Boron (B)
Number of nonmetal atoms = _____	18
Charge on ion if single *halogen* atom = _____	1−
Charge if *alkali metal* ion	1+
Charge if *alkaline earth metal* ion	2+

Two-Way Cards (*without* Notch)

Location of alkali metals in the periodic table = _____	Below H, first column family name = _____
Location of halogens = _____	Next-to-last column family name = _____
Location of noble gases = _____	Last column family name = _____
Location of transition metals = _____	Between tall columns, family name = _____
Term for single halogen atoms with 1− charge = _____	Halide ions definition = _____

Some frequently encountered metals have symbols based on their Latin names. You have previously learned sodium and potassium. If the additional metals presented in the cards below are not firmly in memory in both directions, add these cards to your stack.

Two-Way Cards (*without* Notch)

Copper symbol =	Cu = _____	Mercury symbol =	Hg = _____
Silver symbol =	Ag = _____	Iron symbol =	Fe = _____
Gold symbol =	Au = _____	Lead symbol =	Pb = _____
Tin symbol =	Sn = _____		

PRACTICE **B**

Run the flashcards above until they are initially in memory, then use a full periodic table and your knowledge about the table to answer these questions.

1. State the name of the *family* of the periodic table in which these atoms are found.

    a. Iodine   b. Cesium   c. Gold

    d. Neon   e. Strontium   f. Uranium

2. Without consulting a periodic table, add the metal–nonmetal dividing line to the portion of the periodic table below, then circle the metalloid atoms.

					(H)	He
	B	C	N	O	F	Ne
	Al	Si	P	S	Cl	Ar
Zn	Ga	Ge	As	Se	Br	Kr
Cd	In	Sn	Sb	Te	I	Xe
Hg	Tl	Pb	Bi	Po	At	Rn

3. Since ancient times, copper, silver, and gold have been used as "coinage metals." In part this is because, unlike most elements that are metals, they do not readily corrode or dissolve if they become wet (a desirable characteristic for money).

    In the periodic table, are the coinage metals a family? Should they be?

SUMMARY

Chapter 6 introduces the terminology and fundamentals of chemistry. The flashcards and practice problems will convey facts and relationships that at this point you need firmly in memory.

By the end of this chapter, you should also be able to fill in a blank periodic table with the first 20 and first two and last two columns of atom names and symbols presented in Lesson 6.1.

In a science text, new vocabulary is defined when it is first introduced, but some passages may contain quite a few words with unfamiliar meanings. When that happens, working memory can become overloaded, and comprehension becomes difficult. A study procedure that will help when first reading about a topic is to:

- List words that have meanings you cannot recall from memory.
- Define each word on the basis of what is in the passage or available online.
- Design and practice "two-way" flashcards with the words and meanings.
- Then, read the text passage again.

After the meaning of new terms can be recalled even to a limited extent, your working memory has more space for understanding. If you practice the flashcards over additional days and weeks, the words and their associations are likely to "stick" in long-term memory.

REVIEW QUIZ

Use a periodic table to answer the following questions:

1. Fill in the chart.

Subatomic Particle	Location	Symbol	Mass	Charge
Neutron				
Electron				
Proton				

2. Fill in the blanks.

Protons	Neutrons	Electrons	Atomic Number	Mass Number	Nuclide Symbol	Ion Symbol
8		10		16		
				137		Cs^+
		54			^{137}Ba	
35	46	36				

3. Write the ion symbols for the anions in the table in question 2.

4. Which subatomic particles are not gained or lost during chemical reactions?

5. Add a charge to these symbols to show the *ion* that a single atom of each tends to form.

   a. Mg   b. I   c. Li   d. P   e. S

6. Which family in the periodic table tends not to form ions?

7. Using the list H_2O, Cl_2, Au, S_8, Si, Co, and H_2SO_4, write the formulas for

   a. elements on the list.   b. substances without ionic or covalent bonds.

   c. substance(s) that are diatomic.   d. elements that are metals.

   e. metalloids (semimetals).

8. Which lettered choices below represent isotopes?

   a. 47 p^+ and 60 n^0 b. 46 p^+ and 60 n^0

   c. 47 p^+ and 62 n^0 d. 48 p^+ and 62 n^0

   e. 47 p^+ and 61 n^0

9. Which of these lists contains all nonmetals?

   a. C, N, S, Na, O   b. H, I, He, P, C

   c. F, H, Ne, Si, S   d. Br, H, Al, N, C

10. What weight in pounds (lb.) is equal to 4.10 kg? (454 g/lb.)

11. Without looking at a periodic table, fill in the *symbols* for the atoms in the first three *rows* and the first two and last two *columns*.

PERIODIC TABLE

1A	2A		3A	4A	5A	6A	7A	8A

ANSWERS

Lesson 6.2

Practice A   1a. C   1b. O   1c. Cl   1d. Ca

2a. Potassium   2b. Sodium   2c. Fluorine

3. Neutrons

4.

Particle	Symbol	Charge	Mass	Location
Proton	p^+	1+	1.0 amu	Nucleus
Neutron	n^0	0	1.0 amu	Nucleus
Electron	e^-	1−	<0.001 amu	Outside nucleus

5. See the periodic table.

6.

Atom Name	Symbol	Protons	Electrons	Atomic Number
Sodium	Na	11	11	11
Nitrogen	N	7	7	7
Carbon	C	6	6	6
Phosphorus	P	15	15	15
Fluorine	F	9	9	9

Practice B 1. During chemical reactions, atoms do *not* gain or lose **protons**, which have a positive charge. Atoms can take on a charge to become *ions* by gaining or losing **electrons**, which have a **negative** charge.

2a. **20 protons, 20 electrons**   2b. **20 protons, 18 electrons**

3. …a cation will always have more **protons** than **electrons**.

4a. S = **Sulfur**   4b. Si = **Silicon**   4c. P = **Phosphorus**

5a. Bromine = **Br**   5b. Boron = **B**   5c. Barium = **Ba**

6.

Symbol	Protons	Electrons
O	8	8
O^{2-}	8	10
Mg^{2+}	**12**	10
Al^{3+}	13	10
Au	79	79

Symbol	Protons	Electrons
H^+	1	0
Br^-	35	36
Se^{2-}	34	36
I^-	**53**	54
Cl^-	**17**	**18**

7. Mg^{2+}, Al^{3+}, H^+

Lesson 6.3

Practice A 1.

Protons	Neutrons	Atomic Number	Mass Number	Nuclide Symbol	Nuclide Name
6	6	6	**12**	^{12}C	**Carbon-12**
7	7	**7**	14	^{14}N	**Nitrogen-14**
53	**78**	53	131	^{131}I	Iodine-131
92	**143**	92	235	^{235}U	**Uranium-235**
2	2	2	4	^4He	**Helium-4**

2. **Uranium**. All nuclei with more than 82 protons are radioactive.

Practice B 1.

Protons	Neutrons	Electrons	Atomic Number	Mass Number	Nuclide Symbol	Ion Symbol
90	144	88	90	**234**	^{234}Th	Th^{2+}
94	148	**92**	94	242	^{242}Pu	Pu^{2+}
82	**124**	78	82	206	^{206}Pb	Pb^{4+}
1	0	**0**	1	1	^1H	H^+
1	2	2	1	3	^3H	H^-
38	52	36	**38**	90	^{90}Sr	Sr^{2+}
11	**12**	10	11	23	^{23}Na	Na^+
15	16	18	**15**	31	^{31}P	P^{3-}

2. **Plutonium (Pu)**. No atoms found in nature on Earth have more than 92 protons.

Lesson 6.4

Practice A   1. 1a and 1d.   2. 2a, 2b, and 2c.   3a. 2a.   3b. 1a and 2d.   3c. 2c.

Practice B   1a. NO_3^- 1b. Na^+ 1c. O_2 (one kind of atom, electrically neutral particle)   1d. H_2O (more than one kind of atom, electrically neutral particle)   1e. O_2 1f. Na^+ (contains one atom)

2. See the lesson.   3a. Chemical reaction   3b. Nuclear reaction   3c. Physical change

Lesson 6.5

Practice A   1a. Last column   1b. Column 1 (not including H)   1c. Group 17 (tall column 7A), just before the noble gases
1d. The 10 columns in the middle dip   2a. Br^- 2b. Ra^{2+} 2c. Cs^+ 2d. O^{2-}

Practice B   1a. Halogens   1b. Alkali metals   1c. Transition metals   1d. Noble gases   1e. Alkaline earth metals
1f. Either inner transition metals or actinides

2. See the table in the lesson.

3. In the periodic table, the coinage metals are in the same column, so they can be considered to be in a family. The fact that they share similar chemical behavior (in not reacting with water) is what would be expected within a periodic table family.

Review Quiz

1.

Subatomic Particle	Location	Symbol	Mass	Charge
Neutron	Nucleus	n^0	1.0 amu	0
Electron	Outside nucleus	e^-	<0.001 amu	1−
Proton	Nucleus	p^+	1.0 amu	1+

2.

Protons	Neutrons	Electrons	Atomic Number	Mass Number	Nuclide Symbol	Ion Symbol
8	8	10	8	16	^{16}O	O^{2-}
55	82	54	55	137	^{137}Cs	Cs^+
56	81	54	56	137	^{137}Ba	Ba^{2+}
35	46	36	35	81	^{81}Br	Br^-

3. O^{2-} and Br^-

4. Protons and neutrons

5a. Mg^{2+} 5b. I^- 5c. Li^+ 5d. P^{3-} 5e. S^{2-}

6. Noble gases

7a. Cl_2, Au, S_8, Co, Si   7b. Au, Co   7c. Cl_2 7d. Au, Co   7e. Si

8. **a, c,** and **e**

9. **b**

10. WANTED: ? lb. =   (WANT a single unit.)

   DATA: (4.10 kg)   (Single-unit DATA.)

   454 g = 1 lb.

   SOLVE: ? lb. = 4.10 kg $\cdot$ $\dfrac{1000\ g}{1\ kg}$ $\cdot$ $\dfrac{1\ lb.}{454\ g}$ = **9.03 lb.**

11. See the periodic table.

7

Writing Names and Formulas

Lesson 7.1   Elements and Compounds

Chemical substances are identified by both a name and a chemical formula. To write names and formulas that both identify and differentiate substances, a *system* is required.

Historically, compounds containing carbon and hydrogen are studied in **organic chemistry**, which has its own naming system. In this chapter, we will confine our attention to the naming rules for **inorganic chemistry**, which govern substances not based on carbon and hydrogen.

Different types of inorganic substances have different naming systems. We will begin with the rules for naming elements.

Names and Formulas for Elements

An *element* is an electrically neutral substance that contains only one kind of atom. The **name** of an element is simply the name of its **atoms**.

> **Example:** The element composed of neutral atoms with 20 protons is called **calcium**. Calcium is a metal, and formulas for elements that are metals are written as if they were monatomic elements. The formula for the element calcium is written **Ca**.

Example: Neutral **oxygen** atoms at typical room conditions are most stable in diatomic molecules. For elemental oxygen, the formula is O_2.

Some elements have more than one stable form.

Example: Graphite, diamond, and fullerenes are all stable substances made from neutral, bonded **carbon** atoms, but the elemental state of carbon is generally designated to be graphite. Although graphite particles consist of many carbon atoms bonded together, in chemical reaction equations graphite is represented by the simplified formula **C**.

Note that for elements, the name does not distinguish between monatomic, diatomic, and polyatomic structures. This is only an issue for a few of the elements, but for the millions of chemical compounds, a more systematic naming system is required. Based on the bonds within them, compounds have different naming systems.

Types of Bonds

The two types of bonds encountered most often in substances are *covalent* and *ionic*.

1. In **covalent bonds**, electrons are *shared* between two atoms.

2. In **ionic bonds**, an atom (or group of atoms) has lost one or more electrons compared to its electrically neutral form, and another neutral atom or group has gained one or more electrons. The loss and gain of electrons from neutral particles results in charged particles (ions). Ions are bonded by the attraction of their opposite charges.

3. In *most* cases:

    *   A bond between two *nonmetal* atoms is a covalent bond.

    *   A bond between a *metal* atom and a *nonmetal* atom is an *ionic* bond.

To identify the *type* of bond between two atoms, begin by asking:

*   Are *both* atoms nonmetals? If so, predict the bond is *covalent*.
*   Is one atom a *metal* and the other a *nonmetal*? If so, predict the bond is *ionic*.

The location of the nonmetals in the periodic table is shown in the illustration below. Recall that hydrogen is classified as a nonmetal and that all atoms in the last two columns are nonmetals.

			(H)	He
C	N	O	F	Ne
	P	S	Cl	Ar
		Se	Br	Kr
			I	Xe
			At	Rn

The six noble gases rarely bond. The remaining 12 nonmetal atoms nearly always form covalent bonds when they bond with each other.

──────────▶ **TRY IT**

Q. Using the above rules and a periodic table, below each of the following write whether a bond between the two atoms is likely to be *ionic* or *covalent*.

1. C----H   2. Na----S   3. N----Cl   4. K----F   5. H----H

STOP

Answers:

1. C----H   Both are nonmetals; predict this will be a covalent bond.

2. Na----S   A metal and a nonmetal; predict an ionic bond.

3. N----Cl   Both are nonmetals; predict a covalent bond.

4. K----F   A metal and a nonmetal; predict an ionic bond.

5. H----H   Both are nonmetals; predict a covalent bond.

Types of Compounds

1. If a compound contains *all* covalent bonds, it is classified as a **molecular compound** (also known as a *covalent compound*).

2. If a compound has *one* or more ionic bonds, even if it also has *many* covalent bonds, it will tend to have ionic behavior and is classified as an **ionic compound**.

These rules mean that in most cases:

- A compound with *all nonmetal* atoms is a *molecular* compound.
- A compound that combines *metal* and *nonmetal* atoms is an *ionic* compound.

The above general rules do not cover all types of bonds and compounds and there are exceptions, but these rules will give us a starting point for both naming compounds and writing their formulas.

TRY IT

Q. Using the above rules and a periodic table, label these compounds as *ionic* or *molecular* in the space below each formula.

1. NaCl   2. CH$_4$ 3. H$_2$O   4. HCl

STOP

Answers:

1. NaCl: Na is a metal and Cl is a nonmetal, the bonding is ionic, the compound is ionic.

2. CH$_4$: Both kinds of atoms are nonmetals; the compound is molecular.

3. H$_2$O: Both atoms are nonmetals; the compound is molecular.

4. HCl: Both atoms are nonmetals; predict the compound is molecular.

Physical Properties In some cases, physical properties can indicate whether a compound is molecular or ionic.

- If a compound is in a gas or liquid state at typical room temperature and pressure, it is molecular.
- If a compound is ionic, it is always a solid at typical room conditions.

But a compound that is solid at room conditions can be either ionic or molecular.

Flashcards

Run the new cards for several days in a row, then add them to your "test prep" stack.

One-Way Cards (with Notch)	Back Side—Answers
The formula for elemental oxygen = _____	O$_2$
A metal-to-nonmetal bond is what type?	Usually ionic
A bond between two nonmetals is what type?	Usually covalent
A compound with 10 covalent bonds and one ionic bond is what type of compound?	An ionic compound
A compound with all nonmetal atoms is what type?	Usually a molecular compound
A compound with metal and nonmetal atoms is what type?	Usually an ionic compound
If a compound is a gas or liquid at room conditions, what type of compound is it?	A molecular compound
What is the state of an ionic compound at typical room conditions?	Always in the solid state

Two-Way Cards (*without* **Notch**)

Monatomic means _____	Word meaning "composed of one atom" = _____
Diatomic means _____	Word meaning "composed of two atoms" = _____
Polyatomic means _____	Word meaning "composed of more than one atom" = _____

PRACTICE

For the problems below, use the type of periodic table that you are permitted to view on exams in your course.

On problems with multiple parts, circle the last letter and complete that part in a future study session.

1. After each formula, label the bond as ionic or covalent.

   a. Na ---- I   b. C ---- Cl   c. S ---- O

   d. Ca ---- F   e. C ---- H   f. K ---- Br

2. Label these compounds as ionic or molecular.

   a. CF_4 b. KCl   c. CaH_2

   d. H_2O e. NF_3 f. CH_3ONa

3. Which two families in the periodic table are all nonmetal elements?

Lesson 7.2   Naming Binary Molecular Compounds

(*Note*: This lesson can be done at this point *or after* Lessons 7.3 and 7.4 on naming ionic compounds. Lessons 7.3 and 7.4 do not require that this lesson be completed first. Complete the lessons in the order that these topics are covered in *your* course.)

Molecular and ionic compounds have different naming systems. This lesson will cover the rules for compounds that are molecular.

Binary molecular compounds contain *two* different nonmetals.

Binary molecular compounds that contain *hydrogen* are generally referred to by their historic and familiar but **nonsystematic** names, such as water (H_2O), ammonia (NH_3), and hydrogen sulfide (H_2S). In nearly all other cases, the steps that follow will supply the name used in practice to identify *most* binary molecular compounds.

Steps for Naming Binary Molecular Compounds

1. The name contains two words. The first includes the name of one atom, and the second includes the *root* of the name of the other atom.

2. The *roots* used in naming are H = hydr-, C = carb-, N = nitr-, O = ox-, F = fluor-, P = phosph-, S = sulf-, Cl = chlor-, Se = selen-, Br = brom-, I = iod-, and At = astat-. Not all of these roots are "regular," but their use will become intuitive with practice.

3. The number of atoms of each kind is represented by a Greek prefix. The examples in this text will be limited to these eight prefixes.

Two-Way Cards (*without* a Notch)

mono- =	1 atom
di- =	2 atoms
tri- =	3 atoms
tetra- =	4 atoms
penta- =	5 atoms
hexa- =	6 atoms
hepta- =	7 atoms
octa- =	8 atoms

4. The format for the name is *prefix-atom name* and then *prefix-root*-**ide**.

   Example: The name of Br_2O_5 is dibromine pentaoxide.

5. The following rule takes precedence over the rules below: For binary molecular compounds that contain O atoms, the second word is prefix-*oxide*.

6. The *first* word contains the name of the *atom* that is in the column more to the *left* in the periodic table. If the two atoms are in the same column, the *lower* atom is named first. ("First word: left, then lower.")

7. The second word contains the *root* of the second atom name, with the suffix *-ide* added.

8. In the *first* word, mono- is left off but is assumed to apply if no prefix is given.

9. If a formula contains one oxygen, the second word is *monoxide*. If an *o* or *a* at the end of a prefix is followed by a first letter of an atom or root that is a vowel, the *o* or *a* in the prefix is *sometimes* omitted (both inclusion and omission of the *o* and *a* are allowed, and you may see such names both ways).

To begin learning these steps, prepare two-sided flashcards for the eight molecular prefixes. You may want to use a different color of card or color of ink for these "prefix" cards. Run the cards until you have recalled them all from memory at least once, then try the following problems.

TRY IT

Use a periodic table and the rules above.

Q1. What is the name of CS_2?

Answer:

Carbon is in a column more to the left in the periodic table, so *carbon* is the first word. For one atom, the prefix would be *mono-*, but *mono-* is omitted if it applies to the *first* word. The name's first word is simply *carbon*.

For the second word, *sulfur* becomes *sulfide*. Because there are two sulfur atoms, the name of the compound is **carbon disulfide**.

Q2. What is the name of the combination of four fluorine atoms and two nitrogen atoms?

Answer:

Nitrogen is in the column more to the left in the periodic table, so the first word contains *nitrogen*. Because there are two nitrogen atoms, use the prefix *di-*, so the first word is *dinitrogen*. For the second word, the prefix for four atoms is *tetra-* and the *root*-ide is *fluoride*. The name of the compound is **dinitrogen tetrafluoride**.

Flashcards and Practice

To learn the steps for molecular naming, try these strategies.

- Run the prefix flashcards until you can recall each prefix in both directions quickly and correctly.

- Write a summary in your own words of the shaded rules, then practice writing the summary from memory.

- Prepare the flashcards that follow below. The last flashcard, calling for recall of two examples, may be especially helpful in recalling the steps for naming.

- Answer *parts a–c* of each Practice problem below, applying the rules from memory.

- Wait 1–2 days, write your rules and run your cards again, then finish the Practice.

One-Way Cards (with Notch)	Back Side—Answers
Binary molecular name format = _____	Prefix-atom name *then* prefix-root-ide
For binary molecular names, which atom is named first?	Atom more to the left in the periodic table; lower if in the same column
Which molecular naming rule has precedence?	If a compound has O, the name ends in prefix-oxide

Two-Way Cards (*without* Notch)

Formula for ammonia = ?	Name of NH_3 = ?
Formula for carbon monoxide = ?	Name of CO = ?
Formula for dinitrogen tetrachloride = ?	Name of N_2Cl_4 = ?
Recall two examples of molecular names	Carbon monoxide and dinitrogen tetraoxide

PRACTICE

Answer these in your notebook. Consult a periodic table as needed.

1. Write the formula for these names.

    a. Nitrogen trichloride   b. Sulfur hexafluoride   c. Dinitrogen monoxide

2. Write the name for these nonmetal atom combinations.

    a. Three chlorine and one iodine   b. One oxygen and two chlorine

    c. One bromine and one iodine   d. Eight oxygen and four phosphorus

3. Name these molecular compounds.

    a. SCl_2 b. PI_3

    c. NO   d. I_2O_7

4. Nonmetals often form several stable oxides, including the combinations below. Name that compound!

    a. Six oxygen and four phosphorus

    b. Five oxygen and two nitrogen

    c. NO_2 d. CCl_4 e. SO_3

Lesson 7.3   Naming Ions

Ionic compounds are combinations of *ions*: particles with a charge. In most first-year chemistry courses, you will be asked to memorize the names and symbols for some of the frequently encountered ions. This task is simplified by the patterns for ion charges that are found in the periodic table.

Categories of Ions

1. All ions are either positive (cations) or negative (anions).
   - The charges on cations are most often 1+, 2+, 3+, or 4+.
   - The charges on anions are most often 1−, 2−, or 3−.

2. Ions are either **monatomic** or **polyatomic**.

   - A monatomic ion is composed of a single atom with a charge.

     Examples: Na^+, Al^{3+}, Cl^-, and S^{2-}

   - A polyatomic ion is a particle with two or more covalently bonded atoms and a non-zero electrical charge.

     Examples: OH^-, NH_4^+, and SO_4^{2-}

The Charge of Metal Ions

Metals in their elemental form are electrically neutral, but they have a tendency to lose electrons. All metal atoms can form monatomic ions (single atoms with a charge), though they may form other ions as well. The charge on a metal ion is predicted by these rules.

Rules for Monatomic Metal Ions

1. When a metal atom is written as the first atom in an ionic compound formula, it is *usually* a *monatomic* ion.

2. Monatomic metal ions are always positive ions (cations). Their charges are most often 1+, 2+, 3+, or 4+.

In many cases, the charge on a monatomic metal ion can be predicted from the location of the metal in the periodic table. For example, metals in the *first two* columns of the periodic table form only *one* stable monatomic ion. The charge on that ion is easy to predict.

3. Metal atoms in column 1 (the alkali metals) form only one stable ion: a single atom with a 1+ charge.

4. Metals in column 2 form only one stable ion: a single atom with a 2+ charge.

The charges on metal ions in the remainder of the periodic table are more difficult to predict. Some metals form one stable cation, and others form two.

In these lessons, you will need to know the formulas for the following *six* ions that are frequently encountered.

5. Assume that if Ag or Al is the first atom in a compound formula, it is an ion with a charge of Ag^+ or Al^{3+}.

6. Know that copper and iron form *two* stable monatomic ions: Cu^+ and Cu^{2+}; Fe^{2+} and Fe^{3+}.

PRACTICE **A**

Write a summary in your own words of the six rules above. Then try to complete these problems without consulting a periodic table, but use the table if you are not *certain* where an atom is located.

1. For these atoms, add the *charge* for the *ion* that the atom tends to form.

      a. Ba   b. Al   c. Na   d. Ag   e. K

2. Write symbols for these.

      a. Strontium ion   b. Lithium ion

      c. Aluminum ion   d. Calcium ion

Naming Metal Ions

How a metal ion is named depends on whether the metal forms only one stable ion or forms more than one. Add the following to your metal ion rules.

7. If a metal forms only *one* stable ion, the ion name is the atom name.

      Examples: Na^+ is sodium ion. Al^{3+} is aluminum ion.

8. If a metal forms more than one stable positive ion (ions with different positive charges), the **systematic** name of the ion is the atom name followed by a Roman numeral in parentheses that states the ion's positive charge.

      Example: Fe^{2+} is named iron(II) ion. Fe^{3+} is named iron(III) ion.

9. Add the (Roman numeral) designation *only* for ions of metals that form *more* than one ion. Do *not* use the (Roman numeral) designation in ion names for metals that can form only *one* ion.

      Example: Because copper can form two ions, the name of a copper ion must have a Roman numeral. It will be named either copper(I) or copper(II).

      Silver forms only one ion: Ag^+. Its name is simply *silver ion*. Because only one silver ion is stable, the Roman numeral representing the charge is not included in the name.

Flashcards

One-Way Cards (with Notch)	Back Side—Answers
Define a monatomic ion	One atom with a charge
Define a polyatomic ion	Two or more covalently bonded atoms with an overall charge
A monatomic metal ion always has what kind of charge?	Positive
Ions of column 1 atoms have what charge?	1+
Ions of column 2 atoms have what charge?	2+
When is a (Roman numeral) used in an ion name?	If a metal forms more than one stable ion (different positive charges)

PRACTICE   B

In your notebook, complete a written summary in your own words of the nine metal ion rules, run the flashcards above, then complete these problems. Do not use a periodic table.

1. Write the ion symbols for the following ions.

    a. Copper(II) ion   b. Iron(III) ion   c. Magnesium ion

2. For the following metals, these are the ions that are stable. Name *each* ion.

    a. Ag^+ b. Li^+ c. Cu^+ and Cu^{2+}

3. Lead has two stable ions: Pb^{2+} and Pb^{4+}. Name each ion.

Monatomic Anions

In these lessons, for use in learning to name ionic compounds, the names and symbols of the following eight monatomic anions must be memorized.

- Four *halide* ions: the fluoride, chloride, bromide, and iodide ions (F^-, Cl^-, Br^-, and I^-).
- Two ions in the oxygen family: the oxide (O^{2-}) and sulfide (S^{2-}) ions.
- Two ions in the nitrogen family: the nitride (N^{3-}) and phosphide (P^{3-}) ions.

The name of a monatomic *anion* is the ion's *root* name followed by *-ide*. Check how this rule is applied to the eight ions above.

On a periodic table, note the location of the atoms that form 1–, 2–, and 3– ions.

Group	1A	2A	3B → 2B	3A	4A	5A	6A	7A	8A
	1	2	3–12	13	14	15	16	17	18
Family Name	Alkali metals (Except H)	Alkaline earth metals	Transition metals			Nitrogen family	Oxygen family	Halogens	Noble gases
Charge if Monatomic Ion	1+	2+	Positive, but varies			3−	2−	1−	None

Polyatomic Ions

A polyatomic ion is a particle that has two or more atoms held together by covalent bonds and also has an overall electric charge. In a polyatomic ion, the number of protons is not equal to the number of electrons in the particle.

An example of a polyatomic anion is the hydroxide ion, OH^-. One way to form this ion is to start with a neutral water molecule, $H—O—H$, which has $1 + 8 + 1 = 10$ protons and 10 balancing electrons, and take away an H^+ ion (which has one proton and no electrons). The result is a particle composed of two atoms with a total of 9 protons and 10 electrons: it therefore has an overall *negative one* charge. The negative charge behaves as if it is attached to the oxygen. A structural formula for the hydroxide ion is

$$H—O^-$$

At this point, our interest is the *ratios* in which ions combine. For that purpose, it may help to think of a monatomic ion as a charge that has one atom attached and a polyatomic ion as a charge with several atoms attached.

Learning the Polyatomic Ions

In this text, to practice naming ionic compounds, you will need to learn the names and formulas for the following 10 polyatomic ions.

Two-Way Cards (*without* Notch)

OH^-	Hydroxide ion	PO_4^{3-}	Phosphate ion
NO_3^-	Nitrate ion	SO_4^{2-}	Sulfate ion
MnO_4^-	Permanganate ion	SO_3^{2-}	Sulfite ion
CO_3^{2-}	Carbonate ion	ClO_3^-	Chlorate ion
CrO_4^{2-}	Chromate ion	NH_4^+	Ammonium ion

Suggested ways to learn the ion names and formulas include the following:

1. Practice reciting the ion name and formula.

    Example: "Phosphate ion is P O 4 3 minus."

    "Practice reciting your lines" is an especially effective memory strategy.

2. Copy the listed ions, noting both the spelling of the ion name and which numbers go where when writing each ion formula.

    Example: For nitrate ion, NO_3^- is correct, NO^{3-} is *not* correct.

3. With a folded sheet of paper, cover the ion formulas on the left column and try, from the name on the right, to *recite* the formula. Then *write* the formula on the paper. Then cover the name on the right and, from the formula, see if you can recite, then write, the ion name.

4. Next, make "two-sided" (no notch) flashcards using the ion names and formulas. You may want to use a unique card color to identify these as the *ion* cards. Using the flashcards in both directions, recite what the flashcard answer will be. Carry the cards and practice recall several times over several days.

If additional ions are assigned in your course, they can quickly be learned by this procedure. If problems in this text use additional ions, we will supply names and formulas.

Ions of Hydrogen

Hydrogen has unique characteristics. In many of its compounds, hydrogen bonds covalently, but in compounds classified as *acids*, H^+ ions form when the acid is dissolved in water. In addition, when bonded to metal atoms, hydrogen behaves as a hydride ion (H^-). We will address the special case of hydrogen ions later in this text.

PRACTICE C

Write the rules and run the flashcards for the ion names and symbols, *then* try these problems.

1. In this chart of ions, from memory, add *charges*, *names*, and ion *formulas*.

Symbol	Ion Name	Symbol	Ion Name
	Chloride ion	CO_3	
Br			Radium ion
	Silver ion	MnO_4	
	Hydroxide ion	CrO_4	
Al		K	
ClO_3			Ammonium ion
	Nitrate ion	PO_4	
	Sodium ion		Sulfate ion
F			Sulfide ion
	Calcium ion	Mg	
	Iodide ion		Nitride ion
P		SO_3	

2. Circle the polyatomic ion symbols in the *first* "Symbol" column above.

3. Write the formula(s) for the polyatomic cation(s) in the tables above.

Lesson 7.4   Names and Formulas for Ionic Compounds

If ions have opposite charges, they attract. Ionic compounds are composed of positive ions (cations) combined with negative ions (anions).

An example of an ionic compound is calcium hydroxide. A single neutral unit is composed of three ions: one monatomic calcium ion and two polyatomic hydroxide ions. This unit can be represented by the following diagram.

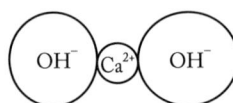

Together, these three ions contain two positive and two negative charges. The charges are said to be *balanced*, and the net charge is zero.

The composition of an ionic compound can be expressed in three ways:

*   By a **name**: in this case, calcium hydroxide
*   By a **standard formula**: $Ca(OH)_2$
*   As **separated ions**: $1\ Ca^{2+} + 2\ OH^-$

As a part of solving many problems, given one type of identification, you will need to write the other two.

Ionic compounds can initially be confusing because their names and standard formulas do not *show* the charges on the ions. To solve problems that involve ionic compounds, a key step will be to translate the name or standard formula into the *separated-ions* format that *shows* the ions with their charges and their ratio in the compound.

Balancing Separated Ions

In matter, there is a powerful tendency for particles with an electric charge to arrange themselves so that the overall *number* of positive and negative *charges* in a collection of particles is balanced. One result is this rule:

> In an ionic compound, the ions must be present in a *ratio* that balances the charges, resulting in electrical neutrality.
>
> In any combination of ions, the *total number* of *positive charges* must equal the *total number* of *negative charges*, so that the overall net charge is *zero*.

When ions combine, only *one ratio* will result in electric neutrality. In problems, you will often need to determine that ratio. Let's learn how to do this with an example.

━━━━━━━━━━▶ **TRY IT**

Q. Find the ratio that balances the charges when S^{2-} and Na^+ combine.

Apply the following steps, then check your answer below.

1. Write the two ion symbols separated by a + sign. Writing the cation (positive ion) first is preferred. Leave space to write a number in front of each ion symbol.

Complete step 1 here:

(STOP)

Answer:

$Na^+ +$ S^{2-}

2. Insert the **coefficients**. Coefficients are numbers written in *front* of ion or particle symbols. Coefficients are a count that shows the ratio in which the particles exist or react. In ionic compounds, that ratio must balance the charges.

When inserting coefficients to balance separated-ion formulas:

(Coefficient *multiplied by* charge of cation) must *balance* (coefficient *multiplied by* charge of anion).

When you are asked to balance ion charges, the only change that you can make, and the one change that *you* must make, is to *write* whole-number *coefficients* in front of the particle symbols. You *cannot* alter the symbols or subscripts or the charge of an ion.

Insert the coefficients that balance charges in your step 1 answer above.

(STOP)

$2\,Na^+ + 1\,S^{2-}$ This is the *separated-ions* formula.

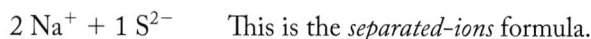

For the charges, $(2 \times 1+ = 2+)$ balances $(1 \times 2- = 2-)$. This means that in the compound, there *must* be *two* sodium ions for every *one* sulfide ion.

3. Reduce the coefficients to the *lowest* whole-number ratios.

(STOP)

2 and 1 are the lowest whole-number ratios.

Try another problem using the steps above.

━━━━━━━━━━▶ **TRY IT**

Q. Insert coefficients so that the charges balance: _____ $Al^{3+} +$ _____ SO_4^{2-}

(STOP)

Answer:

One way to determine the coefficients is to make the coefficient in front of each ion equal to the *number* of the charge of the *other* ion.

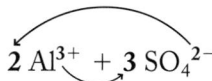

$$2\,Al^{3+} + 3\,SO_4^{2-}$$

For these ions, $(2 \times 3+ = 6+)$ balances $(3 \times 2- = 6-)$.

When balancing charges using this method, however, you often must adjust the *final* coefficients to be the *lowest* whole-number ratios.

▶ TRY IT

Q. Insert proper coefficients: _____ Ba^{2+} + _____ CrO_4^{2-}

Answer:

If balancing produces a ratio of $2\ Ba^{2+}$ + $2\ CrO_4^{2-}$, write the *final* coefficients as

$$1\ Ba^{2+} + 1\ CrO_4^{2-}$$

The separated-ion coefficients *must* be the *lowest* whole-number ratio that results in electrical neutrality.

PRACTICE   A

Insert the lowest whole-number coefficients that balance these separated ions for charge.

1.	Na^+ +	Cl^-	2.	Ca^{2+} +	Br^-
3.	Mg^{2+} +	SO_4^{2-}	4.	Cl^- +	Al^{3+}
5.	Na^+ +	MnO_4^-	6.	In^{3+} +	CO_3^{2-}
7.	Al^{3+} +	PO_4^{3-}	8.	Pb^{4+} +	NO_3^-

From Name to Separated Ions On the basis of the name, we need to be able to write the ratio of the ions in the compound. For a compound composed of two kinds of ions, the first word in the name is the name of the *positive* ion. The second word is the name of the negative ion. To find the ratio of the ions, follow these steps.

1. Write the *symbol* for the positive ion, followed by a + sign, followed by the *symbol* for the negative ion. The ion symbol includes its subscripts and its charge. Leave space for a number in *front* of each symbol.

2. Insert the lowest whole-number coefficients that balance the charges.

Apply these steps to the following problem.

▶ TRY IT

Q. Write the balanced separated-ions formula for aluminum carbonate:

Answer:

1. Write the ion symbols: Al^{3+} + CO_3^{2-}

2. Balance the charges: $\mathbf{2}\,Al^{3+}$ + $\mathbf{3}\,CO_3^{2-}$

The separated-ions formula shows what the name does not: In aluminum carbonate, there must be *two* aluminum ions for every *three* carbonate ions.

When writing separated ions, write the charges *high*, subscripts *low*, and the coefficients at the *same* level as the atom symbols.

PRACTICE   B

If you have not done so today, run your ion flashcards. Then, write balanced *separated-ion* formulas for the ionic compounds below. Try to write the ion formulas from memory, but for the charges on monatomic ions, if you are unsure, consult a periodic table.

1. Sodium hydroxide

2. Aluminum chloride

3. Rubidium sulfite

4. Iron(III) nitrate

5. Copper(II) phosphate

6. Magnesium chloride

Writing Standard Formulas from Names

The standard formula for an ionic compound shows the atoms in each ion in a neutral unit but does not show their charges.

> **Example:** The standard formula for ammonium nitrate is written NH_4NO_3; strontium nitrate is $Sr(NO_3)_2$.

These formulas are referred to by different textbooks in different ways: as "formula unit" formulas, as "empirical formulas based on ions rather than atoms," or simply as the way formulas for ionic compounds have traditionally been written. In this text, we will refer to this formula format as the **standard** formula, but you should use the term for these formulas that is preferred in your course.

In standard formulas for ionic compounds, charges are *hidden*, but charges must balance. During problem solving, it is often necessary to write out a *separated-ions* formula that shows the ion charges.

To write the standard formula for an ionic compound from its *name*, use these steps.

1. On the basis of the name, write the separated-ions formula (including coefficients). Then, to the right draw an arrow, →.

2. After the →, write the two ion symbols, positive ion first, with a small space between them. Include any *subscripts* that are part of a polyatomic ion symbol, but at this point, *leave out* the charges and coefficients.

3. For the symbols after the arrow, put **p**arentheses ( ) around a **p**olyatomic ion *if* its coefficient in the separated-ions formula on the left is greater than 1.

4. Insert *subscripts* after each ion symbol on the right. For the subscript, write the value of the coefficient that is in front of that ion in the separated-ions formula on the left. Omit subscripts of 1.

   For the polyatomic ions, write their separated-ion coefficients on the left side as subscripts *outside* and *after* their parentheses in the standard formula on the right side.

In your notebook, apply those steps to the following example.

▶ TRY IT

Q. Write the standard formula for potassium sulfide.

Answer:

1. Write the separated-ions formula, then an arrow: $2\,K^+ + 1\,S^{2-} \rightarrow$

2. Rewrite the symbols *without* coefficients or charges: $2\,K^+ + 1\,S^{2-} \rightarrow K\;S$

3. Because both ions are monatomic, add no parentheses.

4. The K ion coefficient becomes its standard formula subscript. The sulfide subscript of 1 is omitted as understood.

$$2\,K^+ + 1\,S^{2-} \rightarrow K_2S$$

The standard formula for potassium sulfide is K_2S.

Recite the *three-Ps rule* until you can repeat it from memory:

For ionic compounds, when writing standard formulas, *put parentheses* around *polyatomic* ions—*if* you need more than one.

Apply the rule and the four steps above to the following problem.

▶ TRY IT

Q. Write the standard formula for magnesium phosphate.

Answer:

1. Write the balanced separated ions and →: $3\,Mg^{2+} + 2\,PO_4^{3-} \rightarrow$

2. Rewrite the symbols without coefficients or charges:

$$3\,Mg^{2+} + 2\,PO_4^{3-} \rightarrow Mg\;PO_4$$

3. Because Mg^{2+} is *monatomic*, it is *not* placed in parentheses. Phosphate is *polyatomic* and we need more than **1**, so add ( ): $Mg\,(PO_4)$

4. The separated coefficient of Mg becomes its standard subscript: $Mg_3(PO_4)$
   The phosphate coefficient becomes its standard subscript: $Mg_3(PO_4)_2$
   $Mg_3(PO_4)_2$ is the standard formula for magnesium phosphate.

PRACTICE C

You may use a periodic table. Complete half of the lettered parts today and the rest during your next study session.

1. Circle the polyatomic ions.

   a. Na^+ b. NH_4^+ c. CH_3COO^- d. Ca^{2+} e. OH^-

2. When do you need parentheses? Write the rule from memory.

3. Write standard formulas for these ion combinations.

   a. $2 K^+ + 1 CrO_4^{2-} \rightarrow$ b. $2 NH_4^+ + 1 S^{2-} \rightarrow$

   c. $1 SO_3^{2-} + 1 Sr^{2+} \rightarrow$

Do the remaining problems in your notebook.

4. Balance these separated ions for charge, then write standard formulas.

   a. $Cs^+ +$ $O^{2-} \rightarrow$ b. $CrO_4^{2-} +$ $Ra^{2+} \rightarrow$

   c. $Sn^{4+} +$ $SO_4^{2-} \rightarrow$

5. From these names, write the separated-ions formula, then the standard formula.

   a. Ammonium sulfite $\rightarrow$ b. Potassium permanganate $\rightarrow$

   c. Calcium nitride $\rightarrow$

6. Write the standard formula.

   a. Tin(II) iodide   b. Barium hydroxide   c. Radium nitrate

Converting from Standard Formula to Separated-Ions Formula

To solve problems involving ions in solutions, you will need to be able to convert from the ionic standard formula to the separated-ions formula. This process is the reverse of converting separated ions to standard formulas; however, when going from "standard to separated," some subscripts in the standard formula become coefficients in the separated ions, but others do not.

Example: To represent the ions in sodium phosphate, we write:

$$Na_3PO_4 \rightarrow 3 Na^+ + 1 PO_4^{3-}$$

In going from the standard to the separated-ions formula, the subscript 3 became a coefficient, but the subscripts 1 and 4 did not.

How can we predict which subscripts become coefficients and which do not? The rule is:

> When converting from a standard formula to separated ions, assume that the separated ions have *familiar*, predicted ion formulas.

That's one reason why the frequently encountered ion formulas must be in memory. To separate standard formulas into ions, you must be able to recognize the formulas of the ions inside the standard formula.

➤ TRY IT

Q. Write the separated ions equation for Cu_2CO_3.

Follow these steps:

1. In going from a standard formula to separated ions, decide the *negative* ion's charge and coefficient first.

   Most negative ions have only one likely charge. For the many metal ions that can have two possible positive charges, the charge on the negative ion is needed to identify the charge on the positive ion, so we write the symbol for the negative ion first.

   Do so here:

Answer:

In Cu_2CO_3, the negative ion is CO_3, which always has a 2– charge: CO_3^{2-}.

This step temporarily splits the standard formula into **Cu_2** and 1 CO_3^{2-}.

2. Decide the positive ion's charge and coefficients.

   Given Cu_2 and CO_3^{2-}, the positive ion or ions must include **2** copper atoms *and* must have a total 2+ charge to balance the charge of CO_3^{2-}.

   This means that **Cu_2**, in the separated-ions formula, must be *either* **Cu_2^{2+}** *or* **2 Cu^+**. Both possibilities balance atoms and charge. Which is correct?

   Copper forms two ions: Cu^+ and Cu^{2+}. This suggests that Cu^+ is the ion that forms. Recall also that when metal ions are written as the first ion of two in an ionic compound formula, they are *usually* monatomic ions.

$$Cu_2CO_3 \rightarrow 2\ Cu^+ + 1\ CO_3^{\,2-}$$

   Copper can also be a Cu^{2+} ion, but in the formula above, there is only one carbonate, and carbonate always has a 2– charge. Two Cu^{2+} ions cannot balance the single carbonate.

   The separation equation above is a "best guess," and it does predict the experimentally determined composition of Cu_2CO_3 correctly. But from the standard formula alone, without additional knowledge of ion behavior, it can be difficult to predict with certainty what the separated ions will be.

3. Check to make sure that the charges balance. Make sure that the number of atoms of each kind is the same on both sides. The equation must also make sense going backwards, from the separated-ions formula to the standard formula.

Try another.

▶ **TRY IT**

Q. Write the separated-ions formula for $(NH_4)_2S$.

Answer:

In the standard formula, parentheses are placed around polyatomic ions. When writing the separated ions, a subscript *after* parentheses *always* becomes the polyatomic ion's *coefficient*. You would therefore split the formula:

$$(NH_4)_2S \rightarrow 2\,NH_4 + 1\,S$$

- Assign the charges that these ions prefer:

$$(NH_4)_2S \rightarrow 2\,NH_4^+ + 1\,S^{2-}$$

- Check: In the separated-ions formula, do the charges balance? Going backwards, do the separated ions combine to give the standard formula?

PRACTICE   D

If you have not done so today, run your ion flashcards in both directions, then try these problems. You may use a periodic table.

1. Finish balancing these by adding charges and coefficients.

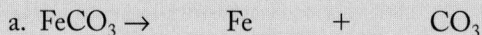

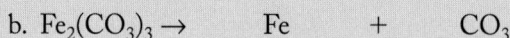

   a. $FeCO_3 \rightarrow \quad Fe \quad + \quad CO_3$

   b. $Fe_2(CO_3)_3 \rightarrow \quad Fe \quad + \quad CO_3$

2. In your notebook, for these ionic compounds, write their separated-ions formula.

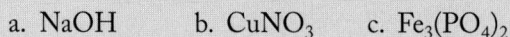

   a. $NaOH$ b. $CuNO_3$ c. $Fe_3(PO_4)_2$

   d. Ag_2CrO_4 e. NH_4Br f. $Mg(OH)_2$

Naming Ionic Compounds

From a standard formula *or* a separated-ions formula, writing the *name* of an ionic compound is easy.

1. If it is not supplied, write the separated-ions formula.

2. On the basis of the separated ions, write the *name* of the cation, then the name of the anion.

Done! In *ionic* compounds, the name ignores the *number* of ions inside. To name the compound, simply name the two ions, positive ion first.

TRY IT

Q. Name Li_3N.

Answer:

$Li_3N \rightarrow 3\ Li^+ + 1\ N^{3-}$; the name is **lithium nitride**.

With practice, you will be able to convert standard formulas to compound names directly, but with ionic compounds, when in doubt, write out the separated-ions formula.

PRACTICE   E

Use a periodic table as needed. If you are unsure of an answer, check it before continuing.

1. Return to Practice D and name each compound.

2. In Practice C, problems 3 and 4, name each compound.

3. Would CBr_4 be named carbon bromide or carbon tetrabromide? Why?

Flashcards

One-Way Cards (with Notch)	Back Side—Answers
In ionic substances, what must be true for charges?	Total positive charges = total negative charges
Term for numbers you insert to balance separated ions = _____	Coefficients
To understand ionic compounds, _____	Write the *separated-ions* formulas
When are ( ) needed in standard formulas?	Put ( ) around polyatomic ions—*if* you need >1

PRACTICE   F

Fill in the blanks in the rows below. Complete half of the rows today and the rest during your next study session. Check your answers after every few rows.

Ionic Compound Name	Separated Ions	Standard Formula
• Name by ion names	• Charges must show	• Positive ion first
• Must be two or more words	• Charges must balance	• Charges balance, but do not show
• Put name of positive ion first	• Coefficients tell ratio of ions	• Put ( ) around polyatomic ions if you need >1

Ionic Compound Name	Separated Ions	Standard Formula
Sodium chloride	$1\ Na^+ + 1\ Cl^-$	NaCl
	$2\ Al^{3+} + 3\ SO_3^{2-}$	$Al_2(SO_3)_3$
Lithium carbonate		
Potassium hydroxide		
	_____ Ag^+ + _____ NO_3^-	
	_____ NH_4^+ + _____ SO_4^{2-}	
		$FeBr_2$
		$Fe_2(SO_4)_3$
Copper(I) chloride		
Tin(II) fluoride		
	_____ Al^{3+} + _____ MnO_4^-	
		K_2CrO_4
		$CaCO_3$
Aluminum phosphide		

SUMMARY

In this chapter, if you have practiced the flashcards over several days, summarized the rules in your own words, and solved the problems, complete the review quiz below and you should be "quiz ready."

By now you may be asking, "Why memorize the molecular prefixes and ion formulas? Why not just look it up?" Science has found that in the *time* it takes to "look up" information, other data from a problem that you need in working memory tends to be forgotten. Plus, "looked up" information takes up one of the 3–5 slots you have in working memory for nonmemorized information, including data specific to a problem. Though memorization takes initial effort, in technical fields and careers the more you "know" in long-term memory, the more problems you can solve. So, "look it up" if needed, but if it is information needed often, move it into memory.

REVIEW QUIZ

Work in your problem notebook. You may use a periodic table but not a calculator.

1. The names of which *type* of compounds identify a count of the atoms in the compound?

2. Write separated-ion formulas for the following.

    a. Ag_2SO_4 b. LiOH   c. K_2CrO_4 d. Rb_3N

3. Write standard formulas for the following.

   a. Sodium chromate   b. Ammonium iodide   c. Aluminum chlorate

4. Name these compounds.

   a. $KClO_3$ b. Br_2O_7 c. Na_2CO_3

5. Which of the compounds in problems 3 and 4 are molecular?

6. Write molecular formulas for these compounds.

   a. Dichlorine heptoxide   b. Sulfur hexabromide

   c. Tetraphosphorus octoxide

7. Name these ionic and molecular compounds.

   a. $CaBr_2$ b. NCl_3 c. $CsOH$ d. $CuCl_2$ e. $RbClO_3$

   f. KI g. MgO h. NO i. NH_4Cl j. P_4S_3

8. Do not use a calculator. Divide 8,523 by 9.

ANSWERS

Lesson 7.1

Practice   1a. Na ---- I   **Ionic**   1b. C ---- Cl   **Covalent**   1c. S ---- O   **Covalent**

1d. Ca ---- F   **Ionic**   1e. C ---- H   **Covalent**   1f. K ---- Br   **Ionic**

2a. CF_4 **Molecular**   2b. KCl   **Ionic**   2c. CaH_2 **Ionic**

2d. H_2O **Molecular**   2e. NF_3 **Molecular**   2f. CH_3ONa **Ionic**

   (All of the ionic compounds contain a metal atom.)

3. Halogens and noble gases.

Lesson 7.2

Practice   1a. NCl_3 1b. SF_6 1c. N_2O

2a. Iodine trichloride   (If in same column, name lower first.)

2b. Dichlorine monoxide   (If includes oxygen, end in oxide. If 1 O, end in monoxide.)

2c. Iodine monobromide   2d. Tetraphosphorus octaoxide *or* octoxide

3a. Sulfur dichloride   3b. Phosphorus triiodide   3c. Nitrogen monoxide

3d. Diiodine heptoxide (or heptaoxide)

4a. Tetraphosphorus hexaoxide (or hexoxide)   4b. Dinitrogen pentoxide (or pentaoxide)

4c. Nitrogen dioxide   4d. Carbon tetrachloride   4e. Sulfur trioxide

Lesson 7.3

Practice A   1a. Ba^{2+} 1b. Al^{3+} 1c. Na^+ 1d. Ag^+ 1e. K^+

2a. Sr^{2+} 2b. Li^+ 2c. Al^{3+} 2d. Ca^{2+}

Practice B   1a. Cu^{2+} 1b. Fe^{3+} 1c. Mg^{2+}

2a. Silver ion   2b. Lithium ion   2c. Copper(I) and copper(II) ion

3. Lead(II) and lead(IV) ion

Practice C   1 and 2.

Symbol	Ion Name
Cl^-	Chloride ion
Br^-	**Bromide ion**
Ag^+	Silver ion
OH^-	Hydroxide ion
Al^{3+}	**Aluminum ion**
ClO_3^-	**Chlorate ion**
NO_3^-	Nitrate ion
Na^+	Sodium ion
F^-	**Fluoride ion**
Ca^{2+}	Calcium ion
I^-	Iodide ion
P^{3-}	**Phosphide ion**

Symbol	Ion Name
CO_3^{2-}	**Carbonate ion**
Ra^{2+}	Radium ion
MnO_4^-	**Permanganate ion**
CrO_4^{2-}	**Chromate ion**
K^+	**Potassium ion**
NH_4^+	Ammonium ion
PO_4^{3-}	**Phosphate ion**
SO_4^{2-}	Sulfate ion
S^{2-}	Sulfide ion
Mg^{2+}	**Magnesium ion**
N^{3-}	Nitride ion
SO_3^{2-}	**Sulfite ion**

   3. The only polyatomic cation is NH_4^+, the ammonium ion.

Lesson 7.4

Practice A   1. $1 Na^+ + 1 Cl^-$ 2. $1 Ca^{2+} + 2 Br^-$ 3. $1 Mg^{2+} + 1 SO_4^{2-}$ 4. $3 Cl^- + 1 Al^{3+}$
5. $1 Na^+ + 1 MnO_4^-$ 6. $2 In^{3+} + 3 CO_3^{2-}$ 7. $1 Al^{3+} + 1 PO_4^{3-}$ 8. $1 Pb^{4+} + 4 NO_3^-$

Practice B   1. $1 Na^+ + 1 OH^-$ 2. $1 Al^{3+} + 3 Cl^-$ 3. $2 Rb^+ + 1 SO_3^{2-}$ 4. $1 Fe^{3+} + 3 NO_3^-$
5. $3 Cu^{2+} + 2 PO_4^{3-}$ 6. $1 Mg^{2+} + 2 Cl^-$

Practice C   1. The polyatomic ions: 1b. NH_4^+ 1c. CH_3COO^- 1e. OH^-
   2. For standard formulas, put parentheses around polyatomic ions if you need more than one.
   3a. K_2CrO_4 3b. $(NH_4)_2S$ 3c. $SrSO_3$
   4a. $2 Cs^+ + 1 O^{2-} \rightarrow Cs_2O$ 4b. $1 CrO_4^{2-} + 1 Ra^{2+} \rightarrow RaCrO_4$ 4c. $1 Sn^{4+} + 2 SO_4^{2-} \rightarrow Sn(SO_4)_2$
   5a. $2 NH_4^+ + 1 SO_3^{2-} \rightarrow (NH_4)_2SO_3$ 5b. $1 K^+ + 1 MnO_4^- \rightarrow KMnO_4$
   5c. $3 Ca^{2+} + 2 N^{3-} \rightarrow Ca_3N_2$
   6. To write standard formulas, write separated ions first.   6a. $1 Sn^{2+} + 2 I^- \rightarrow SnI_2$
   6b. $1 Ba^{2+} + 2 OH^- \rightarrow Ba(OH)_2$ 6c. $1 Ra^{2+} + 2 NO_3^- \rightarrow Ra(NO_3)_2$

Practice D and Practice E, Problem 1

Practice D	Practice E, Problem 1
1a. $FeCO_3 \rightarrow 1 Fe^{2+} + 1 CO_3^{2-}$	**Iron(II) carbonate**
1b. $Fe_2(CO_3)_3 \rightarrow 2 Fe^{3+} + 3 CO_3^{2-}$	**Iron(III) carbonate**
2a. $NaOH \rightarrow 1 Na^+ + 1 OH^-$	**Sodium hydroxide**
2b. $CuNO_3 \rightarrow 1 Cu^+ + 1 NO_3^-$	**Copper(I) nitrate**
2c. $Fe_3(PO_4)_2 \rightarrow 3 Fe^{2+} + 2 PO_4^{3-}$	**Iron(II) phosphate**
2d. $Ag_2CrO_4 \rightarrow 2 Ag^+ + 1 CrO_4^{2-}$	**Silver chromate**
2e. $NH_4Br \rightarrow 1 NH_4^+ + 1 Br^-$	**Ammonium bromide**
2f. $Mg(OH)_2 \rightarrow 1 Mg^{2+} + 2 OH^-$	**Magnesium hydroxide**

Practice E, Problems 2 and 3

2. (3a) **Potassium chromate**   (3b) **Ammonium sulfide**   (3c) **Strontium sulfite**

   (4a) **Cesium oxide**   (4b) **Radium chromate**   (4c) **Tin(IV) sulfate**

3. **Carbon tetrabromide**. (Both are nonmetals, so the compound is molecular. Use prefixes in the names of *molecular* compounds.)

Practice F

Ionic Compound Name	Separated Ions	Standard Formula
Sodium chloride	$1\,Na^+ + 1\,Cl^-$	$NaCl$
Aluminum sulfite	$2\,Al^{3+} + 3\,SO_3^{2-}$	$Al_2(SO_3)_3$
Lithium carbonate	$\mathbf{2\,Li^+ + 1\,CO_3^{2-}}$	$\mathbf{Li_2CO_3}$
Potassium hydroxide	$\mathbf{1\,K^+ + 1\,OH^-}$	$\mathbf{KOH}$
Silver nitrate	$\mathbf{1\,Ag^+ + 1\,NO_3^-}$	$\mathbf{AgNO_3}$
Ammonium sulfate	$\mathbf{2\,NH_4^+ + 1\,SO_4^{2-}}$	$\mathbf{(NH_4)_2SO_4}$
Iron(II) bromide (iron forms two ions)	$\mathbf{1\,Fe^{2+} + 2\,Br^-}$	$FeBr_2$
Iron(III) sulfate	$\mathbf{2\,Fe^{3+} + 3\,SO_4^{2-}}$	$Fe_2(SO_4)_3$
Copper(I) chloride	$\mathbf{1\,Cu^+ + 1\,Cl^-}$	$\mathbf{CuCl}$
Tin(II) fluoride	$\mathbf{1\,Sn^{2+} + 2\,F^-}$	$\mathbf{SnF_2}$
Aluminum permanganate	$\mathbf{1\,Al^{3+} + 3\,MnO_4^-}$	$\mathbf{Al(MnO_4)_3}$
Potassium chromate	$\mathbf{2\,K^+ + 1\,CrO_4^{2-}}$	K_2CrO_4
Calcium carbonate	$\mathbf{1\,Ca^{2+} + 1\,CO_3^{2-}}$	$CaCO_3$
Aluminum phosphide	$\mathbf{1\,Al^{3+} + 1\,P^{3-}}$	$\mathbf{AlP}$

Review Quiz

1. Names for binary molecular compounds

2a. $2\,Ag^+ + 1\,SO_4^{2-}$ 2b. $1\,Li^+ + 1\,OH^-$ 2c. $2\,K^+ + 1\,CrO_4^{2-}$ 2d. $3\,Rb^+ + 1\,N^{3-}$

3a. Na_2CrO_4 3b. NH_4I 3c. $Al(ClO_3)_3$

4a. Potassium chlorate   4b. Dibromine heptoxide (or heptaoxide)   4c. Sodium carbonate

5. Only 4b   6a. Cl_2O_7 6b. SBr_6 6c. P_4O_8

7a. Calcium bromide   7b. Nitrogen trichloride   7c. Cesium hydroxide   7d. Copper(II) chloride

7e. Rubidium chlorate   7f. Potassium iodide   7g. Magnesium oxide   7h. Nitrogen monoxide

7i. Ammonium chloride   7j. Tetraphosphorus trisulfide   8. 947

8

Moles and Balancing Equations

Lesson 8.1   Counting Atoms in Formulas

A step that is central in a variety of chemistry problems is counting *how many* atoms of each kind are represented in a formula. How this is done depends on how the formula is written.

Recall that in a formula, subscripts show the number of atoms of each kind, and if there is no subscript after an atom, a 1 is understood.

Example: H_2SO_4 (sulfuric acid) contains 2 H, 1 S, and 4 O atoms.

To count the atoms of each kind *if* a formula includes parentheses:

For one kind of atom at a time, count the atoms *inside* the parentheses, then multiply by the subscript *outside* the parentheses.

➤ **TRY IT**

Q1. Finish this count of each atom in $Ba(NO_3)_2$. __**1**__ Ba, _____ N, _____ O

Q2. For lead(IV) phosphate, $Pb_3(PO_4)_4$, count each atom: _____ Pb, _____ P, _____ O

STOP

Answers:

1. For N, $(1 \times 2) = 2$. For O, $(3 \times 2) = 6$. Thus, **1 Ba, 2 N, 6 O**.

2. The compound contains **3 Pb**, $(1 \times 4) = $ **4 P**, and $(4 \times 4) = $ **16 O**.

Partial Structural Formulas

Often, chemical formulas are a mix of structural and molecular formulas.

> **Example:** The formula for ethyl alcohol is often written as CH_3CH_2OH. The formula for dimethyl ether is usually written CH_3OCH_3 to show that the O is found in the middle, rather than toward one end as it is in the alcohol. Both formulas could be written as C_2H_6O, but because the different atom arrangements result in different properties, we write the "stretched-out formulas" to distinguish these substances.

In the case where a formula "spreads out the atoms" to show structural information, simply count the *total* for each atom on the basis of the subscript after each atom.

> **Example:** For the *ether* above, $CH_3OCH_3 = $ **2 C, 6 H, 1 O**.

➤ **TRY IT**

Count each kind of atom. Write your answers in the format of the answer to the ether example above.

Q1. $HC_2H_3O_2 =$

Q2. $CH_3COOH =$

STOP

Answers:

Both contain **2 C, 4 H, 2 O** (written in any order). Each formula is a different way to represent acetic acid, the active ingredient in vinegar.

Uppercase versus Lowercase

When counting atoms, and especially when writing formulas by hand, be careful to distinguish uppercase and lowercase letters in atom symbols.

► TRY IT

Copy the following three substance formulas into your notebook. Then, mark the location of this page, close this book, and in your notebook count the atoms in each formula. Write your counts in the format of the ether example earlier.

Q1. $CS_2 =$

Q2. $Cs_2S =$

Q3. $Co(CH_3CO_2)_2 =$

(STOP)

Answers:

1. 1 C, 2 S   2. 2 Cs, 1 S   3. 1 Co, 4 C, 6 H, 4 O

PRACTICE

Do not use a calculator.

1. Copy each formula on separate lines down the page in your notebook. Then, after each formula, write the *number* of each kind of atom and the *full name* for each atom.

   a. HCOOH   b. $CoSO_4$ c. $No(NO_3)_3$ d. Cs_3PO_4

2. Copy each formula into your notebook. After each formula, write the *number* of each kind of atom and the *symbol* for each atom in the format of the Try It answer earlier.

   a. C_2H_5COOH b. $Co_3(PO_4)_2$ c. $Al_2(SO_4)_3$ d. $Pb(C_2H_5)_4$ e. $(NH_4)_3PO_4$

Lesson 8.2   Moles and Molar Masses

Atoms and molecules are extremely small. Visible quantities of a substance must therefore have a very large number of molecules, atoms, or ions.

> **Example:** One drop of water contains about 1,500,000,000,000,000,000,000 (1.5×10^{21}) water molecules.

When solving calculations, rather than writing numbers of this size, chemists use a unit to count large numbers of particles. As we count eggs by the dozen or buy printer paper by the ream (500 sheets), we count chemical particles such as molecules, atoms, and ions by the **mole**. The mole is the metric *base unit* that is used to count small particles.

Just as 1 dozen anythings = 12 individual anythings,

> 1 mole of particles = 6.02×10^{23} individual particles
> (atoms, molecules, or ions).

The number 6.02×10^{23} is called Avogadro's number.

The definition of the mole is based on the isotope carbon-12: exactly 12 grams of ^{12}C contains exactly 1 mole of ^{12}C atoms. Using this definition simplifies the arithmetic in calculations involving other atoms, especially when we convert between grams and moles.

Mole is abbreviated *mol* in the SI system. As with all metric abbreviations, mol is not followed by a period, and no distinction is made between singular and plural when the abbreviation is used.

> When an amount of a substance is large enough to be visible, use moles as the unit to count particles.

Working with Moles

Recall that in exponential notation, when multiplying a *number* by a *number* by an *exponential*, the numbers multiply by the standard rules of arithmetic, but the exponential does not change.

Examples:

Half (exactly) of a mole = $1/2 \times (6.02 \times 10^{23}) = 3.01 \times 10^{23}$ particles

10.0 moles = $10.0 \times (6.02 \times 10^{23}) = 60.2 \times 10^{23} = 6.02 \times 10^{24}$ particles

0.20 mole = $0.20 \times (6.02 \times 10^{23}) = 1.2 \times 10^{23}$ particles

PRACTICE   A

How many particles are in the following? (Answer in scientific notation.)

1. 4.00 mol   2. 0.050 mol

Atomic Mass

In most chemistry calculations, each kind of atom can be treated as if all atoms of that kind have a characteristic mass that is their *atomic mass*, measured in *atomic mass units* (amu; see Lesson 6.2). Atomic masses for the atoms are listed in the Table of Atomic Masses at the back of this book.

To encourage mental arithmetic, the atomic masses in these lessons use fewer significant figures than that used in most textbooks. If you use a different table of atomic masses, your answers may differ slightly from the answers shown here.

Molar Mass

> The **molar mass** of an atom is the mass of 1 mole of its atoms. The units of molar mass are g/mol.

The *number* that represents the atomic mass of an atom in atomic mass units is the *same* as the number that measures the molar mass of the atom in *grams per 1 mole*.

Atomic mass of atom in amu ≡ Molar mass of atom in g/mol

In periodic tables, the consecutive *whole* numbers across rows are the atomic number (the number of protons in the atom), and the numbers with decimal points or in parentheses are the values for *both* the atomic mass in amu *and* the molar mass in g/mol.

> **Examples:** The periodic table at the back of this book shows that *hydrogen* has *one* proton, an atomic mass of **1.008 amu**, and a molar mass of **1.008 g/mol**.
>
> Plutonium (which has 94 protons) is shown with an atomic mass of **(244) amu** for its most stable isotope. One **mole** of the atoms of that isotope has a mass of **244 grams**.

To find the molar mass of a substance that contains *more* than one atom, apply this rule:

The molar mass of a substance is the sum of the molar masses of its atoms.

When you are asked to solve chemistry calculations in homework, quizzes, or tests, you will nearly always be allowed to consult a table that includes the molar masses of atoms. With the values in that table, you can calculate the molar mass of any substance that has a known *formula*.

> **Example:** What is the molar mass of NaOH? Add these molar masses:
>
> Na = 23.$\underline{0}$
>
> O = 16.$\underline{0}$
>
> H = $\underline{\quad 1.008}$
>
> 40.$\underline{0}$08 = **40.0** g/mol NaOH

S.F.: Recall that when *adding* significant figures, because the highest *place* with doubt in the added values is the tenths place, the *sum* has doubt in the tenths place, and the answer must be rounded to that place.

The molar mass supplies an *equality*:

40.0 grams NaOH = 1 mole NaOH

When solving problems, after calculating a molar mass, the *equality* must be written in the DATA. Include the formula for the substance after *both units*: grams and moles.

Molar Masses and Subscripts

To calculate the molar mass from a chemical formula containing *subscripts*, recall that subscripts are exact numbers. Multiplying by a subscript therefore does not change the doubtful digit's *place* in the result.

When calculating a molar mass:

1. First, write a count of each atom symbol.

2. Then, use the following *column format* to keep track of the numbers and the decimal place with doubt as you add the molar masses.

   Example: Find the molar mass of phosphoric acid, H_3PO_4.

$$H_3PO_4 = 3\ H, 1\ P, 4\ O$$

$$1\ mol\ H_3PO_4 = 3\ mol\ H = 3 \times 1.008\ g/mol = \ \ 3.024$$

$$1\ mol\ P = 1 \times 31.0\ g/mol = 31.0$$

$$4\ mol\ O = 4 \times 16.0\ g/mol = \underline{64.0\ \ \ \ }$$

$$98.\underline{0}24 \rightarrow \textbf{98.0 g/mol}$$

3. When done with the column calculation of the molar mass, write in your DATA

   DATA: **# g** formula = **1 mole** formula

   Example: In the example above, in your DATA write

$$98.0\ g\ H_3PO_4 = 1\ mol\ H_3PO_4$$

Let's summarize.

Calculating a Molar Mass

- If you know the *formula* for a substance, you can find its molar mass.
- The units of molar mass are g/mol.
- Use the *column format* to calculate the molar mass.
- When done with the column-format calculation, write your DATA:

   DATA: **# g** formula = **1 mole** formula

The Importance of Molar Mass

Molar mass is the most frequently used conversion in chemistry. Why?

Chemical processes are most easily explained by *counting* the particles, but we don't have machines that can count large numbers of particles directly. What we can easily measure using a balance or scale is the mass of a sample. With the molar mass ratio, we can then convert between the *grams* that we can measure and the particle *counts* that explain chemistry.

Flashcards

In your notebook, in your own words, write a summary of the rules so far in this chapter. Below the rules, draft flashcards that will help you learn the rules and key procedure steps. When done, compare your flashcards to those below. Use the cards that work for you.

One-Way Cards (with Notch)	Back Side—Answers
For the count of an atom inside formula ( ), _____	Multiply count inside ( ) by subscript after
How many C are in $Fe_2(C_2O_2)_3$?	6
To find molar mass from a formula, _____	Add the molar masses of its atoms
After calculating a molar mass, in the DATA write _____	(molar mass) **g** formula = **1 mol** formula

Two-Way Cards (*without* Notch)

6.02×10^{23} is called _____	Avogadro's number = _____
To count a visible number of substance particles, use what units?	When are *moles* used as a unit?
Grams per 1 mole (g/mol) are the units of what quantity?	The units of molar mass = _____

PRACTICE   B

To solve these problems, use the periodic table at the back of this book as needed.

1. What is the mass of 1 mole of helium atoms?

2. Write the molar mass of these single atoms. Include the unit with your answer.

   a. Nitrogen   b. Au   c. Pb

3. How many *oxygen atoms* are represented in each of these formulas?

   a. $Ca_3(PO_4)_2$ b. $Al_2(SO_3)_3$ c. $Co(NO_3)_2$

Complete the remainder of these problems in your notebook.

4. Add *without* a calculator. Round to proper *s.f.*

   24.01 + 86.038 =

5. Add *with* a calculator. Round to proper *s.f.*

   138.0 + 48.0 + 9.072 =

For problems 6 and 7, allow room on your notebook paper for clear and careful work. Use the column format of the H_3PO_4 molar mass calculation earlier. Be sure to do problems 6b, 6f, and 7b; you will need those answers for later problems.

6. Calculate the molar mass for these compounds. Round your answers to proper *s.f.* Write the molar mass equality for the compound below each calculation.

   Without a calculator: a. H_2 b. NaH   c. KSCN

   With a calculator as needed: d. Na_3PO_4 e. $Al_2(SO_4)_3$ f. Barium nitrate

7a. 1 mol H_2S = ? g H_2S 7b. ? g $AgNO_3$ = 1 mol $AgNO_3$

| **Lesson 8.3** | **Grams, Moles, and Particles** |

Knowing how to calculate the grams per 1 mole, we now want to be able to calculate the mass for *any* number of moles of a substance. The problem can be viewed as one of converting units, in this case from moles to grams. An equality, the molar mass, provides the conversion factor.

A **prompt** is a word or two that reminds us of what to do next. In chemistry, certain words or conditions can prompt us to write relationships that are needed to solve problems. In chemistry, the prompt we will use most often is:

The Grams Prompt

In your WANTED and DATA, if you see *grams* of a substance *formula* (or *prefix-grams* such as *kg* or *mg*):

- Calculate the molar mass of that formula.
- Write that molar mass as an *equality* in your DATA.

Example:

DATA: 18.0 g H_2O = 1 mol H_2O

In calculations, a fundamental rule will be:

To convert between grams and moles of a substance, use the molar mass.

The grams prompt will put into your DATA a molar mass conversion that you will need at the SOLVE step.

TRY IT

Solve the following in your notebook using WANTED, DATA, SOLVE and the grams prompt.

Q. Find the mass in grams of 0.25 mole of O_2.

Answer:

WANTED: ? g O_2 =   (First, write the unit *and* substance WANTED.)

DATA: 0.25 mol O_2

32.0 g O_2 = 1 mol O_2 (g O_2 in the WANTED is a grams prompt.)

(If a single unit is WANTED, the DATA contains a single unit to use as the *given* quantity. The rest of the DATA will be written as equalities.)

SOLVE: ? g O_2 = 0.25 mol O_2 • $\dfrac{32.0 \text{ g } O_2}{1 \text{ mol } O_2}$ = **8.0 g O_2**

S.F.: Because 0.25 has 2 *s.f.*, 32.0 has 3 *s.f.*, and 1 is exact, round the answer to 2 *s.f.*

At the WANTED, DATA, and SOLVE steps, always write the *number*, *unit*, and *formula* for each term. Above, by writing the WANTED **g** O_2, you were prompted to write the molar mass conversion that was needed to solve the problem. By listing needed conversions in the DATA, you can focus on arranging conversions without interruption at the SOLVE step.

PRACTICE   A

Try the last lettered part for each numbered problem. If you answer it correctly, go to the last lettered part of the next problem. If you need more practice to feel confident, do another lettered part of the problem. Molar masses for problems 2–4 are found in either problem 1 of this Practice or in Lesson 8.2, Practice B.

1. Working in your notebook, find the molar mass for the following compounds.

    a. H_2SO_4 b. Aluminum nitrate

2. Finish: $? \text{ g NaOH} = 5.5 \text{ mol NaOH} \cdot \dfrac{40.0 \text{ g NaOH}}{1 \text{ mol NaOH}} =$

3. Supply the needed conversion and solve.

    a. $? \text{ g } H_2SO_4 = 4.5 \text{ mol } H_2SO_4 \cdot$ _____ $=$

    b. $? \text{ g AgNO}_3 = 0.050 \text{ mol AgNO}_3$

    (For molar mass, see Lesson 8.2, Practice B, problem 7b.)

4. Use WANTED, DATA, *prompt*, and SOLVE steps in your notebook.

    a. 3.6 moles of H_2SO_4 would have a mass of how many grams?

    b. Find the mass in grams of 2.0×10^{-6} moles of $Al(NO_3)_3$. (Answer in scientific notation. See problem 1b.)

Converting Grams to Moles

If the *grams* of a substance with a known chemical formula are *given* in a problem, how do you find the *moles*? To convert between grams and moles, use the molar mass as a conversion.

▶ TRY IT

Solve the following problem in your notebook.

STOP **Q.** How many moles of O_2 are in 4.00 grams of O_2?

Answer:

WANTED: $? \text{ mol } O_2 =$

DATA: 4.00 g O_2 (See the g O_2? Write the O_2 molar mass.)

32.0 g O_2 = 1 mol O_2

SOLVE: ? mol O_2 = 4.00 g̶ ̶O_2 • $\dfrac{1 \text{ mol } O_2}{32.0 \text{ g̶ } O_2}$ = **0.125 mol O_2**

S.F.: 4.00 has 3 *s.f.*, 1 is exact (infinite *s.f.*), 32.0 has 3 *s.f.*; your answer must be rounded to 3 *s.f.*

PRACTICE   B

Start with the last lettered part for each numbered problem. If your answer is correct, go to the next number. Need more practice? Do another part. Molar masses for these problems were calculated in the two prior Practice sets.

1. Supply conversions and solve. Answer in numbers without exponential terms.

    a. ? mol H_2SO_4 = 10.0 g H_2SO_4 • _____ =

    b. ? mol $Ba(NO_3)_2$ = 65.4 g $Ba(NO_3)_2$

2. Solve in your notebook.

    a. 19.6 g of H_2SO_4 is how many moles?

    b. How many moles are in 51.0 mg of $AgNO_3$? Answer in scientific notation.

Moles and Particles

You will likely need Avogadro's number in calculations that include both a *count* of *invisible* particles (such as molecules, atoms, or ions) and units used to measure *visible* amounts of particles (such as grams or liters).

Let's call this rule

The Avogadro Prompt

If the WANTED or DATA

- includes *both* counts of *invisibly* small particles (such as atoms or molecules) and units used to measure *visible* amounts (such as grams) *or*
- contains any value that includes a *two-digit* power of 10 (**10^{xx}** *or* **10^{-xx}**)

write in your DATA:

1 mol (substance formula) = 6.02×10^{23} particles (substance formula)

▸ **TRY IT**

In your notebook, apply WANTED, DATA, SOLVE plus the grams and Avogadro prompts to solve this problem.

Q. Find the mass in grams of 1.5×10^{22} molecules of H_2O.

Answer:

WANTED: $? \text{ g } H_2O =$

DATA: $1.5 \times 10^{22} \text{ } H_2O$ molecules   (10^{xx} = Avogadro prompt.)

$1 \text{ mol } H_2O = 6.02 \times 10^{23} \text{ } H_2O$ molecules

$1 \text{ mol } H_2O = 18.0 \text{ g } H_2O$ (WANTED unit calls the grams prompt.)

SOLVE:

$$? \textbf{ g } H_2O = 1.5 \times 10^{22} \text{ molec. } H_2O \cdot \frac{1 \text{ mol } H_2O}{6.02 \times 10^{23} \text{ molec. } H_2O} \cdot \frac{18.0 \textbf{ g } H_2O}{1 \text{ mol } H_2O}$$

$$= \frac{1.5 \times 10^{22}}{6.02 \times 10^{23}} \cdot 18.0 \text{ g } H_2O = \textbf{4.5} \times \textbf{10}^{-1} \textbf{ g } H_2O \text{ } or \text{ } \textbf{0.45 g } H_2O$$

Lightly mark the unit cancellation in the conversions above.

There are several ways to do the arithmetic. You may use any method that works, but try doing the *exponential* math without a calculator.

PRACTICE   C

Summarize the two prompts so far in your own words, then solve these problems in your notebook.

1. 3.55 grams of $Cl_2(g)$ (chlorine gas) contain how many molecules of Cl_2?

2. What is the mass in grams of exactly 25 molecules of water? Answer in scientific notation.

Flashcards

Consider adding these cards to your collection. Modify and add your own as needed.

One-Way Cards (with Notch)	Back Side—Answers
To convert between grams and moles, _____	Use the molar mass equality
If *grams* or *prefix-grams* are in WANTED or DATA, write _____	In the DATA: (molar mass) **g** formula = 1 **mol** formula
If WANTED or DATA include 10^{xx}, write _____	In the DATA: 1 mol (formula) = 6.02×10^{23} particles (formula)
If WANTED and DATA mix units measuring visible amounts (g, mol, etc.) and invisibles (atoms, molecules, etc.), write _____	In the DATA: 1 mol (formula) = 6.02×10^{23} particles (formula)

PRACTICE   **D**

Run your flashcards for this chapter to this point, then solve these problems in your notebook.

1. 8.0×10^{24} atoms of aluminum (Al) have a mass of how many grams?

2. How many moles of O_2 are in 6.40×10^{-2} g O_2?

3. Given a sample of dry crystals of a pure molecular substance, what information would you need to know to find the moles of the substance in the sample?

Lesson 8.4   Reactions and Equations

Chemical equations are the language that describes chemical reactions. An example of a **chemical reaction** is the burning of hydrogen gas (H_2) to produce steam (hot H_2O gas). In chemistry, to **burn** something is to react it with oxygen gas (O_2) to form one or more new substances.

In a chemical equation written with *molecular* formulas, the above reaction is represented as

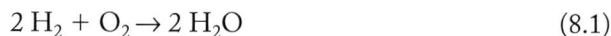

$$2\,H_2 + O_2 \rightarrow 2\,H_2O \tag{8.1}$$

This equation can be read as either "two H two plus one O two react to form two H two O" or "two molecules of hydrogen plus one molecule of oxygen react to form two molecules of water." In an equation with numbers in *front* of some formulas, if no number is shown in front of a formula, a 1 is understood.

The substances on the left side of a reaction equation are termed the **reactants**. The substances on the right side of the arrow are the **products**.

> In chemical reactions, reactants are *used up* and products *form*.

Most chemical reactions are represented by equations using standard formulas, as in Equation 8.1 above. But more information is supplied if an equation is written using structural formulas. An example is

$$\begin{matrix} H & & H \\ | & O{=}O & | \\ H & & H \end{matrix} \rightarrow \begin{matrix} H \\ \diagdown \\ \diagup \\ H \end{matrix} O \quad O \begin{matrix} \diagup H \\ \diagdown \\ \diagdown H \end{matrix} \tag{8.2}$$

▶ **TRY IT**

Q. Compare Equation 8.1 to Equation 8.2. Are they the same reaction?

Answer:

Yes.

By writing the structural formulas, it is easier to see that, in many respects, after the reaction not much has changed. We began with four hydrogen atoms and two oxygen atoms; we end with the same.

> In chemical reactions, bonds break and/or new bonds can form. This changes the substances and their formulas, but the number and kinds of *atoms* stay the same.

The fact that a chemical reaction can neither create nor destroy atoms is called the **law of conservation of atoms**, or the **law of conservation of matter**. In this usage, *conservation* means that what you start with is conserved at the end.

Before, during, and after a reaction, there is also **conservation of mass**: The total mass of the reactants and products does not change at any point during the reaction. Total mass is determined by the number and kinds of atoms in a reaction mixture, which a reaction does not change.

What does change? Because of the new positions of the bonds, after the reaction the products are new substances: They will have formulas, characteristics, and behavior that are different from those of the reactants. In the above reaction, the molecules on the left are explosive when ignited, but the water molecules on the right are quite stable. The oxygen molecules on the left cause many materials to burn. To stop burning, we often use the water on the right.

Reaction Equation Terminology

1. In a reaction equation, **5 H_2O** is called a **term**, and the **5** is called a **coefficient**. It is important to distinguish between subscripts and coefficients.

   - Subscripts are numbers written *after* and *lower than* the atom symbols in a molecule or ion formula. Subscripts count the atoms of each type inside the particle. Subscripts define what the particle is.

   - Coefficients are full-sized numbers written in *front* of a particle formula in an equation.

2. In a reaction equation, if the number and kind of atoms on each side of the arrow is the same, the equation is said to be **balanced**.

3. The coefficients of a *balanced* equation show the exact ratios in which the particles *react* (are used up) and are *formed* in the reaction. Only *one set of ratios* will balance a reaction equation.

   In a balanced equation, writing a coefficient of 1 is optional, and in an equation with coefficients, if no coefficient is shown it is understood to be 1.

4. To balance equations, *atoms* are *counted* on the basis of coefficients and subscripts.

> To count how many of each kind of atom are represented by a *term* in a reaction equation, *first* count the atoms of each kind in the formula, then *multiply* each count by the coefficient.

Example: How many of what atoms are represented by **5 H_2O?**
H_2O has 2 H and 1 O; **5 H_2O** has 10 H and 5 O.

▸ TRY IT

Applying the rule above and the format of the example answer, write the count of each atom represented by these terms.

Q1. 5 CH_4

Q2. 3 CH_3COOH

Q3. 2 $Pb(C_2H_5)_4$

STOP

Answers:

1. CH_4 has 1 C and 4 H; **5 CH_4** has $^5 \pm$ C and ^{20}4 H

2. CH_3COOH has 2 C, 4 H, and 2 O; **3 CH_3COOH** has **6 C, 12 H,** and **6 O**

3. $Pb(C_2H_5)_4$ has 1 Pb, 8 C, and 20 H; **2 $Pb(C_2H_5)_4$** has **2 Pb, 16 C,** and **40 H**

Flashcards

You will need to have the fundamental vocabulary in this lesson firmly in memory. Run the cards for three study sessions in a row, then put them in your "test prep" stack.

One-Way Cards (with Notch)	Back Side—Answers
The law of conservation of mass means _____	During a reaction, total mass stays constant
The law of conservation of atoms means _____	During a reaction, the total number of each kind of atom stays constant
During a chemical reaction, what changes?	Bonds can break, new bonds can form, substance formulas change
What is balanced in a balanced equation?	The number and kind of atoms on each side
To count the atoms of each kind in a term (a formula with a coefficient in front), first _____	Count each atom in the formula, then multiply by the coefficient

Two-Way Cards (*without* Notch)

In reactions, substances used up are called _____	Reactants = _____
In reactions, substances formed are called _____	Products = _____
Numbers inside a formula that count atoms are called _____	Subscripts = _____
In balanced equations, numbers written in front of substance formulas are called _____	Coefficients = _____

PRACTICE

If you are not sure an answer is correct, check it before proceeding to the next question.

1. Label the *reactants* and *products* in this reaction equation. Circle the coefficients.

    $4\,Fe + 3\,O_2 \rightarrow 2\,Fe_2O_3$

2. Copy these terms into your notebook, then count the number of each kind of atom symbol represented by each term. Answer in the format of the previous Try It. Do not use a calculator.

    a. $7\,Na_3PO_4$ b. $3\,Co(OH)_2$ c. $2\,No(NO_3)_2$ d. $5\,Al_2(SO_4)_3$

3. The following equation uses structural rather than molecular formulas.

$$C + C + H{-}H + H{-}H + H{-}H \rightarrow H{-}\underset{\underset{H}{|}}{\overset{\overset{H}{|}}{C}}{-}\underset{\underset{H}{|}}{\overset{\overset{H}{|}}{C}}{-}H$$

    a. Is the equation balanced?

    b. Write the reaction using *molecular* formulas (the type used in problem 1).

    c. In going from reactants to products, what changed? What stayed the same?

    Changes:

    Stays the same:

Lesson 8.5   Balancing Equations

Because coefficients are ratios, if all of the coefficients of a balanced equation are multiplied or divided by the same number, the result is an equation that remains balanced. This means that a balanced equation can be shown with *different* sets of coefficients, as long as the *ratios* among the coefficients are the same.

> **Example**: $2\,H_2 + O_2 \rightarrow 2\,H_2O$ and $H_2 + 1/2\,O_2 \rightarrow H_2O$ are the same balanced equation because the coefficient *ratios* are the same. In the second equation, all of the coefficients have been divided by 2.
>
> In both equations, coefficients that are 1 have been omitted as understood.

Balancing by Trial and Error

The coefficients that balance an equation are not always supplied with the equation. But *if* you are supplied the formulas for the reactants and products, you can determine the coefficients by counting atoms.

> **Balancing Equations**
> - In a balanced equation, the number and kind of atoms must be the *same* in the reactants and products.
> - **Balancing the equation** means *writing* in the equation *coefficients* that make the number and kind of atoms the same on both sides.
> - If all of the formulas are known, the coefficients that balance an equation can always be determined by *trial and error*.

Trial-and-error balancing uses the same rules for all types of equations, and with perseverance it works for all types of equations. Let's learn to balance with an example.

TRY IT

Q. Insert coefficients that balance this equation for the burning of *n*-propanol.

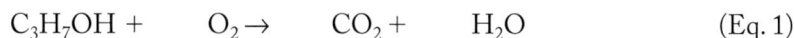

$$C_3H_7OH + \quad O_2 \rightarrow \quad CO_2 + \quad H_2O \qquad\qquad \text{(Eq. 1)}$$

Use these steps:

1. During balancing, you cannot change a formula (including its subscripts). Formulas and subscripts define what the substances are. Balancing does not change what the substances are.

   To balance, *you* must insert numbers into the equation, but the *only* change you can make is to *insert coefficients* that go in *front* of substance formulas.

2. For some equations, one coefficient will be supplied or necessary. If no coefficient is supplied, start the "trial and error" by writing a coefficient of 1 in front of the *most complex* formula (the one with the most atoms or the most different kinds of atoms) on *either* the left or right side of the equation. If two formulas are complex, choose either one.

   Do that step now in Equation 1 above.

Answer:

In Equation 1, the first formula is the most complex, so start with

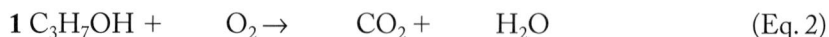

$$\mathbf{1}\,C_3H_7OH + \quad O_2 \rightarrow \quad CO_2 + \quad H_2O \qquad\qquad \text{(Eq. 2)}$$

When writing an already balanced equation, a coefficient that is 1 can be left out as understood, but when trying to figure out what the balanced equation is, writing the 1 is necessary to track which coefficients have been decided.

3. Now that the 1 is in front of C_3H_7OH, let's see what we know. Apply this rule:

> To balance, for each kind of atom on a *side*, find the *total count* for that atom in *all* the terms on that side.

In Equation 2, count the total number of C and H and O atoms to the left of the arrow, fill in the blanks, and then check your answers below.

On left side, number of C = _____, H = _____, O = _____.

On the left side, we have $1 \times 3 = \textbf{3 C}$, and $1 \times (7 + 1) = \textbf{8 H}$, but we have **? O** because there are also O atoms in the O_2 and we have not decided its coefficient yet. But we *do* know the C and H totals on the left, so let's work with those.

4. Insert coefficients on the *other* side of the arrow that must be true if the atoms are balanced.

To start, in Equation 2, decide what one of the coefficients on the right *must be* to have the same number of C atoms on *both* sides. Add that coefficient to Equation 2 on the right side.

The left side has 3 C atoms. Because only CO_2 on the right has C, and CO_2 has 1 C, the only way to have 3 C atoms on the right is to have the CO_2 coefficient be 3.

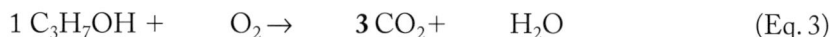

$$1\,C_3H_7OH + \quad O_2 \rightarrow \quad 3\,CO_2 + \quad H_2O \qquad \text{(Eq. 3)}$$

That gives 3 C on both sides. The **C's** are *balanced*.

5. Using Equation 3, decide what the other coefficient on the *right* must be to balance the H atoms. Add that coefficient to Equation 3.

The left has 8 H atoms total. Because only H_2O on the right has H, and it has 2 H's, the only way to have 8 H atoms on the right is to have the H_2O coefficient be **4** ($4 \times 2 = 8$ H). So far, this gives us

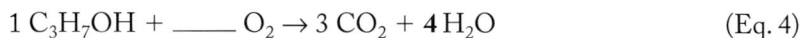

$$1\,C_3H_7OH + \underline{\quad} O_2 \rightarrow 3\,CO_2 + 4\,H_2O \qquad \text{(Eq. 4)}$$

The right side is now *finished*, because each particle has a coefficient. Only the O_2 on the left side lacks a coefficient.

6. Insert the coefficient that *must be* true for the oxygen atoms to balance.

We count the oxygen atoms on the *right* side and get $(3 \times 2) + (4 \times 1) = \textbf{10 O}$. On the left, we see one oxygen atom in propanol, which means we must have a total of **nine** oxygen atoms from O_2. We can get a count of 9 O from the O_2 by writing

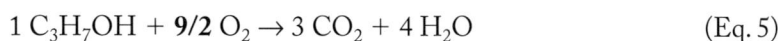

$$1\,C_3H_7OH + \textbf{9/2}\,O_2 \rightarrow 3\,CO_2 + 4\,H_2O \qquad \text{(Eq. 5)}$$

Because each term has an assigned coefficient, the equation *should* be *balanced*.

7. To be *sure* it is balanced, let's check our answer. Count the atoms on each side and fill in the blanks below.

Total left: C = _____, H = _____, O = _____. Total right: C = _____, H = _____, O = _____.

On the left, the total is C = 3, H = 8, O = 10. On the right, the total is C = 3, H = 8, O = 10. When the number for each kind of atom is the same on both sides, the equation is *balanced*.

Fractions as Coefficients

In the above answer, one coefficient was a fraction: 9/2. Fractions are not incorrect when adding coefficients to balance equations, and in some types of problems, the use of fractions to balance equations is required.

Our most frequent use for coefficients, however, will be as *ratios* in chemical reaction calculations. The arithmetic in these calculations will be easier if all coefficients are converted to *lowest whole numbers* by the end of balancing. Because coefficients are ratios, we can multiply all of the coefficients by the same number and still have the same ratios *and* a balanced equation.

When balancing results in only one fraction, the easy way to convert to lowest whole numbers is to multiply each of the coefficients by the denominator of the fraction. In a balanced equation that has fractions with two different denominators, multiplying all coefficients by the lowest common denominator will usually provide lowest-whole-number coefficients.

TRY IT

Q. Using your answer above,

$$1\ C_3H_7OH + 9/2\ O_2 \rightarrow 3\ CO_2 + 4\ H_2O \qquad\qquad (Eq.\ 5)$$

convert the coefficients to lowest whole numbers and insert them here:

$$C_3H_7OH + \underline{\quad} O_2 \rightarrow \underline{\quad} CO_2 + \underline{\quad} H_2O$$

STOP

Answer:

Multiplying each coefficient by the denominator in the fraction 9/2 (the 2) results in

$$\mathbf{2}\ C_3H_7OH + \mathbf{9}\ O_2 \rightarrow \mathbf{6}\ CO_2 + \mathbf{8}\ H_2O \qquad\qquad (Eq.\ 6)$$

But we always check to be *sure* it is balanced. For Equation 6, fill in the blanks.

Total left: C = ____, H = ____, O = ____. Total right: C = ____, H = ____, O = ____.

STOP

On the left, the total is C = 6, H = 16, O = 20. On the right, the total is C = 6, H = 16, O = 20. Balanced.

In the case of Equations 5 and 6, Equation 6 is normally *preferred* because it has lowest-whole-number ratios. Our rule will be

> When balancing an equation, unless other coefficients are specified or required in a problem, convert to *lowest-whole-number* coefficients.

In some cases, you will be asked to convert an already balanced equation to lowest-whole-number ratios.

━━━▭▭▭▭▭▭▭▭▷ **TRY IT**

Q. For $16\,C + 8\,O_2 \rightarrow 16\,CO$, write the lowest-whole-number ratios:

$$C + \quad O_2 \rightarrow \quad CO$$

Answer:

$$\mathbf{2\,C + 1\,O_2 \rightarrow 2\,CO}$$

Finally, when balancing equations, practice your mental math. During balancing, having to go back and forth from paper to calculator keys to display to paper is frustrating at best. For work in the sciences, you need to know your math facts, and balancing is a good place to practice.

PRACTICE   A

Read each *numbered* step below and do every *other lettered* problem. As you go, check your answers. If you need more practice at a step, do a few more lettered problems for that step. Save the rest for your next practice session. Do not use a calculator.

1. A balanced equation must have the same number of each kind of atom on both sides. To determine whether an equation is balanced correctly, count the total number of *one* kind of atom on one side, then count the number of that kind of atom on the *other* side. The left-side and right-side counts for each atom must be equal. Repeat those counts for *each* kind of atom in the equation.

   Using those steps, label each equation below as *balanced* or *unbalanced*.

   a. $2\,Cs + Cl_2 \rightarrow 2\,CsCl$

   b. $4\,HI + O_2 \rightarrow 2\,H_2O + I_2$

   c. $Pb(NO_3)_2 + 2\,LiBr \rightarrow PbBr_2 + LiNO_3$

   d. $BaCO_3 + 2\,NaCl \rightarrow Na_2CO_3 + BaCl_2$

2. In the equations below, one coefficient has been supplied. Use that coefficient to decide one or more coefficients on the *other* side. Then, use your added coefficient(s) to go back and forth, from side to side, filling in the remaining blanks on both sides to balance the equation.

   Remember that balancing is by trial and error. Do what works. If you need help, check your answer after each letter.

   Tip: It helps to balance *last* an atom that is used in two or more different formulas on the same side. Oxygen is the atom most frequently encountered in compounds, so "saving O until last" usually helps in balancing.

   a. $4\,Al + \underline{\quad}\,O_2 \rightarrow \underline{\quad}\,Al_2O_3$ b. $3\,Ca + \underline{\quad}\,N_2 \rightarrow \underline{\quad}\,Ca_3N_2$

   c. $\underline{\quad}\,P_4 + \underline{\quad}\,O_2 \rightarrow 1\,P_4O_6$ d. $\underline{\quad}\,C_3H_8 + \underline{\quad}\,O_2 \rightarrow 9\,CO_2 + \underline{\quad}\,H_2O$

(continued)

e. ____ MgH_2 + ____ $H_2O \rightarrow 1\ Mg(OH)_2$ + ____ H_2

f. $2\ C_2H_6$ + ____ $O_2 \rightarrow$ ____ CO_2 + ____ H_2O

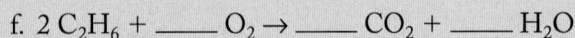

3. Balance these equations. Start by placing a coefficient of **1** in front of the <u>underlined</u> substance. Coefficients *may* be fractions in these equations.

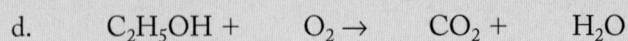

a. K +   <u>F_2</u> $\rightarrow$ KF

b. Cs +   <u>O_2</u> $\rightarrow$ Cs_2O

c. $PCl_3 \rightarrow$ <u>P_4</u> + Cl_2

d.   <u>C_2H_5OH</u> + $O_2 \rightarrow$ CO_2 + H_2O

e. FeS + $O_2 \rightarrow$ <u>Fe_2O_3</u> + SO_2

4. Balance these equations. Start by placing a coefficient of **1** in front of the *most complex formula*. If you get a fraction as you balance, multiply *all* of the existing coefficients by the denominator of the fraction. Repeat this step if you get additional fractions while balancing. Report final coefficients that are lowest-whole-number ratios.

In trial-and-error balancing, be prepared to adjust trial coefficients.

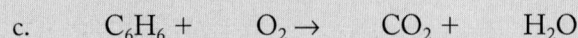

a. Mg + $O_2 \rightarrow$ MgO

b. N_2 + $O_2 \rightarrow$ NO

c. C_6H_6 + $O_2 \rightarrow$ CO_2 + H_2O

d. P_4 + $O_2 \rightarrow$ P_4O_6

e. Al + $HBr \rightarrow$ $AlBr_3$ + H_2

Flashcards

Make any of these flashcards that may help you remember the rules for balancing.

One-Way Cards (with Notch)	Back Side—Answers
During balancing, which numbers do you insert?	Coefficients
How do you usually begin balancing?	Put a 1 in front of a complex formula
What must be balanced in a balanced equation?	The total number of each kind of atom on both sides
How do you avoid fractions as coefficients?	Multiply each known coefficient by the fraction's denominator
What kind of ratios are usually preferred at the end of balancing?	Lowest-whole-number ratios

PRACTICE   B

1. Balance these equations. As final coefficients, write lowest-whole-number ratios.

   a. Al_2O_3 +   HCl → $AlCl_3$ + H_2O

   b. Fe_3O_4 + H_2 →   Fe + H_2O

   c.   C + SiO_2 →   CO +   SiC

   d. N_2 + O_2 + H_2O → HNO_3

   e. $Pb(C_2H_5)_4$ + O_2 →   PbO + CO_2 + H_2O

2. Working in your notebook, write the formulas then balance with lowest-whole-number ratios. (Need formula help? See Lessons 7.2 and 7.4.)

   a. Dinitrogen tetroxide → nitrogen dioxide

   b. Barium carbonate + cesium chloride → cesium carbonate + barium chloride

   c. Silver nitrate + calcium iodide → silver iodide + calcium nitrate

---

SUMMARY

Concepts presented in this chapter include the following:

- A mole is 6.02×10^{23} particles. The molar mass of a substance is its mass in grams *per* 1 mole of substance particles. If you know the formula for a substance, you can calculate its molar mass by adding the molar masses of its atoms. Using the molar mass, and knowing any *one* of the grams, moles, or particles of a substance, you can calculate the other two.
- In chemical reactions, bonds break and new bonds form. Reactants are used up, and products form that have properties that differ from those of the reactants. But in reactions, atoms are conserved: The number and kind of *atoms* is the same before, during, and after the reaction.
- If substance formulas for the reactants and products are known, equations representing reactions can be balanced by trial and error.

To prepare for a quiz and/or test on the material in this chapter:

1. Write the shaded *rules* and *prompts* in your own words.

2. Run chapter flashcards to perfection for at least 3 days.

3. Use the quiz below to identify topics that need further review.

REVIEW QUIZ

1. Find the molar mass of $Co(NO_3)_2$.

2. (Try without a calculator.) Find the mass in grams of 4.00 moles of CH_4.

3. 6.80 grams of NH_3 would be how many moles of NH_3?

4. (Try without a calculator.) 10.0 grams of NaOH is how many moles of NaOH?

5. 2.57 nanograms of S_8 contain how many molecules of S_8?

6. 1.80×10^{23} molecules of CO_2 have what mass in grams?

    a. 0.0840 g CO_2 b. 0.840 g CO_2 c. 8.40 g CO_2

    d. 13.2 g CO_2 e. 9.60 g CO_2

7. Balance these using lowest-whole-number ratios.

    a. Rb_2O + $H_2O \rightarrow$ RbOH

    b. $Mg(NO_3)_2$ + $Na_3PO_4 \rightarrow$ $Mg_3(PO_4)_2$ + $NaNO_3$

    c.   Calcium hydroxide + hydrogen chloride $\rightarrow$ calcium chloride + water

ANSWERS

Lesson 8.1

Practice   1a. HCOOH   2 hydrogen, 1 carbon, 2 oxygen (in any order)   1b. $CoSO_4$ 1 cobalt, 1 sulfur, 4 oxygen

1c. $No(NO_3)_3$ 1 nobelium, 3 nitrogen, 9 oxygen   1d. Cs_3PO_4 3 cesium, 1 phosphorus, 4 oxygen

2a. C_2H_5COOH 3 C, 6 H, 2 O   2b. $Co_3(PO_4)_2$ 3 Co, 2 P, 8 O   2c. $Al_2(SO_4)_3$ 2 Al, 3 S, 12 O

2d. $Pb(C_2H_5)_4$ 1 Pb, 8 C, 20 H   2e. $(NH_4)_3PO_4$ 3 N, 12 H, 1 P, 4 O

Lesson 8.2

Practice A   1. 2.41×10^{24} particles

2. ? particles = 0.050 mol $\times$ (6.02×10^{23} particles/mol) = $\mathbf{3.0 \times 10^{22}}$ **particles**

Practice B   1. 4.00 g

2a. Nitrogen: $\mathbf{14.0\dfrac{g}{mol}}$

2b. Au: $\mathbf{197.0\dfrac{g}{mol}}$

2c. Pb: $\mathbf{207.2\dfrac{g}{mol}}$

3a. 8 O   3b. 9 O   3c. 6 O

4. 110.05 (Round to highest *place* with doubt. See Lesson 3.2).

5. $138.\underline{0} + 48.\underline{0} + 9.072 = 195.072 = \mathbf{195.\underline{1}}$

6a. $H_2 = 2 \times H = 2 \times 1.008 = \mathbf{2.016\ g/mol}$

Multiplying by an exact subscript does not change the *place* with doubt. Adding 1.008 + 1.008 is an option that gives the same result. Below the calculation, write **2.016 g H_2 = 1 mol H_2**

6b. NaH =   6c. KSCN =

    Na = 23.0   K = 39.1

    H = $\underline{1.008}$ S = 32.1

            24.$\underline{0}$08 = **24.0 g/mol**   C = 12.0

Below the calculation, write **24.0 g NaH = 1 mol NaH**   N = $\underline{14.0}$

                                                       97.2 g/mol

                                                  97.2 g KSCN = 1 mol KSCN

6d. Na_3PO_4 = 3 Na, 1 P, 4 O

     $3 \times Na = 3 \times 23.0 = 69.0$

     $1 \times P = 1 \times 31.0 = 31.0$

     $4 \times O = 4 \times 16.0 = \underline{64.0}$

                   164.0 g/mol

   164.0 g Na_3PO_4 = 1 mol Na_3PO_4

6e. $Al_2(SO_4)_3$ = 2 Al, 3 S, 12 O

     $2 \times Al = 2 \times 27.0 = 54.0$

     $3 \times S = 3 \times 32.1 = 96.3$

     $12 \times O = 12 \times 16.0 = \underline{192.0}$

                   342.3 g/mol

   342.3 g $Al_2(SO_4)_3$ = 1 mol $Al_2(SO_4)_3$

6f. Barium nitrate = $Ba(NO_3)_2$ = 1 Ba, 2 N, 6 O

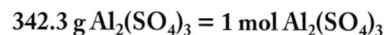

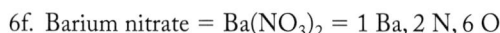

     $1 \times Ba = 1 \times 137.3 = 137.3$

     $2 \times N = 2 \times 14.0 = 28.0$

     $6 \times O = 6 \times 16.0 = \underline{96.0}$

                   261.3 g/mol

   261.3 g $Ba(NO_3)_2$ = 1 mol $Ba(NO_3)_2$

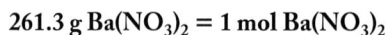

7. This problem asks for the grams per 1 mole for a single substance. That's the molar mass.

7a. H_2S =

     $2 \times H = 2 \times 1.008 = 2.016$

     $1 \times S = 1 \times 32.1 = \underline{32.1}$

                $34.\underline{1}16$ = **34.1** g/mol

   1 mol H_2S = 34.1 g H_2S

7b. $AgNO_3$ =

     $1 \times Ag = 1 \times 107.9 = 107.9$

     $1 \times N = 1 \times 14.0 = 14.0$

     $3 \times O = 3 \times 16.0 = \underline{48.0}$

                   169.9 g/mol

   169.9 g $AgNO_3$ = 1 mol $AgNO_3$

Lesson 8.3

Practice A   1a. H_2SO_4 = 2 H, 1 S, 4 0

     $2 \times H = 2 \times 1.008 = 2.016$

     $1 \times S = 1 \times 32.1 = 32.1$

     $4 \times O = 4 \times 16.0 = \underline{64.0}$

                $98.\underline{1}16$ = **98.1 g/mol**

   98.1 g H_2SO_4 = 1 mol H_2SO_4

1b. Aluminum nitrate = **$Al(NO_3)_3$** = 1 Al, 3 N, 9 O

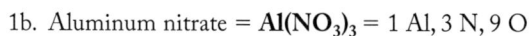

     $1 \times Al = 1 \times 27.0 = 27.0$

     $3 \times N = 3 \times 14.0 = 42.0$

     $9 \times O = 9 \times 16.0 = \underline{144.0}$

                   213.0 g/mol

   213.0 g $Al(NO_3)_3$ = 1 mol $Al(NO_3)_3$

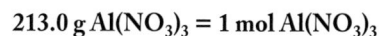

(Note that multiplying by an exact 9—or any exact number—does not change the *place* with doubt.)

2. ? g NaOH = 5.5 ~~mol NaOH~~ • $\dfrac{40.0 \text{ g NaOH}}{1 \text{ ~~mol NaOH~~}}$ = **220 g NaOH**

3a. ? g H_2SO_4 = 4.5 mol H_2SO_4 • $\dfrac{98.1 \text{ g } H_2SO_4}{1 \text{ mol } H_2SO_4}$ = **440 g H_2SO_4**

3b. ? g $AgNO_3$ = 0.050 mol $AgNO_3$ • $\dfrac{169.9 \text{ g } AgNO_3}{1 \text{ mol } AgNO_3}$ = **8.5 g $AgNO_3$**

4a. ? g H_2SO_4 = 3.6 mol H_2SO_4 • $\dfrac{98.1 \text{ g } H_2SO_4}{1 \text{ mol } H_2SO_4}$ = **350 g H_2SO_4**

4b. ? g $Al(NO_3)_3$ = 2.0×10^{-6} mol $Al(NO_3)_3$ • $\dfrac{213.0 \text{ g } Al(NO_3)_3}{1 \text{ mol } Al(NO_3)_3}$ = **4.3×10^{-4} g $Al(NO_3)_3$**

Practice B   1a. $? \text{ mol } H_2SO_4 = 10.0 \text{ g } \cancel{H_2SO_4} \cdot \dfrac{1 \text{ mol } H_2SO_4}{98.1 \text{ g } \cancel{H_2SO_4}} = \textbf{0.102 mol } \mathbf{H_2SO_4}$

        1b. $? \text{ mol } Ba(NO_3)_2 = 65.4 \text{ g } Ba(NO_3)_2 \cdot \dfrac{1 \text{ mol } Ba(NO_3)_2}{261.3 \text{ g } Ba(NO_3)_2} =$

              Answer: If you wrote **0.**250 mol $Ba(NO_3)_2$, go to the head of the class.

              Always write a **0** in front of a decimal point if there is no number in front of the decimal point. This makes the decimal point *visible* when you need this answer for a later step of a lab report or test.

        2a. $? \text{ mol } H_2SO_4 = 19.6 \text{ g } H_2SO_4 \cdot \dfrac{1 \text{ mol } H_2SO_4}{98.1 \text{ g } H_2SO_4} = \textbf{0.200 mol } \mathbf{H_2SO_4}$

        2b. $? \text{ mol } AgNO_3 = 51.0 \text{ mg } AgNO_3 \cdot \dfrac{10^{-3} \text{ g}}{1 \text{ mg}} \cdot \dfrac{1 \text{ mol } AgNO_3}{169.9 \text{ g } AgNO_3} = \mathbf{3.00 \times 10^{-4} \text{ mol } AgNO_3}$

Practice C   Your paper should look like this, but you may omit the comments in parentheses.

        1. WANTED: ? molecules Cl_2 =

        DATA: 3.55 g Cl_2

                  71.0 g Cl_2 = 1 mol Cl_2 (Grams prompt.)

                  1 mol Cl_2 = 6.02×10^{23} molecules Cl_2 (Mix grams and invisibles = Avogadro prompt.)

        SOLVE:

              $? \text{ molecules } Cl_2 = 3.55 \text{ g } Cl_2 \cdot \dfrac{1 \text{ mol } Cl_2}{71.0 \text{ g } Cl_2} \cdot \dfrac{6.02 \times 10^{23} \text{ molecules } Cl_2}{1 \text{ mol } Cl_2} = \mathbf{3.01 \times 10^{22} \text{ molecules } Cl_2}$

        2. WANTED: ? g H_2O =

        DATA: 25 molecules H_2O (Exact.)

                  1 mol H_2O = 6.02×10^{23} H_2O molecules   (Invisible molecules, visible g = Avogadro prompt.)

                  1 mol H_2O = 18.0 g H_2O (WANTED unit = grams prompt.)

        SOLVE:

              $? \text{ g } H_2O = 25 \text{ molecules } H_2O \cdot \dfrac{1 \text{ mol } H_2O \text{ molecules}}{6.02 \times 10^{23} \text{ } H_2O \text{ molecules}} \cdot \dfrac{18.0 \text{ g } H_2O}{1 \text{ mol } H_2O} = \mathbf{7.48 \times 10^{-22} \text{ g } H_2O}$

Practice D   1. In metals, the particles in the "*molecular* formula" are individual atoms, so the metal "molecules" are the same as the metal *atoms*, and the molar mass is the mass of a mole of metal *atoms*.

        WANTED: ? g Al =

        DATA: 8.0×10^{24} Al atoms

                  1 mol Al = 6.02×10^{23} Al atoms   (10^{xx} = Avogadro prompt.)

                  27.0 g Al = 1 mol Al   (Grams prompt in WANTED.)

        SOLVE: $? \text{ g Al} = 8.0 \times 10^{24} \text{ atoms Al} \cdot \dfrac{1 \text{ mol atoms Al}}{6.02 \times 10^{23} \text{ atoms Al}} \cdot \dfrac{27.0 \text{ g Al}}{1 \text{ mol Al}} = \mathbf{360 \text{ g Al}}$

        2. WANTED: ? mol O_2 =

        DATA: 6.40×10^{-2} g O_2

                  32.0 g O_2 = 1 mol O_2 (Grams prompt.)

SOLVE: $? \text{ mol O}_2 = {}^2 64.0 \times 10^{-3} \text{ g O}_2 \cdot \dfrac{1 \text{ mol O}_2}{{}^1 32.0 \text{ g O}_2} = 2.00 \times 10^{-3} \text{ mol O}_2$

3. Methods to find moles of a pure substance include the following: Knowing the mass of the sample in *grams* and its molar mass (*g/mol*), you can convert to its *moles*. Or, knowing the grams and the *formula* of a substance, from the formula you can calculate the g/mol and then the moles in the sample.

Lesson 8.4

Practice

1. ④Fe + ③O₂ → ②Fe₂O₃

       Reactants   Products

2a. Na_3PO_4 has 3 Na, 1 P, and 4 O; 7 Na_3PO_4 has ²¹**3 Na,** ⁷**‡ P,** and ²⁸**4 O**

2b. $Co(OH)_2$ has 1 Co, 2 O, and 2 H; 3 $Co(OH)_2$ has **3 Co, 6 O,** and **6 H**

2c. $No(NO_3)_2$ has 1 No, 2 N, and 6 O; 2 $No(NO_3)_2$ has **2 No, 4 N,** and **12 O**

2d. $Al_2(SO_4)_3$ has 2 Al, 3 S, and 12 O; 5 $Al_2(SO_4)_3$ has **10 Al, 15 S,** and **60 O**

3a. Yes. The same number and kinds of atoms are on each side.

3b. $2 C + 3 H_2 \rightarrow C_2H_6$

3c. Changes: The number of bonds, the bond locations, the molecules, the stored energy, and the appearance and characteristics of the substances involved. Stays the same: The numbers of each kind of atom and the total mass.

Lesson 8.5

Practice A   1a. Balanced   1b. Not balanced   1c. Not balanced   1d. Balanced

2a. **4** Al + **3** $O_2 \rightarrow$ **2** Al_2O_3 2b. **3** Ca + **1** $N_2 \rightarrow$ **1** Ca_3N_2

2c. **1** P_4 + **3** $O_2 \rightarrow$ **1** P_4O_6 2d. **3** C_3H_8 + **15** $O_2 \rightarrow$ **9** CO_2 + **12** H_2O

2e. **1** MgH_2 + **2** $H_2O \rightarrow$ **1** $Mg(OH)_2$ + **2** H_2 (2e is tricky because H is used in more than one formula on each side, rather than the usual O. Save until last the atom that is in two compounds on one or both sides.)

2f. **2** C_2H_6 + **7** $O_2 \rightarrow$ **4** CO_2 + **6** H_2O

3a. **2** K + $\underline{F_2} \rightarrow$ **2** KF   3b. **4** Cs + $\underline{O_2} \rightarrow$ **2** Cs_2O 3c. **4** $PCl_3 \rightarrow \underline{P_4}$ + **6** Cl_2

3d. $\underline{C_2H_5OH}$ + **3** $O_2 \rightarrow$ **2** CO_2 + **3** H_2O 3e. **2** FeS + **7/2** $O_2 \rightarrow \underline{Fe_2O_3}$ + **2** SO_2

4a. **2** Mg + **1** $O_2 \rightarrow$ **2** MgO   4b. **1** N_2 + **1** $O_2 \rightarrow$ **2** NO   4c. **2** C_6H_6 + **15** $O_2 \rightarrow$ **12** CO_2 + **6** H_2O

4d. **1** P_4 + **3** $O_2 \rightarrow$ **1** P_4O_6 4e. **2** Al + **6** HBr $\rightarrow$ **2** $AlBr_3$ + **3** H_2

Practice B   Lowest-whole-number coefficients were requested, so your coefficients must match these exactly.

1a. **1** Al_2O_3 + **6** HCl $\rightarrow$ **2** $AlCl_3$ + **3** H_2O 1b. **1** Fe_3O_4 + **4** $H_2 \rightarrow$ **3** Fe + **4** H_2O

1c. **3** C + **1** $SiO_2 \rightarrow$ **2** CO + **1** SiC   1d. **2** N_2 + **5** O_2 + **2** $H_2O \rightarrow$ **4** HNO_3

1e. **2** $Pb(C_2H_5)_4$ + **27** $O_2 \rightarrow$ **2** PbO + **16** CO_2 + **20** H_2O

2a. **1** $N_2O_4 \rightarrow$ **2** NO_2 2b. $BaCO_3$ + **2** CsCl $\rightarrow$ Cs_2CO_3 + $BaCl_2$

2c. **2** $AgNO_3$ + $CaI_2 \rightarrow$ **2** AgI + $Ca(NO_3)_2$

Review Quiz

1. 182.9 g/mol

2. 64.0 g CH_4

3. 0.400 mol NH_3

4. SOLVE: ? mol NaOH = 10.0 g NaOH • $\dfrac{1 \text{ mol NaOH}}{40.0 \text{ g NaOH}}$ = **0.250 mol NaOH**

5. WANTED: ? molecules S_8

   DATA: 2.57 ng S_8

   256.8 g S_8 = 1 mol S_8 (Any *prefix-grams* = grams prompt.)

   1 mol S_8 = 6.02×10^{23} molecules S_8 (*Invisible atoms*, visible *moles* = Avogadro prompt.)

   SOLVE:

   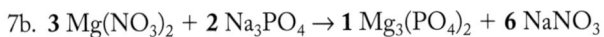

   ? molec. S_8 = ^{1}2.57 ng S_8 • $\dfrac{10^{-9}\,g}{1 \text{ ng}}$ • $\dfrac{1 \text{ mol } S_8}{^{100}256.8 \text{ g } S_8}$ • $\dfrac{6.02 \times 10^{23} \text{ molec. } S_8}{1 \text{ mol } S_8}$ = **6.02×10^{12} molec. S_8**

6. **d. 13.2 g CO_2**

7a. **1** Rb_2O + **1** H_2O → **2** RbOH

7b. **3** $Mg(NO_3)_2$ + **2** Na_3PO_4 → **1** $Mg_3(PO_4)_2$ + **6** $NaNO_3$

7c. 1 $Ca(OH)_2$ + 2 HCl → 1 $CaCl_2$ + 2 H_2O

9

Stoichiometry

Lesson 9.1   Ratios of Reaction

The Meaning of Coefficients

Balanced chemical equations show the *ratios* in which particles (molecules, formula units, or ions) are used up and form. For example, the burning of heptane (a constituent of gasoline) can be represented as

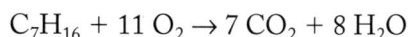

$$C_7H_{16} + 11\,O_2 \rightarrow 7\,CO_2 + 8\,H_2O$$

The equation states that burning 1 molecule of heptane consumes 11 molecules of oxygen. The products are 7 molecules of carbon dioxide and 8 molecules of water. Reactants are used up and products form in known, exact ratios.

What happens if two molecules of heptane burn? Twice as much oxygen must be used up (22 molecules), and twice as much product must be produced: 14 molecules of CO_2 and 16 molecules of H_2O.

In a reaction, the amounts involved can vary, but the *ratios* of the substances used up and formed must stay the same.

PRACTICE A

Do the following problems by "mental math" and write your answers below. Check your answers after each problem.

1. For the balanced equation

$$4\,HBr + O_2 \rightarrow 2\,Br_2 + 2\,H_2O$$

   a. If 16 molecules of HBr react, how many molecules of O_2 must react? How many molecules of Br_2 must form?

   b. If five molecules of O_2 react, how much HBr must react? How much Br_2 must form?

2. For the reaction

$$CS_2 + \quad O_2 \rightarrow \quad CO_2 + \quad SO_2$$

   a. Balance the equation.

   b. If 25 trillion molecules of CS_2 react, how many molecules of O_2 must also be used up? How many molecules of SO_2 must form?

Mole-to-Mole Conversions

In a balanced equation, coefficients are calculated based on *counts* of atoms. Because atoms are not divisible (you may have 1, 2, or 3 atoms, but you cannot have 3.1 atoms), coefficients are exact numbers with no uncertainty.

The balanced equation $2\,H_2 + O_2 \rightarrow 2\,H_2O$ means:

- Two molecules of H_2 used up equals exactly one molecule of O_2 used up.
- Two molecules of H_2 consumed equals exactly two molecules of H_2O formed.
- If one molecule of O_2 reacts, exactly two molecules of H_2O must form.

In this sense, any two terms in a balanced equation are "equal."

In the three "equality statements" above, the coefficient units are individual molecules. Coefficients can also be read as *exact moles*, because once the coefficient ratios from the balanced equation are known, those ratios can be multiplied by *any* number and the equation is still exactly balanced. A mole is simply a very large number.

> When reading an equation, the units attached to coefficients can be read as exact counts of *particles* or exact *moles* of particles.

This means the equation $2\,H_2 + O_2 \rightarrow 2\,H_2O$ can be read as "two *moles* of H_2 plus one *mole* of O_2 react to form 2 *moles* of H_2O," where all numbers are exact.

The three equalities above are therefore true if we substitute *moles* for molecules. For the reaction

$$2\,H_2 + O_2 \rightarrow 2\,H_2O$$

we can write:

2 mol H_2 used up = 1 mol O_2 used up

2 mol H_2 used up = 2 mol H_2O formed

1 mol O_2 used up = 2 mol H_2O formed

These exact equalities can be used as conversions to solve calculations for this reaction.

Reaction Calculations

When dealing with chemical reactions, the question we ask most often is: "If we want or use up this much of one substance, how much of this other substance will be used up or form?" In this type of calculation, the substance *formula* that is WANTED will be different from the substance *formula* that is *given* in the DATA.

To solve reaction calculations by consistent steps, we will apply the following rule.

Rule for Reaction Calculations

For a problem about a chemical reaction, after listing the WANTED and DATA, *if* the WANTED *substance* is not the same as the *given substance*, below the DATA write these sections:

- **Balance:** Write the balanced reaction equation using lowest-whole-number coefficients.
- **Bridge:** Write

  X mol WANTED formula = *Y* mol *given* formula

  Here, *X* and *Y* are the coefficients in front of the WANTED and *given* substances in the balanced equation.

Then use the mole-to-mole bridge conversion at the SOLVE step.

The bridge conversion could be a particle-to-particle ratio from the balanced equation, but most reaction calculations involve visible amounts of a substance. For visible amounts, the arithmetic is easier if we count particles in moles.

TRY IT

In your notebook, apply the rule above to this problem.

Q. Burning ammonia with appropriate catalysts can result in the formation of nitrogen monoxide by the reaction

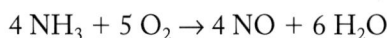

$$4\,NH_3 + 5\,O_2 \rightarrow 4\,NO + 6\,H_2O$$

How many moles of NH_3 are used up to form 8.4 mol H_2O?

(STOP)

Answer:

WANTED: ? mol NH_3 =

DATA: 8.4 mol H_2O

Note the prompt: Because the WANTED substance ≠ the *given* substance (≠ means "does not equal"), add these two sections below your DATA:

Balance: (Supplied)

Bridge: **4 mol NH_3 = 6 mol H_2O**

For the bridge conversion, we could also have written

Bridge: 4 mol NH_3 **used up** = 6 mol H_2O **formed**

which is a more complete description of what the balanced equation is telling us; however, in the interest of time, in the DATA and conversions we usually leave out the words as understood.

Now complete the SOLVE step with our standard conversion rules.

(STOP)

SOLVE:

$$\textbf{? mol } NH_3 = 8.4 \; \cancel{\text{mol } H_2O} \cdot \frac{\textbf{4 mol } NH_3}{\textbf{6 } \cancel{\text{mol } H_2O}} = \textbf{5.6 mol } NH_3$$

S.F.: 8.4 has 2 *s.f.* Coefficients are exact with infinite *s.f.* Round to 2 *s.f.*

Writing a substance formula after each unit in the WANTED and DATA provides us with a *prompt*: In a reaction calculation, *if* the substance WANTED differs from the substance *given*, write the *balance* and *bridge* steps. Those steps get conversions into your DATA that you will need when SOLVING.

PRACTICE **B**

In your notebook, solve using the steps shown in the Try It.

1. In the production of steel, iron(III) oxide can be reduced by carbon. The *unbalanced* equation is

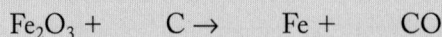

$$Fe_2O_3 + \quad C \rightarrow \quad Fe + \quad CO$$

   a. How many moles of Fe_2O_3 are consumed in reacting with 4.80 mol carbon?

   b. To form 3.6 mol Fe, how many moles of carbon are needed?

2. Burning ammonia can result in the formation of water and elemental nitrogen:

$$4\,NH_3 + 3\,O_2 \rightarrow 2\,N_2 + 6\,H_2O$$

   a. In this reaction, how many moles of O_2 are required to burn 14.0 mol NH_3?

   b. How many moles of O_2 must be used up if 20.0 mol N_2 form?

Lesson 9.2   Conversion Stoichiometry

Perhaps the most frequently encountered type of chemistry calculation is **stoichiometry** (stoy-kee-AHM-et-ree), a term derived from ancient Greek that means "measuring fundamental quantities." There are many types of stoichiometry. For now, we will limit our attention to reactions that *go to completion*: reactions that proceed until at least one reactant is totally used up.

In what we will term "standard" stoichiometry, you are *given* a measured amount of one substance that reacts or is produced, and you WANT to know *how much* of *another* substance in the reaction will react or form. Our steps in this type of stoichiometry are as follows:

- If needed, convert the units of the *given* amount to moles.
- Use the bridge ratio supplied by the balanced equation to convert from the *given* moles to the WANTED moles.
- Convert to other WANTED units if needed.

Let's start with a simple example.

> **TRY IT**

Q. Burning hydrogen gas in air forms steam by the following unbalanced equation:

$$H_2 + \quad O_2 \rightarrow \quad H_2O$$

If 32.0 grams of O_2 is used up in the reaction, how many grams of H_2O form?

In your notebook, write all that you would write *before* the final SOLVE step.

Answer:

WANTED: $? \, g \, H_2O =$ (Single unit WANTED)

DATA: $32.0 \, g \, O_2$ (Single unit *given*)

             $18.0 \, g \, H_2O = 1 \, mol \, H_2O$ (Grams prompt in WANTED unit)

             $32.0 \, g \, O_2 = 1 \, mol \, O_2$ (Grams prompt in DATA)

(Because WANTED $H_2O \neq$ *given* O_2, the balance and bridge steps must be added.)

Balance: $2 \, H_2 + O_2 \rightarrow 2 \, H_2O$

Bridge: $2 \, mol \, H_2O = 1 \, mol \, O_2$ (X mol WANTED $= Y$ mol *given*)

From the equalities above, we could chain conversions to SOLVE, and in most problems we will. But for this first problem, it will be helpful to look at each step individually.

In solving this "grams-to-grams stoichiometry," the first step is to convert from the grams of the *given* substance to its moles. Look at the DATA terms for the given O_2 and see if you can do that conversion "in your head."

$$\text{Moles of } O_2 \text{ given} =$$

STOP

Because 32.0 g O_2 is 1 mol, the oxygen gas used up in the reaction is **1.00 mol O_2**.

If this 1.00 mol O_2 is completely used up, according to the balanced equation how many moles of H_2O will be formed?

$$\text{Moles of } H_2O \text{ formed} =$$

STOP

2.00 mol H_2O will be formed. What is the mass of this quantity of H_2O? Using the molar mass of H_2O in the DATA, try that conversion "in your head" or using "paper-and-pencil" math.

$$\text{Grams of } H_2O \text{ formed} =$$

STOP

Because 1 mol H_2O is 18.0 g, 2.00 mol is $18.0 \times 2.00 = $ **36.0 g H_2O formed**.

This answers the question posed in the problem. If 32.0 g O_2 is used up, 36.0 g H_2O forms.

The above process solved the problem in three steps, solving for the answer at each step. The key step was the mole-to-mole conversion. We needed a ratio that *related* O_2 and H_2O, and the one ratio we knew was the mole-to-mole ratio supplied by the balanced equation. To SOLVE, we began by converting the *given* unit to moles, so that we could apply the mole-to-mole conversion.

Another way to solve the problem (and one that is faster) rather than solving in three separate steps is to chain the conversions, which SOLVES all of the arithmetic at one final step.

TRY IT

In your notebook, for the problem above, SOLVE by the standard method: chain the DATA equalities.

STOP

SOLVE: If you WANT a single unit, start with the single-unit DATA.

$$? \text{ g } H_2O = 32.0 \text{ g } O_2 \cdot \frac{}{\text{g } O_2}$$

STOP

$$? \text{ g } H_2O = 32.0 \text{ g } O_2 \cdot \frac{1 \text{ mol } O_2}{32.0 \text{ g } O_2} \cdot \frac{2 \text{ mol } H_2O}{1 \text{ mol } O_2} \cdot \frac{18.0 \text{ g } H_2O}{1 \text{ mol } H_2O} = \textbf{36.0 g } H_2O$$

Compare this answer to the answer from the Try It above.

In some stoichiometry problems, we will need to solve "one step at a time." But in most "grams and moles" stoichiometry, chaining the conversions will provide an answer more quickly. We will give this method a name:

The Eight Stoichiometry Steps

For a reaction calculation, if the WANTED *formula* is not the same as the *given formula*, write

1. WANTED: **?** *unit* and substance *formula* **=**

2. DATA: (list including prompts)

3. BALANCE: (balance the equation *if* it is unbalanced)

4. BRIDGE: **X mol** WANTED formula **= Y mol** *given* formula

   (*X* and *Y* being the coefficients for the WANTED and *given* substances)

5. SOLVE: If a single unit is WANTED, first write

   ? WANTED unit and formula = # and unit and *given* formula

6. If moles is not the *given unit*, convert to moles of the *given* formula.

7. Convert to moles of WANTED formula using the mole-to-mole ratio.

8. If moles is not WANTED, convert to the unit WANTED.

These steps are simply an extension of the "Rule for Reaction Calculations" in the previous lesson. With practice, the steps will become automatic. A short version is:

The Stoichiometry Prompt

For reaction calculations, if WANTED formula ≠ *given* formula:

- Write the four WDBB steps: WANTED and DATA, **b**alance and **b**ridge.
- Then, write the four SOLVE steps that convert from unit *given* to **mol** *given* to **mol** WANTED to unit WANTED.

During initial reaction calculations, it can help to recite, "If WANTED formula does not equal *given* formula: WDBB, then units to moles to moles to units."

▷ TRY IT

In your notebook, solve the following problem by completing the eight stoichiometry steps.

Q. Sodium burns to form sodium oxide (Na_2O). How many grams of Na_2O can be produced from 2.30 g of Na? The unbalanced equation is

$$Na + \quad O_2 \rightarrow \quad Na_2O$$

Answer:

1. WANTED: $? \text{ g } Na_2O =$ (A single unit is WANTED.)

2. DATA: $2.30 \text{ g } Na$ (The *only* single unit in the DATA—use as *given*.)

   $1 \text{ mol Na} = 23.0 \text{ g Na}$ (g Na in DATA = prompt.)

   $1 \text{ mol Na}_2O = 62.0 \text{ g Na}_2O$ (g Na$_2$O WANTED = prompt.)

(Because this is a *reaction* calculation and the WANTED and *given* substances *differ*, use the eight stoichiometry steps.)

3. Balance: $4 \text{ Na} + 1 \text{ O}_2 \rightarrow 2 \text{ Na}_2O$

4. Bridge: $2 \text{ mol Na}_2O = 4 \text{ mol Na}$ (Moles WANTED to moles *given*.)

If needed, adjust your work and finish.

5. SOLVE: Start with $? \text{ g Na}_2O = 2.30 \text{ g Na} \cdot \dfrac{}{\text{g Na}}$

6. Convert to **moles** *given*. ("Grams and moles—use molar mass.")

$$? \text{ g Na}_2O = 2.30 \text{ } g \text{ } Na \cdot \frac{1 \text{ mol Na}}{23.0 \text{ } g \text{ } Na} \cdot \frac{}{\text{mol Na}}$$

7. Convert to **moles** WANTED using the bridge between *given* and WANTED.

$$? \text{ g Na}_2O = 2.30 \text{ g Na} \cdot \frac{1 \text{ mol Na}}{23.0 \text{ g Na}} \cdot \frac{2 \text{ mol Na}_2O}{4 \text{ mol Na}} \cdot \frac{}{\text{mol Na}_2O}$$

8. Convert to **units** WANTED.

$$? \text{ g Na}_2O = 2.30 \text{ g Na} \cdot \frac{1 \text{ mol Na}}{23.0 \text{ g Na}} \cdot \frac{2 \text{ mol Na}_2O}{4 \text{ mol Na}} \cdot \frac{62.0 \text{ g Na}_2O}{1 \text{ mol Na}_2O}$$

$$= \frac{2.30 \cdot 2 \cdot 62.0}{23.0 \cdot 4} \text{ g Na}_2O = \mathbf{3.10 \text{ g Na}_2O}$$

S.F.: Coefficients and 1 are always exact, 2.30 and 23.0 have 3 *s.f.*, round to 3 *s.f.*

Unit and Substance Formulas: Inseparable

One reason to *always* write *number*, *unit*, and *formula* in each term of your WANTED, DATA, *and* conversions is to avoid errors in unit cancellation. A key rule is:

> Units measuring one substance *cannot* cancel the same units measuring a *different* substance.

> **TRY IT**

Q1. Is the following conversion *legal* or *illegal*? Write your answer, then check below.

$$? \text{ mol HCl} = 32 \text{ g NaOH} \cdot \frac{1 \text{ mol HCl}}{36.5 \text{ g HCl}} =$$

(STOP)

Answer:

Illegal. Because grams of NaOH are not the same as grams of HCl, the units cannot cancel. The substance is an inseparable part of the unit. Writing the substance *formula* after *every* unit (*if* the number and unit apply to only one substance) reduces the chance for improper unit cancellation.

Q2. Is the following conversion *legal* or *illegal*?

$$? \text{ kg NaOH} = 32 \text{ g NaOH} \cdot \frac{1 \text{ kg}}{1000 \text{ g}} =$$

(STOP)

Answer:

Legal, because different formulas are not cancelled. The ratio between grams and kilograms is true for measurements of *all* substances. The molar mass conversion in Q1 was only true for HCl.

> For problems involving more than one substance, if a number and unit apply to only one substance, write the substance formula after the unit.

PRACTICE **A**

Work in your notebook. Use the table of atomic masses or periodic table at the back of this book.

1. Multiply and simplify. Your answers must include a number, unit, and formula.

    a. $5.1 \text{ g Al}_2\text{O}_3 \cdot \dfrac{1 \text{ mol Al}_2\text{O}_3}{102 \text{ g Al}_2\text{O}_3} \cdot \dfrac{3 \text{ mol O}_2}{2 \text{ mol Al}_2\text{O}_3} =$

    b. $1.27 \times 10^{-3} \text{ g Cu} \cdot \dfrac{1 \text{ mol Cu}}{63.5 \text{ g Cu}} \cdot \dfrac{2 \text{ mol Ag}}{1 \text{ mol Cu}} \cdot \dfrac{107.9 \text{ g Ag}}{1 \text{ mol Ag}} =$

2. Phosphorus (P_4) burns in air (O_2) to form the oxide P_4O_{10}.

    a. Write the balanced equation for the reaction.

    b. How many grams of P_4O_{10} would be produced when 62.0 g P_4 burns?

    c. How many moles of O_2 are needed to burn 7.0 mol P_4?

Why Not Go from Grams to Grams in One Conversion?

In a reaction calculation, if grams are WANTED and grams are *given*, why don't we SOLVE in one conversion using a gram-to-gram ratio?

In a reaction, if the substance formulas are known, the *particle* ratios are easy to determine. Simply balance the reaction equation. The coefficients then supply exact and simple mole WANTED to mole *given* ratios. But there is no similar easy way to calculate gram-to-gram ratios.

In reaction calculations, we take the steps we do, in the order we do, so that we can use the particle-to-particle ratio: the only ratio relating the *given* and WANTED substances that is easy to determine.

THE STOICHIOMETRY BRIDGE: MOLES TO MOLES

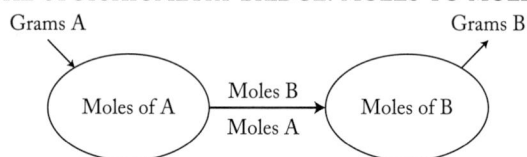

Flashcards

Run these cards until you can do them perfectly, then try the problems below.

One-Way Cards (with Notch)	Back Side—Answers
When reading an equation, the units of coefficients can be read as _____	Counts of particles or moles of particles
When is a mole-to-mole conversion needed in conversions?	In reaction calculations, when WANTED formula ≠ *given* formula
Define *stoichiometry*	A calculation of how much of a substance is used up or forms in a reaction
When are the stoichiometry steps needed?	In reaction calculations, when WANTED formula ≠ *given* formula
What numbers go into a mole-to-mole conversion?	The coefficients of the balanced equation
Recite a short version of the stoichiometry steps	"WDBB, then units to moles to moles to units"

PRACTICE B

Solve problems 1 and 2 in your notebook. For your next study session, run your flashcards again and complete problem 3.

1. Silver metal reacts with concentrated nitric acid (HNO_3: 63.0 g/mol) to produce the red-brown gas nitrogen dioxide. The balanced equation is

$$Ag + 2\,HNO_3 \rightarrow NO_2 + AgNO_3 + H_2O$$

   a. How many grams of nitric acid are required to use up 5.00 g Ag?

   b. How many grams of nitrogen dioxide would be formed?

2. The combustion of iron(II) sulfide can be represented as:

$$FeS + \quad O_2 \rightarrow \quad Fe_2O_3 + \quad SO_2$$

   a. Balance the equation.

   b. Starting with 0.48 mol FeS and excess oxygen, how many grams of Fe_2O_3 (159.6 g/mol) can be formed?

3. Nitrogen dioxide can react with water to form nitric acid and nitrogen monoxide. The balanced equation is

$$3\,NO_2 + H_2O \rightarrow 2\,HNO_3 + NO$$

   a. How many moles of H_2O are required to use up 230. grams of NO_2?

   b. How many milligrams of NO can be formed from 2.4×10^{21} molecules of NO_2?

Lesson 9.3   Limiting Reactants

In a chemical reaction, particles are used up and form *only* in the particle ratios shown by the balanced equation. In any reaction between two substances that goes to completion, for the *reactants* there are two possibilities.

- The reactants are mixed in the exact ratio needed so that each of the reactants is completely used up.
- One of the reactants is completely used up, and some of the other reactant is left over.

Let's take a look at each of these cases.

Stoichiometric Equivalents

Consider this reaction that goes rapidly to completion when ignited:

$$2\,H_2 + O_2 \rightarrow 2\,H_2O$$

▸ TRY IT

Q. If four molecules of H_2 react, how many molecules of O_2 react?

STOP

Answer:

The ratio of H_2 molecules used up to O_2 molecules used up must be 2 to 1. Four reacting H_2 molecules must use up two molecules of O_2.

If one reactant is mixed with another, and both are exactly and completely consumed in the reaction, the reactants initially present are said to be in **stoichiometrically equivalent amounts** or in **stoichiometric ratios**.

▶ TRY IT

Q1. If 4.00 mol H_2 are to be reacted, how many moles of O_2 are needed for stoichiometric equivalence?

Answer:

2.00 mol. Moles is a unit that counts particles. For both reactants to be exactly used up, the initial particle ratio of H_2 to O_2 must be 2 to 1.

Q2. If 4.00 mol H_2 are reacted with 2.00 mol O_2, how much H_2, O_2, and H_2O will be present in the reaction vessel after the reaction is over?

$$H_2 = \qquad O_2 = \qquad H_2O =$$

Answer:

If the coefficients for this reaction are doubled and read as moles, the equation can be read as: "four moles of H_2 react with two moles of O_2 to form four moles of H_2O." This means that after the reaction, present are the **4.00 moles of H_2O** that formed, but **no H_2** and **no O_2**. As products form, reactants are consumed. If the two reactants are mixed in stoichiometrically equivalent amounts, both will be completely used up.

In many reaction calculations, you are asked to solve for a stoichiometric equivalent: the amount of one reactant that is needed to exactly use up another reactant. When the numbers are not as simple as those above, these amounts can be solved by our standard stoichiometry steps.

Limiting Reactants

In many types of lab procedures (such as an acid–base titration), the goal is to mix reactants in amounts that are as close to stoichiometric equivalence as possible. In other experimental situations, two reactants are mixed together in nonstoichiometric ratios. Let's explore what happens when a reaction goes to completion but the reactants were not combined in stoichiometrically equivalent amounts.

▶ TRY IT

Again considering this reaction that goes to completion:

$$2\,H_2 + O_2 \rightarrow 2\,H_2O$$

Q. If *four* molecules of H_2 are reacted with *one* molecule of O_2, how much H_2, O_2, and H_2O will be present after the reaction?

$$H_2 = \qquad O_2 = \qquad H_2O =$$

STOP

Answer:

The one molecule of O_2 reacts with exactly two molecules of H_2. At that point, there is no more O_2 to react, so the reaction must stop. After the reaction, present are the two molecules of H_2O that form, no O_2, and two molecules of H_2 that are left over because they had no O_2 to react with. In this case, the amount of H_2O formed is determined not by the amount of H_2 originally supplied but by the initial amount of the O_2, because O_2 is used up first.

Limiting Reactant Terminology

Many reactions (and all reactions that we are concerned with at this point) go to completion—until one reactant is essentially 100% used up. For those reactions, the following vocabulary, assumptions, and rules will simplify problem solving.

Every reaction that goes to completion has at least one **limiting reactant** (or **limiting reagent**). The limiting reactant is the reactant used up first. This reactant limits and determines both how much of each of the other reactants is used up and how much of each product forms.

A reactant that is *not* completely used up is said to be **in excess**, meaning

- enough is present to use up all of the limiting reactant, and

- some amount remains when the reaction stops.

Which reactant is used up first depends on

- the starting amount of each reactant, and

- the ratios of reaction for the reactants (the reactant coefficients).

For reactions that go to completion, key rules include:

- At least *one* reactant *must* be limiting (totally used up).

- The initial amount of the *limiting reactant* determines how much of each of the other reactants reacts and how much of each product forms.

- The initial amount of the *limiting reactant* must be used as the *given* to calculate the amounts of other substances in the reaction that are used up and form.

- The initial amounts of *reactants in excess* do *not* determine the amounts of other substances that are used up and form. The amount of a reactant *in excess* cannot be used as a *given* to calculate how much of each product forms.

- If a reactant is in excess, its amount is more than enough to use up the limiting reactant, and measurements of its initial amount are not necessary (and can usually be ignored) in calculations.

In the special case of two reactants that are stoichiometrically equivalent, both are limiting, and either reactant amount can be used to calculate how much of each product will form.

Determining the Amounts of the Products

Using the logic of the rules above, in the space provided below, write your answers to the following questions on the basis of mental arithmetic rather than written conversions.

▶ **TRY IT**

Q. Given the balanced equation

$$2\,H_2 + O_2 \rightarrow 2\,H_2O$$

if 10 molecules of H_2 are ignited with 6 molecules of O_2, and the reaction goes to completion,

a. Which reactant is limiting?

b. How much H_2O can be formed?

c. How much of each reactant and product is present in the mixture at the end of the reaction?

$H_2 =$ $O_2 =$ $H_2O =$

STOP

Answer:

a. For 6 molecules of O_2 to react, 12 molecules of H_2 is needed. We don't have that much H_2, so the H_2 is *limiting*. Though the H_2 molecules are initially present in a higher *count*, more H_2 molecules are needed per reaction. In this case, the 10 molecules of H_2 are used up first, preventing the O_2 molecules from being completely used up.

Looked at another way: Reacting 10 molecules of H_2 requires 5 molecules of O_2. We have more than that much O_2, so when we use up all of the H_2, some O_2 remains. The H_2 is therefore limiting (used up ~100%), and there is *excess* O_2: some remains in the mixture at the end of the reaction.

b. If 10 molecules of H_2 are used up, according to the balanced equation 10 molecules of H_2O must form. Because some O_2 remains at the end of the reaction, the 6 molecules of O_2 initially present do *not* determine how much H_2O forms. Only amounts that *react* cause products to form. Only the initial amount of *limiting* reactant predicts the amount of a product that forms.

c. In the mixture after the reaction *no* H_2 remains. Present are the 10 molecules of water formed, plus the 1 molecule of O_2 that did not react.

Flashcards

If you learn new vocabulary *before* starting problems, you will likely need to do fewer problems to recall the concepts you need to know.

Two-Way Cards (*without* Notch)

| Limiting reactant (definition) | In a reaction that goes to completion, the reactant used up first is called the _____ |
| Reactant in excess (definition) | After a reaction that goes to completion, a reactant with some amount left over is called the _____ |

One-Way Cards (with Notch at Top Right)	Back Side—Answers
The reactant with an amount that is ignored in a stoichiometric calculation is called the _____	Reactant in excess
The reactant that determines how much of each product can be formed is called the _____	Limiting reactant
The reactant *amount* that must be *given* to predict other amounts that react and form is the _____	Amount of the limiting reactant
If two reactants are both completely used up, their amounts are said to be _____	Stoichiometrically equivalent

PRACTICE

Answer these using mental math and the balanced equation

$$4\,Fe + 3\,O_2 \rightarrow 2\,Fe_2O_3$$

1. To form four particles of Fe_2O_3:

   a. How many atoms of Fe are needed?

   b. How many molecules of O_2 are needed?

2. How many Fe_2O_3 particles are produced when nine molecules of O_2 are used up?

3. If 20 atoms of Fe are mixed with 20 molecules of O_2 and the reaction goes to completion:

   a. The Fe consumes how much O_2 as it reacts?

   b. How much O_2 is left over?

   c. How many particles of Fe_2O_3 are formed?

   d. Which reactant is in excess?

   e. Which reactant is limiting?

Lesson 9.4 Limiting Reactant Calculations

Lessons 9.1 and 9.2 addressed questions about reaction amounts (stoichiometry) in which the amount to be used as *given* was specified.

In Lesson 9.3, we considered problems in which the reactant (of two or more reactants) used up first needed to be identified. The amount of this limiting reactant controls how much of each of the other reactants is used up and how much of each product forms. From these problems based on simple numbers, we can state these rules:

Limiting Reactant Rules

- If *two* reactant amounts are supplied but the limiting reactant is not identified, to calculate amounts used up and formed, you must *first* determine *which* reactant is limiting.
- The limiting reactant is the one that in the reactant mixture is used up first. When it is completely used up, the reaction must stop.
- The limiting reactant is the one that, when *given*, produces the least amount of each of the reaction products.
- The amount of the limiting reactant must be the *given* when you SOLVE for amounts of other substances used up and formed in the reaction.

Identifying the Limiting Reactant by Stoichiometry

If the particle counts are simple whole numbers, you can often identify the limiting reactant by mental math, as we did in Lesson 9.3. For reactions in which the limiting reactant cannot be determined by mental math, the rules above are the same, but the reactant that is limiting must be determined by calculation. One way to do so is to "SOLVE using each reactant as *given*."

Steps to Calculate Which Reactant Is Limiting

If the limiting reactant cannot be determined by mental math:

1. In two separate stoichiometric SOLVE steps, choose the same unit and *product* formula to be WANTED, but use each of the *two* known reactant amounts as *given*.

2. The SOLVE step that results in the *lowest* amount of the WANTED substance formed has the *limiting* reactant as its *given*. Label this the *limiting* reactant.

Any product chosen will result in the same reactant being identified as limiting; however, *if* the problem asks you to find the amount of *one* of the products that is formed, choose that unit and product as WANTED. The calculation resulting in the *lower* amount of product will answer the question.

Once you determine the limiting reactant, if you are asked to calculate the amounts of *other* products that form, use the amount of the limiting reactant as *given*. The limiting reactant determines how much of *each* product forms.

▶ **TRY IT**

Apply the steps in the box above to the following questions.

Q. If 2.00 g H_2 gas and 4.80 g O_2 gas are mixed and ignited:

  a. Which reactant is limiting?

  b. How many grams of water will form?

Answer:

In reaction calculations, start with WDBB.

  1. WANTED: ? g H_2O =   (A single unit is WANTED.)

  2. DATA: 2.00 **g** H_2 (A single-unit amount supplied.)

             4.80 **g** O_2 (*Another* single-unit amount supplied.)

             1 mol H_2 = 2.016 g H_2

             1 mol O_2 = 32.0 g O_2

             1 mol H_2O = 18.0 g H_2O (There are three grams prompts.)

Strategy: To identify the limiting reactant, choose one product amount as WANTED. Because part (b) asks for the amount of H_2O formed, choose water as the WANTED product. Calculate the WANTED amount of water formed in two separate stoichiometric conversions, each using a different single-unit reactant amount as a *given*.

  3. Balance: **2 H_2 + 1 O_2 → 2 H_2O**

  4. Bridge: (Moles WANTED to moles *given* will vary for the two *givens*.)

  5. SOLVE:

$$? \text{ g } H_2O = 2.00 \text{ g } H_2 \cdot \frac{1 \text{ mol } H_2}{2.016 \text{ g } H_2} \cdot \frac{2 \text{ mol } H_2O}{2 \text{ mol } H_2} \cdot \frac{18.0 \text{ g } H_2O}{1 \text{ mol } H_2O} = 17.9 \text{ g } H_2O$$

$$? \text{ g } H_2O = \boxed{4.80 \text{ g } O_2} \cdot \frac{1 \text{ mol } O_2}{32.0 \text{ g } O_2} \cdot \frac{2 \text{ mol } H_2O}{1 \text{ mol } O_2} \cdot \frac{18.0 \text{ g } H_2O}{1 \text{ mol } H_2O} = 5.40 \text{ g } H_2O$$

The H_2 could produce 17.9 g of H_2O, but the O_2 can only produce 5.40 g of H_2O. In the reactant mixture above, the amount of water that can form is **limited by O_2**. The reaction produces **5.40 g H_2O**. The limiting reactant amount is circled.

Though more grams of oxygen are supplied in the reaction mixture, it is the *moles* (not grams) of the reactants and the *ratios* of the reaction that decide the limiting reactant.

PRACTICE A

1. In your notebook, summarize the rules for limiting reactant calculations in a way that you personally find brief and memorable. Apply your rules to the problems that follow.

2. For the reaction of hydrochloric acid and calcium carbonate, the balanced equation is

$$2\,HCl + CaCO_3 \rightarrow CO_2 + H_2O + CaCl_2$$

If 0.400 mol HCl is reacted with 0.300 mol $CaCO_3$:

   a. Which reactant is limiting?

   b. How many moles of CO_2 can be formed?

3. For the reaction represented by the following unbalanced equation,

$$H_2 + \quad Cl_2 \rightarrow \quad HCl$$

if 14.2 g chlorine gas is reacted with 0.300 mol hydrogen gas:

   a. Which reactant is limiting?

   b. How many moles of HCl can be formed?

   c. How many moles of the "reactant in excess" must be used up to completely consume the limiting reactant?

Identifying Limiting Reactant Problems

How can you tell which problems can be solved by standard stoichiometry and which require finding the limiting reactant first? There are various ways to distinguish these two different types of problems, but in this introductory course, if finding the limiting reactant is necessary, the problem will say so. If it does not, assume that a reaction calculation involving single-unit amounts can be solved by the standard stoichiometry steps.

PRACTICE B

1. Rerun the flashcards for this chapter, practice recall of your rules summary for limiting reactants, and apply your rules to the problems below.

2. For the unbalanced reaction equation

$$H_2SO_4 + \quad NaOH \rightarrow \quad H_2O + \quad Na_2SO_4$$

if 2.00 g NaOH is reacted with 0.0100 mol H_2SO_4:

   a. Which reactant is limiting?

   b. How many grams of water can be formed?

   c. How many grams of the reactant in excess must be used up to completely consume the limiting reactant?

SUMMARY

One of the goals of chemistry is to be able to precisely predict amounts of substances that will be used up and formed in natural and laboratory processes. Stoichiometry uses a known amount of *one* substance to predict how much of *others* will react and form. The coefficients of the balanced equation provide the key whole-number ratios between the *given* and WANTED substance particles.

Coefficients apply to counts of particles, but for visible amounts of a substance it is best to count particles by the mole. We do not have machines to count large numbers of particles directly, but we can measure mass precisely, and mass can be converted to moles on the basis of a substance's molar mass. The steps of "standard" and "limiting reactant" stoichiometry are simply efficient procedures that use a known *given* amount to predict the amounts of other substances used up and formed in chemical reactions.

REVIEW QUIZ

1. For the unbalanced equation

$$NaClO_3 \rightarrow \quad NaCl + \quad O_2$$

   how many moles of O_2 can be obtained from 2.50 mol $NaClO_3$?

2. For the reaction

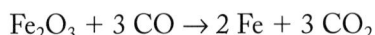

$$Fe_2O_3 + 3\,CO \rightarrow 2\,Fe + 3\,CO_2$$

   if 33.5 g Fe is produced, how many moles of carbon monoxide are used up?

3. How many grams of H_2O gas are produced when 64 mg CH_3OH is burned? The unbalanced equation is

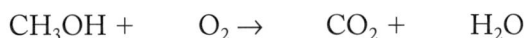

$$CH_3OH + \quad O_2 \rightarrow \quad CO_2 + \quad H_2O$$

4. For the conversion of the iron ore hematite (Fe_2O_3) to iron in a blast furnace, one step is this unbalanced equation:

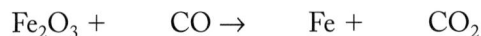

$$Fe_2O_3 + \quad CO \rightarrow \quad Fe + \quad CO_2$$

   If 125 moles of hematite is reacted, how many kilograms of iron can be formed?

5. If 40.2 g H_2 gas is burned with 128 g O_2 gas, how many moles of water form? (Find the limiting reactant first.)

ANSWERS

Lesson 9.1

Practice A   1a. 16 molecules HBr need **4** molecules O_2 (the ratio must be 4 HBr to 1 O_2); 16 molecules of HBr make **8** molecules Br_2 (the ratio must be 4 HBr to 2 Br_2).

1b. **20** molecules HBr react; **10** molecules Br_2 form.

2a. $CS_2 + \mathbf{3}\,O_2 \rightarrow CO_2 + \mathbf{2}\,SO_2$ 2b. 75 trillion molecules O_2; 50. trillion molecules SO_2

Practice B Your paper should look like this, but you may omit the comments in parentheses.

1a. WANTED: $?$ mol **Fe$_2$O$_3$** =   (Single unit WANTED)

    DATA: 4.80 mol **C**   (Single unit *given*)

    (WANTED formula ≠ *given* formula: do balance and bridge steps.)

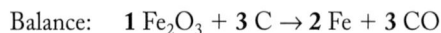

    Balance: **1** Fe$_2$O$_3$ + **3** C → **2** Fe + **3** CO

    Bridge: **1 mol Fe$_2$O$_3$ = 3** mol C   (X mol WANTED = Y mol *given*)

    SOLVE: $?$ mol Fe$_2$O$_3$ = 4.80 mol C • $\dfrac{1\ \text{mol Fe}_2\text{O}_3}{3\ \text{mol C}}$ = **1.60 mol Fe$_2$O$_3$**

1b. WANTED: $?$ mol C =

    DATA: 3.6 mol Fe

    (For a reaction, if formula WANTED ≠ formula *given*, balance and bridge.)

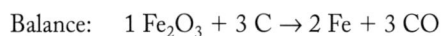

    Balance: 1 Fe$_2$O$_3$ + 3 C → 2 Fe + 3 CO

    Bridge: 3 mol C = 2 mol Fe

    SOLVE: $?$ mol C = 3.6 mol Fe • $\dfrac{3\ \text{mol C}}{2\ \text{mol Fe}}$ = **5.4 mol C**

2. For problem 2, only the SOLVE steps are shown, but your paper should include WANTED, DATA, balance, and bridge steps before SOLVE.

2a. SOLVE: $?$ mol O$_2$ = 14.0 mol NH$_3$ • $\dfrac{3\ \text{mol O}_2}{4\ \text{mol NH}_3}$ = **10.5 mol O$_2$**

2b. SOLVE: $?$ mol O$_2$ = 20.0 mol N$_2$ • $\dfrac{3\ \text{mol O}_2}{2\ \text{mol N}_2}$ = **30.0 mol O$_2$**

Lesson 9.2

Practice A 1a. 0.075 mol O$_2$ 1b. 4.32 × 10^{-3} g Ag

    2a. 1 P$_4$ + 5 O$_2$ → 1 P$_4$O$_{10}$

    2b. 1. WANTED: **$?$ g P$_4$O$_{10}$ =**   (WANT single unit)

        2. DATA: 62.0 g **P$_4$**   (Single unit *given*)

                1 mol P$_4$ = 124.0 g P$_4$ (Grams prompt)

                1 mol P$_4$O$_{10}$ = 284.0 g P$_4$O$_{10}$ (Grams prompt)

        (For a reaction, if WANTED formula ≠ *given* formula, use stoichiometry steps.)

        3. Balance: 1 P$_4$ + 5 O$_2$ → 1 P$_4$O$_{10}$

        4. Bridge: 1 mol P$_4$ = 1 mol P$_4$O$_{10}$

     5–8. SOLVE: (Chain the equalities above)

$$? \text{ g P}_4\text{O}_{10} = 62.0 \text{ g P}_4 \cdot \frac{1 \text{ mol P}_4}{124.0 \text{ g P}_4} \cdot \frac{1 \text{ mol P}_4\text{O}_{10}}{1 \text{ mol P}_4} \cdot \frac{284.0 \text{ g P}_4\text{O}_{10}}{1 \text{ mol P}_4\text{O}_{10}} = \textbf{142 g P}_4\textbf{O}_{10}$$

        S.F.: 62.0 has 3 *s.f.*, 1 is exact, answer must be rounded to 3 *s.f.*

2c. 1. WANTED: ? mol O_2 =

    2. DATA: 7.0 mol P_4

    (If WANTED formula ≠ *given* formula, include the balance and bridge steps.)

    3. Balance: (See part 2a.)

    4. Bridge: 1 mol P_4 = 5 mol O_2 (Bridge must be *mol* WANTED = mol *given*.)

    5–8. SOLVE: (The *given* and WANTED units are moles. Use the one conversion in the DATA to SOLVE.)

$$? \text{ mol } O_2 = 7.0 \text{ mol } P_4 \cdot \frac{5 \text{ mol } O_2}{1 \text{ mol } P_4} = \mathbf{35 \text{ mol } O_2}$$

Practice B   1a. 1. WANTED: ? g HNO_3 =

    2. DATA: 5.00 g Ag   (Single unit is WANTED. This is your single unit *given*.)

        1 mol Ag = 107.9 g Ag   (g Ag in DATA.)

        1 mol HNO_3 = 63.0 g HNO_3 (g HNO_3 in WANTED.)

    3. Balance: (Supplied)

    4. Bridge: 1 mol Ag = 2 mol HNO_3

    5–8. SOLVE: $? \text{ g } HNO_3 = 5.00 \text{ g Ag} \cdot \dfrac{1 \text{ mol Ag}}{107.9 \text{ g Ag}} \cdot \dfrac{2 \text{ mol } HNO_3}{1 \text{ mol Ag}} \cdot \dfrac{63.0 \text{ g } HNO_3}{1 \text{ mol } HNO_3} = \mathbf{5.84 \, g \, HNO_3}$

1b. Showing SOLVE step only:

$$? \text{ g } NO_2 = 5.00 \text{ g Ag} \cdot \frac{1 \text{ mol Ag}}{107.9 \text{ g Ag}} \cdot \frac{1 \text{ mol } NO_2}{1 \text{ mol Ag}} \cdot \frac{46.0 \text{ g } NO_2}{1 \text{ mol } NO_2} = \mathbf{2.13 \, g \, NO_2}$$

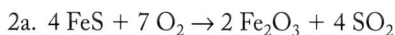

2a. 4 FeS + 7 O_2 → 2 Fe_2O_3 + 4 SO_2

2b. 1. WANTED: ? g Fe_2O_3 =

    2. DATA: 0.48 mol FeS

        1 mol Fe_2O_3 = 159.6 g Fe_2O_3

    (WANTED substance ≠ *given* substance. Use the stoichiometry steps.)

    3. Balance: (See part a.)

    4. Bridge: 4 mol FeS = 2 mol Fe_2O_3

    5–8. SOLVE: $? \text{ g } Fe_2O_3 = 0.48 \text{ mol FeS} \cdot \underset{\text{(step 7)}}{\dfrac{2 \text{ mol } Fe_2O_3}{4 \text{ mol FeS}}} \cdot \underset{\text{(step 8)}}{\dfrac{159.6 \text{ g } Fe_2O_3}{1 \text{ mol } Fe_2O_3}} = \mathbf{38 \, g \, Fe_2O_3}$
                  (step 6)

3a. 1. WANTED: ? mol H_2O =

    2. DATA: 230. g NO_2

        1 mol NO_2 = 46.0 g NO_2

    (WANTED substance ≠ *given* substance. Use the stoichiometry steps.)

    3. Balance: 3 NO_2 + H_2O → 2 HNO_3 + NO

    4. Bridge: 3 mol NO_2 = 1 mol H_2O

    5–8. SOLVE: $? \text{ mol } H_2O = 230. \text{ g } NO_2 \cdot \dfrac{1 \text{ mol } NO_2}{46.0 \text{ g } NO_2} \cdot \dfrac{1 \text{ mol } H_2O}{3 \text{ mol } NO_2} = \mathbf{1.67 \, mol \, H_2O}$

3b. 1. WANTED: ? mg NO =

    2. DATA: 2.4×10^{21} molecules of NO_2 (Only single unit in DATA; must be *given*.)

                    1 mol = 6.02×10^{23} molecules   (Avogadro prompt.)

                    1 mol NO = 30.0 g NO   (**g, mg, kg** of formula = grams prompt.)

    (WANTED substance ≠ *given* substance. Use stoichiometry steps.)

    3. Balance: $3\,NO_2 + H_2O \rightarrow 2\,HNO_3 + NO$

    4. Bridge: 1 *mol* NO = 3 *mol* NO_2

  5–8. SOLVE:

$$? \textbf{ mg NO} = 2.4 \times 10^{21}\,NO_2 \cdot \frac{1\;mol\;NO_2}{6.02 \times 10^{23}\;NO_2} \cdot \frac{1\;mol\;NO}{3\;mol\;NO_2} \cdot \frac{30.0\;\textbf{g NO}}{1\;mol\;NO} \cdot \frac{1\;mg}{10^{-3}g} = \textbf{40. mg NO}$$

Lesson 9.3

Practice   Because coefficients are ratios, they can be doubled to match this problem:

$$8\,Fe + 6\,O_2 \rightarrow \textbf{4 Fe}_2\textbf{O}_3$$

1a. 8 atoms of Fe needed.   1b. 6 molecules of O_2 needed.

2. The ratio is 3 O_2 produce 2 Fe_2O_3, so 9 O_2 produce **6 Fe$_2$O$_3$** particles.

3a. 15 molecules of O_2 3b. 5 molecules of O_2 are left over.   3c. 10 particles of Fe_2O_3 are formed.

3d. O_2 is in excess.   3e. Fe is limiting.

Lesson 9.4

Practice A   2a. WANTED: Which reactant is limiting?

    To find the limiting reactant, find the amount of one product that can be formed by *each* reactant using eight-step stoichiometry. Because we WANT moles of CO_2 in part (b), it is easiest to solve for moles of CO_2 in part (a).

    1. WANTED: ? mol CO_2 =   (A single unit is WANTED.)

    2. DATA: 0.400 mol HCl

               0.300 mol $CaCO_3$

    3. Balance: $2\,HCl + 1\,CaCO_3 \rightarrow 1\,CO_2 + 1\,H_2O + 1\,CaCl_2$

    4. Bridge: (Moles WANTED to moles *given* will vary for the two *givens*.)

    5. SOLVE: $? \text{ mol } CO_2 = 0.400 \text{ mol } \textbf{HCl} \cdot \dfrac{1 \text{ mol } CO_2}{2 \text{ mol HCl}} = \textbf{0.200 mol CO}_2$ forms

$$? \text{ mol } CO_2 = 0.300 \text{ mol } \textbf{CaCO}_3 \cdot \frac{1 \text{ mol } CO_2}{1\;CaCO_3} = \textbf{0.300 mol CO}_2 \text{ forms}$$

    The reactant that *most limits* the amount of product is **HCl**

  2b. WANTED: ? mol CO_2 =

    Use the limiting reactant as *given* to calculate the amount of product that can form. HCl is limiting. This calculation is done in the first SOLVE above: **0.200 mol CO$_2$** forms, then the reaction stops.

3a. To find the limiting reactant, choose a product that is WANTED later in the problem. Moles HCl is WANTED. Find how much of this product can be formed by each reactant.

  1. WANTED: ? mol HCl =   (A single unit is WANTED.)

  2. DATA: 14.2 **g** Cl_2 (One single-unit amount in the DATA.)

               0.300 mol H_2 (A second single-unit amount in the DATA.)

               1 mol Cl_2 = 71.0 g Cl_2

  3. Balance: 1 H_2 + 1 Cl_2 → 2 HCl

  4. Bridge: (Moles WANTED to moles *given* will vary for the two *givens*.)

               (*Strategy*: Identify the limiting reactant.)

  5. SOLVE: ? mol HCl = 14.2 g **Cl_2** • $\dfrac{1 \text{ mol } Cl_2}{71.0 \text{ g } Cl_2}$ • $\dfrac{2 \text{ mol HCl}}{1 \text{ mol } Cl_2}$ = **0.400** mol HCl

               ? mol HCl = 0.300 mol **H_2** • $\dfrac{2 \text{ mol HCl}}{1 \text{ mol } H_2}$ = **0.600** mol HCl

The limiting reactant must be **Cl_2**. When **0.400 mol HCl** forms, the reaction stops.

3b. 0.400 mol HCl is formed [see part (a)].

3c. To calculate the amount of a stoichiometric equivalent, use standard stoichiometry.

  1. WANTED: ? mol H_2 =   [From part (a), H_2 is the *reactant* in excess.]

  2. DATA: 14.2 **g** Cl_2 (The *given* is the known amount of the limiting reactant.)

               1 mol Cl_2 = 71.0 g Cl_2

  3. Balance: 1 H_2 + 1 Cl_2 → 2 HCl

  4. Bridge: 1 mol H_2 = 1 mol Cl_2

  5. SOLVE: ? mol H_2 = 14.2 g Cl_2 • $\dfrac{1 \text{ mol } Cl_2}{71.0 \text{ g } Cl_2}$ • $\dfrac{1 \text{ mol } H_2}{1 \text{ mol } Cl_2}$ = **0.200 mol H_2**

You started with 0.300 mol H_2. You use up 0.200 mol H_2. This confirms H_2 is *in excess*: some is left over when the reaction stops.

Practice B   2a. 1. WANTED: ? **g** H_2O =

  2. DATA: 2.00 **g** NaOH

               0.0100 mol H_2SO_4

               1 mol NaOH = 40.0 g NaOH

               1 mol H_2O = 18.0 g H_2O (Two grams prompts.)

  3. Balance: **1** H_2SO_4 + **2** NaOH → **2** H_2O + **1** Na_2SO_4

  4. Bridge: (Moles WANTED to moles *given* will vary for the two *givens*.)

  5. SOLVE: ? g H_2O = 2.00 g **NaOH** • $\dfrac{1 \text{ mol NaOH}}{40.0 \text{ g NaOH}}$ • $\dfrac{\textbf{2} \text{ mol } H_2O}{\textbf{2} \text{ mol NaOH}}$ • $\dfrac{18.0 \text{ g } H_2O}{1 \text{ mol } H_2O}$ = **0.900** g H_2O

               ? g H_2O = 0.0100 mol **H_2SO_4** • $\dfrac{\textbf{2} \text{ mol } H_2O}{\textbf{1} \text{ mol } H_2SO_4}$ • $\dfrac{18.0 \text{ g } H_2O}{1 \text{ mol } H_2O}$ = **0.360** g H_2O

The limiting reactant is **H_2SO_4**

2b. The mass of water formed is based on the limiting reactant: the 0.360 g H_2O calculated in part (a).

2c. 1. WANTED: ? g NaOH =   (NaOH is the reactant in excess.)

    2. DATA: 0.0100 mol H_2SO_4 (The *given* is the amount of the limiting reactant.)

                     1 mol NaOH = 40.0 g NaOH

    (Because the limiting reactant is known, use standard stoichiometry.)

    3. Balance: **1** H_2SO_4 + **2** NaOH → **2** H_2O + **1** Na_2SO_4

    4. Bridge: **1** mol H_2SO_4 = **2** mol NaOH

  5–8. SOLVE: ? g NaOH = 0.0100 mol H_2SO_4 • $\dfrac{\textbf{2 mol NaOH}}{\textbf{1 mol } H_2SO_4}$ • $\dfrac{\text{40.0 g NaOH}}{\text{1 mol NaOH}}$ = **0.800 g NaOH**

Review Quiz

1. 3.75 mol O_2

2. 0.901 mol CO

3. 0.072 g H_2O

4. ? kg Fe = 125 mol Fe_2O_3 • $\dfrac{\textbf{2 mol Fe}}{\textbf{1 mol } Fe_2O_3}$ • $\dfrac{\text{55.8 g Fe}}{\text{1 mol Fe}}$ • $\dfrac{\text{1 kg}}{10^3 \text{g}}$ = **14.0 kg Fe**

5. **8.00 mol H_2O**   (O_2 is limiting)

10

Molarity

Lesson 10.1   Ratio Units

In prior chapters, the WANTED unit in word problems has been a *single*-unit amount such as grams, moles, or molecules. In this chapter, we will learn to solve word problems in which the answer unit is a *ratio* of two units, such as density in *g/mL* or molar mass in *g/mol*.

Conversions were solved for ratio units in Lesson 4.5. To review briefly:

- When solving for a unit that is a ratio, either the top or bottom *given* unit may be converted first. The *order* in which conversions are multiplied does not affect the answer.

- If a unit and label to the right of the equals sign *matches* a unit and label WANTED, in both what it is and where it is (top or bottom), (circle) it and leave it alone.

- If a unit and label *in* or *after* the *given* unit is *not* what you WANT and where you want it, put it where it will cancel, and convert until it matches the unit WANTED.

PRACTICE   **A**

For additional review, see Lesson 4.5.

1. Complete the conversions and SOLVE.

    a. $\dfrac{? \text{ m}}{\text{s}} = 95 \dfrac{\text{km}}{\text{hr}} \cdot \dfrac{\text{m}}{\rule{2cm}{0.4pt}} \cdot \dfrac{\rule{2cm}{0.4pt}}{\text{min}} \cdot \rule{2cm}{0.4pt} =$

    b. $\dfrac{? \text{ g}}{\text{L}} = 7.20 \dfrac{\text{cg}}{\text{mL}} \cdot \dfrac{\rule{2cm}{0.4pt}}{\text{cg}} \cdot \dfrac{\rule{2cm}{0.4pt}}{\text{L}} =$

(continued)

2. Chain legal conversions in any order and SOLVE.

   a. $? \dfrac{\text{inches}}{\text{second}} = 15.0 \dfrac{\text{feet}}{\text{minute}} \cdot$

   b. $? \dfrac{\text{mg}}{\text{L}} = \dfrac{0.025\ \text{g}}{\text{mL}} \cdot$

Rules for WANTED Units

For a word problem, when writing the WANTED answer unit, we need to distinguish whether a single unit or a ratio unit is WANTED. Let us review and add to the list of rules for writing WANTED units.

1. A ratio between quantities can be represented by the word *per* or by a slash mark (/) in a fraction.

    Example: These are different ways to write the same ratio:

    $$\text{grams per mL} \quad and \quad \text{g/mL} \quad and \quad \frac{\text{g}}{\text{mL}} \quad and \quad \text{g} \cdot \text{mL}^{-1}$$

2. In a ratio with two kinds of units, if no number is written between the *per* or slash mark and the unit after it, the ratio is understood to be "units *per* ONE unit."

    Example: "? meters *per* second" means "? meters *per* ONE second."

3. We will call a WANTED answer unit a *ratio unit* if it consists of one kind of unit *per one* of one other kind of unit.

    Example: If a problem asks for "miles/hour," it is asking for the "miles traveled *per one* hour": a *ratio unit*.

4. We will call an answer unit a *single* unit if it has one kind of base unit and no denominator.

    Example: If a problem asks you to find miles or kilograms or dollars, it is asking for a *single* unit.

TRY IT

After each question below, write only the unit that is WANTED, and then label the unit as a *single* unit or a *ratio* unit.

Q1. On a car loan, if you pay $4,250 over 18 months, how much do you pay per year?

Q2. Averaging 24 km/hr, how many minutes would it take to bicycle 5.0 miles?

Answers:

1. The WANTED unit is "dollars per one year" (a ratio unit).

2. The WANTED unit is minutes (a single unit).

5. Powers (exponents) and metric prefixes that are part of a unit do not affect the *count* of the *kinds* of units in a ratio unit or single unit.

   Example: The following are all the same unit:

$$\text{mg per cm}^3 \quad and \quad \text{mg/cm}^3 \quad and \quad \frac{\text{mg}}{\text{cm}^3} \quad and \quad \text{mg} \cdot \text{cm}^{-3}$$

   Each has one *kind* of base unit on top (grams) and one of one kind of base unit on the bottom (meters), so each represents a *ratio* unit. Prefixes and powers are ignored in a count of the kinds of units.

   Example: If a problem asks for a volume in cm^3, it wants a single unit because it is asking for one *kind* of base unit (meters).

6. If a problem asks for a "unit per *more than one* of one other unit," it WANTS a *single* unit.

   Example: If a problem asks for "grams per 100 milliliters," it is asking for a single unit (in this case, grams).

A ratio unit must be a unit *per one* of *one other* unit.

PRACTICE **B**

After each unit, write whether it would be a *single* or a *ratio* WANTED unit.

1. m/s   2. dm^3 3. miles per gallon   4. g/mol

5. m^2 6. dollars per 3 liters   7. $ng \cdot cm^{-3}$

Writing WANTED Units

To SOLVE calculations, we begin by writing "WANTED: ? (answer unit) = ." When writing single and ratio WANTED units, use the following rules.

1. If a ratio unit is WANTED, write the unit as a fraction with a top and a bottom.

   Example: If an answer unit in a problem is miles/hour, to begin write

$$\text{WANTED: ?} \; \frac{\textbf{mi.}}{\textbf{hr}} =$$

   Do *not* write

$$\text{WANTED: ? miles/hour } or \text{ ? mph}$$

   The slash mark (/) is an easy way to *type* ratios and conversion factors, but when solving with conversions, *writing* ratio answer units as a *fraction*, with a clear numerator and denominator, will be a guide in arranging the conversions that SOLVE.

TRY IT

Q. If a baseball is pitched at 85 miles per hour, what is its speed in meters per second?

Write here what you *begin* by writing:

Answer:

$$\text{WANTED:} \quad ? \frac{\mathbf{m}}{\mathbf{s}} = \quad or \ ? \frac{\text{meters}}{\text{second}} =$$

Write a WANTED ratio unit as a fraction with one unit on *top* of the other. During calculations, *abbreviations* of familiar units are preferred.

2. If a calculation WANTS a single unit, write "WANTED: ? (single unit) = "

   Example: If the answer unit is cubic centimeters, write

   $$\text{WANTED:} \quad ? \, \mathbf{cm}^3 =$$

   Single units are numerator units ("on top"). Single units have an understood 1 as a denominator and are written without a denominator.

3. If a problem asks for a "unit per *more than one* other unit," it WANTS a *single* unit.

   Example: If a problem asks for the "miles traveled in 27 hours," it is asking for miles. Write

   $$\text{WANTED:} \, ? \, \mathbf{mi.} =$$

   A ratio unit *must* be one unit per *one* of *one other* unit.

4. Units can have labels attached. When a unit is measuring a chemical substance, units *must* have the formula attached. Labels do not affect whether the *units* of measurement are single or a ratio.

5. Watch for words such as *each* and *every* that mean *per one*. Watch for words that *can* be *translated* to mean *per one*.

To summarize:

When Writing the WANTED Unit

- If a problem WANTS a unit *per one* of *one* other unit, write the WANTED unit as a "top-over-bottom fraction."
- If a single unit is the answer unit, write "WANTED: (single unit) = ."

PRACTICE **C**

In your notebook, for each problem below, write "WANTED: (answer unit) = ." Do not finish the problem.

1. If a car travels 270 miles in 6 hours, what is its average speed?

2. What would be the volume in milliliters of 14.0 grams of silver? The density of silver is 10.5 g/mL.

Flashcards

Make the cards you need. Add a 10.1 (for Lesson 10.1) in the bottom right corner of the front of each card. You will need these flashcards for the next several chapters.

One-Way Cards (with Notch)	Back Side—Answers
If "unit X *per* one of unit Y" is WANTED, write WANTED: _____	? $\dfrac{\text{Unit } X}{\text{Unit } Y} =$
If "unit X *per* more than one of unit Y" is WANTED, write WANTED: _____	? Unit $X =$
If "# unit X per one of unit Y" is DATA, write in DATA: _____	# Unit X = 1 Unit Y
If "unit X/unit Y" is WANTED, write WANTED: _____	? $\dfrac{\text{Unit } X}{\text{Unit } Y} =$
If "# unit X/unit Y" is DATA, write in DATA: _____	# Unit X = 1 Unit Y

PRACTICE   D

In your notebook, for each problem below, write "WANTED: (answer unit) = ." Do *not* do more of the problem. Check your answers as you complete each problem.

1. A sample of copper has 0.281 mol Cu and a volume of 2.0 mL. Find its density in grams per cubic centimeter.

2. The density of iron is $7.87 \text{ g} \cdot \text{mL}^{-1}$. What volume in milliliters is occupied by one Fe atom?

3. The $50 American Eagle gold coin has a mass of 33.9 grams. If the coin is pure gold, how many moles of gold does the coin contain?

4. The density of aluminum is 2.7 g/mL. What would be the mass of 3.5 mL of Al?

5. An atom has a mass of 1.67×10^{-24} grams. What is the molar mass of the atom?

Lesson 10.2   Answer Units That Are Ratios

Recall these rules (from Chapter 5) for listing the DATA in word problems that can be solved with conversions:

1. Write DATA as *single* units *or* as *equalities*.

   In problems that can be solved using conversions, most of the DATA will be two measurements that are equal or are equivalent, written in the DATA as equalities.

2. If you WANT a *single* unit, in most cases *one* amount in the DATA will be "single unit." This single-unit DATA is used as the *given* at the SOLVE step. The rest of the DATA will be equalities that will be conversion factors.

Add this rule:

3. If you WANT a *ratio* unit, *all* of the DATA will be equalities.

PRACTICE **A**

1. Write the DATA section for problem 1 in Practice C of Lesson 10.1.

2. For problems 1, 2, and 5 in Practice D of Lesson 10.1, write the WANTED and DATA sections that you would use to SOLVE the problem. Put each answer at the top of a new page in your notebook (so you have room to SOLVE these later). Each problem includes at least one *prompt* that *must* be written in the DATA.

3. Did all of the DATA sections in your answers for this practice set follow the shaded rule 3 above?

Choosing a *Given* Quantity

At the SOLVE step, you begin by writing what you wrote at the WANTED step. Next you must choose a *given* quantity: the first term in your chain of conversions. A key rule is:

When Choosing a *Given*

- If you WANT a single unit, start with a single unit.
- If you WANT a ratio, start with a ratio.

The "law of dimensional homogeneity" requires that at the SOLVE step:

- If a single-unit amount is alone on the left side of the equal sign, a single-unit amount must result after unit cancellation on the right side.

- If a ratio unit is on the left side of the equal sign, a ratio must start and be the result on the right side.

Otherwise, the two sides cannot be *equal*. This law of dimensions simplifies choosing the *given* quantity in problems that can be solved with conversions.

When solving for a *single* unit, usually only one item of DATA will be a single unit, and by choosing it as *given*, your conversions will be "right-side up."

When solving for a *ratio* unit, it does not matter which DATA ratio you choose as your *given*, but it is possible to arrange your *given* so that it will never cancel to give the WANTED unit because your *given* is algebraically "upside down" in relation to the WANTED unit.

If this happens, it will become apparent as you do your conversions. One solution is to start over with your *given* inverted (flipped over); however, the following methods will help you to pick a *given* ratio that is *right-side up* the *first* time.

Method 1: Match One *Given* Unit with One WANTED Unit

To SOLVE for a ratio unit, one method to choose and arrange the *given* ratio is:

- Pick as your *given* an equality from your DATA that *includes* one of the units that is WANTED.

- Write that unit in your *given* ratio where it needs to be (on the top or bottom) to match the WANTED unit.

By doing this, the other unit of your *given* is the only one that needs converting, and your conversions will be right-side up.

▷ TRY IT

In your notebook, write only the WANTED and DATA (do not write the SOLVE step).

Q. If a car is traveling 45.0 feet per second, what is its speed in miles per hour? (5,280 feet = 1 mile)

Answer:

WANTED: $\dfrac{? \text{ mi.} =}{\text{hr}}$

DATA: 45.0 ft. = 1 s

5,280 ft. = 1 mi.

(When a ratio unit is WANTED, all of the DATA will be in equalities.)

Now, using method 1, after the equal sign below, write your *given ratio,* then stop.

SOLVE: $\dfrac{? \text{ mi.} =}{\text{hr}}$

(STOP)

SOLVE: $\dfrac{? \textbf{ mi.}}{\text{hr}} = \dfrac{\textbf{1 mi.}}{5{,}280 \text{ ft.}}$

(The only WANTED unit in the DATA is *miles,* so choose as a *given* the conversion equality that includes miles, and arrange the conversion so that *miles* is where you WANT it [in this case on top] in the answer.)

Now solve for the WANTED unit.

(STOP)

SOLVE: $\dfrac{? \textbf{ mi.}}{\text{hr}} = \dfrac{1 \text{(mi.)}}{5{,}280 \text{ ft.}} \cdot \dfrac{45.0 \text{ ft.}}{\text{s}} \cdot \dfrac{60 \text{ s}}{1 \text{ min}} \cdot \dfrac{60 \text{ min}}{1 \text{(hr)}} = 30.7 \dfrac{\textbf{mi.}}{\textbf{hr}}$

PRACTICE   B

Use method 1 above to arrange the *given,* and solve all three problems below.

1. In 1988, Florence Griffith-Joyner set a new women's world record in the 200-meter dash with a time of 21.34 seconds. What was her average speed in miles per hour?

   (Use 1.609 km = 1 mi. Assume the distance is 200.0 m, very carefully measured in a certified world-record race.)

2. For an unknown substance, 0.0500 mole has a mass of 9.01 grams.

   a. To find the molar mass, what unit is WANTED?

   b. Calculate the molar mass of the substance.

3. Complete problem 5 in Practice D of Lesson 10.1. The WANTED and DATA sections are in your notebook from Practice A of *this* lesson.

Method 2: Arrange the *Given* on the Basis of Labels

Here is a second way to pick a *given* ratio that is right-side up. In your WANTED and DATA:

- Include complete and descriptive *labels* after each unit—words that describe what the unit is measuring.
- Pick and arrange the *given* to match a *label* in the WANTED unit.

> ▷ **TRY IT**

Q. If a 250 mL solution contains 0.020 lb. of dissolved NaCl, how many moles of NaCl are dissolved per liter of solution? (Use 1 lb. = 454 g and 58.5 g NaCl/mol)

Using the method 2 steps, *start* this problem. Work as far as writing the *given* unit at the SOLVE step. Choose as your *given* a DATA equality that includes the word *solution*. Arrange that equality so that *solution* is in the same place in the *given* fraction (on the top or bottom) as the word *solution* in the WANTED unit. Then stop.

STOP

Answer:

WANTED: $\dfrac{?\ \text{mol NaCl}}{\text{L } solution} =$

DATA: 250 mL NaCl *solution* = 0.020 lb. NaCl   (Two measures of one object.)

1 lb. NaCl = 454 *g NaCl*   (Put formulas after units to see prompts.)

58.5 g NaCl = 1 mol NaCl

(When a ratio unit is WANTED, all of the DATA will be equalities.)

SOLVE: $\dfrac{?\ \text{mol NaCl}}{\text{L } solution} = \dfrac{0.020\ \text{lb. NaCl}}{250\ \text{mL } solution}$

(Only one item of DATA has *solution*. Because solution is on the bottom in the WANTED unit, put it on the bottom in the *given* unit. Finish SOLVING from here.)

STOP

$$\frac{?\ \text{mol NaCl}}{\text{L } solution} = \frac{0.020\ \text{lb. NaCl}}{250\ \text{mL } solution} \cdot \frac{454\ g}{1\ \text{lb.}} \cdot \frac{1\ \text{mol NaCl}}{58.5\ \text{g NaCl}} \cdot \frac{1\ \text{mL}}{10^{-3}\ \text{L}} = \mathbf{\frac{0.62\ mol\ NaCl}{L\ solution}}$$

By using the label *solution* to arrange the *given*, the answer units are right-side up.

To summarize:

> **When a Ratio Unit Is WANTED, Two Methods to Arrange the *Given* Are:**
> - put one *unit* and *label* of a DATA equality where it is in the WANTED unit; or
> - put one *label* in a DATA equality where it is in the WANTED unit.

For most problems, one or both of these methods will arrange your *given* ratio right-side up. You may use any method you prefer. We will learn additional ways to arrange *given* ratios in Chapter 11.

PRACTICE **C**

Use method 2 on problem 1, and then on problem 2 if you need more practice.

1. 10.0 g of NaOH is dissolved to make 1,250 mL of solution. Calculate the concentration of the solution in moles of NaOH per liter of solution (40.0 g NaOH/mol).

2. A water bath absorbs 24 calories of heat from a reaction that forms 0.88 g of carbon dioxide. What is the heat released by the reaction, in kilocalories of heat per mole of CO_2?

Ratios Represented by Negative Powers

When solving with conversions, units in the form $A \cdot B^{-1}$ must be treated as A/B: as a *ratio* unit.

Example: Treat $g \cdot mol^{-1}$ as g/mol.

- If $g \cdot mol^{-1}$ is WANTED, write

$$\text{WANTED} = ? \ \frac{g}{mol}$$

- If DATA includes $18.0 \ g \cdot mol^{-1}$, write

$$\text{DATA:} \ 18.0 \ g = \mathbf{1} \ mol$$

The general rule is $A \cdot B^{-x} = A/(B^x)$.

Example: In conversion calculations, treat $g \cdot cm^{-3}$ as g/cm^3.

Flashcards

Add to your collection any flashcards below that you cannot answer quickly. Practice until you can answer each card correctly, then do the following problems.

One-Way Cards (with Notch)	Back Side—Answers
In a word problem, if a ratio unit is WANTED, all DATA will be in _____	Equalities (two units equal or equivalent in the problem)
For a *given*, if you WANT a single unit, _____	Start with a single unit
For a *given*, if you WANT a ratio, _____	Start with a ratio
When a ratio unit is WANTED, how can you choose and arrange a *given* ratio?	Match a *given* unit and label, or just a label, with its position in the WANTED unit
If "unit $X \cdot Y^{-x}$" is WANTED, write WANTED: _____	$? \ \text{Unit} \ X = \dfrac{}{\text{Unit} \ Y^x}$
If "# unit $X \cdot Y^{-x}$" is DATA, write in DATA: _____	# Unit X = 1 Unit Y^x

PRACTICE D

1. Rewrite these in the slash (/) format.

   a. $0.47 \text{ mg} \cdot \text{mol}^{-1}$ b. $9.8 \text{ m} \cdot \text{s}^{-2}$ c. $7.2 \text{ kg} \cdot \text{dm}^{-3}$

(SOLVE problems 2 and 3 using either method 1 or 2 to arrange the *given* unit.)

2. If a bar of lead with a volume of 24.0 mL contains 1.31 mol Pb, what is its density in $\text{g} \cdot \text{cm}^{-3}$?

3. Complete problem 2 in Practice D of Lesson 10.1. The WANTED and DATA sections are in your notebook from Practice A of this lesson.

Lesson 10.3   Molarity Calculations

Solution Terminology

Many substances dissolve to some extent in other substances. The resulting mixture is termed a **solution**. When a larger amount of one substance in its liquid state has smaller amounts of one other substance dissolved within it, the liquid that dissolves a substance is the **solvent**, and the substance dissolved is the **solute**.

> **Example:** When table salt (NaCl) is mixed with water, it dissolves. The solute NaCl is said to be **soluble** in water, the solvent.

When two liquids are mixed and dissolve in each other, the liquid with the larger original volume is generally considered to be the solvent.

> **Example:** "80 proof vodka" is a mixture with a volume consisting of 40% ethanol (the solute) and 60% water (the solvent).

Any liquid can be a solvent, but water is the solvent most commonly used in chemistry and in everyday applications. When substances are dissolved in water, the result is termed an **aqueous** solution (from the Latin *aqua*, for "water"). In these lessons, if the solvent for a solution is not specified, you should assume the solvent is water.

Preparing Solutions

The composition of a solution is measured by its **concentration**. Concentration can be measured by a variety of units, but in chemistry calculations, unless otherwise specified, you should assume that *concentration* refers to **molar concentration**: the *moles of dissolved solute per liter of solution.*

Calibrated glassware such as **burets, pipettes,** and **volumetric flasks** can accurately measure solution volumes. If a solution has a known concentration of a dissolved substance, by measuring the volume of a sample we can convert to a *count* of the solute particles in the sample. Counting particles is the key to understanding chemical processes and solving chemistry calculations.

To make an aqueous solution of known molarity, a substance is weighed, placed in a volumetric flask, and completely dissolved in distilled water (water with minerals and impurities removed). The quantity of water is then increased, with mixing, until a precisely marked volume on the neck of the flask is reached. After thorough mixing, the dissolved particles are evenly distributed, and each sample will have the same *concentration* of the solute.

In preparing a molar solution, the amount of water *added* is not measured and is not used to calculate the solution concentration. What *is* measured carefully and used in calculations is the volume of the *mixture* of the dissolved solute and water: the volume of the *solution*.

Expressing Molarity

The concentration of a solution, measured in *moles per liter*, is termed the **molarity** of the solution. Moles per liter can be written as *mol/L* or *mol • L⁻¹* or can be abbreviated by a capital **M**.

A solution labeled "0.50 M HCl" is referred to as a "0.50 **molar** HCl solution," and these terms mean that the solution (abbreviated soln.) contains 0.50 mole of dissolved HCl per liter of solution volume.

Brackets [ ] are used as shorthand for the molar concentration of a substance: [NaCl] is read as "the concentration of NaCl" or as "the molarity of the NaCl."

Flashcards

Cover one column, then the other, and write a check mark if you can quickly recall the vocabulary term or its definition below. For any rows without check marks on both sides, make and practice the card. In science courses, these are terms and meanings that you will need to recall automatically.

Two-Way Cards (*without* Notch)

Solution (definition)	A liquid with one or more dissolved substances
Aqueous solution (definition)	Water that contains one or more dissolved substances
Examples of glassware used to measure solution volumes	Pipettes, burets, volumetric flasks
Solvent (definition)	A liquid that can dissolve substances
Solute (definition)	A substance dissolved in a solvent
Molar concentration (definition)	Moles of solute per 1 L of solution

Rules for Concentration Calculations

Solution concentration is a *ratio* unit: moles of dissolved substance *per* one liter of solution. Molarity calculations can be solved with conversions that apply these rules:

1. When reading for the WANTED unit and DATA, if you see
   * **concentration** *or* **brackets [ ]** *or*
   * a capital **M** *or* **molar** *or* **molarity** *or*
   * **mol/L** *or* **mol • L⁻¹**

treat all of these terms as meaning "**moles solute** *per* **1 liter of solution.**"

The abbreviation **M** or word ***molar*** *cannot* be used in *most* WANTED or DATA or conversion terms. Concentration has ratio units, and ratio units must be written as a *fraction* when solving with conversions.

2. If mol/L is WANTED, write the WANTED unit as a "top-over-bottom" fraction.

> **Example:** If a problem WANTS the "*concentration* of NaCl" or "*molarity* of NaCl" or [NaCl], write
>
> WANTED: **?** $\dfrac{\text{mol NaCl}}{\text{L NaCl soln.}} =$ (Concentration is a *ratio* unit.)

▶ TRY IT

In your notebook, apply the rules above and WANTED, DATA, SOLVE.

Q. If 2.98 g KCl is dissolved to make 250. mL of solution, find the [KCl].

Answer:

WANTED: **?** $\dfrac{\text{mol KCl}}{\text{L KCl soln.}} =$ (Rule 2.)

DATA: 2.98 *g KCl* = 250. mL KCl soln. (Two measures of the *same* solution.)

74.6 g KCl = 1 mol KCl (Grams prompt in DATA.)

(SOLVING for a *ratio* unit, all of the DATA will be equalities.)

SOLVE: (Want a ratio? Start with a ratio.)

? $\dfrac{\text{mol KCl}}{\text{L KCl soln.}} = \dfrac{2.98 \text{ g KCl}}{250. \text{ mL KCl soln.}} \cdot \dfrac{1 \text{ mol KCl}}{74.6 \text{ g KCl}} \cdot \dfrac{1 \text{ mL}}{10^{-3} \text{ L}} = 0.160 \dfrac{\text{mol KCl}}{\text{L KCl soln.}}$

Those conversions started with a *given* that placed *solution* on the bottom, but your conversions, if same-side up, may be in *any* order.

---

3. If you see *mol/L* or any of its equivalent abbreviations in the DATA, write "DATA: # mol (formula) = 1 L (formula) soln."

> **Example:** If you see *0.25 M NaOH* or *0.25 molar NaOH* or *0.25 mol/L NaOH*, write:
>
> DATA: 0.25 mol NaOH = 1 L NaOH soln.

▶ TRY IT

Apply rule 3 to this problem.

Q. How many grams of NaOH are required to make 150. mL of a 0.300 M NaOH solution?

Answer:

WANTED: ? g NaOH =   (You want a single unit.)

DATA: 150. mL NaOH soln.   (Single-unit DATA.)

0.300 mol NaOH = 1 L NaOH soln.   (Rule 3 above.)

40.0 g NaOH = 1 mol NaOH   (See WANTED unit.)

(If needed, adjust your work and finish.)

SOLVE: (Want a single unit? Start with a single unit.)

$$? \text{ g NaOH} = 150. \text{ mL NaOH soln.} \cdot \frac{10^{-3}L}{1 \; mL} \cdot \frac{0.300 \text{ mol NaOH}}{1 \text{ L NaOH soln.}} \cdot \frac{40.0 \text{ g NaOH}}{1 \text{ mol NaOH}}$$

$$= \textbf{1.80 g NaOH}$$

Let's summarize these rules for aqueous solution calculations as follows.

The M Prompt

- Treat mol/L = M = [ ] = molar = mol $\cdot$ L^{-1} as **mol solute** *per* **1 L soln.**
- If mol/L is WANTED, write

    WANTED: ? $\dfrac{\text{mol (formula)}}{\text{L (formula) soln.}}$ =

- If mol/L is DATA, write

    DATA: # mol (formula) = 1 L (formula) soln.

In addition to the M prompt, use these fundamental rules at the SOLVE step:

- If you WANT a single unit, *start* with a single unit as your *given.*
- If you WANT a ratio, start with a ratio: a top over bottom *given.*

Learning the M Prompt

The rules for solution calculations will be needed for the remainder of this course and for work in many scientific careers. After the rules can be recalled from memory, working problems will promote both long-term retention of their content and an intuitive sense of when to apply the rules. Try these steps.

1. In your notebook, summarize the "M Prompt" in your own words. Practice until you can write your M prompt from memory.

2. Use the flashcards below to practice recalling the *parts* of the M Prompt.

3. Work the problems in Practice A to apply the prompt.

Flashcards

One-Way Cards (with Notch)	Back Side—Answers
Treat M = concentration = [ ] = molar = molarity = mol/L = mol • L^{-1} as _____	Moles of solute *per* 1 liter of solution
If [X] or "molarity of X" is WANTED, write WANTED: _____	$\dfrac{?\ \text{mol } X}{\text{L } X \text{ soln.}} =$
If "0.50 M X" is in a problem, write _____	DATA: 0.50 mol X = 1 L X *soln.*
If "0.50 mol • L^{-1} X" is DATA, write _____	DATA: 0.50 mol X = 1 L X *soln.*
See "[X] = 0.50 mol/L"? Write _____	DATA: 0.50 mol X = 1 L X *soln.*
If you get stuck on a complex problem, _____	Add detail to WANTED and DATA labels

PRACTICE A

Check your answers after each problem.

1. Write your version of the M prompt from memory.

2. Find the moles of solute in 100. mL of 0.40 M KBr.

3. Name the solute and solvent in problem 2.

4. What is the molarity of an NaCl solution that has 2.34 g NaCl dissolved in 400. mL of solution volume?

5. How many moles of dissolved potassium hydroxide are present in 0.20 L of 0.60 mol • L^{-1} KOH?

Labeling Solution Conversions

In the problems above, ratios were given a *complete* label: number, unit, and substance. Because we rely on units and labels to SOLVE problems, complete labels are important.

Often, however, parts of labels are omitted as "understood." For example, in a problem about an NaOH solution, all of the measurements are about *one* substance formula: NaOH. In such cases, "NaOH solution" should be indicated once in the problem, but after that, volume units may be "mL soln." or just "mL" with the NaOH and "solution" after volume units omitted as understood.

This shortcut needs to be used carefully. When we encounter problems that involve *two* substances, or volumes of *both* a gas and a liquid, we will need labels that clearly identify what each unit is measuring.

When problems are complex, write complete labels.

PRACTICE   B

Problem 4 must be done and checked; it has important information discussed in its answer.

1. How many grams of HCl are needed to prepare 500. mL of 0.200 M HCl?

2. If 0.050 mol NaBr is dissolved to make 250 mL of solution, find the [NaBr].

3. How many milliliters of 0.100 molar KCl contain 4.48 g of dissolved KCl?

4. At typical room temperatures, liquid water has a density of 1.00 g/mL. What is the concentration of this liquid water in moles per liter?

SUMMARY

Writing WANTED, DATA, prompts, and labels takes time, but it is likely to improve your success at problem solving. As one example, writing the WANTED unit first

- prompts the writing of conversions that may not be specified in the problem but will be needed to SOLVE;
- indicates when to stop conversions and "do the math";
- specifies how the DATA should look (such as: if a ratio unit is WANTED, DATA is all equalities);
- helps in choosing the *given* at the SOLVE step (such as: Want a ratio? Start with a ratio); and
- suggests ways to arrange the *given* so that conversions are right-side up.

By using a methodical approach, you are more likely to know what to do at each step for a wide variety of calculations in the sciences.

REVIEW QUIZ

You may use a calculator and a periodic table.

1. A sample of the metal uranium has a volume of 47.0 mL and a density of 18.95 g/mL. How many moles of uranium are in the sample?

2. If 3.65 g of HCl is dissolved to make 250. mL of solution, what is the [HCl]?

3. How many moles of HNO_3 are present in 100. mL of 0.35 M HNO_3 solution?

4. How many grams of KOH are required to make 500. mL of 0.150 M KOH solution?

5. If 150 mg NaOH is dissolved to form 50.0 mL of NaOH solution, find the [NaOH].

6. If 0.10 grams of water require 6.0 hr to evaporate, how many molecules evaporate per minute?

ANSWERS

Lesson 10.1

Practice A   1a. $? \dfrac{m}{s} = 95 \dfrac{km}{hr} \cdot \dfrac{10^3 \,(m)}{1\ km} \cdot \dfrac{1\ hr}{60\ min} \cdot \dfrac{1\ min}{60\,(s)} = \mathbf{26 \dfrac{m}{s}}$

(In the *given*, because kilometers is not the unit WANTED on top, it is put where it cancels and converted to the unit WANTED on top. Next, as hours is on the bottom, but seconds is WANTED, hours is converted to the seconds WANTED on the bottom.)

1b. $? \dfrac{g}{L} = 7.20 \dfrac{cg}{mL} \cdot \dfrac{10^{-2}(g)}{1\ cg} \cdot \dfrac{1\ mL}{10^{-3}(L)} = 7.20 \times 10^{-2+3} = 7.20 \times 10^1 = \mathbf{72.0 \dfrac{g}{L}}$

In problem 1b, your conversions may be different (such as 1 g = 100 cg) as long as you arrive at the same answer.

2a. Your conversions may be in a different order.

$? \dfrac{inches}{second} = 15.0 \dfrac{feet}{minute} \cdot \dfrac{12\,(inches)}{1\ foot} \cdot \dfrac{1\ minute}{60\,(seconds)} = \mathbf{3.00 \dfrac{inches}{second}}$

2b. $? \dfrac{mg}{L} = 0.025 \dfrac{g}{mL} \cdot \dfrac{1\,(mg)}{10^{-3}\ g} \cdot \dfrac{1\ mL}{10^{-3}(L)} = \mathbf{2.5 \times 10^4 \dfrac{mg}{L}}$

Practice B   1. m/s **ratio**   2. dm^3 **single**   3. miles per gallon **ratio**   4. g/mol **ratio**

5. m^2 **single**   6. dollars per 3 liters **single**   7. $ng \cdot cm^{-3}$ **ratio**

Practice C   1. WANTED: $? \dfrac{mi.}{hr} =$ (You must write a "top-over-bottom fraction.")

Any speed unit could be chosen, but for this DATA, miles per hour is easy.

2. WANTED: ? mL Ag =

If the substance is known, add the formula to every unit.

Practice D   1. WANTED: $? \dfrac{g\ Cu}{cm^3\ Cu} =$

2. WANTED: $? \dfrac{mL\ Fe}{atom\ Fe} =$

The question can be read as asking for "mL Fe *per* individual Fe atom." (See rule 5.)

3. WANTED: $? \dfrac{mol\ Au}{coin} =$

Moles is the unit used to count large numbers of particles.

4. WANTED: ? g Al =

The mass is WANTED. Milligrams or kilograms could be chosen as WANTED, but given the density in g/mL, grams is the easiest mass unit to find.

5. Hint: What are the *units* of molar mass?

WANTED: $? \dfrac{g}{mol} =$ (In this problem, we don't know the atom symbol.)

Lesson 10.2

Practice A   1. DATA: 270 mi. = 6 hr

2. From Lesson 10.1, Practice D:

   1. WANTED: $? \dfrac{\text{g Cu}}{\text{cm}^3 \text{ Cu}} =$

       DATA: 0.281 mol Cu = 2.0 mL Cu   (Two measurements of the same sample.)

           63.5 g Cu = 1 mol Cu   (The WANTED unit has g Cu)

           1 mL = 1 cm^3 (Including metric definitions is optional.)

   2. WANTED: $? \dfrac{\text{mL Fe}}{\text{atom Fe}} =$

       DATA: 7.87 g Fe = 1 mL Fe   (Two measurements of same sample.)

           55.8 g Fe = 1.0 mol Fe   (WANTED unit included g Fe, a grams prompt.)

           1 mol Fe = 6.02 × 10^{23} atoms Fe   (Units include invisible atoms, visible grams = Avogadro prompt.)

   5. WANTED: $? \dfrac{\text{g}}{\text{mol}} =$

       DATA: 1 atom = 1.67 × 10^{-24} g   (Two measures of same sample.)

           1 mol atoms = 6.02 × 10^{23} atoms   (10^{-xx} = Avogadro prompt.)

3. All WANTED are ratios; all DATA are equalities.

Practice B   In all three problems below, a *given* was chosen that matched the *top* unit WANTED.

   1. WANTED: $? \dfrac{\text{mi.}}{\text{hr}}$ (Write WANTED *X per Y* units as a *ratio*.)

       DATA: 200.0 m = 21.34 s   (Equivalent: two measures of one event. See Lesson 5.2.)

           1.609 km = 1 mi.

(*Strategy*: If you start your *given* by putting the *miles* DATA on top, all that remains is to convert *kilometers* on the bottom to *hours*. The **arrows** above show a path through the DATA from kilometers to hours, giving the needed conversions in order. Use of arrows may be a helpful technique.)

       SOLVE: (Your conversions may be in a different order but should all be right-side up compared to these. Want a ratio? Start with a ratio.)

$$? \frac{\text{mi.}}{\text{hr}} = \frac{1\ \text{mi.}}{1.609\ \text{km}} \cdot \frac{1\ \text{km}}{10^3\ \text{m}} \cdot \frac{200.0\ \text{m}}{21.34\ \text{s}} \cdot \frac{60\ \text{s}}{1\ \text{min}} \cdot \frac{60\ \text{min}}{1\ \text{hr}} = \mathbf{20.97\ \frac{mi.}{hr}}$$

   S.F.: 1 and definitions are exact and have infinite *s.f.*

   2. WANTED: $? \dfrac{\text{g}}{\text{mol}} =$ (The unit of molar mass is g/mol.)

       DATA: 0.0500 mol = 9.01 g

       SOLVE: (You WANT the ratio g/mol. Your DATA is a ratio of grams and moles. The fundamental rule is: *Let the units tell you what to do.*)

$$? \frac{\text{g}}{\text{mol}} = \frac{9.01\ \text{g}}{0.0500\ \text{mol}} = \mathbf{180.\ \frac{g}{mol}}$$ (When the units on the two sides match, stop conversions and do the math.)

   3. SOLVE: $? \dfrac{\text{g}}{\text{mol}} = \dfrac{1.67 \times 10^{-24}\ \text{g}}{1\ \text{atom}} \cdot \dfrac{6.02 \times 10^{23}\ \text{atoms}}{1\ \text{mol}} = \mathbf{1.01\ \dfrac{g}{mol}}$

Practice C   In each Practice C problem, DATA *labels* are used in to arrange the *given* term.

1. WANTED: $\dfrac{?\ \text{mol NaOH}}{\text{L of solution}} =$ (Write WANTED *X per Y* units as a *ratio*.)

   DATA: 10.0 g NaOH = 1,250 mL soln.   (Equivalent: two measures of the same soln.)

   40.0 g NaOH = 1 mol NaOH

   SOLVE: (Below, "solution" is placed to match the WANTED unit, but your conversions may be in any order.)

   $\dfrac{?\ \text{mol NaOH}}{\text{L soln.}} = \dfrac{10.0\ \text{g NaOH}}{1,250\ \text{mL soln.}} \cdot \dfrac{1\ \text{mL}}{10^{-3}\ \text{L}} \cdot \dfrac{1\ \text{mol NaOH}}{40.0\ \text{g NaOH}} = \textbf{0.200 mol NaOH}\over \textbf{L soln.}$

   S.F.: All of the nonexact numbers have 3 *s.f.*; the answer is rounded to 3 *s.f.*

   Note that the second conversion is true for all substances and processes. The first and last conversions need a substance formula after the unit because that ratio of numbers and units is not always true.

2. WANTED: $\dfrac{?\ \text{kilocalories heat}}{\text{mol CO}_2} =$ (Write WANTED *X per Y* units as a *ratio*.)

   DATA: 24 calories heat = 0.88 *g* CO$_2$ (Equivalent: see Lesson 5.2.)

   44.0 *g* CO$_2$ = 1 mol CO$_2$ (Grams prompt.)

   SOLVE: (Pick as your *given* any DATA equality that puts a heat term on top or a CO$_2$ term on the bottom. Your conversions may be in a different order but must all be right-side up compared to these.)

   $\dfrac{?\ \text{kilocalories heat}}{\text{mol CO}_2} = \dfrac{24\ \text{calories heat}}{0.88\ \text{g CO}_2} \cdot \dfrac{1\ \text{kilocalorie}}{10^3\ \text{calories}} \cdot \dfrac{44.0\ \text{g CO}_2}{1\ \text{mol CO}_2} = \dfrac{\textbf{1.2 kilocalories heat}}{\textbf{mol CO}_2}$

Practice D   1a. 0.47 mg/mol   1b. 9.8 m/s^2 1c. 7.2 kg/dm^3

2. WANTED: $\dfrac{?\ \text{g Pb}}{\text{cm}^3\ \text{Pb}} =$

   DATA: 1.31 mol Pb = 24.0 mL Pb   (Two measurements of the same sample.)

   207.2 g Pb = 1 mol Pb   (WANTED unit included g Pb)

   1 mL = 1 cm^3 (Including this metric definition is *optional.*)

   SOLVE: $\dfrac{?\ \text{g Pb}}{\text{cm}^3\ \text{Pb}} = \dfrac{207.2\ \text{g Pb}}{1\ \text{mol Pb}} \cdot \dfrac{1.31\ \text{mol Pb}}{24.0\ \text{mL Pb}} \cdot \dfrac{1\ \text{mL}}{1\ \text{cm}^3} = \dfrac{\textbf{11.3 g Pb}}{\textbf{cm}^3\ \textbf{Pb}}$

3. For WANTED and DATA, see problem 2 in Practice A of this lesson.

   SOLVE: (Want a ratio? Start with a ratio. Your conversions may be in any order.)

   $\dfrac{?\ \text{mL Fe}}{\text{atom Fe}} = \dfrac{1\ \text{mL Fe}}{7.87\ \text{g Fe}} \cdot \dfrac{55.8\ \text{g Fe}}{1\ \text{mol Fe}} \cdot \dfrac{1\ \text{mol atoms}}{6.02 \times 10^{23}\ \text{atoms}} = \textbf{1.18} \times \textbf{10}^{-23}\ \dfrac{\textbf{mL Fe}}{\textbf{atom Fe}}$

Lesson 10.3

Practice A   Your answers should look like these, but with the (comments) omitted.

2. WANTED: ? mol KBr =

   DATA: 100. mL of KBr solution

   0.40 mol KBr = 1 L KBr solution   (M prompt.)

   SOLVE: (WANT a single unit? Start with the single unit in your DATA.)

   $?\ \textbf{mol KBr} = 100.\ \text{mL KBr soln.} \cdot \dfrac{10^{-3}\ \text{L}}{1\ \text{mL}} \cdot \dfrac{0.40\ \textbf{mol KBr}}{1\ \text{L KBr soln.}} = \textbf{0.040 mol KBr}$

3. The solute is KBr, and if the solvent is not specified, assume it is water.

4. WANTED: $\underline{? \text{ mol NaCl} =}$ (M prompt)
$\qquad\qquad\;\;$ L soln.

DATA: 2.34 *g NaCl* = 400. mL soln.   (Two measures of the *same* solution.)

$\qquad\qquad$ 58.5 g NaCl = 1 mol NaCl

(If you want a ratio unit, all of the DATA will be in equalities.)

SOLVE: (You want a ratio. Start with a ratio. Your conversions may be in any order if they are the "same side up" as these.)

$$\frac{?\text{ mol NaCl}}{\text{L soln.}} = \frac{2.34\text{ g NaCl}}{400.\text{ mL soln.}} \cdot \frac{1\text{ mol NaCl}}{58.5\text{ g NaCl}} \cdot \frac{1\text{ mL}}{10^{-3}\text{ L}} = \mathbf{0.100\ \frac{mol\ NaCl}{L\ soln.}}$$

5. WANTED: ? mol KOH =

DATA: 0.20 L KOH soln.

$\qquad\qquad$ 0.60 mol KOH = 1 L KOH soln.   (mol $\cdot$ L^{-1} = mol/L = M prompt)

SOLVE: (WANT a single unit? Start with a single unit.)

$$? \text{ mol KOH} = 0.20\text{ L KOH soln.} \cdot \frac{0.60\text{ mol KOH}}{1\text{ L KOH soln.}} = \mathbf{0.12\ mol\ KOH}$$

Practice B   1. WANTED: ? *g* HCl =

DATA: 500. mL HCl solution

$\qquad\qquad$ 0.200 mol HCl = 1 L HCl solution   (M prompt.)

$\qquad\qquad$ 36.5 g HCl = 1 mol HCl   (Grams prompt in WANTED unit.)

SOLVE: (WANT a single unit? Start with a single unit.)

$$? \text{ g HCl} = 500.\textbf{ mL HCl} \cdot \frac{10^{-3}\text{ L}}{1\textbf{ mL}} \cdot \frac{0.200\text{ mol HCl}}{1\text{ L}} \cdot \frac{36.5\text{ g HCl}}{1\text{ mol HCl}} = \mathbf{3.65\ g\ HCl}$$

("Soln." and "HCl soln." are omitted from some volume units as understood.)

2. Showing the SOLVE step only:

SOLVE: $\dfrac{?\text{ mol NaBr}}{\text{L}} = \dfrac{0.050\text{ mol NaBr}}{250\text{ mL}} \cdot \dfrac{1000\text{ mL}}{1\text{ L}} = \mathbf{0.20\ \dfrac{mol\ NaBr}{L}}$

3. WANTED: ? mL KCl soln. =

DATA: 4.48 **g** KCl (Because you WANT a single unit, start with this as your *given*.)

$\qquad\qquad$ 0.100 moles KCl = 1 liter of solution   (M prompt.)

$\qquad\qquad$ 74.6 g KCl = 1 mol KCl   (Grams prompt.)

SOLVE: (The arrows trace the path from the *given* to the WANTED unit.)

$$? \text{ mL KCl soln.} = 4.48\text{ g KCl} \cdot \frac{1\text{ mol KCl}}{74.6\text{ g KCl}} \cdot \frac{1\text{ L soln.}}{0.100\text{ mol KCl}} \cdot \frac{1\text{ mL}}{10^{-3}\text{ L}} = \mathbf{601\ mL\ KCl}$$

4. WANTED: $\dfrac{?\ \text{mol}\ H_2O}{L\ H_2O} =$

DATA: 1.00 g liquid H_2O = 1 mL liquid H_2O

18.0 g H_2O = 1 mol H_2O (Grams prompt.)

SOLVE: $\dfrac{?\ \text{mol}\ H_2O}{L\ H_2O} = \dfrac{1\ \text{mol}\ H_2O}{18.0\ \text{g}\ H_2O} \cdot \dfrac{1.00\ \text{g}\ H_2O(\ell)}{1\ \text{mL}\ H_2O(\ell)} \cdot \dfrac{1\ \text{mL}}{10^{-3}\ \text{L}} = \mathbf{55.6\ mol\ H_2O}\ \mathbf{L\ H_2O}$

As calculated in this answer, pure liquid water has a concentration of *about* **55** moles per liter. For substances dissolved in water, even for those that are very soluble in water, the highest-concentration solutions usually have a limit of about **20** moles of solute per liter.

If you calculate a concentration for an aqueous solution that is higher than 20 mol/L, check your work.

Review Quiz

1. WANTED: ? mol U =   (Single unit WANTED.)

DATA: 47.0 mL U   (Single unit *given.*)

18.95 g U = 1 mL U

238.0 g U = 1 mol U   (The density included g U.)

SOLVE: ? mol U = 47.0 mL U $\cdot$ $\dfrac{18.95\ \text{g}\ U}{1\ \text{mL}\ U} \cdot \dfrac{1\ \text{mol}\ U}{238.0\ \text{g}\ U}$ = **3.74 mol U**

2. [HCl] = 0.400 M   3. 0.035 mol HNO_3 4. 4.21 g KOH   5. 0.075 M NaOH

6. 9.3×10^{18} molecules/min (Use grams and Avogadro prompts.)

11

Dimensions

Lesson 11.1   Units and Dimensions

To this point, we have focused on calculations that ask one question about one set of DATA. We also need to be able to solve calculations that ask *several* questions about one set of DATA. To do so, let's take a closer look at quantities, measurements, and units.

The Fundamental Quantities

The quantities of the universe that can be *measured* are termed the **physical quantities**, and measurements record the **dimensions** of the quantities. The physical quantities can be divided into two types: fundamental and derived.

All measurements are based on the **fundamental quantities**. In chemistry, the fundamental quantities studied are distance, mass, time, electrical charge, temperature, and the count of entities. An additional fundamental quantity, luminosity, is studied in physics.

A measurement must have a **magnitude** (a number) and a *unit*. All measurement systems begin with a definition for a *base unit* that measures each fundamental quantity.

Each *fundamental* quantity is measured by a *single base unit* to the *first* power.

The quantities studied in chemistry and their base units are given below.

Fundamental Quantities	Base Unit Used in Chemistry
distance	**meter**
mass	**gram**
time	**second**
count of particles	**mole**
temperature	kelvin
electric charge	coulomb

In many types of problems, it is essential to label each unit with the dimension it is measuring. A key rule is:

> To label a unit with its dimension, look at the unit *and* its exponent, but ignore any prefix.

If a prefix is added to a base unit, the unit is no longer termed a "base" unit, but the same fundamental quantity is being measured.

> **Example:** The fundamental quantity *distance* is measured by the base unit *meter*. Millimeter, kilometer, and centimeter are not base units, but they all measure *distance* because if their prefixes are ignored, their units are *meter* to the *first* power.

Prefixes can be ignored because they represent exponential terms (such as *kilo-* = $\times 10^3$) that are part of the magnitude of a measurement rather than its unit.

PRACTICE   A

1. At this point in the course, you need to know the names of the four quantities in **bold** in the table of fundamental quantities *and* which base unit is used to measure that quantity in chemistry. To learn these, draw this table in your notebook:

Fundamental Quantity	Base Unit Used in Chemistry

Practice until from memory you can fill in the blanks with the information in **bold** in the table of fundamental quantities.

2. Which of these is *not* a measure of a fundamental quantity: kilograms, cubic centimeters, or millimeters?

3. Write the fundamental quantity measured by each of these units.

    a. kilograms   b. nanoseconds

    c. millimoles   d. decimeters

Derived Quantities

Derived quantities measure physical quantities having definitions that *either* include more than one fundamental quantity *or* consist of a fundamental quantity with a power *not* equal to one.

> **Example:** The derived quantity *volume* is *defined* in terms of the fundamental quantity distance, cubed. All geometric formulas for volume are based on a *distance* multiplied three times.
>
> $$\text{Volume of a cube} = \textbf{side} \times \textbf{side} \times \textbf{side} = (\textbf{side})^3$$
>
> $$\text{Volume of a cylinder} = \boldsymbol{\pi} \times \textbf{radius} \times \textbf{radius} \times \textbf{height} = \boldsymbol{\pi}\textbf{r}^2\textbf{h}$$
>
> **Example:** The derived quantity *density* is defined as the fundamental quantity mass *divided* by the derived quantity volume.

A derived quantity is measured in a unit that is based on the units that measure the quantities in its definition.

> **Example:** Volume can be measured in cubic meters (m^3), cubic centimeters (cm^3), and cubic decimeters (dm^3). Ignoring the prefixes, all three of those units are the distance base unit meters, cubed.
>
> **Example:** Because *density* is defined as *mass/volume*, the units used to measure density must be mass *units* over volume *units*, such as g/cm^3 and kg/m^3.

The table below lists some of the derived quantities that are frequently measured in chemistry.

Derived Quantity	Dimension	SI Base Units	Units Often Used in Chemistry
volume	**distance cubed**	meter^3	$\textbf{cm}^3$ **(= mL)**
			$\textbf{dm}^3$ **(= L)**
speed (velocity)	**distance over time**	meters/second	**m/s**
density	**mass over volume**	kg/m^3	$\textbf{g/cm}^3$ **(= g/mL)**
concentration	**particles per volume**	mol/m^3	**mol/L**
molar mass	**mass per particle count**	kg/mol	**g/mol**
area	distance squared	meter^2	m^2, cm^2
acceleration	velocity over time	m/s^2	m/s^2
pressure	(mass × accel.)/area	kg/(m • s^2)	pascal, mm Hg, atm
frequency	1/time	1/s	1/s (= hertz)
energy	mass × acceleration × distance	(kg • m^2)/s^2	joules, calories

The SI units of the "official" metric system differ in some cases from the units generally used in chemistry. In addition, when measuring derived quantities, some combinations of base units are frequently abbreviated with a single word.

Examples

- The volume unit *cubic decimeter* (dm^3) is called a liter (L).

- The volume unit *cubic centimeter* (cm^3) is called a milliliter (mL).

- The unit of molar concentration *moles per cubic decimeter* (mol/dm^3) is abbreviated as *mol/L* or as *molar*.

- *Hour* is equivalent to *3,600 seconds*.

- *Joule* (which measures energy) is an equivalent for *(kg • m^2)/s^2*

PRACTICE   B

1. Draw the table below in your notebook, then practice until you can draw and fill in the five quantities that are in **bold** in the table of derived quantities, their definitions, and one to two units that measure them.

Derived Quantity	Dimension	Units Often Used in Chemistry

Note that if you know the definition, the units column is easy: substitute the base unit that measures each fundamental quantity in the definition. Similarly, the unit of a derived quantity, expressed in base units, indicates what the definition must be.

2. Write a metric base unit used in chemistry to measure the following.

    a. time   b. mass

    c. distance   d. volume

3. Which *unit* measures the molar concentration of solutions?

4. Label each of these units with the quantity it is measuring. Choose from distance, volume, mass, density, time, speed, solution concentration, or molar mass.

    a. liters   b. cubic meters

    c. kilograms   d. decimeters

    e. grams/mole   f. kg/L

5. Which units in problem 4 are measures of a fundamental quantity?

Intensive and Extensive Quantities

Physical quantities can be classified as **intensive** and **extensive**.

An **extensive** physical quantity has a value that is proportional to the *size* of the system being measured. For extensive quantities, what is measured is an *amount*. Examples of extensive quantities include distance, mass, area, time, volume, and energy.

A unit that contains a *single* metric base unit (with or without a prefix) that has a positive power of 1, 2, or 3 always measures an *amount* and therefore measures an extensive quantity. Other quantities that are extensive are measured by complex units.

An **intensive** quantity has a value that does *not* depend on the amount of substance being measured. Pressure, density, molar mass, temperature, and concentration are examples of intensive quantities. When measuring those quantities, the amount of substance being measured does not affect the value (magnitude) of the measurement.

A quantity that is a ratio of two extensive quantities is an intensive quantity, but other types of derived quantities can also be intensive.

Dimensional Analysis

The fundamental and derived quantities together are termed *dimensions*. The technique of using dimensions and their units to solve problems is called **dimensional analysis**. Use of the WANTED unit or dimension as a guide to arranging the given unit to solve a problem is an application of the law of *dimensional homogeneity*. The analysis of units and their dimensions can be a powerful tool for solving science problems.

Flashcards

Add cards that you cannot quickly answer to your "daily" collection. Practice until you can do each correctly, then try the problems in Practice C. Run your cards as described for two more days, then put them in your "quiz and test prep" stack.

One-Way Cards (with Notch)	Back Side—Answers
Name six fundamental quantities	Distance, mass, time, temperature, particles, charge
What are the two parts of a measurement?	A magnitude and a unit
What quantity is based on distance cubed?	Volume
Define an extensive quantity	A quantity that measures an amount
Define an intensive quantity	A quantity independent of amount
Name three intensive quantities	Concentration, pressure, density, temperature, or molar mass
Which quantity is measured by distance/time?	Speed or velocity
Which quantity is mass per unit of volume?	Density

PRACTICE C

1. From memory, fill in the blanks in the table.

Derived Quantity	Dimension	Units Often Used in Chemistry
	mass over volume	
		liters
concentration		
		meters/second
	mass per particle count	

2. Label each of the following with the derived quantity the unit is measuring.

   a. gallons

   b. nanoseconds

   c. g/mL

   d. feet/second

   e. mm

   f. dL

   g. moles dissolved per liter

   h. km/hr

3. Which quantities in problem 1 are extensive?

4. Which units in problem 2 are measuring intensive quantities?

5. Which units in problem 2 are measuring fundamental quantities?

Lesson 11.2   Using Dimensions to Arrange Conversions

To solve with conversions for a ratio unit, in Chapter 10 we discussed two ways to pick a *given* fraction that is right-side up: starting with a *given* fraction that has either a *unit* or a *label* where you WANT it in the answer.

A third way to pick a *given* ratio that is right-side up is to arrange the *given dimension* to match the WANTED *dimension*.

> **Example:** If mg/cm^3 is WANTED and "1 L = 24 g" is in the DATA, because you WANT a *mass* unit *over* a *volume* unit, you can use as your *given* the mass-to-volume equality in the DATA, with the mass placed on top.

$$\text{SOLVE: ?}\ \frac{\text{mg}}{\text{cm}^3} = \frac{24\,\text{g}}{1\,\text{L}} \cdot$$

> This *given* will arrange your conversions right-side up.

By the law of dimensional homogeneity, each dimension in the *given* must either be the *same* as the "top and bottom" dimensions WANTED or a dimension that can be *converted* to the dimension WANTED.

━━━━━▶ **TRY IT**

Q. In 0.500 second, if a baseball travels 60.5 feet from the pitcher's mound to home plate, what is its average speed in miles per hour?

For this problem, write the WANTED and DATA sections, then at the SOLVE step, write the WANTED and *given* terms. Arrange the dimensions of the *given* equality to match the dimensions in the WANTED fraction. Then stop.

🛑

Answer:

WANTED: $\dfrac{? \text{ mi.}}{\text{hr}} =$ (This unit is distance over time.)

DATA: 0.500 s = 60.5 ft.

SOLVE: $\dfrac{? \text{ mi.}}{\text{hr}} = \dfrac{60.5 \text{ feet}}{0.500 \text{ s}} \cdot$

(Because you WANT distance over time, if you arrange the equality in your DATA as a distance unit over a time unit, your conversions are probably arranged "right-side up." Finish solving from here.)

🛑

SOLVE: $\dfrac{? \text{ mi.}}{\text{hr}} = \dfrac{60.5 \text{ feet}}{0.500 \text{ s}} \cdot \dfrac{1 \text{ mi.}}{5{,}280 \text{ ft.}} \cdot \dfrac{60 \text{ s}}{1 \text{ min}} \cdot \dfrac{60 \text{ min}}{1 \text{ hr}} = \mathbf{\dfrac{82.5 \text{ mi.}}{hr}}$

When a ratio unit is WANTED, different methods may lead you to pick different ratios as your *given*. That's okay. When solving for a ratio, the order of the conversions does not matter, and there are multiple ways to arrange conversions right-side up. That said, choosing your *given* fraction on the basis of dimensions often results in conversions that are easier to arrange.

PRACTICE

1. If a sample of gas at a given temperature and pressure has a density of 0.860 mg/mL, what is its density in $kg \cdot L^{-1}$? (Use the WANTED dimension to arrange the *given* dimension.)

2. A 16 fluid ounce (fl. oz.) can of soft drink contains 52 g of sugar. Convert the concentration of sugar to $mg \cdot cm^{-3}$ (12 fl. oz. = 355 mL). (Use the WANTED dimensions to arrange the *given* dimensions.)

3. At a given temperature and pressure, 1 mole of He gas has a volume of 22.4 liters. What is the He density in grams per liter? (Use *any* method you choose to arrange the *given* fraction, then solve.)

Lesson 11.3   Ratios versus Two Related Amounts

In solving with conversions, we have divided DATA into two types: single-unit amounts and ratios (which we list as equalities). That distinction is all that is needed to solve most conversion calculations, but in some problems three DATA divisions are needed: single-unit amounts, ratios, and equivalencies. Let's explore when this is the case.

━━━▶ TRY IT

Q. Fill in the blanks to answer parts (a), (b), and (c) for these two problems. Use 1.61 km = 1 mi.

1. A cyclist travels 6.0 miles in exactly 1 hour.	2. A cyclist is traveling at a speed of 6.0 miles per hour.
a. What is her speed in km/hr?	a. What is her speed in km/hr?
WANTED:	WANTED:
DATA:	DATA:
SOLVE:	SOLVE:
b. How many minutes did the cyclist travel?	b. How many minutes did the cyclist travel?

c. How do these two problems differ?

(STOP)

Answer:

For part (a), the WANTED, DATA, and SOLVE steps should be identical. For part (b), the columns differ.

- From the wording of problem 1, we know three quantities: the amount of distance the cyclist traveled, the amount of time required, and the average speed for the trip. We know *two amounts* and *one ratio*.

- From the wording of problem 2, we know the speed of the trip, but *not* the distance or time of the trip. Problem 2 supplies one ratio, but *no amounts*.

- Problem 1 gives us more information.

We can answer question (b) for problem 1, but not for problem 2. How can we indicate that the miles-to-hours equalities in the two problems are different?

Two Types of DATA Equalities

To SOLVE for a ratio unit, we have been listing DATA as an equality in two cases: when two *amounts* are proportional in a problem (as in problem 1 above), and when two quantities are not amounts but are proportional (as in problem 2). For problems in which it is necessary to distinguish these two DATA types, we will adopt the following terminology:

- A **ratio** represents two quantities that are proportional.
- An **equivalence** is the special case of two *amounts* that are also a ratio.

Single-Unit Amounts

In an amount, "how much" is present and/or the "size" of an object determines the *magnitude* of the measurement.

> **Examples:** mm, km, g, kg, dm^3 (liters), cm^3 (mL), m^2, s, min, and mol are all among the units that measure *amounts*.

A quantity measured by a single base unit is always an amount (though some amounts are also measured by complex units). An amount may be a fundamental or derived quantity. A single unit measuring an amount may have a prefix and may be raised to a positive power, but it must have *one* kind of *base* unit in the numerator and *no unit* in the denominator. This represents *no change* in our past definition of single-unit measurements.

Ratios

When a number with its unit (which may have a label attached) is "directly proportional" to another number and unit, it means that if one number goes up by a multiple, the other number goes up by the same multiple. Two such quantities can be related in words by *per* and written as *equal* in DATA. The equality can be written as a *ratio* or *fraction* or *conversion factor* (all have equivalent meanings in this usage) that has an algebraic value of one.

By this definition, *ratios* include the following:

1. Metric-prefix definitions.

   > **Example:** 1 kilometer $= 10^3$ meters

2. Conversions relating two units that measure the same dimension.

   > **Example:** 1 minute $= 60$ seconds

3. Some derived quantities *defined* as ratios that express a relationship between two units but do not convey amounts in a problem.

   > **Examples**: *Concentration, speed, density,* and *molar mass* are ratios that convey the relationship between two units in a problem.

This represents *no change* in our past definitions and practices in listing DATA.

Equivalencies

An *equivalency* is the special case of two single-unit *amounts* that are measurements of the same object or process *and* are directly proportional. These amounts can be used separately as single-unit DATA, *and* they can also be used as a ratio in a conversion factor. An equivalency is especially important in solving problems because it supplies *three* items of DATA: two single-unit amounts and a ratio.

Identifying Ratios and Equivalencies

For problems in these lessons in which we need to distinguish ratios and equivalencies, after a DATA equality that is "2 amounts and a ratio" we will write **(2A+R)**. After an equality that is a ratio *not* based on two *amounts* in a problem we will write **(R)**.

▷ TRY IT

Q. In the first Try It in Lesson 11.3, find the *miles-to-hours* equality in the DATA for problem 1 and for problem 2. After the equality that is *not* two amounts, write an **(R)** for ratio. After the equality composed of two single-unit amounts that can also be used as a ratio, write **(2A+R)**.

[STOP]

Answer:

For problem 1:   DATA: 6.0 mi. = 1 hr **(2A+R)**

For problem 2:   DATA: 6.0 mi. = 1 hr **(R)**

Distinguishing Ratios from Two Related Amounts

The distinction between a ratio and an equivalency often requires a careful analysis of the wording of a problem. Some practical tips are:

- When a quantity in a problem can be read as *per one* or is written with a slash mark (/), it nearly always represents a *ratio* **(R)** rather than two related amounts.
- DATA that is a molarity, density, speed, molar mass, or unit conversion is, by definition, a ratio that is not two amounts.

▷ TRY IT

In your notebook, for the problem below, write the DATA section *only*. Include any equalities called by prompts. Then, after each *equality*, by the rules above write (2A+R) or (R).

Q. If 0.24 g NaOH is dissolved to make 250 mL of solution, find the [NaOH].

[STOP]

Answer:

DATA: $0.24\ g\ NaOH = 250$ mL soln. **(2A+R)**

   40.0 g NaOH = 1 mol NaOH **(R)**

The first equality is composed of two *amounts* that have been measured for this solution. The second equality, a molar mass that will be needed as a conversion, is a ratio between grams and moles that is always true for NaOH but is not based on the amounts present in this problem.

PRACTICE

For each part below, write the DATA section *only*. Include prompts but omit metric-prefix definitions. Then *label* each equality in the DATA as a ratio (**R**) or as two single-unit amounts and a ratio (**2A+R**).

1. An English to metric system volume conversion is "12 fl. oz. = 355 mL." What is the number of milliliters per fluid ounce?

2. A can of soda is labeled "12 fl. oz. = 355 mL." How many liters of soda does the can hold?

3. A sample of neon gas has a density of 0.842 gram per liter at room temperature and pressure. What would be the volume of 1 mole of the gas at these conditions?

4. A water bath absorbs 24 calories of heat from a reaction that forms 0.88 g CO_2. What is the heat released by the reaction in kilocalories per mole of CO_2?

Lesson 11.4   Solving Problems with Parts

To this point, for calculations in which a single unit was WANTED, one item of DATA had a single unit, and that is your *given* quantity. In some problems, however, you will need to SOLVE for a single unit when *all* of the DATA is equalities. Let's learn to do so with an example.

> ━━━━▶ **TRY IT**
>
> In your notebook, SOLVE *just* part (a) below.
>
> **Q.** An aqueous solution with a volume of 32 fluid ounces (fl. oz.) contains 0.050 pounds (lb.) of dissolved table salt (NaCl). (Use 454 g = 1 lb. and 12.0 fl. oz. = 355 mL)
>
> a. Find the volume of the solution in liters.
>
> b. How many moles of NaCl does the solution contain?
>
> c. What is the [NaCl]?
>
> **Answer:**
>
> Your paper should include:
>
> a. WANTED: ? L NaCl soln. =   (Label units with formulas.)
>
>   DATA: 32 fl. oz. NaCl soln. = 0.050 lb. NaCl
>
>   454 *g NaCl* = 1 lb. NaCl
>
>   12.0 fl. oz. soln. = 355 mL soln.
>
>   58.5 g NaCl = 1 mol NaCl   (g NaCl = grams prompt.)
>
>   SOLVE: ? L NaCl soln. =
>
> Which measurement in the DATA should be chosen as the *given*?

You may have chosen the correct *given* by intuition, but let's use this example to develop rules for picking a *given* from several possibilities.

Rules for Selecting the *Given*

In the Try It above, a single unit is WANTED in part (a), but the DATA is all equalities. To handle such cases, apply this rule.

> By the law of dimensions, to convert to an amount, you must be *given* an amount. You cannot choose as a *given* one part of a ratio *unless* the ratio is *also* two equivalent *amounts*.

To apply the rule to problems, use these steps.

When All DATA Is Equalities but a Single Unit Is WANTED

1. Label each *equality* in the DATA as a ratio (**R**) or as two related single-unit amounts *and* a ratio (**2A+R**).

2. As the *given*, choose *one* single-unit amount from those that are 2A+R.

3. Most often, only one unit in the 2A+R equality can be converted to the WANTED unit. This unit will either measure the *dimension* WANTED or be convertible to the dimension WANTED. Choose this unit as your *given*.

► TRY IT

Q. For the first Try It problem of Lesson 11.4, apply steps 1–3 to SOLVE part (a).

Answer:

Your paper should include:

DATA: 32 fl. oz. soln. = 0.050 lb. NaCl (**2A+R**) (Two measures of one soln.)

454 *g NaCl* = 1 lb. NaCl (**R**) (An English–metric conversion.)

12.0 fl. oz. = 355 mL (**R**) (An English–metric conversion.)

58.5 g NaCl = 1 mol NaCl (**R**) (Molar mass is a ratio, not amounts.)

SOLVE:

(Your *given* must be one of the two amounts in the 2A+R equality. Which one? The WANTED unit is liters: a *volume*. In the 2A+R equality, 32 fluid ounces is a volume. Conversions within the same dimension can always be accomplished.)

$$? \text{ L soln.} = 32 \text{ fl. oz. soln.} \cdot \frac{355 \text{ mL}}{12.0 \text{ fl. oz.}} \cdot \frac{10^{-3} \text{ L}}{1 \text{ mL}} = \boxed{0.947 \text{ L soln.}}$$

That answers part (a), but note that we rounded to three significant figures. Why?

Significant Figures in Problems with Parts

In a calculation that has *multiple steps* or *multiple parts*, it is generally considered a good practice to

- carry one more digit than called for by significant figures until the *final* step;
- then, at the *final* step, round to correct *s.f.* on the basis of the supplied problem DATA.

This method limits variation in the final answer due to rounding in the parts.

TRY IT

Q. Now let's SOLVE part (b) of the first Try It of Lesson 11.4 in a manner similar to part (a). In your notebook write:

b. WANTED: ? mol NaCl =

   DATA: Same as for part (a)

   SOLVE: ? mol NaCl =

Choose your *given* amount and SOLVE part (b).

Answer:

(To SOLVE for moles, a single-unit amount, you must use as a *given* one of the two 2A+R amounts. Which one? Pounds can be converted to grams and grams to moles. A tip: If two parts of a problem ask for single-unit amounts and the first part uses one side of the 2A+R equality as *given*, usually the second part will use the other side.)

$$\text{SOLVE:} \quad \text{? mol NaCl} = 0.050 \text{ lb. NaCl} \cdot \frac{454 \text{ g}}{1 \text{ lb.}} \cdot \frac{1 \text{ mol NaCl}}{58.5 \text{ g NaCl}} = \boxed{0.388 \text{ mol NaCl}}$$

(Carry an extra *s.f.* until the last step.)

Solving for a Ratio in Problems with Parts

Note that the answers above were $\boxed{\text{boxed}}$ for parts (a) and (b). This is one way to indicate answers from early parts that can be used as DATA for a later part.

TRY IT

Q. With that note in mind, see if you can SOLVE part (c) in one conversion.

Answer:

c. WANTED: $? \dfrac{\text{mol NaCl}}{\text{L soln.}} =$ (M prompt)

   DATA: See part (a), plus $\boxed{\text{boxed}}$ answers to parts (a) and (b).

(Apply the fundamental rule of conversions: *Let the units tell you what to do.*)

SOLVE: [The units say: divide the part (b) moles by the part (a) liters.]

$$? \frac{\text{mol NaCl}}{\text{L soln.}} = \frac{0.388 \text{ mol NaCl}}{0.947 \text{ L soln.}} = \textbf{0.41} \frac{\textbf{mol NaCl}}{\textbf{L soln.}}$$

(*S.F.*: Two original measurements were 2 *s.f.*; round the *final* answer to 2 *s.f.*)

You could also have solved starting from the problem's original data, but using the earlier answers avoided the need to repeat some steps.

To summarize:

Hints for Multiple Questions about One Set of DATA

1. Box the answer to each part.

2. For later parts, see if you can use answers from earlier parts.

PRACTICE

1. From memory, be able to answer these two questions in your own words.

   a. List three steps to take to SOLVE for a single-unit amount when the DATA is all equalities.

   b. List two hints that help in solving multiple questions about one problem scenario.

2. In the 1972 Olympics, Melissa Belote set a women's world record in the 200-meter backstroke with a time of 139.19 seconds. (Use 1.609 km = 1 mi. Assume the distance is 200.0 m)

   a. What was the length of the race in miles?

   b. What was the time of the race in hours?

   c. What was her average speed in miles per hour?

3. 11.7 g NaCl is dissolved to make 250. mL of aqueous solution.

   a. How many liters of solution were prepared?

   b. How many moles of NaCl were used?

   c. What is the [NaCl] in the solution?

4. 0.0250 mol KCl are dissolved in 750. mL of solution.

   a. How many grams of KCl are in the solution?

   b. How many liters of solution were prepared?

   c. What is the [KCl]?

SUMMARY

When solving conversions for ratio units, dimensions can help in choosing a *given* fraction and arranging it "right-side up." Solving for single units, dimensions are the best guide to choosing which part of an equivalency to use as a *given* amount. But dimensions will be of greatest assistance for problems in upcoming chapters that require equations.

Given a unit, you need to recall quickly from memory which dimension it measures. "Unit-to-dimension" flashcards will help, but you should also practice writing the tables for the fundamental and derived quantities and their units. The patterns in tables organize information conceptually. Seeing the patterns will help you to know intuitively which facts and procedures are most likely to be needed to solve different types of problems.

REVIEW QUIZ

1. Label each unit with the quantity it is measuring.

Unit	Quantity Being Measured
kilometers	
liters	
moles dissolved per liter	
ms	
dm^3	
m/s	

Unit	Quantity Being Measured
kg/L	
centigrams	
moles	
day	
grams/mole	
mg/mL	

2. List the units in problem 1 that measure fundamental quantities.

3. If the speed of an energy wave is measured to be 1.80×10^7 km • min^{-1}, what is its speed in m/s?

4. A 2.0 L (67.6 fl. oz.) bottle of a soft drink contains an artificial sweetener with a concentration of 4.5 mg per fluid ounce.

   a. Write an equality from this DATA that represents two related amounts.

   b. Write an equality from this DATA that represents a ratio that is not two measured amounts.

   c. Calculate the milligrams of sweetener found in 0.25 L of the drink.

5. An 18-carat gold bar has a volume of 125 cm^3 and weighs 4.27 lb. (1 kg = 2.20 lb.). Find

   a. The mass of the bar in grams.

   b. The volume of the bar in liters.

   c. The density of the bar in grams/liter.

ANSWERS

Lesson 11.1

Practice A   1. See table in lesson.

2. Cubic centimeters. (Units of fundamental quantities must be *first* power.)

3a. mass   3b. time   3c. count of particles   3d. distance

Practice B   1. See table in lesson.

2. These are a few of the possibilities:

a. time: **seconds, minutes, hours, days, years, centuries**

b. mass: **grams, kg, mg**

c. distance: **meters, km, cm**

d. volume: **L, mL, dL, m^3, cm^3, dm^3**

3. Moles of dissolved solute per liter of solution (mol/L)

4. a. liters: **volume**   b. cubic meters: **volume**   c. kilograms: **mass**   d. decimeters: **distance**

e. grams/mole: **molar mass**   f. kg/L: **density** (A mass unit over a volume unit)

5. **kilograms** measures mass; **decimeters** measures distance

Practice C   1.

Derived Quantity	Dimension	Units Often Used in Chemistry
density	mass over volume	**g/mL** or **any mass unit over a volume unit**
volume	**distance cubed**	liters
concentration	**particles per volume**	**moles per liter soln.**
speed (or velocity)	**distance/time**	meters/second
molar mass	mass per particle count	**grams/mole**

2a. gallons: **volume**

2b. nanoseconds: **time**

2c. g/mL: **density**

2d. feet/second: **speed** or **velocity**

2e. mm: **distance**

2f. dL: **volume**

2g. moles/liter: **concentration**

2h. km/hr: **speed** or **velocity**

3. Volume is the only extensive quantity.

4. **2c, 2d, 2g,** and **2h** are all ratios of two units that measure an extensive quantity.

5. **2b** (nanoseconds) and **2e** (deciliters)

Lesson 11.2

Practice   In some answers, steps are omitted. Your calculations should always include WANTED, DATA, and SOLVE. SOLVE conversions may be in a different order but must be "same-side up" as these. To assist with readability, some units that should be (circled) or cancelled on your paper may not be marked in these answers.

1. WANTED: $? \dfrac{kg}{L} =$ ($kg \cdot L^{-1} = kg/L$)

   DATA: $0.860 \ mg = 1 \ mL$

   SOLVE: (You WANT mass over volume. As *given*, place mass over volume.)

   $$\frac{? \ kg}{L} = \frac{0.860 \ mg}{1 \ mL} \cdot \frac{10^{-3} \ g}{1 \ mg} \cdot \frac{1 \ \textcircled{kg}}{10^3 \ g} \cdot \frac{1 \ mL}{10^{-3} \ \textcircled{L}} = 8.60 \times 10^{-4} \ \frac{kg}{L}$$

2. WANTED: $\dfrac{? \ mg \ sugar}{L \ drink} =$ ($mg \cdot mL^{-1} = mg/mL =$ a ratio unit is WANTED.)

   DATA: $16 \ fl. \ oz. \ drink = 52 \ g \ sugar$ (Two measures of the same drink.)

   $355 \ mL = 12 \ fl. \ oz.$

   SOLVE: $\dfrac{? \ mg \ sugar}{cm^3 \ drink} = \dfrac{52 \ g \ sugar}{16 \ fl. \ oz.} \cdot \dfrac{12 \ fl. \ oz.}{355 \ mL} \cdot \dfrac{1 \ mL}{1 \ cm^3} \cdot \dfrac{1 \ mg}{10^{-3} \ g} = 110 \ \dfrac{mg \ sugar}{cm^3 \ drink}$

3. SOLVE: $\dfrac{? \ g \ He}{L \ He} = \dfrac{1 \ mol \ He}{22.4 \ L \ He} \cdot \dfrac{4.00 \ g \ He}{1 \ mol \ He} = 0.179 \ \dfrac{g \ He}{L \ He}$

Lesson 11.3

Practice
1. DATA: $12 \ fl. \ oz. = 355 \ mL$ **(R)**   This ratio is not measuring an object or process.

2. DATA: $12 \ fl. \ oz. = 355 \ mL$ **(2A+R)**   Both sides are measuring the amount of soda.

3. DATA: $0.842 \ g \ Ne = 1 \ mol \ Ne.$ **(R)**   (Density is defined as a ratio.)

   $20.2 \ g \ Ne = 1 \ mol \ Ne$ **(R)**   (Molar mass is defined as a ratio.)

   This problem has no measured amounts.

4. DATA: $24 \ calories \ heat = 0.88 \ g \ CO_2$ **(2A+R**—two amounts, same reaction.**)**

   $44.0 \ g \ CO_2 = 1 \ mol \ CO_2$ **(R)**

Lesson 11.4

Practice
1. See rules and hints in the lesson.

2a. WANTED: $? \ mi. =$

   DATA: $200.0 \ m = 139.19 \ s$ (2A+R)

   $1.609 \ km = 1 \ mi.$ (R)

   SOLVE: $? \ miles = 200.0 \ meters \cdot \dfrac{1 \ km}{10^3 \ meters} \cdot \dfrac{1 \ mile}{1.609 \ km} = \boxed{0.12430 \ \textbf{miles}}$

2b. SOLVE: $? \ hr = 139.19 \ s \cdot \dfrac{1 \ min}{60 \ s} \cdot \dfrac{1 \ hr}{60 \ min} = \boxed{0.0386639 \ \textbf{hours}}$

2c. DATA: See part (a) plus the boxed answers.

   SOLVE: The units say to divide miles by hours.

   $\dfrac{? \ \textbf{mi.}}{\textbf{hr}} = \dfrac{part \ (a) \ answer}{part \ (b) \ answer} = \dfrac{0.12430 \ \textbf{mi.}}{0.0386639 \ \textbf{hr}} = 3.215 \ \dfrac{\textbf{mi.}}{\textbf{hr}}$

3a. WANTED: ? L soln. =

    DATA: 11.7 *g NaCl* = 250. mL NaCl soln.   (**2A+R**)

             58.5 g NaCl = 1 mol NaCl   (**R**)   (Molar mass is a ratio by definition.)

    SOLVE: (Pick a 2A+R single-unit *given* that converts to L WANTED.)

$$? \text{ L soln.} = 250. \text{ mL soln.} \cdot \frac{10^{-3} \text{ L}}{1 \text{ mL}} = \boxed{0.2500 \text{ L soln.}}$$

3b. SOLVE: $? \text{ mol NaCl} = 11.7 \text{ g NaCl} \cdot \dfrac{1 \text{ mol NaCl}}{58.5 \text{ g NaCl}} = \boxed{0.2000 \text{ mol NaCl}}$

3c. WANTED: $\dfrac{? \text{ mol NaCl}}{\text{L soln.}} =$ (Units of molarity = mol/L)

    DATA: See part (a) plus boxed answers.

    SOLVE: $\dfrac{? \text{ mol NaCl}}{\text{L soln.}} = \dfrac{\text{part (a) answer}}{\text{part (b) answer}} = \dfrac{0.2000 \text{ mol NaCl}}{0.2500 \text{ L soln.}} = \mathbf{0.800} \dfrac{\textbf{mol NaCl}}{\textbf{L soln.}}$

    (*S.F.*: Original measurements were 3 *s.f.*; round the *final* answer to 3 *s.f.*)

4a. SOLVE: ("Grams and moles, use molar mass.")

$$? \text{ g KCl} = 0.0250 \text{ mol KCl} \cdot \frac{74.6 \text{ g KCl}}{1 \text{ mol KCl}} = \boxed{1.865 \text{ g KCl}}$$

4b. SOLVE: $? \text{ L soln.} = 750. \text{ mL soln.} \cdot \dfrac{10^{-3} \text{ L}}{1 \text{ mL}} = \boxed{0.7500 \text{ L soln.}}$

4c. SOLVE: $\dfrac{? \text{ mol KCl}}{\text{L soln.}} = \dfrac{0.0250 \text{ mol KCl}}{0.750 \text{ L soln.}} = \mathbf{0.0333} \dfrac{\textbf{mol KCl}}{\textbf{L soln.}}$

Review Quiz

1.

Unit	Quantity Being Measured	Unit	Quantity Being Measured
kilometers	**distance**	kg/L	**density**
liters	**volume**	centigrams	**mass**
moles dissolved per liter	**molar concentration**	moles	**count of particles**
ms	**time**	day	**time**
dm^3	**volume**	grams/mole	**molar mass**
m/s	**speed**	mg/mL	**density**

2. Kilometers, ms, centigrams, moles, day

3. SOLVE: $\dfrac{? \text{ m}}{\text{s}} = \dfrac{1.80 \times 10^7 \text{ km}}{1 \text{ min}} \cdot \dfrac{10^3 \text{ m}}{1 \text{ km}} \cdot \dfrac{1 \text{ min}}{60 \text{ s}} = \mathbf{3.00 \times 10^8} \dfrac{\textbf{m}}{\textbf{s}}$

4a. 2.0 L = 67.6 fl. oz.   4b. 4.5 mg = 1 fl. oz.   4c. 38 mg sweetener

5. SOLVE: $\dfrac{? \text{ g Au}}{\text{L Au}} = \dfrac{\textbf{part (a) answer}}{\textbf{part (b) answer}} = \dfrac{\textbf{1,941 g Au}}{\textbf{0.1250 L Au}} = \mathbf{1.55 \times 10^4} \dfrac{\textbf{g Au}}{\textbf{L Au}}$

12

Concentration Calculations

Lesson 12.1   Dilution

Converting between Milliliters and Liters

When a simple operation is needed often, it is best to practice until it is automated (so you can do it "in your head"). For calculations involving solutions, it is especially helpful to be able to convert quickly between milliliters (mL) and liters (L). Let's review the logic of a "quick rule" for these conversions.

1. Milli- means $\times 10^{-3}$, so $\times 10^{-3}$ can be *substituted* for **milli-**.

   Example: $? \, \mathbf{L} = 47 \, \mathbf{mL} = 47 \times 10^{-3} \, \mathbf{L}$ or $\mathbf{0.047 \, L}$

2. Recall that 1 L = 1000 mL, so 4.2 L = 4,200 mL

3. In fixed-decimal notation, when converting between mL and L:

   - Move the decimal *three* times.
   - Make the mL number 1000 times larger than the L number.

PRACTICE **A**

1. Fill in the blanks.

   a. 35 mL = 35 × _____ L = _____ L   b. 125 mL = 125 × _____ L = _____ L
                      ∧   ∧   ∧   ∧
   Notation: Exponential   Fixed   Exponential   Fixed
                                   decimal   decimal

2. Answer without exponentials.

   a. 2.500 L = _____ mL   b. 15 mL = _____ L   c. 0.77 L = _____ mL

3. In *fixed* notation, when converting between mL and L, the decimal point in the L should always

   be _____ places to the _____ of the mL value.

4. Given milliliters or liters, write the other in fixed notation.

   a. 150 mL =   b. 33 L =   c. 9.21 mL =

   d. 0.45 L =   e. 0.833 mL =   f. 0.0655 L =

Dilution Terminology

To mix a solution, a *solute* is dissolved in a *solvent* in which the solute is *soluble*. To **dilute** a solution, more *solvent* is added. As it is diluted, the solution becomes **less concentrated**: The average distance between the dissolved particles increases.

When a solution is diluted, its volume increases, and the concentration of its solute decreases, but the number of particles of the solute stays the same. That's the key relationship in dilution:

In dilution, for the solute, *moles* **Concentrated** = *moles* **Diluted**.

Solving Dilution by Inspection

Dilution involves *four* measurements: an initial volume and molarity and a final volume and molarity. In dilution calculations, you know three of those values and you are asked to find the fourth.

If the numbers in a dilution involve simple multiples, calculations can often be solved "by inspection."

Dilution for Easy Multiples

In a dilution calculation, if an initial *volume* or *concentration* is changed by a *multiple*, the final value for the *other* quantity will be WANTED. To find that final value, multiply the quantity's initial value by *1/multiple*.

Examples

- If the *volume* of a solution is *doubled* (**2×**) by adding more solvent, the *concentration* of the solute is cut in *half* (**1/2**).

- If the volume of a solution is *quadrupled*, the solute concentration becomes *1/4* of its original value.

- If a solute concentration WANTED is **1/10th** the [original], add solvent until the volume is **10** times higher.

(Recall that [original] means "concentration of the original solution.")

To apply the dilution rule for easy multiples, the steps are as follows:

1. Find the two supplied numbers that have the same *unit*.

2. Find an easy *multiple* that takes you from the *initial* to the *final* value for that unit.

3. Multiply the *known* initial value for the *other* unit by *1/multiple*.

> **TRY IT**

Q. To 250 mL of a 0.45 M solution of glucose (a biologically important sugar), distilled water is added until the volume is 750 mL. What is the new [glucose]?

Answer:

The total volume increased from 250 mL to 750 mL: it *tripled*. The final [glucose] will be *1/3* the 0.45 M original = **0.15 M** glucose.

PRACTICE   B

1. In your notebook, write a summary that, recalled from memory, will guide you in SOLVING simple dilution problems. Practice recalling your summary.

   Apply your summary from memory to SOLVE the next three problems. Use pencil and paper for math if needed, but no calculator.

2. If 100. mL of 2.0 M KCl is diluted to a 400. mL total volume, what is the final [KCl]?

3. To dilute 250. mL of 1.00 M HCl to a concentration of 0.200 M, what must be the final volume of the solution?

4. If distilled water is added to increase the volume of a 0.60 M NaOH from 3.0 liters to 9.0 liters, what is the new [NaOH]?

The Dilution Equation

A dilution problem that cannot be solved by inspection can be SOLVED using conversions or an equation, but the equation method is generally faster.

In a dilution problem, we will label measurements for the concentrated solution with a **C** and those for the diluted solution with a **D**. In dilution, because the moles of solute stays constant, we can write:

$$\text{liters } \mathbf{C} \cdot \frac{\text{moles } \mathbf{C}}{\text{liter } \mathbf{C}} = \boxed{\text{moles } \mathbf{C} = \text{moles } \mathbf{D}} = \text{liters } \mathbf{D} \cdot \frac{\text{moles } \mathbf{D}}{\text{liter } \mathbf{D}}$$

A way to rewrite this equation is

$$\text{Volume}_{\mathbf{C}} \times \text{Molarity}_{\mathbf{C}} = \text{Volume}_{\mathbf{D}} \times \text{Molarity}_{\mathbf{D}}$$

which can be abbreviated with symbols as the **dilution equation**:

In dilution: $V_C \times M_C = V_D \times M_D$

and memorized by recitation:

"In dilution, volume times molarity equals volume times molarity."

A restriction on the use of equations is that the units must be **consistent**. Concentration will nearly always be in mol/L (M), but volumes may be in liters or milliliters. The equation works for any volume units, but all of the volumes must be converted to the *same unit*.

Steps for Solving with the Dilution Equation

Let's learn to use the dilution equation with an example.

▬▬▬▶ **TRY IT**

Q. To 225 mL of an aqueous 0.200 M KOH solution, water is added until the total volume is 4.00 L. What is the resulting [KOH]?

Complete the following steps in your notebook.

1. List the WANTED unit and DATA in the usual manner.

2. If a problem involves dilution, write

   SOLVE: In dilution: $\mathbf{V_C \times M_C = V_D \times M_D}$

3. Label each item in the WANTED and DATA with a *symbol* from the equation.

   a. First, label each item in the WANTED and DATA as being a volume (**V**) or a concentration (**M**), on the basis of its *unit*. Molarities (**M**) will be written as ratios if WANTED and as equalities in the DATA.

   b. Then label each **V** as either $\mathbf{V_C}$ or $\mathbf{V_D}$. Mark each **M** as $\mathbf{M_C}$ or $\mathbf{M_D}$. The solution with the higher molarity *or* lower volume is the more concentrated (**C**).

c. Once you have one solution identified as **C** or **D**, the other must be the opposite. *Each* of the four symbols in the equation must be used *once* and *only* once in labeling the WANTED and DATA. It often helps to mark the WANTED unit as **C** or **D** *last*, by process of elimination.

Finish those steps for this problem, then check below.

Answer:

WANTED: ? $\dfrac{\text{mol KOH}}{\text{L KOH soln.}}$ $\mathbf{M_D}$

DATA: 225 mL KOH soln. $\mathbf{V_C}$

0.200 mol KOH = 1 L KOH soln. $\mathbf{M_C}$

4.00 L KOH soln. $\mathbf{V_D}$

SOLVE: In dilution: $\mathbf{V_C \times M_C = V_D \times M_D}$

4. Convert the DATA to consistent units. (*If* the two volume *units* differ, either convert both to milliliters *or* both to liters.)

You can work in liters: ~~225 mL KOH soln.~~ $\mathbf{V_C}$ 0.225 **L** KOH soln.

Or in milliliters: ~~4.00 L KOH soln.~~ $\mathbf{V_D}$ 4.00×10^3 **mL** KOH soln.

One volume must be converted so that both volumes have the *same* unit, but which unit you convert to will not affect the final answer.

5. Using algebra, SOLVE the equation in *symbols* for the *symbol* WANTED.

WANTED: ? $\mathbf{M_D} = \dfrac{V_C \times M_C}{V_D}$

Do not plug numbers into an equation until *after* you have solved for the WANTED symbol. When numbers include units, symbols move faster and with fewer mistakes.

6. Plug in the numbers with their units and SOLVE.

When using *conversions*, you must write **M** as moles *per* liter because you use **M** as a ratio. With the dilution *equation*, however, use **M** to abbreviate moles/liter so that the equation will SOLVE more quickly.

$$? M_D = \frac{V_C \times M_C}{V_D} = \frac{225 \text{ mL KOH } \mathbf{C} \times 0.200 \text{ M KOH } \mathbf{C}}{4.00 \times 10^3 \text{ mL KOH } \mathbf{D}}$$

$$= 0.0113 \text{ M KOH soln. diluted}$$

7. *Check* your answer. *Round* the problem numbers to make the change an easy multiple. Then estimate the answer in your head or by pencil-and-paper math. Compare your estimate to the calculated answer. They should be close.

Try that step, then check below.

If 200 mL is increased to 4,000 mL, the increase is by a factor of 20. The concentration should therefore be cut by 1/20. 0.200 M × 1/20 = *0.010 M*. This is close to the answer above. Check!

A Caution about Equations, Units, and Labels

In step 6 above, the *units* cancel properly to give the WANTED unit, but the **C** and **D** *labels* with each quantity seem not to cancel. Actually, units and labels must cancel in equations. If we had written out M_C as the full "mol KOH/L KOH **C** soln.," the units and labels would have cancelled properly, but the equation would not have solved as quickly. If you abbreviate mol/L as **M**, the equation solves quickly, but the label cancellation may not work.

The bottom line: To SOLVE quickly, our rule will be to abbreviate mol/L as **M** when you substitute into the dilution equation, but do so carefully. You will not have label cancellation as a check on your work.

Learning the Dilution Steps

To summarize:

Using the Dilution Equation

If a dilution calculation cannot be solved with simple multiples:

1. List the WANTED and DATA.

2. At SOLVE, write the dilution equation.

3. Label the WANTED and DATA with the four equation symbols.

4. If needed, convert to consistent volume units.

5. SOLVE the dilution equation for the WANTED symbol in *symbols*.

6. Plug in DATA values, use M for mol/L, and SOLVE.

7. Round the dilution to an easy multiple and check your answer.

The Dilution Prompt

Dilution calculations often do not include the word *dilution*. How do you recognize that a problem is *about* dilution?

Dilution Prompt

- *If* a problem involves two solutions, before and after, that contain the same dissolved substance, or
- *If* the WANTED and DATA contain two volumes (usually in mL or L) and two concentrations (M) of the same substance,

see if the wording can be adjusted to describe "dilution." If so, use a dilution method to solve.

Flashcards

Make these cards, run them to perfection today, then do the problems in Practice C. Repeat running the cards to perfection on two more days before placing them in the "quiz and test prep" stack.

One-Way Cards (with Notch)	Back Side—Answers
The fundamental rule of dilution is _____	*Moles* of *solute* are not changed by dilution
In dilution calculations, if *volume* or *concentration* is changed by a multiple, _____	The other value is multiplied by 1/multiple
The dilution equation in words = _____	Volume times molarity = volume times molarity
The dilution equation in symbols = _____	$V_C \times M_C = V_D \times M_D$
When solving with equations, label the WANTED and DATA with _____	The *symbols* in the equation

PRACTICE C

1. Given liters, write milliliters; given milliliters, write liters. Answer in fixed notation.

   a. 250 mL =   b. 0.62 L =   c. 3.5 mL =

For problems 2–5, try to apply the steps to SOLVE from memory.

2. If 50.0 mL of 3.00 M NaOH is diluted

   a. to 500. mL total volume, what is the [NaOH]?

   b. to 0.750 L total volume, what is the [NaOH]?

3. To what volume must 20.0 mL of 2.5 M KCl be diluted to make 0.24 M KCl?

4. To make 250 mL of 0.65 M NaCl starting from 2.00 M NaCl, how many milliliters of the concentrated solution are required?

5. If 6.2 mL of an HCl solution is added to water to make 0.250 L of 0.15 mol/L HCl, what was the initial [HCl]?

Lesson 12.2   Ion Concentrations

All ionic compounds are solids at room temperature. In a solid, the ions have minimal freedom of motion, but when an ionic solid is melted or is dissolved in water, the ions **dissociate:** they separate and move about freely. The electrical charges on the ions can "flow." This ability of the charges to move means that when ions are melted or dissolved in water, they can conduct electricity. A substance that separates into ions when dissolved in water is termed an **electrolyte**.

Unless otherwise noted, for the portion of an ionic substance that dissolves in water, you should assume that *all* of the ions are separated particles. In dilute solutions, this "ideal solution behavior" is generally close to true.

Some ionic compounds dissolve only slightly in water, while others dissolve 100% in dilute solutions. In either case, dissolving an ionic solid in water can be written as if it were a simple chemical reaction. As the solid dissolves, the original formula units of the solid are *used up* as the *separated* ions form.

When an ionic solid dissolves 100% in water, *all* of its nondissociated ions are used up and are assumed to form 100% separated ions.

Example: Na_3PO_4 is highly soluble in water: In dilute solutions, the ionic solid dissolves 100% and its ions also separate by close to 100%. This reaction can be expressed as

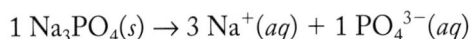

$$1 \, Na_3PO_4(s) \rightarrow 3 \, Na^+(aq) + 1 \, PO_4^{3-}(aq)$$

The products of the dissociation can be predicted by applying the rules for converting standard ionic formulas to balanced separated ions (see Lesson 7.4).

- In the example, *one* particle was used up and *four* particles form. One formula unit of an ionic compound, when it dissolves, will always produce *two or more* ions. In chemical reactions, the total *number* of particles in the reactants and products will often differ, but the count for each kind of *atom* must be the *same* on both sides.

- If an ionic solid completely dissolves, the concentration of the *nondissociated* ions [labeled in the dissociation equation by (s)] becomes 0 M. The nondissociated ions are completely used up.

To simplify problem solving, use this rule:

In a *calculation* for an ionic substance dissolved in water, begin by writing the balanced equation for the forming of its separated ions.

Calculating Moles of Ions

For an ionic solid dissolved in water, one measure often needed is the *moles* of each separated ion in the solution. If the moles of solid that were dissolved is known, the moles of ions formed are based on the simple whole-number ratios of the balanced equation.

TRY IT

For this problem, do needed math by mental arithmetic.

Q. The ionic solid $BaCl_2$ dissolves 100% in dilute aqueous solutions. If 0.40 mol solid $BaCl_2$ dissolves:

    a. Write the balanced equation for ion separation.

    b. How many moles of Cl^- ions are present in the solution?

    c. How many moles of nondissociated $BaCl_2$ are present in the solution?

Answers:

    a. The balanced equation for dissolving is

$$1\ BaCl_2(s) \rightarrow 1\ Ba^{2+}(aq) + 2\ Cl^-(aq)$$

    b. For every *1 mole* of $BaCl_2$ formula units used up as it dissolves and separates, *2 moles* of Cl^- ions form. Dissolving 0.40 mol $BaCl_2$ will therefore result in **0.80 mol** Cl^- in the solution.

    c. **0 mol** of nondissociated $BaCl_2$ is in the solution. All of the nondissociated $BaCl_2$ formula units are used up as $BaCl_2$ dissolves.

PRACTICE   A

1. Briefly describe what happens at the particle level when an ionic solid dissolves in water.

2. As 0.50 mol of solid Na_2SO_4 is added to water, it dissolves 100%. In the resulting solution, what will be the number of *moles* of

    a. Na^+ ions?   b. SO_4^{2-} ions?

    c. Na_2SO_4 particles?   d. Ions in the solution?

Ion Concentrations

When ionic solids dissolve 100% in water, the simple coefficient ratios that predict *moles* used up and formed also predict the *mol/L* of both the *solid* that dissolves and *ions* that form. Why?

- Because coefficients are *ratios*, all coefficients can be multiplied or divided by the same number.

- The total solution *volume* is essentially the same before and after an ionic substance dissolves. This means the coefficients read in moles can all be divided by the number of liters of solution that exist before and after dissolving to give mole *per liter* (concentration) ratios.

To summarize:

> ### In Calculations for an Ionic Solid Dissolved in Water
>
> 1. Write the balanced equation for separation of the solid into its ions.
>
> 2. The coefficients can be read as particles, moles, or moles *per liter*.

Ion Concentration Calculations

When describing the concentration of particles in a solution, chemists often use a shortcut. The solution concentration is usually represented by the moles per liter of solid, nondissociated particles that dissolved in the water, even though all of the nondissociated particles have been used up. We use this "concentration as mixed" because it is brief and because it is easily converted to the actual concentrations of the particles in the solution.

TRY IT

Q. In a bottle of solution labeled 0.15 M K_2CrO_4, what is the

a. $[K^+]$?   b. $[CrO_4^{2-}]$?   c. $[K_2CrO_4]_{as\ mixed}$?

Answer:

To find moles or [ions] when ionic solids dissolve, begin by writing the balanced equation for separation:

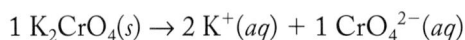

$$1\ K_2CrO_4(s) \rightarrow 2\ K^+(aq) + 1\ CrO_4^{2-}(aq)$$

According to the coefficients, for every 0.15 mole per liter of K_2CrO_4 formula units mixed:

a. **0.30** mol/L of K^+ forms.   b. **0.15** mol/L of CrO_4^{2-} forms.

c. The $[K_2CrO_4]$ is *labeled* as **"0.15 M,"** and "0.15 M K_2CrO_4" represents the moles of K_2CrO_4 per liter that were dissolved to *make* the solution, so 0.15 M is the "$[K_2CrO_4]$ **as mixed**." However, in the solution are *zero* nondissociated particles of K_2CrO_4.

Writing "0.15 M K_2CrO_4" is quicker than writing "0.30 M K^+ *and* 0.15 M CrO_4^{2-}," so the "[solid] as mixed" is usually how the concentration is expressed.

The *REC* Steps

In dilute aqueous solutions, soluble ionic solids dissolve and dissociate close to 100%. Other compounds we will study later in the course dissolve in water but dissociate only *slightly*.

To develop a consistent method to calculate ion concentrations in both types of dissociation, we will "write the *REC* steps." The *REC* steps are simply the balanced equation for a dissociation reaction, with the particle concentrations written beneath each particle.

The *REC* Steps

To SOLVE calculations for solutions of dissociated ions, write the following:

- **R**: The balanced *reaction* equation. After the equation, write
- **E**: The *extent* of the reaction (such as "goes 100%"). Below each particle, write
- **C**: The *concentration* of each particle based on the balanced equation.

Example: To find the concentrations of the particles in a 0.10 M solution of Na_3PO_4, write

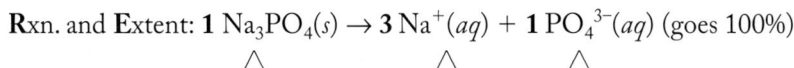

Rxn. and Extent: 1 $Na_3PO_4(s) \rightarrow$ **3** $Na^+(aq) +$ **1** $PO_4^{3-}(aq)$ (goes 100%)

$\wedge \qquad\qquad \wedge \qquad\qquad \wedge$

Concentrations: **0.10 M** ($\rightarrow$ **0 M**)   **0.30 M**   **0.10 M**

The *REC* steps show the [solid] both as mixed and after dissociation.

→ TRY IT

In your notebook, write the *REC* steps to answer the following problem.

Q. In a solution labeled 0.25 M $Ca(NO_3)_2$, what are the following concentrations?

a. $[Ca(NO_3)_2]_{as\ mixed} =$ b. $[Ca^{2+}] =$ c. $[NO_3^-] =$

STOP

Answer:

Because this reaction is a solid separating into ions, and particle concentrations are WANTED, write the *REC* steps.

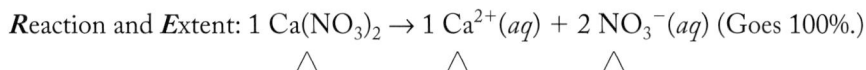

Reaction and **E**xtent: 1 $Ca(NO_3)_2 \rightarrow$ 1 $Ca^{2+}(aq) +$ 2 $NO_3^-(aq)$ (Goes 100%.)

$\wedge \qquad\qquad \wedge \qquad\qquad \wedge$

Concentrations: **0.25 M** ($\rightarrow$ **0 M**)   **0.25 M**   **0.50 M**

Flashcards

Add these to your collection. Practice until perfect, then try the problems below.

One-Way Cards (with Notch)	Back Side—Answers
Coefficients can be read as _____	Particles or moles—or mol/L if all reactants and products are in the same volume
To find [ions] for substances that separate into ions, _____	Write the *REC* steps
The *REC* steps are _____	Write the balanced Reaction equation, Extent, and Concentrations

Two-Way Cards (*without* Notch)	
Dissociation (definition)	A substance separating into smaller particles
Electrolyte	A substance or mixture that is composed of separated ions and conducts electricity

PRACTICE   B

Assume these solids dissolve 100%. SOLVE in your notebook using the *REC* steps. Save one problem for your next study session.

1. In a 0.30 M solution of radium nitrate:

   a. $[Ra^{2+}] = ?$ b. $[NO_3^-] = ?$

2. In a 0.60 molar solution of sodium carbonate:

   a. $[Na^+] = ?$ b. $[CO_3^{2-}] = ?$

3. In a solution of potassium phosphate, $[K^+] = 0.45$ M.

   a. $[PO_4^{3-}] = ?$ b. $[K_3PO_4]_{\text{as mixed}} = ?$

4. In a solution of aluminum sulfate, if the sulfate ion concentration is 0.036 M:

   a. $[Al^{3+}] = ?$ b. $[Al_2(SO_4)_3]_{\text{as mixed}} = ?$

Lesson 12.3   Fractions and Percentages

Fractions and Decimal Equivalents

A fraction is a ratio: one quantity divided by another. In math, a fraction can be any ratio, but in science, "fraction" often (but not always) refers to a *part* of a larger total: a smaller quantity over a larger quantity.

For the next two lessons that review the math of fractions, we will number each *rule* in **bold** for ease of reference.

Let's begin with three rules.

1. By definition,

$$\text{Fraction} = \frac{\text{quantity A}}{\text{quantity B}}$$

and generally in chemisty,

$$\text{Fraction} = \frac{\text{part}}{\text{total}} = \frac{\text{smaller number}}{\text{larger number}}$$

2. To find the **decimal equivalent** of a fraction, divide the numerator by the denominator.

Example: For the fraction 1/2, its decimal equivalent is **0.50**

(In these lessons, if a *fraction* has numbers without units, such as 1/2 or 2/5, assume for significant figures that the numbers are exact.)

In chemistry calculations, any number in the form 0.XX... (a value between zero and one) may be referred to as a fraction.

3. Fraction = any value between zero and one.

TRY IT

Q. (Use a calculator if needed.) For 5/8, what is its decimal equivalent?

Answer:

The decimal equivalent is **0.625**.

Percentages and Decimal Equivalents

A percentage can be defined in a number of ways. In equation form:

4. Percentage $= \dfrac{Part}{Total} \times 100\% = fraction \times 100\% = $ (decimal equivalent) $\times 100\%$

Example: 1/2 = 50% because 1/2 = 0.50 × 100% = 50%

A percentage must be converted to a decimal equivalent before it can be used in most calculations, but the conversion between a percentage and its decimal equivalent is easy mental arithmetic. Because a percentage is a decimal equivalent multiplied by 100%:

5. The decimal point in a percentage is always two places to the right of the decimal in its decimal equivalent.

Example: 0.50 = 50%

In all percentage calculations, "100" is exact and does not affect the uncertainty (and significant figures) in a calculated answer.

TRY IT

Q1. If the decimal equivalent is 0.075, the percentage =

Answer:

0.075 × 100% = **7.5%**

Q2. If the percentage = 85%, its decimal equivalent =

Answer:

85%/100% = **0.85**

In scientific calculations, percentages are often measurements (such as "4.5% KCl by mass") that have uncertainty. We will address "percentages with chemistry attached" in the next lesson, but unless otherwise noted, you should assume that "number percentages without units attached" (such as 25%) are exact.

PRACTICE A

1. In your notebook, summarize rules 1–5 into a form that fits on "one-way" (notched) flashcards. Make the cards and write a "12.3" on each at the bottom right corner. You may also want the cards for Lessons 12.3 and 12.4 to be in a different color.

   Practice your cards, then complete the following problems trying not to look back at the cards or the lesson. For this Practice set, do any needed math with pencil and paper, not a calculator.

2. 34.8% has what decimal equivalent?

3. If the decimal equivalent is 0.013, what is the percentage?

4. Write the decimal equivalent of 0.25%

5. Write 9.5/100,000 as a decimal equivalent and a percentage.

Percentages may be WANTED or DATA. Let's consider those cases one at a time.

When a Numeric Percentage Is WANTED

6. If a "top-over-bottom" *fraction* is *given* and a percentage is WANTED, divide to find the decimal equivalent, then convert to its percentage.

TRY IT

Q. 7/8 is what percentage?

Answer:

7/8 (divide) = 0.875 = 87.5%

With practice, you can solve many numeric percent calculations using mental arithmetic by writing the *fraction*, then its *decimal equivalent*, then its percentage.

▶ **TRY IT**

🛑 STOP

Q. (Try without a calculator.) 2/5 is what percentage?

Answer:

2/5 = 0.40 = **40%** (exactly)

---

7. If *numbers* are *given* and a percentage is WANTED:
  • Write the *fraction* definition equation.
  • Substitute the numbers.
  • Calculate the decimal equivalent.
  • Write the percentage.

▶ **TRY IT**

🛑 STOP

Q. 25 is what percentage of 400?

Answer:

WANTED: Percentage   (Find the fraction first.)

$$? = \text{fraction} = \frac{\text{part}}{\text{total}} = \frac{\text{smaller number}}{\text{larger number}} = \frac{25}{400} = \mathbf{0.0625}$$

Percentage = 6.25%

In general, the sequence is:

8. To find a %, write smaller/larger, then decimal, then %.

PRACTICE   B

1. Make a "one-way" notched flashcard designed above for rule 8. Practice with the card. Add it to the other cards for this lesson, then shuffle the cards. Practice until perfect, then do the problems below. Use a calculator as needed.

2. 3/5 is what decimal equivalent and what percentage?

3. What percentage of 25 is 7?

4. Twelve is what percentage of 24,000?

When a Numeric Percentage Is DATA

If a "number percent" (a percentage without a unit or substance attached) is supplied as DATA, SOLVE using

> **9.** $X\%$ *of* $Y =$ (decimal equivalent) *times* Y

► TRY IT

Q. 3.5% of 12,000 grams is how many grams?

STOP

Answer:

3.5% of 12,000 g = **0.035** $\times$ 12,000 g = **420 g**

The problem includes a unit, but the supplied percentage does not, so rule 9 applies.

Recalling the Rules

You don't need to know the numbers of the rules above, but you need to be able to recall and apply the rules automatically. Finish designing cards for any rules in this lesson that cannot yet be recalled automatically. Practice the cards until given the front, you can recite the back. Then complete the practice below.

PRACTICE   C

If you answer these "short-answer" questions on a sheet of paper beside the question, you can "re-quiz" yourself on these when fractions are a part of later topics. Assume all numbers are exact.

For problems 1–5, use pencil and paper if needed, but *not* a calculator.

1. How much is 25% of 80?

2. 16.7% has what decimal equivalent?

3. 12% of 200 = ?

4. Write the decimal equivalent of 0.6%.

5. 45/10,000 has what decimal equivalent and what percentage?

For problems 6–9, use a calculator.

6. How much is 0.0450% of 7,500?

7. 3/8 has what decimal equivalent and is what percentage?

8. 1.8 is what percent of 45?

9. 0.25% of 12,400 = ?

Lesson 12.4 — Concentration in Mass Percent

Mass and Weight Percent Solutions

In chemistry, solution concentration is most often measured as a molarity, but concentration may also be expressed as a **mass percent** (also called **percent by mass**).

> **Example:** In a *5% by mass* NaCl solution, 5 g of NaCl are dissolved per 100 g of solution.

In the health sciences, solution concentration may be expressed as a **percent by weight** or **percent (weight/weight)** or **percent (w/w)**, abbreviated as **% (w/w)**. Each of these terms has the same meaning as a *mass percent* or *percent by mass* or **% (m/m)**. In solution calculations, "weight" is generally measured in mass units.

To summarize:

> **1.** For calculations in which solution concentration is measured using percentages, the terms
> - *mass percent* and *percent by mass* and *% (m/m)* and
> - *percent by weight* or *percent (weight/weight)* or *percent (w/w)* or *% (w/w)*
>
> all have the same meaning, and their calculations are solved in the same way.

The Importance of Labels

Percent mass calculations can involve up to three measured entities: the dissolved *solute*, the *solvent*, and their combination: the *solution*. A key to percent mass calculations is:

> **2.** Label each unit with either the solute *formula*, the solvent *name*, or *"soln."*

In a problem about one solute, units measuring the solution may be labeled with "formula solution" (such as "KOH soln.") or just "soln." with the formula omitted as understood. Measurements of "just the solute" must be labeled with the solute formula.

When a Mass Percent Is WANTED

When percentage calculations have chemistry labels attached, the rules are similar to those of numeric percentages in the prior lesson. In math:

$$\textbf{Percentage} = \frac{Part}{Total} \times 100\% = fraction \times 100\%$$

3. In chemistry, by definition,

$$\textbf{Mass percent} = \frac{\textit{mass of solute dissolved}}{\textit{total mass of solution}} \times 100\% = \textit{mass fraction} \times 100\%$$

4. If a *mass percent* is WANTED, FIND the mass *fraction* **first**. Write

WANTED: ? mass percent =   (Find *fraction* first.)

FIND: mass fraction = ? $\frac{\text{g solute}}{\text{g soln.}}$ =   (Calculate decimal, then %.)

A mass fraction is a fraction made of two masses: g solute (the *part*) and g soln. (the *total*). Finding the decimal equivalent of the mass *fraction* first, rather than trying to substitute numbers into the mass percent definition, will assist in using conversions.

A mass *fraction* will be a decimal equivalent with a value between zero and one (**0.XX…**). A mass *percent* will be a value between 0% and 100%.

Steps If a mass percent or any of its equivalent terms is WANTED:

a. Write the WANTED and FIND terms above.

b. List the DATA.

c. Using conversions, SOLVE the FIND question.

d. Convert the FOUND decimal equivalent to the WANTED percentage with units attached.

▶ TRY IT

Q. If 25.0 grams of KOH is dissolved to make 750 grams of aqueous solution, what is its concentration in mass percent?

STOP

Answer:

WANTED: ? mass percent =   (Find the *fraction* first.)

FIND: mass fraction = ? $\frac{\text{g solute}}{\text{g soln.}}$ = ? $\frac{\text{g KOH}}{\text{g soln.}}$ =   (Rule 4.)

DATA: 25.0 g KOH = 750 g soln.

SOLVE:

FIND: $? \dfrac{\text{g KOH}}{\text{g soln.}} = \dfrac{25.0 \text{ g KOH}}{750 \text{ g soln.}} = \mathbf{0.033} \dfrac{\text{g KOH}}{\text{g soln.}} = \textit{fraction}$

? mass percent = fraction × 100% = **0.033** × 100% = **3.3% KOH by mass**

Some special rules apply to mass percent calculations.

> **5.** In percentage concentration calculations, if moles are not mentioned, you will not need a molar mass. If *moles* is a unit in the WANTED or DATA, write the solute molar mass in the DATA.

Most chemistry relationships are defined in terms of moles, but percentage concentration is defined based on mass. You will need a g/*mol* ratio only if *moles* appear as a unit in the WANTED or DATA.

> **6.** In chemistry, if a concentration is expressed as a percentage, assume it is a *mass* percent unless otherwise noted. (Health science courses may use different conventions.)

PRACTICE   **A**

First learn the rules, then do the problems.

1. Based on the Try It above, how do the steps to find a mass percent differ from the usual steps of solving for a ratio unit?

2. If 6.0 grams of glucose is dissolved to make 150 grams of glucose solution, find the percentage of glucose by mass.

3. If 0.050 mol NaOH is dissolved to make 200. g of solution, find the % (w/w) NaOH (40.0 g NaOH/mol).

When Mass Percent Is DATA

A percentage can be defined as the count of one component part *per* 100 total components. In general, if a percentage of a substance is supplied in a problem, in your DATA you would write either

> A. *X*% means: *X* units of *part* = **100** units of *total; or*
>
> B. *X*% means: (decimal equivalent) units of *part* = **1** unit of *total.*

In these lessons, we will usually write statement A, but statements A and B are mathematically equivalent.

> **Example:** If a problem says "50% of the dots are blue," there are 50 blue dots *per* 100 total dots. Write:
>
> DATA: 50% blue dots means: 50 blue dots = **100 total** dots.

If a *mass* percent is supplied without units, it is assumed to be the number of *grams* of solute per 100 total *grams* of *solution*.

> **7.** When a mass percent is supplied in a problem, in your DATA write:
>
> DATA: *X*% by mass means: *X* g solute *per* **100** g soln.

Example: If "the solution is 3% HCl by mass," write:

DATA: 3% HCl by mass means: **3 g HCl = 100 g soln.**

By rule 6, if a percentage concentration is expressed without a dimension, assume it is a *mass* percent unless otherwise noted.

Example: If "the solution is *5% NaI*," write:

DATA: 5% NaI means: 5 **g** NaI = 100 **g** soln.

Using these rules, problems that state a percentage concentration can be solved using WANTED, DATA, and SOLVE.

► TRY IT

Q. In 2.0% KCl, how many grams of KCl are in 150 g of solution?

Answer:

WANTED: ? g KCl

DATA: 2% KCl means: **2.0** g KCl = **100** g soln.   (Rules 6 and 7.)

150 g soln.

SOLVE: ? g KCl = 150 g soln. • $\dfrac{2.0\ \text{g KCl}}{100\ \text{g soln.}}$ = **3.0 g KCl**

Each unit must be labeled as measuring the *solute* or *soln.*

PRACTICE   B

1. To learn rule 7, make this one-way (notched) flashcard:

   | See 5% NaCl? Write: | 5 g NaCl = 100 g soln. |

Practice until seeing the front, you can recall the back. Then complete the remaining problems.

2. If a solution is 4.7% HBr, how many grams of HBr are in 100. g of solution? (Solve without a calculator.)

3. If 2.50% of an aqueous solution is potassium iodide, how many grams of the solution contain 12.5 g of KI?

4. In 3.50% glucose, how many grams of glucose are in 225 grams of solution?

Dilute and Nondilute Solutions

If no solvent for a solution is identified, assume the solvent is water. A *dilute* aqueous solution is generally considered to be one in which the solute concentration is *less* than either 1% or 0.2 molar.

> **8.** Dilute solution ≈ <1% or <0.2 M

When calculating with percentage concentrations, we often need to know the solution density. Because a dilute (~1%) aqueous solution is 99% water by mass, and water has a density of about 1.00 g/mL at or near room temperature, we can apply the following *approximation*.

> **9.** The *dilute* prompt: If the solution is said to be *dilute*, add to your DATA:
>
> 1 mL *dilute* soln. ≈ 1.00 mL H_2O solvent ≈ 1.00 g H_2O solvent ≈ 1.00 g soln.

In a solution that is *not* dilute, the mass of the solute becomes significant, and the mass of the solvent and the solution differ significantly. Use this rule:

> **10.** If calculations involve *solvent amounts*, write:
>
> a. g solute + g solvent = g solution.
>
> b. Attach labels that distinguish the *solute*, the *solvent*, and the *solution*.

TRY IT

Q. In 100. g of 12% by mass KCl:

a. What is the mass of the KCl in the solution?

b. What is the mass of the water in the solution?

Answer:

a. The mass of the KCl is 12% of the 100. g soln. = 0.12 × 100. g = **12 g KCl**.

b. Because water is the *solvent*, write:

g solute + g solvent = g soln.

12 g KCl + ? g H_2O = 100 g soln.

? g H_2O = 100. g soln. − 12 g KCl = **88 g H_2O**

In a solution that is not dilute, calculations often require the *solution density*. For conversions to be arranged properly, *both* units in the density must be labeled "soln."

> **11.** The units of solution *density* are mass of *solution* per volume of *solution*.

Example: If "the solution density is 1.15 g/mL," write:

DATA: 1.15 g *soln.* = 1 mL *soln.*

► TRY IT

Q. If the density of a 3.31 M KOH solution is 1.14 g/mL, what is its concentration in mass percent? (56.1 g KOH/mol)

STOP

Answer:

WANTED: ? mass percent =   (Find mass *fraction* first.)

FIND: $\dfrac{? \text{ g solute}}{\text{g soln.}} = \dfrac{? \text{ g KOH}}{\text{g soln.}} =$ (Rule 4.)

DATA: 3.31 mol KOH = 1 L soln.   (M prompt.)

1.14 g *soln.* = 1 mL *soln.*   (*Solution* density.)

56.1 g KOH = 1 mol KOH   (g KOH in FIND.)

SOLVE: (Want a ratio? Start with a ratio.)

$$\frac{? \text{ g KOH}}{\text{g soln.}} = \frac{56.1 \text{ g KOH}}{1 \text{ mol KOH}} \cdot \frac{3.31 \text{ mol KOH}}{1 \text{ L soln.}} \cdot \frac{10^{-3} \text{ L soln.}}{1 \text{ mL soln.}} \cdot \frac{1 \text{ mL soln.}}{1.14 \text{ g soln.}}$$

$$= \mathbf{0.163} \ \frac{\text{g KOH}}{\text{g soln.}} = \text{decimal fraction; \% } = \mathbf{16.3\% \ by \ mass \ KOH}$$

(Your conversions may be in a different order.)

PRACTICE   C

Design flashcards for rules 8–11, practice, then try these problems. You may want to save one problem for your next practice session.

1. How many milliliters of dilute 0.50% KCl solution contain 1.6 g of KCl?

2. Express the concentration of a relatively dilute 0.080 M HCl solution as a mass percent.

3. To make 500. g of an aqueous solution that is 6.0% NaCl by mass, how many grams of water would be needed?

4. In 44% H_2SO_4 with a density of 1.34 g/mL, how many moles of H_2SO_4 are in 80. mL of solution? (98.1 g H_2SO_4/mol)

Rules for Percent (Weight/Volume)

In the health sciences, solution concentration may be expressed as **percent (weight/volume)**, abbreviated as **% (w/v)**. For % (w/v) calculations:

---

12. Unless otherwise specified, assume the unit of mass and weight (**w**) is grams and the unit of volume (**v**) is milliliters.

13. For % (**w/v**), substitute "**mL** soln." for "**g** soln." in % (**w/w**)—which is mass percent.

Examples

If *% (w/v)* is WANTED, write

FIND: ? (w/v) fraction = ? $\dfrac{\text{g solute}}{\text{mL soln.}}$ =   (Find decimal, then %.)

If **X% (w/v)** is DATA, write: **X% (w/v)** means: *X* g solute = 100 **mL** soln.

TRY IT

Q. If a problem includes "1.5% (w/v) KI," in your DATA, write:

Answer:

1.5% (w/v) KI means: 1.5 **g** *KI* = 100 **mL** *soln.*

One implication of the dilution approximation is that in a *dilute* solution, the *grams* and *milliliters* of the *water* and *solution* all have the same *numeric value*.

---

14. In *dilute* solutions: % (**w/v**) = % (**w/w**) = mass percent.

PRACTICE   D

Design flashcards for rules 12–14, practice, and then complete these problems.

1. The **normal saline** solution used in intravenous drips is 0.90% (w/v) NaCl. How many grams of NaCl would be contained in 125 mL of normal saline?

2. The *normal saline* solution is 0.90% (w/v) NaCl. Calculate the molarity of normal saline.

SUMMARY

In this chapter, we move from suggesting flashcards toward having you design your own. Learning to self-test on new topics is an essential skill.

Flashcards are especially effective during initial learning because they parallel the way memory works. The brain has what science calls an "atomic structure": Each brain cell (a neuron) can store a small element of knowledge, and each of your ~100 billion neurons can connect to hundreds of others. Hearing or reading a known word or seeing an image activates linked cells that can then be accessed by working memory.

When designing flashcards, connect symbols, words, and images so that when prompted by one, the others come to mind. From your effort at recall and thought about meaning, associations are constructed in memory. If you learn fundamentals with flashcards first, then apply new knowledge to distinctive problems, your working memory will have room to associate new knowledge with different problem types, and you will learn to solve each problem type more quickly.

REVIEW QUIZ

You may use a calculator and a periodic table. On the multiple-choice problems, SOLVE—*then* find your answer in the choices provided.

1. If 200. mL of 0.75 M NaCl is diluted to a volume of 600. mL, what is the diluted [NaCl]?

2. To what volume must 50.0 mL of 2.0 M KOH be diluted to make 0.24 mol $\cdot$ L^{-1} KOH?

    a. 600. mL   b. 420 mL   c. 60. mL   d. 104 mL   e. 24 mL

3. In a 0.20 M solution of ammonium sulfate, the [NH$_4^+$] is:

    a. 0.10 M   b. 0.20 M   c. 0.30 M   d. 0.40 M   e. 0.60 M

4. In a solution of K$_3$PO$_4$, if the concentration of potassium ion is 0.45 M, what is the concentration of the phosphate ion?

5. 3/8 has what decimal equivalent and what percentage?

6. 21 is what percentage of 84?

7. 0.75% of 16,800 = ?

8. A solution with a mass of 200. grams contains 4.0 grams of NaOH. What is the % NaOH in the solution?

9. If a dilute solution is 1.60% KCl, how many milliliters would be required to supply 2.00 grams of KCl?

10. A 10.0% by mass aqueous solution of H$_2$SO$_4$ has a density of 1.07 g/mL. How many milliliters would be needed to supply 26.8 g H$_2$SO$_4$?

    a. 0.400 mL   b. 250. mL   c. 287 mL

    d. 28.7 mL   e. 25.0 mL

ANSWERS

Some answers are *partial*. Complex *calculations* must include WANTED, DATA, and SOLVE.

Lesson 12.1

Practice A   1a. 35 mL = 35 × 10^{-3} L = **0.035 L**   1b. 125 mL = 125 × 10^{-3} L = **0.125 L**

2a. 2.500 L = **2,500.** mL   2b. 15 mL = **0.015 L**   2c. 0.77 L = **770** mL

3. **Three** places to the **left**.   4a. 150 mL = **0.15 L**   4b. 33 L = **33,000 mL**

4c. **0.00921 L**   4d. **450 mL**   4e. **0.000833 L**   4f. **65.5 mL**

Practice B   2. Because the volume has been *quadrupled*, the concentration must be cut to **1/4** the original amount. The final [KCl] = **0.50 M**.

3. Because [final] has been cut to **1/5** [original], volume must become *five times* higher. Final volume = 5 × 250. mL = **1,250 mL**.

4. The volume is tripled, so [NaOH] is cut by 1/3 to **0.20 M**.

Practice C   1a. 250 mL = **0.25 L**   1b. 0.62 L = **620 mL**   1c. 3.5 mL = **0.0035 L**

2a. **0.300 M NaOH**. If volume increases by 10×, the solution is 1/10th as concentrated.

2b. WANTED: [NaOH] = ? M NaOH = ? $\dfrac{\text{mol NaOH}}{\text{L NaOH soln.}}$ = M_D

   DATA: 50.0 **mL** V_C (**C**, lower of two volumes.)

   3.00 mol NaOH = 1 L soln. M_C (Paired with 50.0 mL, which is **C**.)

   ~~0.750 L~~ 750. mL soln. V_D (Higher volume = diluted.)

   SOLVE: Dilution: $V_C × M_C = V_D × M_D$ Solve for WANTED *symbol*.

$$? M_D = \frac{V_C × M_C}{V_D} = \frac{50.0\ \cancel{mL} × 3.00\ M\ NaOH}{750.\ \cancel{mL}} = 0.200\ \textbf{M NaOH diluted}$$

Check: 50 to 800 mL makes V about 15 times larger, so the [new] will be about 1/15× original = ~1/15 × 3 M = ~3/15 M = ~1/5 M = ~0.20 M. Check!

3. WANTED: ? mL KCl soln. = V_D (Choose mL as consistent volume unit.)

   DATA: 20.0 mL KCl V_C (Goes with 2.5 M, which is C.)

   2.5 mol KCl = 1 L KCl M_C ([Higher])

   0.24 mol KCl = 1 L KCl M_D

(Strategy: Labeling DATA C or D *first* helps in labeling WANTED C or D.)

   SOLVE: $V_C × M_C = V_D × M_D$ SOLVE for WANTED symbol.

$$? \textbf{mL}\ V_D = \frac{V_C × M_C}{M_D} = \frac{20.0\ mL × 2.5\ M}{0.24\ M} = 210\ \textbf{mL KCl soln. D}$$

Check: From 2.5 M **C** to 0.24 M **D** is *about* 1/10th, so the V_D must increase the V_C by *about* 10 times. 10 × 20 mL = 200 mL. Pretty close.

4. WANTED: $? \mathbf{mL}$ NaCl soln. $= \mathbf{V_C}$ (Given a choice, pick a unit to match DATA.)

    DATA: 250 mL soln. $\mathbf{V_D}$ (D because it is paired with 0.65 M, which is D.)

                0.65 mol NaCl = 1 L $\mathbf{M_D}$ ([Lower])

                2.00 mol NaCl = 1 L $\mathbf{M_C}$ ([Higher])

    SOLVE: $V_C \times M_C = V_D \times M_D$

$$? \text{ mL NaCl soln.} = V_C = \frac{V_D \times M_D}{M_C} = \frac{250 \text{ mL} \cdot 0.65 \text{ M}}{2.00 \text{ M}} = \mathbf{81 \text{ mL NaCl soln.}}$$

Check: The 0.65 M must be about tripled to reach 2.00 M, so the new V must be about 1/3 of the original 250 mL = about 80 mL.

5. SOLVE: In dilution: $V_C \times M_C = V_D \times M_D$

$$[\text{HCl}] = ? \mathbf{M_C} = \frac{V_D \times M_D}{V_C} = \frac{250 \text{ mL} \cdot 0.15 \text{ M}}{6.2 \text{ mL}} = \mathbf{6.0 \text{ M HCl}}$$

Check: Increasing V from 6 mL to 250 mL is ~40× increase, so the new M is ~1/40th [original]. If the original was 6 M, 1/40th is about 0.15 M. Check.

Lesson 12.2

Practice A   1. The ions in the solid separate and move about freely in the solution.

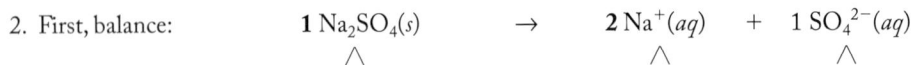

      2. First, balance: $1 \text{ Na}_2\text{SO}_4(s)$ $\rightarrow$ $2 \text{ Na}^+(aq)$ $+$ $1 \text{ SO}_4^{2-}(aq)$

                                          $\wedge$ $\wedge$ $\wedge$

        Then, based on coefficients: 0.50 moles *used up* $\rightarrow$ **1.0 mol** $+$ **0.50 mol** *formed*

        In solution are: a. **1.0 mol Na$^+$** ions   b. **0.50 mol SO$_4^{2-}$** ions   c. **no Na$_2$SO$_4$** particles

        d. **1.5 total** moles of ions

Practice B   1. Radium nitrate $= \mathbf{1 \text{ Ra}^{2+}} + \mathbf{2 \text{ NO}_3^-} = \mathbf{1 \text{ Ra(NO}_3)_2}$ (Goes 100%.)

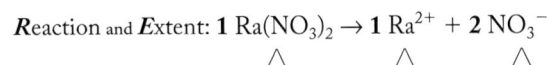

        *R*eaction and *E*xtent: $1 \text{ Ra(NO}_3)_2 \rightarrow 1 \text{ Ra}^{2+} + 2 \text{ NO}_3^-$

                                  $\wedge$ $\wedge$ $\wedge$

        *C*oncentrations: 0.30 M ($\rightarrow$ 0 M) $\rightarrow$ **0.30 M** $+$ **0.60 M** formed

      2. *R*eaction and *E*xtent: $1 \text{ Na}_2\text{CO}_3 \rightarrow 2 \text{ Na}^+ + 1 \text{ CO}_3^{2-}$ (Goes 100%.)

                                    $\wedge$ $\wedge$ $\wedge$

        *C*oncentrations: 0.60 M ($\rightarrow$ 0 M) $\rightarrow$ **1.2 M** $+$ **0.60 M**

      3. *R*eaction and *E*xtent: $1 \text{ K}_3\text{PO}_4(aq) \rightarrow \mathbf{3 \text{ K}^+} + 1 \text{ PO}_4^{3-}$ (Goes 100%.)

                                      $\wedge$ $\wedge$ $\wedge$

        *C*oncentrations: **0.15 M** ($\rightarrow$ 0 M) $\rightarrow$ 0.45 M $+$ **0.15 M** formed

        a. **3 K$^+$** are formed for every **one** PO$_4^{3-}$. **[PO$_4^{3-}$] = 0.15 M**

        b. The ratio of K$_3$PO$_4$ to K$^+$ is 1 to 3, so **[K$_3$PO$_4$]$_{\text{used up}}$ = 0.15 M.**

      4. *R*eaction and *E*xtent: $1 \text{ Al}_2(\text{SO}_4)_3 \rightarrow \mathbf{2 \text{ Al}^{3+}} + \mathbf{3 \text{ SO}_4^{2-}}$ (Goes 100%.)

                                      $\wedge$ $\wedge$ $\wedge$

        *C*oncentrations: **0.012 M** ($\rightarrow$ 0 M) $\rightarrow$ **0.024 M** $+$ 0.036 M formed

        a. **[Al^{3+}] = 0.024 M**

        b. [Al$_2$(SO$_4$)$_3$]$_{\text{as mixed}}$ is 1/3 of the given [SO$_4^{2-}$] = 1/3 (0.036 M) = **0.012 mol/L**

Lesson 12.3

Practice A   2. 34.8% = **0.348**   3. 1.3%   4. 0.0025

5. To divide by 100,000, move the decimal to the left five times.

Decimal equivalent = **0.000095** or **9.5 × 10⁻⁵** = **0.0095%** or **9.5 × 10⁻³ %**.

Practice B   2. 3/5 = **0.60 (exact)**; percentage = **60% (exact)**

3. Fraction = $\dfrac{\text{smaller}}{\text{larger}}$ = $\dfrac{7}{25}$ = 0.28; % = decimal equivalent × 100 = **28%**

4. Fraction = $\dfrac{\text{part}}{\text{total}}$ = $\dfrac{12}{24{,}000}$ = $\dfrac{12}{24 \times 10^3}$ = 0.5 × 10⁻³; % = decimal × 10² % = **0.050%**

Practice C   1. X% of Y = (decimal equivalent) times Y; 25% of 80 = 0.25 × 80 = **20**

2. 16.7% = **0.167**   3. 12% of 200 = 0.12 × 200 = **24**

4. Decimal equivalent = percent/100% = 0.6%/100% = **0.006**

5. 45/10,000 = **0.0045**; percentage = **0.45%**

6. 0.00045 × 7,500 = **3.4**   7. 3/8 = 0.375 = **37.5%**

8. 0.040 = **4.0%**   9. 0.0025 × 12,400 = **31**

Lesson 12.4

Practice A   1. Before you solve for the %, solve for the ratio (the fraction) inside the percent definition. After finding the ratio (fraction), you must convert the fraction's decimal value to a percentage.

2. WANTED: ? mass % glucose = (Find mass *fraction* first.)

FIND: $\dfrac{?\ \text{g glucose}}{\text{g soln.}}$ = (Write fraction, then decimal, then %; rule 4.)

DATA: 6.0 g glucose = 150 g soln.   (Two measures of one solution.)

SOLVE: $\dfrac{?\ \text{g glucose}}{\text{g soln.}}$ = $\dfrac{6.0\ \text{g glucose}}{150\ \text{g soln.}}$ = **0.040** = fraction = **4.0%** glucose by mass

3. DATA: 0.050 mol NaOH = 200. g soln.   (Two measures of one solution.)

40.0 g NaOH = 1 mol NaOH   (Rule 5: moles in DATA.)

SOLVE: $\dfrac{?\ \text{g NaOH}}{\text{g soln.}}$ = $\dfrac{0.050\ \text{mol NaOH}}{200\ \text{g soln.}}$ · $\dfrac{40.0\ \text{g NaOH}}{1\ \text{mol NaOH}}$ = **0.010** = fraction = **1.0% (w/w) NaOH**

Practice B   2. WANTED: ? g HBr =

DATA: 4.7 g HBr = 100 g soln.   (Rule 7.)

100 g soln.

SOLVE: ? g HBr = 100 g soln. · $\dfrac{4.7\ \text{g HBr}}{100\ \text{g soln.}}$ = **4.7 g HBr**

3. SOLVE: ? g soln. = 12.5 g KI · $\dfrac{100\ \text{g soln.}}{2.50\ \text{g Kl}}$ = **500. g soln.**

4. SOLVE: ? g glucose = 225 g soln. · $\dfrac{3.50\ \text{g glucose}}{100\ \text{g soln.}}$ = **7.88 g glucose**

Practice C   1. SOLVE: ? mL KCl soln. = 1.6 g KCl • $\dfrac{100 \text{ g soln.}}{0.50 \text{ g KCl}}$ • $\dfrac{1 \text{ mL soln.}}{1.00 \text{ g soln.}}$ = **320 mL KCl soln.**

2. DATA: 0.080 *mol* HCl = 1 L soln.   (M prompt.)

         1 mL *dilute* soln. ≈ 1.00 mL water ≈ 1.00 g H_2O ≈ 1.00 g soln.

         36.5 g HCl = 1 mol HCl   (Moles is in DATA.)

     SOLVE: $\dfrac{? \text{ g HCl}}{\text{g soln.}}$ = $\dfrac{36.5 \text{ g HCl}}{1 \text{ mol HCl}}$ • $\dfrac{0.080 \text{ mol HCl}}{1 \text{ L soln.}}$ • $\dfrac{10^{-3} \text{ L soln.}}{1 \text{ mL soln.}}$ • $\dfrac{1 \text{ mL soln.}}{1.00 \text{ g soln.}}$

             = **0.0029** $\dfrac{\text{g HCl}}{\text{g soln.}}$ = fraction; **% = 0.29% HCl by mass**

3. WANTED: ? g water =   (Hint: Find g NaCl first.)

     SOLVE: **? g NaCl** = 500 g soln. • $\dfrac{6.0 \text{ g NaCl}}{100 \text{ g total soln.}}$ = 30. g NaCl   What must be the mass of the *water*?

         g solute + g solvent = g soln.; 30 g NaCl + **?** g H_2O = 500 g soln.

         ? g H_2O = 500. g soln. − 30. g NaCl = **470. g H_2O**

4. WANTED: ? mol H_2SO_4 =

     DATA: 1.34 g soln. = 1 mL soln.

         44% H_2SO_4 means: 44 g H_2SO_4 = 100 g soln.   (Rules 6 and 7.)

         98.1 g H_2SO_4 = 1 mol H_2SO_4 (Rule 5: Moles is WANTED.)

         80. mL soln.

     SOLVE: ? mol H_2SO_4 = 80. mL soln. • $\dfrac{1.34 \text{ g soln.}}{1 \text{ mL soln.}}$ • $\dfrac{44 \text{ g } H_2SO_4}{100 \text{ g soln.}}$ • $\dfrac{1 \text{ mol } H_2SO_4}{98.1 \text{ g } H_2SO_4}$ = **0.48 mol H_2SO_4**

Practice D   1. SOLVE: ? g NaCl = 125 mL soln. • $\dfrac{0.90 \text{ g NaCl}}{100 \text{ mL soln.}}$ = **1.1 g NaCl**

2. DATA: 0.90% (w/v) NaCl means: 0.90 g NaCl = 100 **mL** soln.   (Rules 12 and 13.)

         58.5 g NaCl = 1 mol NaCl   (Moles is in WANTED unit.)

     SOLVE: $\dfrac{? \text{ mol NaCl}}{\text{L soln.}}$ = $\dfrac{1 \text{ mol NaCl}}{58.5 \text{ g NaCl}}$ • $\dfrac{0.90 \text{ g NaCl}}{100 \text{ mL soln.}}$ • $\dfrac{1 \text{ mL soln.}}{10^{-3} \text{ L soln.}}$ = **0.15** $\dfrac{\text{mol NaCl}}{\text{L soln.}}$

Review Quiz

     1. 0.25 M NaCl   2. **b.** 420 mL soln.   3. **d.** 0.40 M   4. $[PO_4^{3-}]$ = 0.15 M

     5. 0.375 = 37.5%   6. 25%   7. 126   8. 2.0% NaOH by mass

     9. 125 mL KCl soln.   10. **b.** 250. mL H_2SO_4 soln.

13

Ionic Equations and Precipitates

Lesson 13.1   Ionic Compound Solubility

Prerequisites

If you have any difficulty with the questions in the following Practice set, review your ion flashcards from Lesson 7.3 and the naming rules in Lesson 7.4 before continuing with this chapter.

PRACTICE A

Answer these in your notebook.

1. Write the separated-ion formulas for these ionic compounds. Include coefficients showing the ion ratios.

    a. $PbCl_2$ b. $Mg(OH)_2$ c. K_3PO_4

2. Write standard formulas for these ionic compounds.

    a. Sodium nitrate   b. Ammonium carbonate   c. Aluminum bromide

3. Name these compounds.

    a. K_2CrO_4 b. NH_4I c. $FeSO_4$

Solubility

In chemistry, a major topic of interest is aqueous solubility—the extent to which a substance dissolves in water. When solubility is discussed, you should assume the solvent is water unless otherwise noted.

All ionic compounds dissolve in water to *some* extent. The following definitions are generally accepted:

* If less than 0.10 mol/L of a substance dissolves, it is termed either **slightly soluble** or **insoluble**.
* If 0.10 mol/L or *more* dissolves at room temperature, the substance is termed **soluble**.

A Solubility Scheme

The solubility of an ionic compound can often be predicted from its name or formula and a **solubility scheme**—a set of rules to predict the solubility of ion combinations. The scheme below is *limited* but predicts the solubility of many of the ionic compounds encountered in first-year chemistry.

This scheme is **hierarchical**: Higher rules take *precedence* over those beneath them. This means that you should base your prediction for the solubility of a compound on the *first* rule that applies, starting from the top.

You will need to commit this scheme to memory:

Positive Ions	**Negative** Ions	**Solubility**
1. (alkali metals)$^+$, NH_4^+	NO_3^-, CH_3COO^-	**Soluble**
2. Pb^{2+}, Ag^+	CO_3^{2-}, PO_4^{3-}, S^{2-}, CrO_4^{2-}	**Insoluble**
3.	Cl^-, Br^-, I^-	**Soluble**

Exceptions to these rules include the following: column 2 sulfides and aluminum sulfide decompose in water; $AgCH_3COO$ is moderately soluble; copper(I) halides are insoluble. But in these lessons, we won't assign problems about exceptions.

> **Example:** According to this scheme, is NaCl soluble in water? Because NaCl combines a metal and a nonmetal atom, it is predicted to be ionic, composed of Na^+ and Cl^- ions. By rule 1, all compounds that contain Na^+ ions are *soluble*. [From experience, we can confirm that NaCl (table salt) dissolves to a substantial extent in water.]

If only one ion of the two in a compound is in this table, presume that the compound will follow the rule for the ion that is in the table. This will not always be accurate but is a best guess. With solubility, there are exceptions.

Using the Scheme to Make Predictions

When using any solubility scheme, if you are unsure of which ions are in a compound, you should write out the *separated-ion* formula that shows the ion charges.

━━━▶ TRY IT

For the questions below, write your answer and your solubility scheme reasoning.

Q1. Is $Ba(NO_3)_2$ soluble?

STOP

Answer:

1. $Ba(NO_3)_2 \rightarrow Ba^{2+} + 2\,\mathbf{NO_3^-}$. All nitrates are **soluble** by rule 1.

Q2. Is $PbCl_2$ soluble or insoluble?

Q3. Is $Cu(NO_3)_2$ soluble or insoluble?

STOP

2. $PbCl_2 \rightarrow \mathbf{Pb^{2+}} + 2\,Cl^-$. Compounds containing Pb^{2+} ion by rule 2 are **insoluble**. Compounds containing chloride ion by rule 3 are soluble, but rule 2 takes precedence. $PbCl_2$ is **insoluble**.

3. $Cu(NO_3)_2 \rightarrow Cu^{2+} + 2\,NO_3^-$. The above solubility scheme makes no prediction for Cu^{2+}, but based on rule 1 that all nitrates are soluble, predict **soluble**. If only one ion in a pair is in the table, base your prediction on the rule for that ion.

Devote special attention to rule 1. In a hierarchical table, the higher the rule, the more likely it is to be used. Some important generalizations:

1. If a compound contains one or more *alkali metal* atoms, each of those atoms will be a 1+ ion, and the compound will be *soluble*.

2. If the *name* of a compound includes the words *ammonium* or *nitrate* or *acetate*, or a formula includes any of those ions, predict the compound will be *soluble*.

The scheme above does not cover all ion combinations. In this text, if problems include cases not covered by the scheme, we will identify whether the combination is soluble or insoluble.

PRACTICE   B

Write the solubility scheme until you can do so from memory. Design flashcards that summarize the two shaded rules and practice their recall. Then, to solve these problems, use a scheme written from memory.

1. Write the names and formulas for the four ions (or families of ions) that, if present in a compound, mean the compound is always predicted to be soluble.

(continued)

2. Label each ion combination as soluble or insoluble and state a reason for your prediction (in a form similar to the Try It answers in this lesson).

   a. $K^+ + S^{2-}$

   b. $Sr^{2+} + Cl^-$

   c. $Ca^{2+} + CO_3^{2-}$

   d. $Ag^+ + CrO_4^{2-}$

3. Write the ions found in these combinations, then label the ion combination as soluble or insoluble and state a reason for your prediction.

   a. Lead(II) bromide $\rightarrow Pb^{2+} + Br^-$: **insoluble** by rule 2 for Pb^{2+} (example)

   b. Barium carbonate $\rightarrow$

   c. Sodium hydroxide $\rightarrow$

   d. $MgBr_2 \rightarrow$

   e. Silver nitrate $\rightarrow$

   f. Ammonium hydroxide $\rightarrow$

   g. $Fe_3(PO_4)_2 \rightarrow$

   h. $Pb(CH_3COO)_2 \rightarrow$

4. Label each compound as soluble or insoluble and state a reason for your prediction.

   a. $NiCrO_4$

   b. $RbBr$

   c. Fe_2S_3

5. Name two ions that in combination with Ag^+ form soluble compounds.

6. Are all compounds containing phosphate ions insoluble?

Lesson 13.2   Total and Net Ionic Equations

Mixing Ions

When all or part of an ionic compound dissolves in water, the portion that dissolves is said to *dissociate*, meaning that the ions that dissolve *separate*.

When aqueous solutions of different soluble ionic compounds are mixed, new combinations of positive and negative ions are possible, and reactions among the ions can occur. One type of reaction is **precipitation**.

Example: When dissolved sodium chloride (table salt) is added to dissolved silver nitrate, a white cloud of small solid particles forms immediately as the two solutions are mixed. Over several minutes, the particles from the cloud settle, leaving a clear solution with a layer of solid on the bottom.

A solid, insoluble compound that is formed when some solutions of soluble ionic compounds are mixed is termed a **precipitate**.

Total Ionic Equations

In the example above, the precipitate is silver chloride (AgCl). This reaction can be written as

$$NaCl(aq) + AgNO_3(aq) \rightarrow AgCl(s) + NaNO_3(aq) \tag{13.1}$$

After each formula is a notation of its physical **state** (also called its **phase**). The (aq) means that the substance or particle is in the aqueous phase. The (s) after AgCl is an abbreviation for *solid*: the state of a precipitate.

A precipitation reaction is often written using standard formulas with state notations as above; however, when ionic compounds are dissolved, their ions separate and move about freely. To understand a precipitation, it is best to write each compound that is (aq) as *separated ions*. For the reaction above, this is

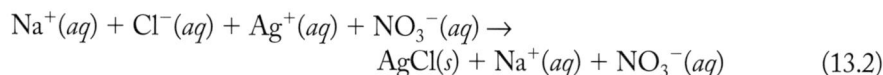

$$Na^+(aq) + Cl^-(aq) + Ag^+(aq) + NO_3^-(aq) \rightarrow$$
$$AgCl(s) + Na^+(aq) + NO_3^-(aq) \tag{13.2}$$

This format is termed the *total ionic equation* for the reaction.

The *total* **ionic equation** shows dissolved ions as *separated* ions in an (aq) state and a precipitate as a *standard* formula followed by (s) meaning *solid* state.

From equations 13.1 and 13.2, it can be seen that to form the precipitate, Cl^- ions from one solution reacted with Ag^+ ions from the other.

The important general rule is:

To understand the reactions of ionic compounds, if some ions are separated before or after reaction, write the equation with the separated ions separated.

Net Ionic Equations

The total ionic equation (13.2) shows that before *and* after the reaction, the Na^+ and NO_3^- ions are the *same* dissolved and separated ions: Those ions are not changed by this reaction.

A **spectator ion** is present during a reaction but has the same formula and state before and after the reaction.

Total ionic equations include spectators so that we can see *all* of the ions present; however, as in mathematical equations, terms that are the same on each side of an equation can be cancelled.

> A *net* **ionic equation** shows only the reactants and products that *change* their formula or state (those that *react*) in a reaction.

In other words,

> In a *net* ionic equation, the *spectator ions* are *left out*.

To write a net ionic equation, starting from the total ionic equation,

- *cancel* the ions with the same formula and state on both sides, and
- write the equation without the cancelled spectators.

TRY IT

Q. In your notebook, starting from the total ionic equation (13.2), write the *net* ionic equation.

Answer:

Cancel the spectators:

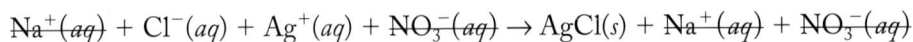

$$\cancel{Na^+(aq)} + Cl^-(aq) + Ag^+(aq) + \cancel{NO_3^-(aq)} \rightarrow AgCl(s) + \cancel{Na^+(aq)} + \cancel{NO_3^-(aq)}$$

Net ionic equation:

$$Ag^+(aq) + Cl^-(aq) \rightarrow AgCl(s) \tag{13.3}$$

> All reaction equations must balance for *atoms* and *charge*.

This means that the number of each kind of atom must be the same, *and* the overall net charge must be the same, on both sides of all equations. In the total and net ionic equations above, the overall charge is zero on both sides.

Equation Writing: Assume (*aq*)

To save space and time, for the remainder of these lessons, when discussing *aqueous solutions*, if an equation formula shows no state, assume the omitted state is (*aq*). When *writing* ionic equations, if a particle is aqueous, you *may* leave out the (*aq*) as understood, but you must include symbols such as (*s*) for solid and (*g*) for gas for particles *not* in the (*aq*) state.

1. Distill the rules above into flashcards. For vocabulary, make "two-sided" cards that associate the word and its meaning in both directions.

2. For these total ionic equations, circle the precipitate, cross out the spectators, and in the space below each write the *net* ionic equation.

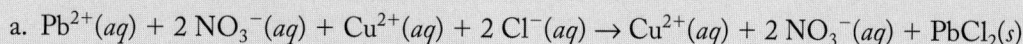

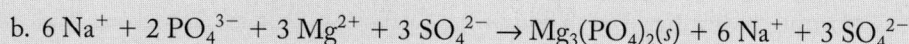

   a. $Pb^{2+}(aq) + 2\,NO_3^-(aq) + Cu^{2+}(aq) + 2\,Cl^-(aq) \rightarrow Cu^{2+}(aq) + 2\,NO_3^-(aq) + PbCl_2(s)$

   b. $6\,Na^+ + 2\,PO_4^{3-} + 3\,Mg^{2+} + 3\,SO_4^{2-} \rightarrow Mg_3(PO_4)_2(s) + 6\,Na^+ + 3\,SO_4^{2-}$

Leaving Out the Spectators

In the laboratory, a test substance or solution may be *labeled* with a *single* ion, such as Ag^+ or OH^-, but in such cases, other ions must also be present so that the *net* charge of the test substance is zero. All substances must have an overall net charge of zero. "Leaving out the nonreactive spectators" on labels is a way to emphasize the ion that is likely to *react*. The presence of other ions that balance the charge is "understood."

> **Example:** If a dropper bottle containing a solution of dissolved ions is labeled "CO_3^{2-}," the solution must also contain positive ions that are soluble when combined with carbonate ions, such as alkali metal ions or ammonium ion.

1. If a dropper bottle of a clear aqueous solution is labeled "Pb^{2+}," name two ions that might also be in the bottle.

2. A test solution labeled "CrO_4^{2-}" will appear to be "water with a yellow dye" because dissolved chromate ions give their solutions a yellow color. Name six other ions that might also be present in the solution.

Lesson 13.3 Balancing Precipitation Equations

Balancing Standard Formulas

Precipitation reactions described using standard formulas can be balanced by our regular trial-and-error methods. To balance the equation, ignore the *(states)* and express final coefficients as lowest-whole-number ratios.

TRY IT

Q. Balance this precipitation equation.

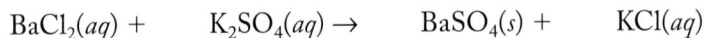

$$BaCl_2(aq) + \quad K_2SO_4(aq) \rightarrow \quad BaSO_4(s) + \quad KCl(aq)$$

Answer:

$$\textbf{1}\,BaCl_2(aq) + \textbf{1}\,K_2SO_4(aq) \rightarrow \textbf{1}\,BaSO_4(s) + \textbf{2}\,KCl(aq)$$

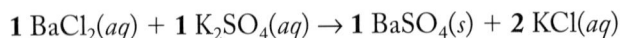

(To review balancing, see Lesson 8.5.)

During balancing, you should write coefficients that are **1** to keep track of which particles have been balanced. But if you see an equation in which some formulas have coefficients but others do not, assume the equation is balanced and that an omitted coefficient is 1.

PRACTICE   **A**

Balance these precipitation equations.

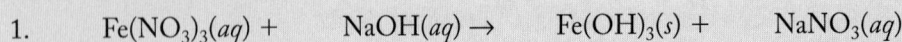

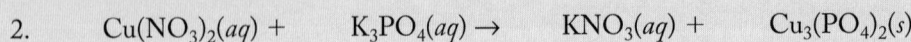

1. $Fe(NO_3)_3(aq) + \quad NaOH(aq) \rightarrow \quad Fe(OH)_3(s) + \quad NaNO_3(aq)$

2. $Cu(NO_3)_2(aq) + \quad K_3PO_4(aq) \rightarrow \quad KNO_3(aq) + \quad Cu_3(PO_4)_2(s)$

Converting Standard to Net Ionic Equations

To predict whether two solutions will form a precipitate when mixed, one step involves balancing dissolved compounds written as separated ions.

Example: When lead(II) nitrate is mixed with sodium bromide, white crystals of lead bromide form. In standard formulas, the balanced equation is:

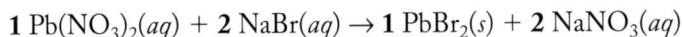

$$\textbf{1}\,Pb(NO_3)_2(aq) + \textbf{2}\,NaBr(aq) \rightarrow \textbf{1}\,PbBr_2(s) + \textbf{2}\,NaNO_3(aq)$$

Showing dissolved compounds as separated ions, the balanced equation can be represented as:

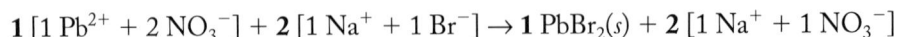

$$\textbf{1}\,[1\,Pb^{2+} + 2\,NO_3^-] + \textbf{2}\,[1\,Na^+ + 1\,Br^-] \rightarrow \textbf{1}\,PbBr_2(s) + \textbf{2}\,[1\,Na^+ + 1\,NO_3^-]$$

Note the following in the example:

- Compounds that are (*aq*) are written inside brackets [ ] as separated ions, matching how those ions are found in the solutions.

- Precipitates are labeled (*s*) and are *not* shown as separated ions, matching how they are found after the reaction.

- Coefficients in front of standard formulas become coefficients in front of the *brackets* that enclose separated ions.

To convert from a precipitation represented with standard formulas to the format that includes separated ions, follow these steps.

1. Below each standard formula that is (*aq*), write its separated-ions formula *inside brackets*. For a precipitate, retain the *standard* formula followed by (*s*).

2. Balance the equation written with standard formulas.

3. Place the standard-formula coefficients in *front* of each bracket and the precipitate.

▭▭▭▶ TRY IT

Q. In your notebook, write this precipitation equation as a balanced equation with dissolved compounds in brackets.

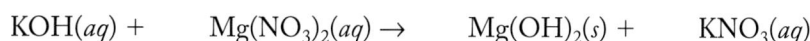

$$\text{KOH}(aq) + \text{Mg(NO}_3)_2(aq) \rightarrow \text{Mg(OH)}_2(s) + \text{KNO}_3(aq)$$

STOP

Answer:

By the step numbers:

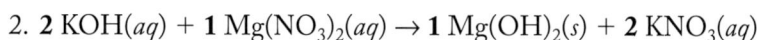

1. $[1\,\text{K}^+ + 1\,\text{OH}^-] + [1\,\text{Mg}^{2+} + 2\,\text{NO}_3^-] \rightarrow \text{Mg(OH)}_2(s) + [1\,\text{K}^+ + 1\,\text{NO}_3^-]$

2. $\mathbf{2}\,\text{KOH}(aq) + \mathbf{1}\,\text{Mg(NO}_3)_2(aq) \rightarrow \mathbf{1}\,\text{Mg(OH)}_2(s) + \mathbf{2}\,\text{KNO}_3(aq)$

3. $\mathbf{2}\,[1\,\text{K}^+ + 1\,\text{OH}^-] + \mathbf{1}\,[1\,\text{Mg}^{2+} + 2\,\text{NO}_3^-] \rightarrow \mathbf{1}\,\text{Mg(OH)}_2(s) + \mathbf{2}\,[1\,\text{K}^+ + 1\,\text{NO}_3^-]$

From the format with separated ions, it is easy to convert to the total and net ionic equations.

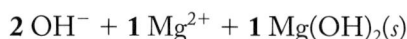

4. To write the *total* ionic equation, take out the brackets as you would take out parentheses in algebra: Multiply each coefficient inside a bracket by the coefficient in front of the bracket.

5. To write the *net* ionic equation, take out the spectators.

▭▭▭▶ TRY IT

Q. In your notebook, below your step 3 answer of the Try It above, write the total ionic equation, and below it write the net ionic equation.

STOP

Answer:

Total ionic equation:

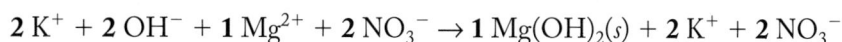

$$\mathbf{2}\,\text{K}^+ + \mathbf{2}\,\text{OH}^- + \mathbf{1}\,\text{Mg}^{2+} + \mathbf{2}\,\text{NO}_3^- \rightarrow \mathbf{1}\,\text{Mg(OH)}_2(s) + \mathbf{2}\,\text{K}^+ + \mathbf{2}\,\text{NO}_3^-$$

To write the net ionic equation:

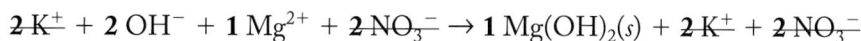

$$\mathbf{2}\,\cancel{\text{K}^+} + \mathbf{2}\,\text{OH}^- + \mathbf{1}\,\text{Mg}^{2+} + \cancel{\mathbf{2}\,\text{NO}_3^-} \rightarrow \mathbf{1}\,\text{Mg(OH)}_2(s) + \cancel{\mathbf{2}\,\text{K}^+} + \cancel{\mathbf{2}\,\text{NO}_3^-}$$

Net ionic equation:

$$\mathbf{2}\,\text{OH}^- + \mathbf{1}\,\text{Mg}^{2+} + \mathbf{1}\,\text{Mg(OH)}_2(s)$$

Now, without a calculator, write the *count* of the atoms of each kind, and the *net* charge, on each side of the *total* ionic equation. Is the equation balanced?

STOP

Total ionic equation: 2 K, 8 O, 2 H, 1 Mg, 2 N atoms on each side. Zero net charge on both sides. Balanced.

Now write the *count* of each atom and the *net* charge on each side of the *net* ionic equation. Is the equation balanced?

STOP

On each side, 2 O, 2 H, 1 Mg and zero net charge. Balanced.

PRACTICE   B

1. For the equation balanced in Practice A, problem 1 of this lesson:

   a. Write the balanced equation in the format that uses brackets to represent compounds that are (*aq*) but precipitates as standard formula followed by (*s*).

   b. Write the *total* ionic equation.

   c. Write the net ionic equation.

2. For the equation balanced in Practice A, problem 2 of this lesson:

   a. Write the total ionic equation.

   b. Write the net ionic equation.

Balancing Separated-Ion Precipitation Equations

A precipitation equation showing separated ions can be balanced without balancing the standard equation first. The steps are:

   1. Add coefficients *inside* the brackets to balance *charge*.

   2. Add coefficients in *front* of the brackets to balance the *atoms* needed to form the precipitate.

▶ **TRY IT**

Q. Balance this equation, then write the total and net ionic equations.

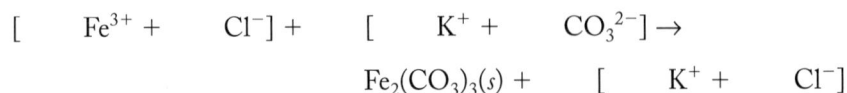

$$[\quad Fe^{3+} + \quad Cl^-] + \quad [\quad K^+ + \quad CO_3^{2-}] \rightarrow$$
$$Fe_2(CO_3)_3(s) + \quad [\quad K^+ + \quad Cl^-]$$

Answer:

First balance charge inside the brackets:

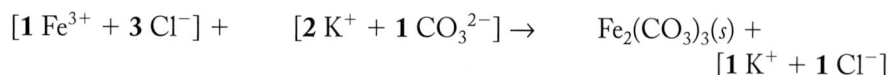

$$[\mathbf{1}\ Fe^{3+} + \mathbf{3}\ Cl^-] + \qquad [\mathbf{2}\ K^+ + \mathbf{1}\ CO_3{}^{2-}] \rightarrow \qquad Fe_2(CO_3)_3(s) +$$
$$[\mathbf{1}\ K^+ + \mathbf{1}\ Cl^-]$$

To form one precipitate particle, two iron(III) ions and three carbonate ions are needed.

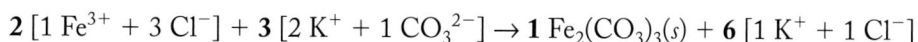

$$\mathbf{2}\,[1\ Fe^{3+} + 3\ Cl^-] + \mathbf{3}\,[2\ K^+ + 1\ CO_3{}^{2-}] \rightarrow \mathbf{1}\ Fe_2(CO_3)_3(s) + \mathbf{6}\,[1\ K^+ + 1\ Cl^-]$$

Take out the brackets to write the *total* ionic equation:

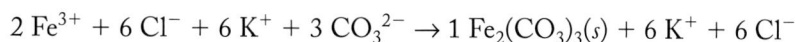

$$2\ Fe^{3+} + 6\ Cl^- + 6\ K^+ + 3\ CO_3{}^{2-} \rightarrow 1\ Fe_2(CO_3)_3(s) + 6\ K^+ + 6\ Cl^-$$

Check: On each side, 2 Fe, 6 Cl, 6 K, 3 C, and 9 O atoms and zero net charge. Balanced.

Omit the spectators to write the *net* ionic equation:

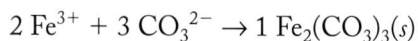

$$2\ Fe^{3+} + 3\ CO_3{}^{2-} \rightarrow 1\ Fe_2(CO_3)_3(s)$$

Check: On each side, 2 Fe, 3 C, and 9 O; zero net charge. Balanced.

PRACTICE   C

Writing on this page, balance charge inside the brackets and supply coefficients in front of the brackets. Then, under each equation, write the *total* ionic equation.

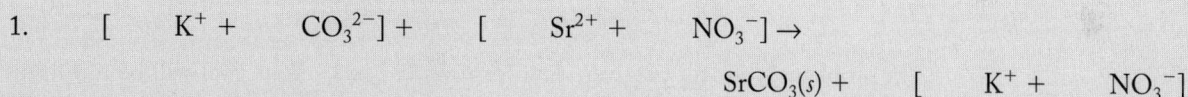

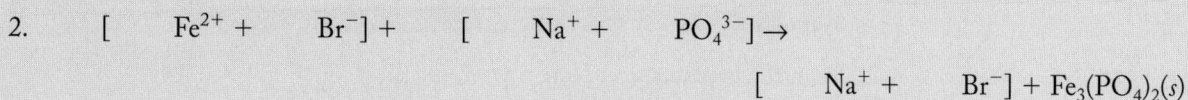

1.   [K^+ + $CO_3{}^{2-}$] +   [Sr^{2+} + $NO_3{}^-$] $\rightarrow$

   $SrCO_3(s)$ +   [K^+ + $NO_3{}^-$]

2.   [Fe^{2+} + Br^-] +   [Na^+ + $PO_4{}^{3-}$] $\rightarrow$

   [Na^+ + Br^-] + $Fe_3(PO_4)_2(s)$

Converting Standard Formulas to Ionic Equations

A precipitation can be represented by equations with ionic compound

1. names,

2. standard formulas, or

3. separated ions.

The summary of the reaction can be written as a

4. total ionic equation and

5. net ionic equation.

Given any *one* of those first *three* equation types, you need to be able to write all five. Let's organize this task in a table.

Example: A completed table will look like this.

1	Names	cesium sulfide(*aq*) + iron(II) nitrate(*aq*) → iron(II) sulfide(*s*) + cesium nitrate(*aq*)
2	Balanced standard formulas	$1\,Cs_2S(aq) + 1\,Fe(NO_3)_2(aq) \rightarrow 1\,FeS(s) + 2\,CsNO_3(aq)$
3	All as balanced separated formulas	$1\,[2\,Cs^+ + 1\,S^{2-}] + 1\,[1\,Fe^{2+} + 2\,NO_3^-] \rightarrow 1\,[1\,Fe^{2+} + 1\,S^{2-}] + 2\,[1\,Cs^+ + 1\,NO_3^-]$
	Total ionic equation	$2\,Cs^+ + 1\,S^{2-} + 1\,Fe^{2+} + 2\,NO_3^- \rightarrow 1\,FeS(s) + 2\,Cs^+ + 2\,NO_3^-$
	Net ionic equation	$1\,S^{2-}(aq) + 1\,Fe^{2+}(aq) \rightarrow 1\,FeS(s)$

In problems, you will be given data for rows 1, 2, *or* 3. Your goal will be to fill in the remaining data. The table is simply a puzzle in which you put together steps that you have practiced previously. These additional steps will help.

1. Based on the data you are given, fill in the *brackets* in row 3. The separated-ion formulas are the key to writing names and standard formulas.

2. From the brackets, complete rows 1 and 2.

3. If a *product* is an *insoluble* ion, add an (*s*) after its row 2 standard formula.

4. The coefficients in row 2 and in front of the brackets in row 3 must be the same.

5. In the *total* ionic equation, write the row 3 separated formulas for ions that are separated, but the row 2 standard formula for the *precipitate*.

► TRY IT

Q. When aqueous solutions of barium chloride and ammonium sulfate are mixed, insoluble barium sulfate and soluble ammonium chloride are formed. Complete the table below for this reaction. (Use your notebook where you need more room.)

1	Names	
2	Balanced standard formulas	
3	All as balanced separated formulas	
	Total ionic equation	
	Net ionic equation	

STOP

Answer:

1	Names	barium chloride + ammonium sulfate → barium sulfate(s) + ammonium chloride
2	Balanced standard formulas	$1\ BaCl_2 + 1\ (NH_4)_2SO_4 \rightarrow 1\ BaSO_4(s) + 2\ NH_4Cl$
3	All as balanced separated formulas	$1\ [1\ Ba^{2+} + 2\ Cl^-] + 1\ [2\ NH_4^+ + 1\ SO_4^{2-}] \rightarrow$ $1\ [1\ Ba^{2+} + 1\ SO_4^{2-}] + 2\ [1\ NH_4^+ + 1\ Cl^-]$
	Total ionic equation	$1\ Ba^{2+} + 2\ Cl^- + 2\ NH_4^+ + 1\ SO_4^{2-} \rightarrow 1\ BaSO_4(s) + 2\ NH_4^+ + 2\ Cl^-$
	Net ionic equation	$1\ Ba^{2+}(aq) + 1\ SO_4^{2-}(aq) \rightarrow 1\ BaSO_4(s)$

PRACTICE   D

1. When aqueous solutions of $AlCl_3$ and NaOH are mixed, insoluble aluminum hydroxide and soluble sodium chloride form. Complete the table below for this reaction.

1	Names	
2	Balanced standard formulas	
3	All as balanced separated formulas	
	Total ionic equation	
	Net ionic equation	

Lesson 13.4 | Predicting Precipitation

Predicting Precipitate Formulas

When solutions of two different reactants are *mixed*, the ions in the original solutions can attract new partners that have opposite charges. In some cases this attraction results in *reactions* to form new compounds. In other cases it does not.

When solutions of ionic compounds are mixed, formation of a precipitate is one type of reaction that can occur. Whether precipitation occurs depends on the solubility of the *new* possible combinations. In the problems in the previous lesson, the precipitate and other *products* were identified. We can also predict, given reactant solutions, whether a precipitate will form, and if so what its formula will be, by applying this rule:

> When solutions of two soluble ionic compounds are *mixed*, if a new combination is *possible* that is **insoluble**, it **will** precipitate.

▶ TRY IT

Q. Aqueous solutions of $Ca(NO_3)_2$ and Na_2CO_3 are mixed. Will a precipitate form, and if so, what is its formula?

1. To start, fill in the first three rows of the table *only* for the two *reactants*. At this point, fill in coefficients *inside* the brackets, but not outside.

1	Names	
2	Balanced standard formulas	
3	All as balanced separated formulas	[+] + [+] →

STOP

Answer:

Your table should match the one below.

1	Names	calcium nitrate + sodium carbonate →
2	Balanced standard formulas	$Ca(NO_3)_2$ + Na_2CO_3 →
3	All as balanced separated formulas	$[1\,Ca^{2+} + 2\,NO_3^-]$ + $[2\,Na^+ + 1\,CO_3^{2-}]$ →

2. When two solutions are mixed, two *new* combinations are possible. Each positive ion can attract the negative ion in the other reactant.

In the underlined blanks for rows 1 and 3 of the table below, write the new *possible products*. One way to do this: In the row 1 products, switch the places of the *negative* ion *names,* and in row 3, inside the brackets for the products, give each cation its new *negative* ion partner.

1	Names	(trade places) calcium nitrate + sodium carbonate → calcium _____ + sodium _____
2	Balanced standard formulas	$Ca(NO_3)_2 + \quad Na_2CO_3 →$
3	All as balanced separated formulas	$[1\ Ca^{2+} + 2\ NO_3^-] + \quad [2\ Na^+ + 1\ CO_3^{2-}] →$ $[\quad Ca^{2+} + \qquad] + \quad [\quad Na^+ + \qquad]$

STOP

Your paper should look like this:

1	Names	calcium nitrate + sodium carbonate → calcium **carbonate** + sodium **nitrate**
2	Balanced standard formulas	$Ca(NO_3)_2 + \quad Na_2CO_3 →$
3	All as balanced separated formulas	$[1\ Ca^{2+} + 2\ NO_3^-] + \quad [2\ Na^+ + 1\ CO_3^{2-}] →$ $[\quad Ca^{2+} + \mathbf{CO_3^{2-}}] + \quad [\quad Na^+ + \mathbf{NO_3^-}]$

3. Now finish the first three rows with formulas and coefficients. Start with the coefficients inside the brackets.

STOP

1	Names	calcium nitrate + sodium carbonate → calcium **carbonate** + sodium **nitrate**
2	Balanced standard formulas	$\mathbf{1}\ Ca(NO_3)_2 + \mathbf{1}\ Na_2CO_3 → \mathbf{1\ CaCO_3} + \mathbf{2\ NaNO_3}$
3	All as balanced separated formulas	$\mathbf{1}\ [1\ Ca^{2+} + 2\ NO_3^-] + \mathbf{1}\ [2\ Na^+ + 1\ CO_3^{2-}] →$ $\mathbf{1}\ [1\ Ca^{2+} + 1\ CO_3^{2-}] + 2\ [1\ Na^+ + 1\ NO_3^-]$
4	Solubility	_____ + _____ → _____ + _____

4. In this type of experiment, each of the two *reactant* solutions always contains a *soluble* ionic compound. Some solutions may have a *color* due to the dissolved ions, but the *reactants* contain no solids.

> In precipitation experiments, the two *reactants* are always *soluble*.

In row 4 above, in the first two blanks, write *soluble*.

5. Based on solubility rules, in row 4 write *soluble* or *insoluble* beneath each *product*.

6. In rows 2 or 3, circle the combinations that are actually present in the solution.
   - Circle both *reactants* in **row 3**, because each reactant must consist of *soluble, separated* ions.
   - If a *product* is *insoluble*, circle it in **row 2**, where it has the *solid* formula of a precipitate.
   - If a *product* is *soluble*, circle it in **row 3**, because if the *new* combination is soluble, its ions will remain *separated*.

7. Will there be a precipitate in this case? If so, what is its standard formula?

Complete steps 4–7 in the spaces in the table above.

🛑

1	Names	calcium nitrate + sodium carbonate → calcium carbonate + sodium nitrate
2	Balanced standard formulas	$1 \, Ca(NO_3)_2 + 1 \, Na_2CO_3 \rightarrow \boxed{1 \, CaCO_3} + 2 \, NaNO_3$
3	Balanced separated formulas	$\boxed{1 \, [1 \, Ca^{2+} + 2 \, NO_3^-]} + \boxed{1 \, [2 \, Na^+ + 1 \, CO_3^{2-}]} \rightarrow$ $1 \, [1 \, Ca^{2+} + 1 \, CO_3^{2-}] + \boxed{2 \, [1 \, Na^+ + 1 \, NO_3^-]}$
4	Solubility	soluble + soluble → **insoluble** + **soluble**

Calcium carbonate ($CaCO_3$) is **insoluble** by the rule for carbonates, so it *will* precipitate.

Sodium nitrate ($NaNO_3$) is also a possible new combination, but it is soluble, and soluble combinations do not precipitate. The sodium and nitrate ions will remain dissolved in the solution.

> If ions are mixed but *both* of the new products are *soluble*, no precipitation reaction will take place.

These rules and steps will become intuitive quickly—with practice. Let's summarize.

To Predict Products in Precipitation Reactions

1. Write a four-row table: names; balanced *standard* formulas; balanced *separated* formulas; and solubility.

2. Fill in the *reactants* in rows 1–3, then trade partners in rows 1 and 3 to write the new possible *products*.

3. Finish rows 1–3, *balancing* atoms and charges as you go.

4. In row 4, write the solubility of each compound. The rules are:
   • The two *reactants* are always soluble, separated ions.
   • If a new product is possible that is **insoluble**, it *will* precipitate.
   • If a new product is possible that is soluble, its ions will remain separated.

5. Circle the formulas that are actually present in the reactants and products.

Apply the rules to the following problem.

▶ TRY IT

Q. When lead(II) nitrate and potassium iodide solutions are mixed, will a precipitate form? If so, write its standard formula.

In your notebook, draw and fill in the four-row table for this reaction.

Answer:

Your table should look like this:

1	Names	lead(II) nitrate + potassium iodide → lead(II) iodide + potassium nitrate
2	Balanced standard formulas	$1\,Pb(NO_3)_2 + 2\,KI \rightarrow \boxed{1\,PbI_2} + 2\,KNO_3$
3	Balanced separated formulas	$1\,[1\,Pb^{2+} + 2\,NO_3^-] + 2\,[1\,K^+ + 1\,I^-] \rightarrow$ $\quad 1\,[1\,Pb^{2+} + 2\,I^-] + 2\,[1\,K^+ + 1\,NO_3^-]$
4	Solubility	soluble + soluble → **insoluble** + **soluble**

The precipitate is PbI_2.

PRACTICE   A

1. Label these combinations as soluble or insoluble.

   a. $Mg^{2+} + CO_3^{2-}$ b. $Pb^{2+} + Br^-$

   c. Silver carbonate   d. Ammonium chloride

   e. Na_2S f. $CuCl_2$

2. When solutions of $AgNO_3$ and Na_2CrO_4 are mixed, a brick-red precipitate forms. Complete a four-row table for this reaction.

Finding Total and Net Ionic Equations

From our four-row table, we can also write total and net ionic equations to describe a predicted precipitate.

▶ TRY IT

Q. Earlier in this lesson, you completed the four-row table for the mixing solutions $Ca(NO_3)_2$ and Na_2CO_3, shown below.

Using the circled formulas representing the particles actually present in the reactants and products, fill in the total ionic equation and then the net ionic equation.

1	Names	calcium nitrate + sodium carbonate → calcium carbonate 1 sodium nitrate
2	Balanced standard formulas	$1\,Ca(NO_3)_2 + 1\,Na_2CO_3 \rightarrow \boxed{1\,CaCO_3} + 2\,NaNO_3$
3	Balanced separated formulas	$\boxed{1\,[1\,Ca^{2+} + 2\,NO_3^-]} + \boxed{1\,[2\,Na^+ + 1\,CO_3^{2-}]} \rightarrow$ $1\,[1\,Ca^{2+} + 1\,CO_3^{2-}] + \boxed{2\,[1\,Na^+ + 1\,NO_3^-]}$
4	Solubility	soluble + soluble → insoluble + soluble
5	**Total ionic equation**	
6	**Net ionic equation**	

Answer:

Check that both the row 5 and row 6 equations are balanced for atoms and charge.

STOP

Your completed rows should be:

5	Total ionic equation	$1\,Ca^{2+} + 2\,NO_3^- + 2\,Na^+ + 1\,CO_3^{2-} \rightarrow 1\,CaCO_3(s) + 2\,Na^+ + 2\,NO_3^-$
6	Net ionic equation	$1\,Ca^{2+}(aq) + 1\,CO_3^{2-}(aq) \rightarrow 1\,CaCO_3(s)$

The table method provides a complete view of the precipitation process. With practice, you will be able to answer questions about precipitation automatically or by writing only parts of the table. That's the goal.

PRACTICE B

1. Design flashcards for the shaded rules in the past two lessons. Add cards for additional steps or strategies you find helpful.

2. In your notebook, for the earlier Try It that formed lead iodide, add and fill in rows 5 and 6 for the total and net ionic equations.

3. When potassium hydroxide and cobalt(II) nitrate solutions are mixed, an intense blue precipitate of cobalt(II) hydroxide forms. Knowing the precipitate, complete a six-row table, including the total and net ionic equations.

4. Combining solutions of magnesium chloride and sodium nitrate:

   a. Name the two new combinations that are possible.

   b. Which of the new combinations will be soluble in aqueous solutions?

   c. Which of the new combinations will precipitate?

SUMMARY

In this chapter, our focus was a procedure to predict whether two solutions of dissolved ions will form a precipitate when mixed. With focused attention, effort, and practice, your brain learns to sequence facts and the steps to apply them intuitively.

Attention is where learning begins. Every waking second, your senses are bombarded with sounds, sights, and other sensations, but when you focus your attention on a *portion* of those sensory elements, you move those elements into working memory. Sustained attention increases the opportunity for working memory to identify the cues, context, and concepts needed to fit new information into long-term memory.

Distraction is the enemy of attention. To maximize the effectiveness of your study time, try to find a quiet (cell-phone free) space where others are focused on study as well (such as the tables in a library)—an environment that supports focused attention and minimizes distraction.

REVIEW QUIZ

1. Label these combinations as soluble or insoluble.

   a. $Li^+ + CO_3^{2-}$ b. $Na^+ + Br^-$

   c. Magnesium chromate   d. Ammonium phosphate

   e. BaS   f. $Pb(CH_3COO)_2$

2. Balance:

   a. $Ca(NO_3)_2(aq) +$ $KOH(aq) \rightarrow$ $Ca(OH)_2(s) +$ $KNO_3(aq)$

   b. $AgNO_3 +$ $Na_2S \rightarrow$ $Ag_2S(s) +$ $NaNO_3$

3. Balance with coefficients inside and in front of brackets, then write total and net ionic equations.

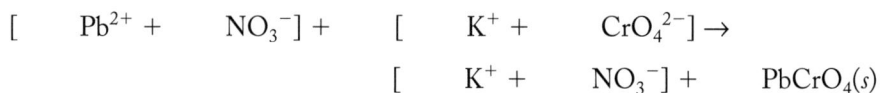

   $[\quad Pb^{2+} +\quad NO_3^-] + \quad [\quad K^+ +\quad CrO_4^{2-}] \rightarrow$

   $[\quad K^+ +\quad NO_3^-] +\quad PbCrO_4(s)$

4. Write total and net ionic equations for this reaction. If there is a precipitate, name it.

   $K_2S(aq) + CuCl_2(aq) \rightarrow$

ANSWERS

Lesson 13.1

Practice A   1a. $1\ Pb^{2+} + 2\ Cl^-$ 1b. $1\ Mg^{2+} + 2\ OH^-$ 1c. $3\ K^+ + 1\ PO_4^{3-}$ 2a. $NaNO_3$

2b. $(NH_4)_2CO_3$ 2c. $AlBr_3$ 3a. Potassium chromate   3b. Ammonium iodide   3c. Iron(II) sulfate

Practice B   1. Compounds that contain any alkali metal ions (such as Na^+ and K^+) or ammonium ion (NH_4^+), nitrate ion (NO_3^-), or acetate ion (CH_3COO^-) are always soluble.

2a. $K^+ + S^{2-}$: **Soluble**. By rule 2, sulfides are insoluble, but by rule 1 alkali metal ions are soluble, and rule 1 has precedence.   2b. $Sr^{2+} + Cl^-$: **Soluble** by rule 3 for chloride.   2c. $Ca^{2+} + CO_3^{2-}$: **Insoluble** by rule 2 for carbonate.   2d. $Ag^+ + CrO_4^{2-}$: **Insoluble** by rule 2 for chromate and/or silver ion.

3b. $Ba^{2+} + CO_3^{2-}$: **Insoluble** by rule 2 for carbonate.   3c. $Na^+ + OH^-$: **Soluble** by rule 1 for alkali metal ions.

3d. $Mg^{2+} + Br^-$: **Soluble** by rule 3 for bromide.   3e. $Ag^+ + NO_3^-$: **Soluble** by rule 1 for nitrate.

3f. $NH_4^+ + OH^-$: **Soluble** by rule 1 for ammonium ion.   3g. $Fe^{2+} + PO_4^{3-}$: **Insoluble** by rule 2 for phosphate.

3h. $Pb^{2+} + CH_3COO^-$: **Soluble** by rule 1 for acetate.

4a. **Insoluble** by rule 2 for chromates.   4b. **Soluble** by rule 1 for alkali metal ions.

4c. **Insoluble** by rule 2 for sulfide.   5. The anions covered by rule 1: **nitrate and acetate**.

6. **No.** When the phosphate anion is combined with either alkali metal or ammonium cations, the phosphate is soluble. Rule 1 has precedence.

Lesson 13.2

Assume (aq) if no state is shown. Coefficients of 1 may be omitted.

Practice A   1. $Pb^{2+}(aq) + \cancel{2\ NO_3^-(aq)} + \cancel{Cu^{2+}(aq)} + 2\ Cl^-(aq) \rightarrow \cancel{Cu^{2+}(aq)} + \cancel{2\ NO_3^-(aq)} + \boxed{PbCl_2(s)}$

Net ionic equation: $Pb^{2+}(aq) + 2\ Cl^-(aq) \rightarrow PbCl_2(s)$

2. $\cancel{6\,Na^+}$ + $2\,PO_4^{3-}$ + $3\,Mg^{2+}$ + $\cancel{3\,SO_4^{2-}}$ → $\boxed{Mg_3(PO_4)_2(s)}$ + $\cancel{6\,Na^+}$ + $\cancel{3\,SO_4^{2-}}$

Net ionic equation: $3\,Mg^{2+}(aq) + 2\,PO_4^{3-}(aq) \rightarrow Mg_3(PO_4)_2(s)$

Practice B   1. Pb^{2+} is soluble in lead(II) nitrate and lead(II) acetate.

2. Ions soluble with the chromate anion include Li^+, Na^+, K^+, Rb^+, Cs^+, Fr^+, and NH_4^+.

Lesson 13.3

Practice A   1. **1** $Fe(NO_3)_3(aq)$ + **3** $NaOH(aq) \rightarrow$ **1** $Fe(OH)_3(s)$ + **3** $NaNO_3(aq)$

2. **3** $Cu(NO_3)_2$ + **2** $K_3PO_4 \rightarrow$ **6** KNO_3 + **1** $Cu_3(PO_4)_2(s)$

Practice B   1a. $1\,[1\,Fe^{3+} + 3\,NO_3^-] + 3\,[1\,Na^+ + 1\,OH^-] \rightarrow 1\,Fe(OH)_3(s) + 3\,[1\,Na^+ + 1\,NO_3^-]$

1b. Total ionic equation: $1\,Fe^{3+} + 3\,NO_3^- + 3\,Na^+ + 3\,OH^- \rightarrow 1\,Fe(OH)_3(s) + 3\,Na^+ + 3\,NO_3^-$

1c. Net ionic equation: $1\,Fe^{3+}(aq) + 3\,OH^-(aq) \rightarrow 1\,Fe(OH)_3(s)$

2a. Total ionic equation: $3\,Cu^{2+} + 6\,NO_3^- + 6\,K^+ + 2\,PO_4^{3-} \rightarrow 6\,K^+ + 6\,NO_3^- + 1\,Cu_3(PO_4)_2(s)$

2b. Net ionic equation: $3\,Cu^{2+} + 2\,PO_4^{3-} \rightarrow 1\,Cu_3(PO_4)_2(s)$

Practice C   1. $1\,[2\,K^+ + 1\,CO_3^{2-}] + 1\,[1\,Sr^{2+} + 2\,NO_3^-] \rightarrow 1\,SrCO_3(s) + 2\,[1\,K^+ + 1\,NO_3^-]$

Total ionic equation: $2\,K^+ + 1\,CO_3^{2-} + 1\,Sr^{2+} + 2\,NO_3^- \rightarrow 1\,SrCO_3(s) + 2\,K^+ + 2\,NO_3^-$

2. $3\,[1\,Fe^{2+} + 2\,Br^-] + 2\,[3\,Na^+ + 1\,PO_4^{3-}] \rightarrow 6\,[1\,Na^+ + 1\,Br^-] + 1\,Fe_3(PO_4)_2(s)$

Total ionic equation: $3\,Fe^{2+} + 6\,Br^- + 6\,Na^+ + 2\,PO_4^{3-} \rightarrow 6\,Na^+ + 6\,Br^- + 1\,Fe_3(PO_4)_2(s)$

Practice D

1	Names	aluminum chloride + sodium hydroxide → aluminum hydroxide + sodium chloride
2	Balanced standard formulas	$1\,AlCl_3 + 3\,NaOH \rightarrow 1\,Al(OH)_3(s) + 3\,NaCl$
3	All as balanced separated formulas	$1\,[1\,Al^{3+} + 3\,Cl^-] + 3\,[1\,Na^+ + 1\,OH^-] \rightarrow 1\,[1\,Al^{3+} + 3\,OH^-] +$ $3\,[1\,Na^+ + 1\,Cl^-]$
	Total ionic equation	$1\,Al^{3+} + 3\,Cl^- + 3\,Na^+ + 3\,OH^- \rightarrow Al(OH)_3(s) + 3\,Na^+ + 3\,Cl^-$
	Net ionic equation	$1\,Al^{3+} + 3\,OH^- \rightarrow Al(OH)_3(s)$

Lesson 13.4

Practice A   1a. **Insoluble**   1b. **Insoluble**   1c. **Insoluble**   1d. **Soluble**   1e. **Soluble**   1f. **Soluble**

2. 1	Names	**silver nitrate + sodium chromate → silver chromate + sodium nitrate**
2	Balanced standard formulas	**2 AgNO₃ + 1 Na₂CrO₄ →** $\boxed{1\,Ag_2CrO_4}$ **+ 2 NaNO₃**
3	All as balanced separated formulas	$\boxed{2\,[1\,Ag^+ + 1\,NO_3^-]} + 1\,\boxed{[2\,Na^+ + 1\,CrO_4^{2-}]} \rightarrow$ $1\,[2\,Ag^+ + CrO_4^{2-}] + 2\,\boxed{[1\,Na^+ + 1\,NO_3^-]}$
4	Solubility	soluble + soluble → **insoluble** + **soluble**

Practice B	2. 5	Total ionic equation	**1** Pb^{2+} + **2** NO_3^- + **2** K^+ + **2** $I^- \rightarrow$ **1** $PbI_2(s)$ + **2** K^+ + **2** NO_3^-
	6	Net ionic equation	**1** $Pb^{2+}(aq)$ + **2** $I^-(aq) \rightarrow$ **1** $PbI_2(s)$

3. The problem identifies the precipitate (ppt.) as cobalt(II) hydroxide.

1	Names	**potassium hydroxide + cobalt(II) nitrate → potassium nitrate + cobalt(II) hydroxide**
2	Balanced standard formulas	$2\,KOH + 1\,Co(NO_3)_2 \rightarrow 2\,KNO_3 + \boxed{1\,Co(OH)_2}$
3	All as balanced separated formulas	$\boxed{2\,[1\,K^+ + 1\,OH^-]} + \boxed{1\,[1\,Co^{2+} + 2\,NO_3^-]} \rightarrow$ $\boxed{2\,[1\,K^+ + 1\,NO_3^-]} + 1\,[1\,Co^{2+} + 2\,OH^-]$
4	Solubility	soluble + soluble → **soluble + insoluble**
5	Total ionic equation	$2\,K^+ + 2\,OH^- + 1\,Co^{2+} + 2\,NO_3^- \rightarrow 2\,K^+ + 2\,NO_3^- + 1\,Co(OH)_2(s)$
6	Net ionic equation	$1\,Co^{2+}\,(aq) + 2\,OH^-(aq) \rightarrow 1\,Co(OH)_2(s)$

4a. magnesium chloride + sodium nitrate → **magnesium nitrate + sodium chloride**

4b. **Both** new combinations are soluble.   4c. **Neither** will precipitate.

Review Quiz

1a. **Soluble**   1b. **Soluble**   1c. **Insoluble**   1d. **Soluble**   1e. **Insoluble**   1f. **Soluble**

2a. $1\,Ca(NO_3)_2(aq) + 2\,KOH(aq) \rightarrow 1\,Ca(OH)_2(s) + 2\,KNO_3(aq)$

2b. $2\,AgNO_3 + 1\,Na_2S \rightarrow 1\,Ag_2S(s) + 2\,NaNO_3$

3. $1\,[1\,Pb^{2+} + 2\,NO_3^-] + 1\,[2\,K^+ + 1\,CrO_4^{2-}] \rightarrow 2\,[1\,K^+ + 1\,NO_3^-] + 1\,PbCrO_4(s)$

Total ionic equation: $1\,Pb^{2+} + 2\,NO_3^- + 2\,K^+ + 1\,CrO_4^{2-} \rightarrow 2\,K^+ + 2\,NO_3^- + 1\,PbCrO_4(s)$

Net ionic equation: $1\,Pb^{2+} + 1\,CrO_4^{2-} \rightarrow 1\,PbCrO_4(s)$

4.

1	Names	potassium sulfide + copper(II) chloride → potassium chloride + copper(II) sulfide
2	Balanced standard formulas	$1\,K_2S + 1\,CuCl_2 \rightarrow 2\,KCl + \boxed{1\,CuS}$
3	All as balanced separated formulas	$\boxed{1\,[2\,K^+ + 1\,S^{2-}]} + \boxed{1\,[1\,Cu^{2+} + 2\,Cl^-]} \rightarrow$ $\boxed{2\,[1\,K^+ + 1\,Cl^-]} + 1\,[1\,Cu^{2+} + 1\,S^{2-}]$
4	Solubility	soluble + soluble → soluble + insoluble
5	Total ionic equation	$2\,K^+ + 1\,S^{2-} + 1\,Cu^{2+} + 2\,Cl^- \rightarrow 2\,K^+ + 2\,Cl^- + 1\,CuS(s)$
6	Net ionic equation	$1\,Cu^{2+}(aq) + 1\,S^{2-}(aq) \rightarrow 1\,CuS(s)$

The precipitate is **copper(II) sulfide**.

14

Acid–Base Neutralization

Lesson 14.1   Ions in Acids and Bases

Chemical particles can be classified as being *acids*, *bases*, or *"acid–base neutral"* (and some particles can act as both acids and bases). Acids and bases may be ions or they may be compounds that are ionic or molecular. When dissolved in water, however, all acids and bases react to form *ions* to at least some extent.

For acids and bases, a variety of definitions exist, each helpful in certain types of reactions. In this chapter, for the limited purpose of studying the reactions between acids and bases in aqueous solutions, we will adopt the following definitions:

- An **acid** is a substance that forms H^+ ions when dissolved in water.
- A **base** is a substance that can react with (use up) H^+ ions.

The Structure of H^+: A Proton

A neutral hydrogen atom contains one proton and one electron. A small percentage of naturally occurring H atoms also contain one or two neutrons, but neutrons do not affect the chemical reactions in which an atom participates.

An **H^+ ion** is a hydrogen atom without an electron, so in most cases an H^+ ion is a single *proton*. The terms *H^+ ion* and *proton* are both used to describe the active particle in an acid.

> The ion formed by acids in water = H^+ = a proton.

In aqueous solutions, the proton released by an acid is nearly always found attached to a water molecule, forming a **hydronium ion (H_3O^+)**, but the proton behaves in most respects as if it were a free H^+ ion. For now, to simplify our work, we will use H^+ as the symbol for the ion present in acid solutions.

Acids

Acids can be categorized as *strong* or *weak*.

> A **strong acid** separates into ions essentially 100% when it dissolves in water.

Example: Nitric acid (HNO_3) dissociates 100% in water to form H^+ ions and nitrate ions.

$$1\ HNO_3 \xrightarrow{H_2O} 1\ H^+ + 1\ NO_3^- \qquad \text{(The reaction goes ~100\%)}$$

Other strong acids include:

- HCl, *hydrochloric* acid, which ionizes in water to form an H^+ ion and a chloride ion (Cl^-).
- H_2SO_4, *sulfuric* acid, which is used in lead-acid car batteries. It is termed a **diprotic acid** because each neutral H_2SO_4 molecule can ionize to form *two* H^+ ions.

> **Weak acids** dissociate only slightly in water.

Example: Acetic acid, the active ingredient in vinegar, is a weak acid. The formula for acetic acid can be written as CH_3COOH or as $HC_2H_3O_2$. Acetic acid is soluble in water but ionizes only slightly: About 1% of its molecules form an H^+ ion and an acetate anion (written as CH_3COO^- or $C_2H_3O_2^-$).

$$1\ CH_3COOH \xrightarrow{H_2O} 1\ H^+ + 1\ CH_3COO^- \qquad \text{(The reaction goes \textbf{~1\%})}$$

For these four acids that are encountered frequently, you will need to know their names, formulas, and acid strengths:

- Strong acids: hydrochloric acid (HCl), nitric acid (HNO_3), and sulfuric acid (H_2SO_4)
- Weak acid: acetic acid (CH_3COOH *or* $HC_2H_3O_2$)

Bases

Bases can also be categorized as strong or weak. By our definitions, in a **strong base**, for every one base particle that dissolves, at least one hydroxide ion (OH^-) forms.

> **Examples:** Strong bases include NaOH (sodium hydroxide) and KOH (potassium hydroxide).

Many other particles can act as bases, such as ammonia (NH_3), carbonate ions (CO_3^{2-}), and bicarbonate ions (HCO_3^-), but these particles are weaker bases than NaOH and KOH. In weak bases, a lower percentage of the particles cause OH^- ions to be formed in water.

Identifying Acidic Hydrogens

Hydrogen atoms bonded to other atoms can be divided into two types.

- **Acidic** hydrogen atoms react with hydroxide ions.
- **Nonacidic** hydrogen atoms do not react with hydroxide ions.

Many particles contain both acidic *and* nonacidic hydrogen atoms.

> **Examples:** In CH_3COOH (acetic acid), the H atom at the *end* of the formula reacts with NaOH, but the other three H atoms do not. The H at the end is an *acidic* hydrogen, and the other H atoms are *nonacidic* hydrogens.
>
> In ammonium ion (NH_4^+), one hydrogen is acidic, but the other three are not.

A particle with one or more acidic hydrogens can be either a strong or a weak acid (both types of acid react with hydroxide ions).

From the standard formula for an acid, can you predict which hydrogens will be acidic and which will not? In most cases, the rules are:

- If a substance formula can be separated into H^+ ions and *familiar anions*, the H is acidic.

> **Examples:** HCl, HNO_3, and H_2SO_4 can each ionize to form H^+ ions and familiar anions (Cl^-, NO_3^-, and SO_4^{2-}). Their H atoms are acidic.

- If one or more H atoms is written at the *front* of a formula, while other H atoms are not, the H atoms at the front will be *acidic* and the others will *not*.

Example: Acetic acid is often written as $HC_2H_3O_2$. Only the H in front is acidic.

- The H at the *end* of a –COOH group (also written –CO_2H) is acidic.

  Examples: In C_6H_5COOH, only the H at the end is acidic. In $C_3H_7CO_2H$, only the H at the end is acidic.

- In a formula, if one or more H atoms are written after a *metal* atom but before other atoms, each of those H atoms is acidic.

  Examples: In $KHC_8H_4O_4$, the *first* H (and only the first) is acidic. In $NaH_2C_6H_5O_7$, only the two hydrogens after Na are acidic.

- If a substance with only one H reacts with hydroxide ions, the H is acidic.

PRACTICE A

1. Design flashcards for the shaded rules and vocabulary above. Practice with the cards before beginning the following problems.

2. Draw an arrow pointing toward the acidic hydrogens in each compound. Then, after each formula, write the number of acidic hydrogens in the compound.

   a. NaH_2PO_4 b. $C_{12}H_{25}COOH$ c. $H_2C_4H_4O_6$

   d. H_3PO_4 e. $KHC_8H_4O_4$ f. $NaHSO_4$

Identifying Acids and Bases

From a formula, can you tell whether a particle is an acid or a base? It is not always easy to determine, but one rule is this:

To Identify Whether a Compound Is an Acid or a Base

Write the equation for the substance separating into *familiar* ions.

- Compounds that ionize to form H^+ ions are **acids**.
- Compounds that contain OH^- ions can act as **bases**.

PRACTICE **B**

In each reaction below, assume that each compound on the left is added to water and separates to form ions. Write a balanced equation showing the ions formed, then state whether the initial reactant is an *acid* or a *base*. For review of separation into ions, see Lesson 7.4.

1. $LiOH \xrightarrow{H_2O}$

2. $HClO_3 \rightarrow$

3. $Ca(OH)_2 \rightarrow$

4. $HC_2H_3O_2 \rightarrow$

Lesson 14.2   Balancing Hydroxide Neutralization

Acids and bases *react* with each other. In a **neutralization** reaction, as an acid and a base are mixed together, both are used up in simple whole-number particle ratios. In the reaction, both the acid and the base are said to be **neutralized**.

Acids can be neutralized by many different types of bases, including carbonates, sulfites, and ammonia, but in this chapter we will limit our study to reactions between acids and strong hydroxide bases.

When solutions of an acid and a hydroxide are mixed, the neutralization forms *liquid water*. The ions that react can be represented as

$$H^+ + OH^- \rightarrow H\!-\!OH(\ell) \qquad \text{(goes } \sim 100\%) \qquad (14.1)$$

Water can be written as H_2O or HOH or $H\!-\!OH$, but $H\!-\!OH$ helps to emphasize the reaction that takes place between an acid and hydroxide ions.

> **Example:** Hydrochloric acid solution mixed with sodium hydroxide solution forms water and dissolved sodium chloride.
>
> $$HCl(aq) + NaOH(aq) \rightarrow H\!-\!OH(\ell) + NaCl(aq) \qquad (14.2)$$
>
> Equation 14.2 is one way that this reaction is represented, but in aqueous solutions, HCl, NaOH, and NaCl all exist as separated ions. An equation that better represents the reacting particles is:
>
> $$H^+ + Cl^- + Na^+ + OH^- \rightarrow H\!-\!OH(\ell) + Na^+ + Cl^- \qquad (14.3)$$
>
> Note in equation 14.3 that the sodium and chloride ions are *spectator* ions: The reaction does not change their formula or state.

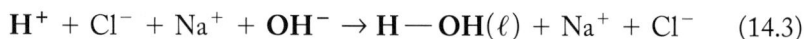

In equation 14.3 and for all *solution* reactions in these lessons, assume unless otherwise noted that H_2O is liquid (ℓ) and other substances and ions are *aqueous* (*aq*).

Partial versus Complete Neutralization

In the reaction between an acid and a base, if *either* the acid or the base (or both) is strong, the reaction will go *to completion*: The limiting reactant is completely used up.

In compounds that contain more than one *acidic hydrogen* or *basic group,* it is often possible to mix the acid and base so that some of the acidic or basic groups are neutralized, but others are not. But in these lessons, unless otherwise noted, you should assume that *neutralization* means *exact* and *complete* neutralization: that the acid and base are mixed in stoichiometrically equivalent amounts so that *all* of the acidic hydrogens and hydroxide ions are completely used up.

Predicting the Products of Hydroxide Neutralization

Often in neutralization calculations, the formulas for *products* are not supplied, but our interest is nearly always limited to the ratio of reaction between the acid and the base. To find that ratio, you generally need to know only *one* reaction *product.* In acid–hydroxide reactions, one product is always water.

> If an acid reacts with OH^- ions, assume one product is $H - OH$.

For acid–hydroxide reactions, this rule means:

* For every one H^+ ion used up, exactly one OH^- ion must be used up.
* The overall ratio of H^+ to OH^- ions reacting to $H - OH$ formed must be 1 to 1 to 1.

► TRY IT

Q. Add coefficients in front of these reactants. Assume that all of the hydroxide ions are reactive.

$$HNO_3 + \quad Al(OH)_3 \rightarrow$$

STOP

Answer:

Because one of the products must be $H - OH$, you may be able to supply the left-side coefficients by inspection:

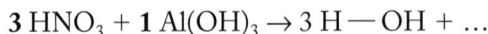

$$\mathbf{3}\,HNO_3 + \mathbf{1}\,Al(OH)_3 \rightarrow 3\,H - OH + \ldots$$

Balancing by inspection is always acceptable.

A more rigorous analysis would be to apply the rule that "to understand the reaction of ions in solution, write the separated-ion formulas." Because nitric acid is strong, it ionizes 100%. Given that all of the hydroxide ions can react with the H^+ ions, the equation can be balanced using these formulas:

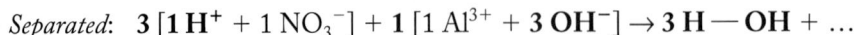

Standard: $\quad HNO_3 + Al(OH)_3 \rightarrow H - OH + \ldots$

Separated: $\quad \mathbf{3}\,[\mathbf{1}\,H^+ + 1\,NO_3^-] + \mathbf{1}\,[1\,Al^{3+} + 3\,OH^-] \rightarrow 3\,H - OH + \ldots$

Balanced: $\quad \mathbf{3}\,HNO_3 + \mathbf{1}\,Al(OH)_3 \rightarrow \mathbf{3}\,H - OH + \ldots$

The coefficients that you supply in *front* of the *brackets* must result in an *equal* number of H^+ and OH^- ions.

The ions that are *not* H^+ and OH^- will often be nonreacting spectators, but in some cases additional reactions will occur that change the final products. However, knowing that one product must be water, we can balance the *reactant* ratios correctly. That will be all we need to solve most neutralization calculations.

PRACTICE

1. Design flashcards for the shaded rules and vocabulary presented after Practice A of Lesson 14.1. Master the cards before starting the problems below.

2. Supply lowest-whole-number coefficients for the ratio of reaction between the acid and the base.

   a. HNO_3 + $KOH \rightarrow$

   b. KOH + $H_2SO_4 \rightarrow$

   c. H_2SO_4 + $Al(OH)_3 \rightarrow$

3. In your notebook, write these as equations that use standard formulas, then supply lowest-whole-number ratios of reaction. Assume all acidic hydrogens and hydroxide ions react.

   a. Cesium hydroxide + $H_2SO_4 \rightarrow$

   b. $Ca(OH)_2$ + nitric acid $\rightarrow$

   c. Barium hydroxide reacting with sulfuric acid

   d. Hydrochloric acid reacting with magnesium hydroxide

Lesson 14.3   Neutralization Calculations

Solutions and Stoichiometry

One goal in neutralization calculations is to predict *how much* of an acid or base will react. A calculation of *how much* in a reaction is stoichiometry. At the point where the *moles* of an acid and base have been mixed in the exact ratio that matches their coefficients in their balanced equation, both reactants are limiting, and the reactants have been mixed in stoichiometrically *equivalent* amounts (see Lesson 9.3).

For two particles reacted in stoichiometrically equivalent amounts, calculations can be solved using conversion stoichiometry. If a single-unit amount is WANTED and an amount of either reactant is supplied in the DATA, that amount is the *given* quantity for stoichiometric conversions.

The Stoichiometry Prompt

For reactions, if WANTED substance ≠ *given* substance:

• Write WANTED and DATA, balance, and bridge (WDBB).
• Convert "? unit WANTED = # *unit given*" to **mol** *given* to **mol** WANTED to unit WANTED.

The quick version of the stoichiometry steps is: WDBB, then units to *moles* to *moles* to *units*. (For review, see Lesson 9.2.) In calculations involving aqueous solutions, you will also need to apply:

The M Prompt

In the WANTED and DATA, translate *M, molar, molarity, concentration,* and *brackets* [ ] into *moles per 1 liter*. Write moles per liter as a ratio when it is WANTED and as an equality in DATA.

The Importance of Labels

To solve stoichiometry, careful *labeling* is essential.

• In WANTED, DATA, and conversions, after each unit write a *label*: the substance formula and/or words that identify what the unit is measuring.
• The label is an inseparable part of the unit. A unit that measures *one* substance cannot cancel the same unit measuring a *different* substance.

TRY IT

Q. How many milliliters of 0.0500 M H_2SO_4 are required to neutralize 25.0 milliliters of 0.220 M KOH? The unbalanced equation is

$$H_2SO_4 + \quad KOH \rightarrow \quad H_2O + \quad K_2SO_4$$

If you get stuck, read a *portion* of the answer until you are unstuck, then try again.

Answer:

WANTED: ? mL **H_2SO_4** soln.   (You want a single unit.)

DATA: 0.0500 mol H_2SO_4 = 1 L H_2SO_4 (M prompt.)

25.0 mL **KOH** soln.   (The single unit *given*.)

0.220 mol KOH = 1 L KOH   (M prompt.)

For a reaction calculation in which the WANTED and *given* substances differ, use the stoichiometry steps. Start with WDBB. For calculations in which all volumes are *solution* volumes, you may label one volume (ml or L) for each substance as "soln," but leave off "soln." as understood in other volume labels in the DATA and conversions.

Balance: **1** H_2SO_4 + **2** KOH → **2** H_2O + **1** K_2SO_4

Bridge: **1 mol** H_2SO_4 = **2 mol** KOH

SOLVE: (Convert units of *given* substance to moles of *given* to moles of WANTED to WANTED units of the WANTED substance.)

$$? \, \textbf{mL} \, \textbf{H}_2\textbf{SO}_4 = 25.0 \, \text{mL KOH} \cdot \frac{10^{-3} \, \text{L}}{1 \, \text{mL}} \cdot \frac{0.220 \, \text{mol KOH}}{1 \, \text{L KOH}} \cdot \frac{1 \, \text{mol H}_2\text{SO}_4}{2 \, \text{mol KOH}}$$

$$\cdot \frac{1 \, \text{L H}_2\text{SO}_4}{0.500 \, \text{mol H}_2\text{SO}_4} \cdot \frac{1 \, \text{mL}}{10^{-3} \, \text{L}}$$

$$= \textbf{55.0 mL H}_2\textbf{SO}_4 \, \textbf{soln.}$$

Take the Paper You Need

A stoichiometric calculation may require an entire sheet of paper. That's okay. Paper recycles. Trees grow. Take the time and paper needed for careful work. For problems with large numbers of conversions, you may want to solve on graph paper that is turned sideways (landscape format) to provide more room for conversions.

PRACTICE

1. Potassium hydrogen phthalate ($KHC_8H_4O_4$, abbreviated as KHPht) is a solid organic acid with a molar mass of 204.2 g/mol. How many milliliters of 0.0750 M KOH solution are required to neutralize 0.300 g of this acid? The balanced equation can be written as:

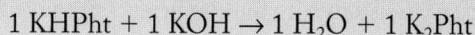

$$1 \, \text{KHPht} + 1 \, \text{KOH} \rightarrow 1 \, \text{H}_2\text{O} + 1 \, \text{K}_2\text{Pht}$$

Lesson 14.4   Titration

Titration Terminology

Titration is an experimental technique used to study reactions carried out in aqueous solutions. Calibrated **burets** precisely measure the amounts of a solution added as a reaction takes place. When the **endpoint** is reached, both of the reactants being titrated are completely used up.

An **indicator** is a dye added to the solution where the reaction is taking place. The indicator changes color at the precise endpoint of the titration. The endpoint is an *equivalence point* at which the *moles* of two reacting particles have reacted in a simple whole-number ratio.

In titration, the *endpoint* is an *equivalence point* in moles.

In an acid–base titration, an acid or base solution from a buret may be added to a solution of its opposite or added to a solid acid or base. If a solid acid or base is initially insoluble, it will dissolve as the titration proceeds.

If either the acid or base is *weak*, the opposite solution must be strong for an acid–base indicator to show a sharp endpoint, thus careful selection of an indicator is required.

Acid–Hydroxide Titration Calculations

If either a strong or weak acid is reacted with a hydroxide, because OH^- ion is a strong base, the reaction goes to completion.

> At an acid–hydroxide endpoint, **moles H^+** from acid = **moles OH^-** from base.

For an acid with *one* acidic hydrogen, at the endpoint the moles of acid in a titrated sample will equal the moles of OH^- supplied by the base.

If an acid particle contains *more* than one acidic hydrogen, neutralization produces a series of equivalence points. For many of these **polyprotic** acids, by carefully selecting an appropriate indicator it may be possible to titrate to an equivalence point for different acidic hydrogens in the acid formula.

At the endpoint, you should assume that the moles of acid and base have been mixed in the exact ratio shown by the coefficients of the balanced neutralization equation (at the uncertainty indicated by the significant figures in the data). At that *equivalence* point, calculations can be solved by the steps of conversion stoichiometry.

> For reaction calculations based on measurements at an *equivalence point,* or a *titration endpoint,* or a point of *exact neutralization,* use conversion stoichiometry.

Stoichiometry without Product Formulas

Until now, in all of our stoichiometric calculations, the reactant *and* product formulas have been provided. Knowing those formulas, all equations can be balanced, and balancing supplies the ratio that converts *given* to WANTED measurements. But in acid–base calculations, often *product* formulas are not supplied. Without knowing the products, can you find the ratio of reaction?

In the case of acid–hydroxide reactions, we know *one* of the products: H — OH. Using the balancing strategies in Lesson 14.2, the coefficients for the two *reactants* (the *acid* and *base)* can be determined, and those coefficients supply the key mole-to-mole conversion needed in stoichiometry.

━━━━━━━━▶ **TRY IT**

Apply the acid–hydroxide balancing and stoichiometry steps to this example.

Q. How many milliliters of 0.200 M sodium hydroxide solution are required to neutralize all three of the acidic hydrogens in 2.34 g of solid arsenic acid? (141.9 g H_3AsO_4/mol)

(STOP)

Answer:

(Your paper should look like this, minus the parenthetical comments.)

WANTED: ? mL **NaOH** soln.

DATA: 0.200 mol NaOH = 1 L NaOH   (M prompt.)

2.34 *g* **H_3AsO_4**   (Single unit *given*.)

141.9 g H_3AsO_4 = 1 mol H_3AsO_4 (Grams prompt.)

(Include substance formulas after all units.)

Balance: **1** H_3AsO_4 + **3** NaOH → **3** H—OH + …

Bridge: **1** mol H_3AsO_4 = **3** mol NaOH

SOLVE: (Want a single unit?)

? mL NaOH = 2.34 g H_3AsO_4 · $\dfrac{1\ \text{mol } H_3AsO_4}{141.9\ \text{g } H_3AsO_4}$ · $\dfrac{3\ \text{mol NaOH}}{1\ \text{mol } H_3AsO_4}$

· $\dfrac{1\ \text{L NaOH}}{0.200\ \text{mol NaOH}}$ · $\dfrac{1\ \text{mL}}{10^{-3}\ \text{L}}$ = **247 mL NaOH soln.**

What's the logic of these steps? We want to know the number of milliliters of base that will neutralize a known amount of acid.

- From the grams of the acid and its g/mol, we can find the *moles* of acid in the sample.
- From the balanced equation, we can find the *moles* of base needed to neutralize the acid.
- From the moles of base and its mol/L, we can find its solution volume.

With stoichiometry, we can precisely *predict* amounts that react and form.

PRACTICE

Complete problems 1 and 3 and then problem 2 if you need more practice.

1. Design flashcards for the shaded rules and vocabulary after Lesson 14.2. Practice with the cards before beginning the problems below.

2. How many milliliters of 0.200 potassium hydroxide are required to titrate 25.0 milliliters of 0.145 M hydrochloric acid?

3. In the titration of an H_3PO_4 solution, an indicator is selected that changes color at the precise point when all three of the phosphoric acid hydrogens have been neutralized. How many milliliters of 0.500 M NaOH will be required to titrate 37.5 milliliters of 0.200 M H_3PO_4?

SUMMARY

In Chapter 14, we applied both stoichiometry and molarity rules to reactions that take place in solutions. Had you forgotten any of those rules from prior chapters?

In learning, some forgetting is inevitable, but the consequences of forgetting depend on how well you learned initially. Cognitive scientist Daniel Willingham explains[1] that if you practice "spaced overlearning" (see Lesson 1.1), you are more likely to remember what you learn. Just as important, if forgetting has occurred, you will be able to "relearn" much more quickly, because knowledge "overlearned for several days" is still present in long-term memory to some extent.

Material studied intensely, but for only a day or two, is more likely to be forgotten. In addition, after a few weeks, material that was "crammed" tends to take nearly as long to learn a second time as it took the first.

During a career in rapidly changing sciences, fundamentals learned in introductory courses will be encountered repeatedly. If you practice recall of new facts and procedures and their application over *several* days, followed by occasional review, you construct stronger memory—and relearn faster when needed.

---

[1] Willingham, D. T. (2015). Do Students Remember What They Learn in School? *American Educator, 6, 7.*

REVIEW QUIZ

1. Write the number of acidic hydrogens in each of these acids.

   a. H_2SO_4 b. C_2H_5COOH c. $KHC_8H_4O_4$

2. Write balanced equations showing the ions formed when these substances dissolve and ionize, then label each initial reactant as an *acid* or a *base*.

   a. $HI \rightarrow$

   b. $CsOH \rightarrow$

   c. $CH_3COOH \rightarrow$

3. Supply ratios of reaction assuming that all acidic hydrogens and hydroxide ions react.

   a. $H_2SO_4 +$ $NaOH \rightarrow$

   b. $Mg(OH)_2 +$ $HCl \rightarrow$

   c. $H_3PO_4 +$ $LiOH \rightarrow$

4. Rewrite these equations using standard formulas for ionic compounds, then add coefficients that assume all of the acidic hydrogens and hydroxide ions react.

   a. Nitric acid + barium hydroxide

   b. Aluminum hydroxide + hydrochloric acid

5. A 0.100 g sample of solid oxalic acid ($H_2C_2O_4$, 90.03 g $\cdot$ mol^{-1}) is titrated by 0.0500 molar sodium hydroxide, and the chosen indicator changes color at the point that both acidic hydrogens are neutralized. How many milliliters of NaOH solution are required?

ANSWERS

Lesson 14.1

Practice A   2a. NaH_2PO_4 **Two**   2b. $C_{12}H_{25}COOH$ **One**   2c. $H_2C_4H_4O_6$ **Two**

2d. H_3PO_4 **Three**   (forms familiar $PO_4{}^{3-}$)   2e. $KHC_8H_4O_4$ **One**   2f. $NaHSO_4$ **One**

Practice B   1. $LiOH \rightarrow Li^+ + OH^-$; **base**

2. $HClO_3 \rightarrow H^+ + ClO_3{}^-$; **acid** (This anion is chlorate.)

3. $Ca(OH)_2 \rightarrow Ca^{2+} + 2\ OH^-$; **base**

4. $HC_2H_3O_2 \rightarrow H^+ + C_2H_3O_2{}^-$; **acid** (This is acetic acid.)

Lesson 14.2

Practice   2a. **1** HNO_3 + **1** $KOH \rightarrow$ **1** H—OH + ...   2b. **2** KOH + **1** $H_2SO_4 \rightarrow$ **2** H—OH + ...

2c. **3** H_2SO_4 + **2** $Al(OH)_3 \rightarrow$ **6** H—OH + ...   3a. **2** $CsOH$ + **1** $H_2SO_4 \rightarrow$ **2** H—OH + ...

3b. **1** $Ca(OH)_2$ + **2** $HNO_3 \rightarrow$ **2** H—OH + ...   3c. **1** $Ba(OH)_2$ + **1** $H_2SO_4 \rightarrow$ **2** H—OH + ...

3d. **2** HCl + **1** $Mg(OH)_2 \rightarrow$ **2** H—OH + ...

Lesson 14.3

Practice   1. WANTED: ? mL KOH soln.   (Single unit WANTED.)

DATA: 0.300 g KHPht   (Single unit *given*.)

204.2 g KHPht = 1 mol KHPht

0.0750 mol KOH = 1 L KOH   (M prompt.)

Balance: See problem.

Bridge: **1** mol KHPht = **1** mol KOH

SOLVE: ? mL KOH = 0.300 g KHPht $\cdot \dfrac{1\ \text{mol KHPht}}{204.2\ \text{g KHPht}} \cdot \dfrac{1\ \text{mol KOH}}{1\ \text{mol KHPht}} \cdot \dfrac{1\ \text{L KOH}}{0.0750\ \text{mol KOH}} \cdot \dfrac{1\ \text{mL}}{10^{-3}\ \text{L}}$

= **19.6 mL KOH soln**.

Lesson 14.4

Practice A   2. WANTED: ? mL KOH solution

DATA: 25.0 mL HCl soln.   (Single-unit *given*.)

0.145 mol HCl = 1 L HCl   (M prompt.)

0.200 mol KOH = 1 L KOH   (M prompt.)

Balance: 1 HCl + 1 KOH $\rightarrow$ 1 H_2O + ...

Bridge: 1 mol HCl = 1 mol KOH

SOLVE: ? mL KOH = 25.0 mL HCl $\cdot \dfrac{10^{-3}\ \text{L}}{1\ \text{mL}} \cdot \dfrac{0.145\ \text{mol HCl}}{1\ \text{L HCl}} \cdot \dfrac{1\ \text{mol KOH}}{1\ \text{mol HCl}} \cdot \dfrac{1\ \text{L KOH}}{0.200\ \text{mol KOH}} \cdot \dfrac{1\ \text{mL}}{10^{-3}\ \text{L}}$

= **18.1 mL KOH solution**

3. WANTED: ? mL NaOH soln.   (You want a single unit.)

    DATA: 0.200 mol H_3PO_4 = 1 L H_3PO_4 soln.   (M prompt.)

              37.5 mL H_3PO_4 (The single unit *given*.)

              0.500 mol NaOH = 1 L NaOH   (M prompt.)

Balance: **1** H_3PO_4 + **3** NaOH → **3** H_2O + ...

Bridge: **1 mol** H_3PO_4 = **3 mol** NaOH

SOLVE:

? mL NaOH = 37.5 mL H_3PO_4 • $\dfrac{10^{-3}\text{ L}}{1\text{ mL}}$ • $\dfrac{0.200\text{ mol }H_3PO_4}{1\text{ L }H_3PO_4}$ • $\dfrac{3\text{ mol NaOH}}{1\text{ mol }H_3PO_4}$ • $\dfrac{1\text{ L NaOH}}{0.500\text{ mol NaOH}}$ • $\dfrac{1\text{ mL}}{10^{-3}\text{ L}}$

       = **45.0 mL NaOH soln.**

Review Quiz

1a. H_2SO_4 **2**   1b. C_2H_5COOH **1**   1c. $KHC_8H_4O_4$ **1**

2a. HI → H^+ + I^-; HI is an **acid**   2b. CsOH → Cs^+ + OH^-; CsOH is a **base**

2c. CH_3COOH → H^+ + CH_3COO^-; CH_3COOH is acetic **acid**

3a. **1** H_2SO_4 + **2** NaOH →   3b. **1** $Mg(OH)_2$ + **2** HCl →   3c. **1** H_3PO_4 + **3** LiOH →

4a. 2 HNO_3 + 1 $Ba(OH)_2$ →   4b. 1 $Al(OH)_3$ + 3 HCl →

5. ? mL NaOH soln. = 0.100 g $H_2C_2O_4$ • $\dfrac{1\text{ mol }H_2C_2O_4}{90.03\text{ g }H_2C_2O_4}$ • $\dfrac{2\text{ mol NaOH}}{1\text{ mol }H_2C_2O_4}$ • $\dfrac{1\text{ L NaOH}}{0.0500\text{ mol NaOH}}$

                • $\dfrac{1\text{ mL}}{10^{-3}\text{ L}}$ = **44.4 mL NaOH soln.**

15

Redox Reactions

Lesson 15.1   Oxidation Numbers

In Chapters 13 and 14, we studied two types of reactions that can occur when particles are mixed:

- In *precipitation*, ions from two soluble compounds are mixed and form a compound that is insoluble in water.

- In *hydroxide neutralization*, a proton (H^+) from an acid combines with a hydroxide ion from a base to form products that include water.

In a **redox reaction**, electrons transfer. *Redox* is a combination of <u>red</u>uction and <u>ox</u>idation.

- **Oxidation** is the *loss* of electrons.
- **Reduction** is the *gain* of electrons.

A redox reaction occurs when a reactant loses one or more of its electrons and another reactant gains the transferred electrons.

The original meaning of *oxidation* was "reaction with oxygen." Elemental oxygen (O_2) is the primary reactive component of the air around us. In reactions such as the *burning* of substances and *rusting* of iron, elemental oxygen removes electrons from atoms, causing them to become **oxidized**. In modern usage, the term *oxidation* is applied to any reaction in which particles lose electrons.

The term *reduction* reflects a discovery made in ancient times: When some types of minerals are heated with substances that tend to lose electrons, the mass of the minerals is seemingly *reduced* as metals form. Because metals can be bent, shaped, and sharpened, they have been of value throughout human history in uses from jewelry to weaponry.

Assigning Oxidation Numbers

Oxidation numbers are values related to electric charge that can be assigned to *each atom* in a chemical particle. Oxidation numbers can assist in tracking the transfer of electrons that occurs during redox reactions.

Oxidation numbers are similar in many respects to the charges on ions, but they are *not* the same. To distinguish oxidation numbers from ion charges, chemists have adopted these conventions:

- The charge on an *ion* is shown as a superscript *after* the symbol. The number of charges is *followed* by a + or −, and values of 1 are omitted as understood.

- In **oxidation number diagrams**, an oxidation number (O.N.) is written above and below each atom symbol. The + or − sign is written *first*, and all numbers, including 0 and 1, are shown.

  Example: The oxidation number diagram for the chlorate ion is:

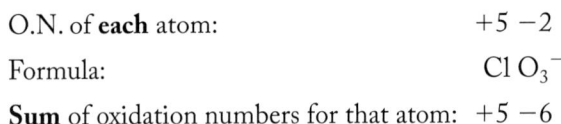

  O.N. of **each** atom: $+5 \; -2$

  Formula: $Cl \, O_3{}^-$

  Sum of oxidation numbers for that atom: $+5 \; -6$

Some sets of rules for assigning oxidation numbers are more complex than others, and those rules will predict behavior for a larger number of substances. In these lessons, we will use the following *limited* set of rules that predict redox behavior for *most* chemical particles.

Rules for Assigning Oxidation Numbers

1. Assign each atom in an *element* an O.N. of *zero*.

Formulas for elements have *one kind* of atom and no charge.

  Examples: Element formulas include C, Na, O_2, S_8, and Cl_2 (for more on elements, see Lesson 7.2).

2. Assign each atom in a *monatomic ion* an O.N. equal to the *charge* on the ion.

  Examples:

  For Al^{3+}, Al is assigned an O.N. of +3.

  For S^{2-}, S is assigned an O.N. of −2.

3. For particles *not* elements or monatomic ions:

   a. Above *each* hydrogen (H) atom, write an O.N. of +1.

   b. Above each oxygen (O) atom, write an O.N. of −2.

   c. Write an O.N. of +1 above *each* atom located in column 1 of the periodic table and +2 above each atom in column 2.

Example: For NaOH, write:

O.N. for *each* atom: +1 −2 +1
Formula: Na O H

4. Assign the remaining oxidation numbers so that the *sum of all* the oxidation numbers equals the *overall charge* on the particle.

▶ **TRY IT**

Q. Write oxidation numbers above each atom in the hydrogen sulfide ion: HS⁻

O.N. for *each* atom:

Formula: H S⁻

Answer:

O.N. for *each* atom: +1 −2
Formula: H S⁻

By rule 3a, write an oxidation number above H of +1. Because the sum of the oxidation numbers must be the 1− charge on the ion, the oxidation number of S must be −2.

Calculating Oxidation Numbers

When solving problems, what is most important is the oxidation number of *each individual* atom, but to find those values for particles that have *subscripts*, the sum for *all* the atoms of that kind must be considered. To find sums and individual values, we complete an **oxidation number diagram**. The steps are:

1. Write the symbol for the particle, including its charge if any.

2. Write the *rule-defined* O.N. for each atom above each symbol.

3. Below each symbol, write the known O.N. *sum*, obtained from multiplying the O.N. for *each* atom by its subscript.

4. Fill in numbers *below* each atom so that the *sum* of bottom numbers equals the overall charge on the particle.

5. Divide each sum by the atom subscript to get the O.N. for *each* individual atom. Write that *individual O.N. above* each remaining atom symbol.

➤ TRY IT

Q. Fill in this diagram for NO_2. Consult the rules and steps above.

Each:

Formula: N O_2

Sum:

STOP

Answer:

Assign each oxygen an O.N. of -2; for two oxygens, the *sum* is -4.

Each: −2

Formula: N O_2

Sum: −4   [(individual O.N.) × (subscript) = sum]

(Note the "column arrangement" of $-2 \times 2 = -4$)

The numbers on the bottom must add up to the zero charge on the particle, so the *sum* for all of the N atoms must be **+4**. Because we have only one nitrogen, its *individual* oxidation number (on top) is also **+4**.

Each: +4 −2

Formula: N O_2

Sum: +4 −4   (Must sum to equal charge on particle.)

Check that for each atom, "*each* times *subscript* equals *sum*."

Try one more.

➤ TRY IT

Q. What is the oxidation number for each carbon atom in $C_2O_4^{2-}$?

STOP

Answer:

Each: +3 −2

Formula: C_2 O_4^{2-}

Sum: +6 −8   (Must add up to −2.)

First assign -2 to each oxygen and -8 as the four-oxygen sum. The O.N. sums must sum to the $2-$ charge on the ion, so the two C must sum to $+6$ and *each* C atom must be **+3**.

Check that in the numbers for each atom, "*each* times *subscript* = sum."

In a sense, oxidation numbers ask the question: If each atom in a particle were an ion, what would be its charge? In reality, the distribution of charge inside most particles is more complex than oxidation numbers indicate, but oxidation numbers provide a simple model to track the *transfer* of electrons between atoms during reactions. These O.N. values will be especially helpful when we balance redox equations.

PRACTICE

1. Write the oxidation number (O.N.) of *each* atom for the atoms specified in these particles. You *may* look back at the rules and examples above.

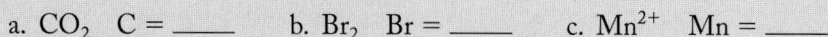

   a. CO_2 C = _____   b. Br_2 Br = _____   c. Mn^{2+} Mn = _____

2. In your notebook, distill the shaded rules 1–4 in the section above into a form that you find easy to write and remember. Practice writing your summary from memory.

3. Write the *oxidation number diagrams* for these particles.

   a. $CaBr_2$ b. NO_3^-

   c. Na_2CrO_4 d. $Cr_2O_7^{2-}$

4. Using rules recalled from memory, write the oxidation number (O.N.) for the individual atoms specified in these particles.

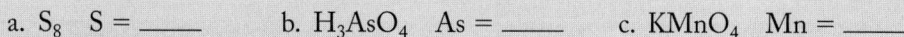

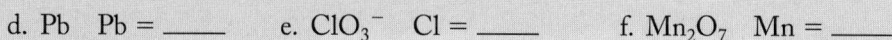

   a. S_8 S = _____   b. H_3AsO_4 As = _____   c. $KMnO_4$ Mn = _____

   d. Pb   Pb = _____   e. ClO_3^- Cl = _____   f. Mn_2O_7 Mn = _____

Lesson 15.2   Half Reactions and Balancing Charge

In a redox reaction, the number of electrons lost by one reactant must equal the number gained by another reactant, but this balanced transfer is hidden when a redox equation is written. One way to see the electrons lost and gained is to separate the reaction into two parts.

A redox **half-reaction** shows the *electrons* lost or gained.

Half reactions separate a redox reaction into its two components: the oxidation and the reduction.

Every redox reaction is a result of *two* half-reactions: one that loses electrons, and the other that gains the same number of electrons.

> **Example:** When lead metal is placed in a solution of silver ions, lead atoms dissolve as silver metal forms. The reaction can be represented as

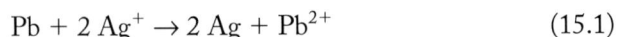

$$Pb + 2\,Ag^+ \rightarrow 2\,Ag + Pb^{2+} \tag{15.1}$$

The reaction is a result of two *half*-reactions. Each lead atom loses two electrons, and each silver ion gains one electron. To balance the loss and gain of electrons, the half reactions are written as

$$Pb \rightarrow Pb^{2+} + \mathbf{2\,e^-} \qquad \text{(One Pb is oxidized.)} \tag{15.2}$$

$$\mathbf{2\,e^-} + 2\,Ag^+ \rightarrow 2\,Ag \qquad \text{(Two Ag}^+ \text{ are reduced.)} \tag{15.3}$$

In the above example, note that:

- In half reactions, electrons can be reactants (to the left of the arrow) or products.
- If you "add" equations 15.2 and 15.3 on each side of the arrows, the two electrons in each equation cancel and the result is equation 15.1.
- When breaking reactions into half-reactions, the half-reaction coefficients *may* or *may not* be *lowest*-whole-number ratios. The half reaction is balanced so that atoms balance, charges balance, *and* the number of electrons is the *same* on opposite sides of the arrows.

In a half reaction, each electron symbol represents a 1− charge. To balance a half reaction, electrons are added to balance charge.

To Balance Half-Reactions:

1. First *supply coefficients* that balance the atoms.

2. Then *supply electrons* (e^-) to balance charge.

> ━━━━━▶ **TRY IT**

Q. Balance this half reaction:

$$Br_2 \rightarrow \qquad Br^-$$

🛑

Answer:

Because there are two bromine atoms on the left, a coefficient of 2 must be inserted on the right.

$$Br_2 \rightarrow \mathbf{2}\,Br^-$$

That balances atoms, but the charge is not balanced. The right side has a total charge of 2−, but the left is neutral. To balance charge, two electrons must be added to the left.

$$\mathbf{2\,e^-} + Br_2 \rightarrow 2\,Br^-$$

The total charge on each side is 2−. Atoms and charge are now balanced.

PRACTICE   A

When adding half reactions, coefficients are often *not* lowest whole numbers, but do supply lowest whole numbers on these.

1. Add coefficients and electrons to balance these half reactions.

a. $Co \rightarrow Co^{2+}$ b. $O_2 \rightarrow O^{2-}$

c. $Cu^{2+} \rightarrow Cu^+$ d. $S_8 \rightarrow S^{2-}$

2. Balance these half reactions.

a. $H^+ \rightarrow H_2$ b. $I^- \rightarrow I_2$

c. $P_4 \rightarrow P^{3-}$ d. $Sn^{4+} \rightarrow Sn^{2+}$

Balancing Atoms and Charge

In redox equations, spectator ions are often left out. When spectators are omitted, the counts for each atom must be the same on both sides, and total charge must be the same on both sides, but the total charge is not required to be zero.

Example: In the equation

$$Zn + 2\,Ag^+ \rightarrow Zn^{2+} + 2\,Ag$$

on the left, the total charge is 2+, and on the right, the total charge is 2+. Both atoms and charges balance.

By supplying coefficients, redox equations can be balanced by trial and error. The steps are: First balance the atoms, then balance the charge.

TRY IT

Balance both of these equations with lowest-whole-number coefficients.

Q1. $Mg + H^+ \rightarrow Mg^{2+} + H_2$

(Recall that it often helps to *start* with a 1 in front of one of the "most complex" formulas, then adjust by trial and error.)

Answer:

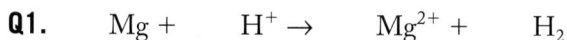

$$\mathbf{1}\,Mg + \mathbf{2}\,H^+ \rightarrow \mathbf{1}\,Mg^{2+} + \mathbf{1}\,H_2$$

Q2. $Al + H^+ \rightarrow Al^{3+} + H_2$

Answer:

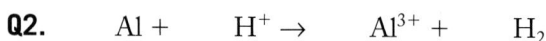

$$2\,Al + 6\,H^+ \rightarrow 2\,Al^{3+} + 3\,H_2$$

Trial-and-error balancing is easy in some cases and time consuming in others. In a later lesson we will learn a way to speed redox balancing, but when you are asked to balance what looks to be a simple equation, try trial and error first.

PRACTICE B

Working in your notebook, balance by trial and error. In each equation, all of the coefficients will be a 1, 2, or 3. Do not use a calculator.

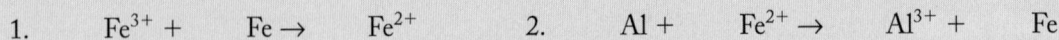

1. $Fe^{3+} + Fe \rightarrow Fe^{2+}$ 2. $Al + Fe^{2+} \rightarrow Al^{3+} + Fe$

3. $ClO_3^- \rightarrow Cl^- + O_2$

Lesson 15.3   Oxidizing and Reducing Agents

In redox reactions, one reactant must be oxidized, and another reactant must be reduced. The particle *oxidized* contains an *atom* that *loses* some of its electrons. This particle is termed the **reducing agent (RA)** because, by giving its electrons to another particle, it is the agent *causing* the other particle to be reduced.

The reactant particle *reduced* contains an atom that *gains* electrons. This particle is termed the **oxidizing agent (OA)** because, by taking electrons from another particle, it is the agent causing the other particle to be oxidized.

The following definitions must be committed to memory.

Redox Definitions

- **Oxidation** is the loss of electrons. **Reduction** is the gain of electrons.
- An **oxidizing agent** is a particle that *accepts* electrons in a reaction; it removes electrons from another atom.
- A **reducing agent** is a particle that *loses* electrons in a reaction; it donates electrons to another atom.

In a redox reaction:
- One reactant must be an OA and one must be an RA.
- The **reducing agent** is **oxidized** as it loses electrons.
- The **oxidizing agent** is **reduced** as it gains electrons.

TRY IT

Consulting the definitions above, write answers to the following. For the reaction

$$Cu + 2\,Ag^+ \rightarrow 2\,Ag + Cu^{2+}$$

Q1. Which symbol represents silver metal?

STOP

Answer:

Ag. By definition, a symbol for one kind of metal atom without a charge (or "with a charge of zero") is the formula for the elemental form of a metal.

Ag^+ represents a silver *ion*. An ion must be part of an ionic compound. Compounds that contain metal *ions* do not bend, shape, or conduct electricity in the way that metals do.

Q2. Is the copper metal being oxidized or reduced?

STOP

Answer:

The neutral Cu metal atom becomes a Cu^{2+} ion, and it must lose two electrons to do so. As it loses electrons, the neutral Cu is being oxidized.

Q3. Is Ag^+ being oxidized or reduced?

STOP

Answer:

Ag^+ becomes Ag in the reaction. To do so, it must gain an electron. Particles that gain electrons in a reaction are being reduced. In addition, because the Cu reactant is being oxidized, the Ag^+ reactant must be being reduced.

Q4. Is the Ag^+ acting as an oxidizing agent or a reducing agent?

STOP

Answer:

Because Ag^+ is being reduced, it must be an oxidizing agent. By accepting the electron from neutral copper, Ag^+ is the agent that causes Cu to be oxidized.

PRACTICE   A

1. In your notebook, design flashcards for the vocabulary and rules in this lesson, make and practice the cards, then answer the remaining questions in this lesson. Try not to look back at the cards or the lesson.

2. Write a label below each *reactant* identifying it as an oxidizing agent (OA) or reducing agent (RA), then circle the reactant particle being *oxidized*.

   a. $Sn^{4+} + Ni \rightarrow Ni^{2+} + Sn^{2+}$ b. $Mg + 2\,H^+ \rightarrow H_2 + Mg^{2+}$

Identifying Agents Using Oxidation Numbers

Knowing the following two rules can help to "anchor" your knowledge of redox behavior.

- Alkali metals are strong reducing agents: They tend to form +1 ions.
- Elemental chlorine (Cl_2) is a strong oxidizing agent: It tends to form chloride ion (Cl^-).

In redox reactions that are complex, to identify the oxidizing and reducing agents, it is often necessary to track the transfer of electrons using oxidation numbers. Apply these steps.

1. On the written reaction, balanced or unbalanced, write the O.N. for each atom above each atom. Coefficients, if provided in equations, do not affect the values for oxidation numbers in a particle.

2. Underline the *two* atoms that have their oxidation numbers *changed* going from reactants to products.

3. The reactant particle with the atom losing electrons (going toward a higher, more positive O.N.) is the reducing agent.

4. The reactant particle with the atom gaining negative charges (electrons) is the oxidizing agent.

▶ TRY IT

For this redox reaction,

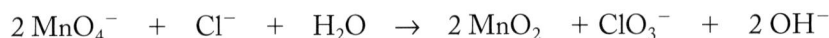

$$2\,MnO_4^- \;+\; Cl^- \;+\; H_2O \;\rightarrow\; 2\,MnO_2 \;+\; ClO_3^- \;+\; 2\,OH^-$$

Q1. Assign oxidation numbers to each atom.

Q2. Write the formula for the reactant particle with an atom being oxidized.

Q3. Write the formula for the reactant particle that is the oxidizing agent.

Answers:

1. Assign oxidation numbers:

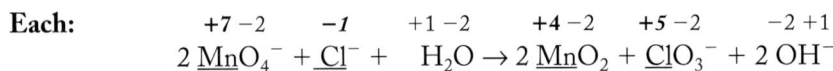

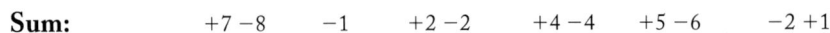

Each:
$$\overset{+7\,-2}{2\,\underline{Mn}O_4^-} + \overset{-1}{\underline{Cl}^-} + \overset{+1\,-2}{H_2O} \rightarrow \overset{+4\,-2}{2\,\underline{Mn}O_2} + \overset{+5\,-2}{\underline{Cl}O_3^-} + \overset{-2\,+1}{2\,OH^-}$$

Sum: $+7\,-8$ -1 $+2\,-2$ $+4\,-4$ $+5\,-6$ $-2\,+1$

2. The **Cl** is -1 in the reactants and $+5$ in the products. The Cl^- ion is therefore losing six electrons and is the reactant being **oxidized**.

3. The Mn atom is $+7$ on the left and $+4$ on the right, so each Mn atom is gaining three electrons. MnO_4^- contains the atom being reduced, so $\mathbf{MnO_4^-}$ ion is the **oxidizing agent**.

PRACTICE **B**

1. In your notebook, assign oxidation numbers to each atom. Then label one reactant as an oxidizing agent (OA), one reactant as a reducing agent (RA), and circle the reactant particle being *reduced*.

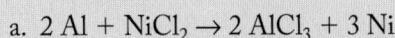

   a. $2\,Al + NiCl_2 \rightarrow 2\,AlCl_3 + 3\,Ni$

   b. $4\,As + 3\,HClO_3 + 6\,H_2O \rightarrow 4\,H_3AsO_3 + 3\,HClO$

2. Label one reactant as an oxidizing agent (OA), one reactant as a reducing agent (RA), and circle the reactant particle being *oxidized*.

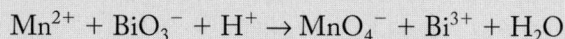

$$Mn^{2+} + BiO_3^- + H^+ \rightarrow MnO_4^- + Bi^{3+} + H_2O$$

Lesson 15.4   Balancing Based on Oxidation Numbers

The first step in understanding a chemical reaction is balancing the reaction equation. Trial and error can balance all equations, but for many redox reactions, trial and error can be a slow process. Oxidation numbers can speed balancing by suggesting coefficients.

In a *balanced* redox reaction, the number of electrons lost by the reducing agent must equal the number gained by the oxidizing agent.

By assigning oxidation numbers, the gain and loss of electrons for the two reacting particles can be identified. Four coefficients can then be determined that balance the electron gain and loss, and final balancing can be completed by trial and error.

The use of oxidation numbers for redox balancing is best learned by example.

▶ **TRY IT**

Q. In your notebook, write this redox equation (leaving some room between the terms) and complete the following steps to balance the equation.

$$NaClO_3 + HI \rightarrow I_2 + H_2O + NaCl$$

1. Above the atom symbols, write the oxidation numbers for *each atom* in the equation. Write sums below if needed.

Answer:

Each: +1 +5 −2   +1 −*1*   0   +1 −2   +1 −*1*

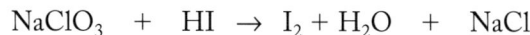

$$NaClO_3 \quad + \quad HI \ \rightarrow \ I_2 + H_2O \quad + \quad NaCl$$

Sum: +1 +5 −6

2. Underline the *two* atoms on *each* side that *change* their *individual* atom oxidation numbers in going from reactants to products.

Each Cl changes from +5 to −1. Each I goes from −1 to zero.

3. For each of those two atoms, draw arrows connecting the atom on one side to the *same* atom on the other side. Draw one arrow above and the other below the equation. Label each arrow with the *electron change* that must take place in going from one oxidation number to the other.

STOP

Each: $\overbrace{}^{\text{gained 6 e}^-}$

$\underset{+1+5-2}{} \quad \underset{+1-1}{} \quad \underset{0}{} \quad \underset{+1-2}{} \quad \underset{+1-1}{}$

$\mathrm{Na\underline{Cl}O_3 + H\underline{I} \rightarrow \underline{I}_2 + H_2O + Na\underline{Cl}}$

$\underbrace{}_{\text{lost 1 e}^-}$

One arrow must lose electrons, and the other must gain electrons.

To go from +5 to −1, each chlorine atom must gain six negative electrons. To go from −1 to zero, each iodine atom must lose one electron.

4. Calculate the lowest common *multiple* (LCM) by which each loss or gain can be multiplied so that the *total* electrons lost and gained are *equal*. Write each multiplier (such as 3×) next to the value for each electron change.

STOP

To get an equal number of electrons transferred, multiply the **I** arrow by 6× and the **Cl** arrow by 1×.

Each: $\overbrace{}^{\text{gained 6 e}^-\ (\mathbf{1x})}$

$\underset{+1+5-2}{} \quad \underset{+1-1}{} \quad \underset{0}{} \quad \underset{+1-2}{} \quad \underset{+1-1}{}$

$\mathrm{Na\underline{Cl}O_3 + H\underline{I} \rightarrow \underline{I}_2 + H_2O + Na\underline{Cl}}$

$\underbrace{}_{\text{lost 1 e}^-\ (\mathbf{6x})}$

5. Rewrite the original equation. In front of each particle connected by an arrow, write the number in the *multiplier* for that arrow as a trial coefficient.

STOP

$$\mathbf{1}\,\mathrm{NaClO_3} + \mathbf{6}\,\mathrm{HI} \rightarrow \mathbf{6}\,\mathrm{I_2} + \mathrm{H_2O} + \mathbf{1}\,\mathrm{NaCl}$$

The two arrows supply *four* trial coefficients.

In this reaction, the multipliers supply a 1-to-6 ratio for the reactants. The logic? Based on the arrows, every Cl atom reactant gains six electrons. Because every one I⁻ releases one electron, the I⁻ release must happen six times more often for the electron transfer to *balance*.

6. The **subscript tweak**. For each of the *atoms* connected by an arrow, if the atom *subscript* is *not* 1, divide the trial coefficient in front of its *particle* by the subscript of that atom.

STOP

One of the 4 "arrowed atoms" has a subscript. For the iodine atom with a subscript of 2, divide its trial coefficient by 2.

$$1\,\mathrm{NaClO_3} + 6\,\mathrm{HI} \rightarrow {}^{3}\!\!\not{6}\,\mathrm{I_2} + \underline{}\,\mathrm{H_2O} + 1\,\mathrm{NaCl}$$

7. Write coefficients that finish balancing by trial and error.

STOP

$$1\,\mathrm{NaClO_3} + 6\,\mathrm{HI} \rightarrow 3\,\mathrm{I_2} + \underline{\mathbf{3}}\,\mathrm{H_2O} + 1\,\mathrm{NaCl}$$

8. *Check*: 1 Na, 1 Cl, 3 O, 6 H, and 6 I atoms on each side. Neutral charge on each side. The equation is balanced. Done!

Oxidation number balancing provides four *trial* coefficients that are good hints, but they are often *not* final answers.

In *all* methods of balancing, modify coefficients if needed, by trial and error, until the atoms and overall charge are the same on each side.

The method will become easier with practice.

PRACTICE

Do not use a calculator.

1. Consulting the steps above as needed, balance this equation.

$$KI + O_2 + H_2O \rightarrow I_2 + KOH$$

2. Add flashcards for the two shaded rules in this lesson to your collection. Then, in your notebook, write a summary of the eight steps in this lesson. Learn your summary, then use it to balance these redox equations. If you need help, peek at the answer and try again.

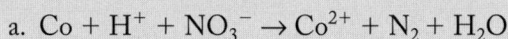

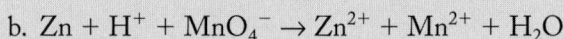

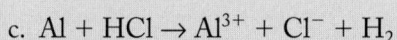

   a. $Co + H^+ + NO_3^- \rightarrow Co^{2+} + N_2 + H_2O$

   b. $Zn + H^+ + MnO_4^- \rightarrow Zn^{2+} + Mn^{2+} + H_2O$

   c. $Al + HCl \rightarrow Al^{3+} + Cl^- + H_2$

3. Write the formula for the reactant that is the reducing agent in each of the equations in problem 2.

4. Write the formula for the reactant that is being oxidized in each of the equations in problem 2.

5. Balance the equation in Lesson 15.3, Practice B, problem 2.

SUMMARY

In this chapter, we have practiced two study strategies that research says will help in learning science.

Strategy 1: Master mental arithmetic. Balancing redox equations is one of many activities in science that can be rewarding if you have mastered your math facts but frustrating if you need a calculator. To keep your mental math sharp, look for opportunities to solve problems with simple numbers by "pencil and paper" or mental math, reserving the calculator for a check if needed.

Strategy 2: Take advantage of the "testing effect." Measurements of learning have found that re-reading, underlining, and highlighting are not as effective at building memory as "self-testing" activities such as flashcards. In addition, by making an effort to learn facts and procedures before starting practice problems, you can treat each problem as a practice quiz. Challenging your brain to recall knowledge is especially effective at strengthening memory.

The goal of study is to add to and organize your long-term memory.

REVIEW QUIZ

You may use a periodic table. Do not use a calculator.

1. Define oxidation.

2. Define a reducing agent.

3. Find the oxidation number of:

    a. The nitrogen atom in an ammonium ion: NH_4^+. N = _____

    b. The sulfur atom in a sulfite ion: SO_3^{2-}. S = _____

    c. Each chlorine atom in dichlorine heptoxide. Cl = _____

4. Balance these half-reactions with lowest-whole-number coefficients.

    a. $Pb^{2+} \rightarrow$ Pb^{4+} b. $Cl_2 \rightarrow$ Cl^-

5. Write a label below each *reactant* identifying it as an oxidizing agent (OA) or reducing agent (RA), then write the formula for the particle being reduced.

    a. $Ag^+ + Pb \rightarrow Ag + Pb^{2+}$ b. $I^- + Fe^{3+} \rightarrow I_2 + Fe^{2+}$

6. Balance the problem 5a equation.

7. Balance the problem 5b equation.

8. Balance: $FeCl_2 + KMnO_4 + HCl \rightarrow MnCl_2 + FeCl_3 + H_2O + KCl$

9. In problem 8, which reactant particle is the reducing agent?

ANSWERS

Lesson 15.1

Practice   1a. **C = +4**

           Each: **+4** −2

           Formula: C O_2

           Sum: +4 −4   (Must sum to zero.)

    1b. **Br = zero**. Br_2 is an element.   1c. **Mn = +2**. The O.N. of a monatomic ion is equal to its charge.

    3a. *Each*:   +2 −1

         Formula: $CaBr_2$

         Sum: +2 −2   (Must sum to zero.)

    3b. *Each*:   +5 −2   3c. *Each*:   +1 +6 −2   3d. *Each*:   +6 −2

         Formula: NO_3^- Formula: Na_2CrO_4 Formula: $Cr_2O_7^{2-}$

         Sum: +5 −6   *Sum*:   +2 +6 −8   *Sum*:   +12 −14

    4a. **S = zero**. S_8 is an element. Atoms in elements have an O.N. of zero.

    4b. **As = +5**   4c. **Mn = +7**   4d. **Pb = 0**   4e. **Cl = +5**   4f. **Mn = +7**

Lesson 15.2

Practice A   Coefficients of 1 *may* be omitted (but must be included *during* balancing).

    1a. $1\,Co \rightarrow 1\,Co^{2+} + 2\,e^-$ 1b. $1\,O_2 + 4\,e^- \rightarrow 2\,O^{2-}$ 1c. $1\,e^- + 1\,Cu^{2+} \rightarrow 1\,Cu^+$

    1d. $1\,S_8 + 16\,e^- \rightarrow 8\,S^{2-}$

    2a. $2\,H^+ + 2\,e^- \rightarrow 1\,H_2$ 2b. $2\,I^- \rightarrow 1\,I_2 + 2\,e^-$ 2c. $12\,e^- + 1\,P_4 \rightarrow 4\,P^{3-}$ 2d. $2\,e^- + Sn^{4+} \rightarrow Sn^{2+}$

Practice B   1. $2\,Fe^{3+} + 1\,Fe \rightarrow 3\,Fe^{2+}$ 2. $2\,Al + 3\,Fe^{2+} \rightarrow 2\,Al^{3+} + 3\,Fe$ 3. $2\,ClO_3^- \rightarrow 2\,Cl^- + 3\,O_2$

Lesson 15.3

Practice A   2a. $\text{Sn}^{4+} + \widehat{\text{Ni}} \rightarrow \text{Ni}^{2+} + \text{Sn}^{2+}$

             OA   RA

As Sn^{4+} changes to Sn^{2+}, it gains two electrons so it is being reduced and is the oxidizing *agent*. As Ni loses two electrons, it is being *oxidized*. Because Ni donates electrons to Sn^{4+}, Ni is a reducing agent.

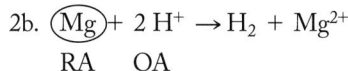

   2b. $\widehat{\text{Mg}} + 2\,\text{H}^+ \rightarrow \text{H}_2 + \text{Mg}^{2+}$

       RA   OA

Mg on the left loses two electrons as it forms Mg^{2+}: It is being *oxidized*. Because it is giving away electrons, it is the reducing agent. That means H^+ must be the oxidizing agent. Two H^+ ions gain two electrons to form neutral hydrogen.

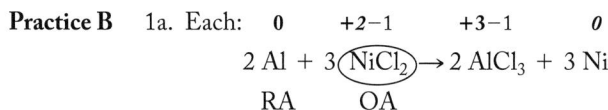

Practice B   1a. Each: **0**   **+2**−1   **+3**−1   **0**

$$2\,\text{Al} + 3\,\widehat{\text{NiCl}_2} \rightarrow 2\,\text{AlCl}_3 + 3\,\text{Ni}$$

          RA   OA

Neutral Al and Ni atoms are elements. Atoms in elements have O.N. = 0. NiCl_2 and AlCl_3 can both be considered to be *ionic* compounds because they combine metal and nonmetal atoms. All of their atoms are monatomic ions. The oxidation number for each atom is its charge as an ion.

Al goes from neutral to 3+, losing three electrons, being oxidized, and acting as a reducing agent. NiCl_2 must therefore be the oxidizing agent being reduced. The Ni^{2+} gains two electrons as it forms Ni.

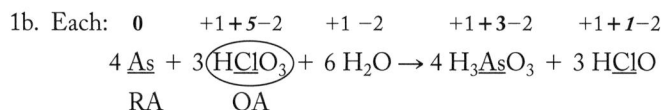

   1b. Each: **0**   +1**+5**−2   +1 −2   +1**+3**−2   +1**+1**−2

$$4\,\underline{\text{As}} + 3\,\widehat{\text{HClO}_3} + 6\,\text{H}_2\text{O} \rightarrow 4\,\text{H}_3\underline{\text{As}}\text{O}_3 + 3\,\text{H}\underline{\text{Cl}}\text{O}$$

       RA   OA

The As loses three electrons, is oxidized, and acts as a reducing agent. The **Cl** atom goes from a +5 O.N. to +1, gaining four electrons, being reduced, and acting as the oxidizing agent.

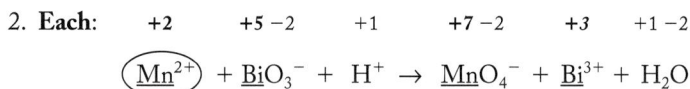

   2. **Each**:   **+2**   **+5** −2   +1   **+7**−2   **+3**   +1 −2

$$\widehat{\text{Mn}^{2+}} + \underline{\text{B}}\text{iO}_3^- + \text{H}^+ \rightarrow \underline{\text{Mn}}\text{O}_4^- + \underline{\text{Bi}}^{3+} + \text{H}_2\text{O}$$

Mn^{2+} loses five electrons, acting as a reducing agent. The Bi atom gains two electrons. The particle with Bi is therefore being reduced and is an oxidizing agent.

Lesson 15.4

Practice   Answers are given at *selected steps* of the balancing process.

Problem 1:

   4.   ⌐**(2x)** lost 1 e⁻ ⟶

       +1**−1**   *0*   +1−2   **0**   +1**−2**+1

$$\text{K}\underline{\text{I}} + \underline{\text{O}}_2 + \text{H}_2\text{O} \rightarrow \underline{\text{I}}_2 + \text{K}\underline{\text{O}}\text{H}$$

            ⌐**(1x)** gained 2 e⁻ ⟶

   5. Initial trial coefficients: **2** KI + **1** O_2 + **?** H_2O → **2** I_2 + **1** KOH

   6. *Subscript tweak*: Fractions are okay at this point.

$$2\,\text{KI} + {}^{1/2}\!\!\not{1}\,\text{O}_2 + \not{?}\,\text{H}_2\text{O} \rightarrow {}^1\!\not{2}\,\text{I}_2 + 1\,\text{KOH}$$

   7. Finish by trial and error. The 2 K on the left require 2 K on the right: Adjust the *trial* KOH coefficient. With 2 H on the right, one water is needed on the left.

$$2\,\text{KI} + 1/2\,\text{O}_2 + \underline{\mathbf{1}}\,\text{H}_2\text{O} \rightarrow 1\,\text{I}_2 + \mathbf{2}\,\text{KOH}$$

This equation is balanced, but lowest-whole-number coefficients are preferred. To eliminate the fraction, multiply all terms by the fraction's denominator.

$$4 \, KI + 1 \, O_2 + 2 \, H_2O \rightarrow 2 \, I_2 + 4 \, KOH$$

8. *Check:* Based on the *second* step 7 equation: 4 K, 4 I, 4 O, and 4 H atoms on each side. Zero charge on each side. Balanced.

Trial coefficients were adjusted at step 7 for the K atoms to balance. Oxidation numbers give good *hints*, but always adjust by trial and error if needed.

Problem 2a:

4.

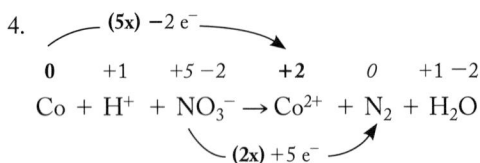

$$\overset{0}{Co} + \overset{+1}{H^+} + \overset{+5\ -2}{NO_3^-} \rightarrow \overset{+2}{Co^{2+}} + \overset{0}{N_2} + \overset{+1\ -2}{H_2O}$$

6. *Tweak:* $5 \, Co + \underline{\quad} \, H^+ + 2 \, NO_3^- \rightarrow 5 \, Co^{2+} + {}^1\!2 \, N_2 + \underline{\quad} \, H_2O$

7. The 6 O on the left require 6 H_2O on the right, so 12 H^+ must be on the left.

$$5 \, Co + 12 \, H^+ + 2 \, NO_3^- \rightarrow 5 \, Co^{2+} + 1 \, N_2 + 6 \, H_2O$$

8. *Check:* 5 Co, 12 H, 2 N, and 6 O atoms on each side. +10 charge on each side. Balanced.

Problem 2b: $5 \, Zn + 16 \, H^+ + 2 \, MnO_4^- \rightarrow 5 \, Zn^{2+} + 2 \, Mn^{2+} + 8 \, H_2O$

Problem 2c: $2 \, Al + 6 \, HCl \rightarrow 2 \, Al^{3+} + 6 \, Cl^- + 3 \, H_2$

3. For 2a: **Co**   For 2b: **Zn**   For 2c: **Al**

4. The particle being oxidized is the reducing agent. Answers are the same as in problem 3.

5.

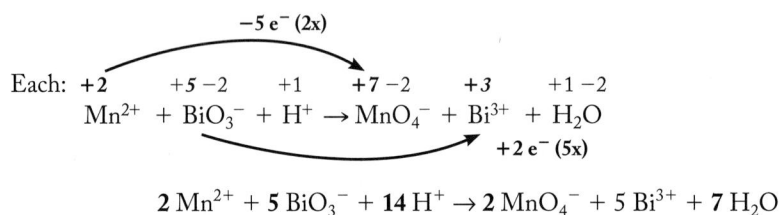

$$Each: \ \overset{+2}{Mn^{2+}} + \overset{+5\ -2}{BiO_3^-} + \overset{+1}{H^+} \rightarrow \overset{+7\ -2}{MnO_4^-} + \overset{+3}{Bi^{3+}} + \overset{+1\ -2}{H_2O}$$

$$2 \, Mn^{2+} + 5 \, BiO_3^- + 14 \, H^+ \rightarrow 2 \, MnO_4^- + 5 \, Bi^{3+} + 7 \, H_2O$$

Review Quiz

1. Oxidation is the loss of electrons.

2. A reducing agent donates electrons and loses some of its electrons.

3a. **N = −3**   3b. **S = +4**   3c. In Cl_2O_7, **Cl = +7**

4a. $1 \, Pb^{2+} \rightarrow 1 \, Pb^{4+} + 2 \, e^-$ 4b. $2 \, e^- + 1 \, Cl_2 \rightarrow 2 \, Cl^-$

5a. $Ag^+ + Pb \rightarrow Ag + Pb^{2+}$ 5b. $I^- + Fe^{3+} \rightarrow I_2 + Fe^{2+}$
    OA   RA   (Ag^+ is being reduced.)   RA   OA   (Fe^{3+} is being reduced.)

6. $2 \, Ag^+ + 1 \, Pb \rightarrow 2 \, Ag + 1 \, Pb^{2+}$

7. $2 \, Fe^{3+} + 2 \, I^- \rightarrow 2 \, Fe^{2+} + 1 \, I_2$

8. $5 \, FeCl_2 + 1 \, KMnO_4 + 8 \, HCl \rightarrow 1 \, MnCl_2 + 5 \, FeCl_3 + 4 \, H_2O + 1 \, KCl$

9. **$FeCl_2$** (*Or* the **Fe^{2+}** ion.)

16

Gas Laws I: Solving Equations

Lesson 16.1   Gas Fundamentals

Chemistry is most often concerned with matter in three states: gas, liquid, and solid. The gas state is in many respects the easiest to study because, by most measures, gases have similar and highly predictable behaviors.

Gas quantities can be measured using four variables: pressure, volume, temperature, and number of gas particles. Symbols for these variables are P, V, T, and n. Gas volumes are measured in units including liters (L, dm^3) and milliliters (mL, cm^3). Counts of gas particles are measured in moles (mol). Let's take a look at the units that measure *pressure* and *temperature*.

Gas Pressure

In an experiment, a glass tube 100 centimeters in length is sealed at one end and filled with mercury (symbol Hg), a dense, silver-colored metal that is a liquid at room temperature. The open end of the filled tube is covered, the tube is turned over, and the covered end is placed under the surface of additional mercury in a partially filled beaker. The tube end under the mercury in the beaker is then uncovered.

What happens? In all experiments at or near standard atmospheric pressure, the result is the same. The top of the mercury in the tube quickly falls from the top of the tube until it is about 76 centimeters above the surface of the mercury in the beaker. There, the mercury

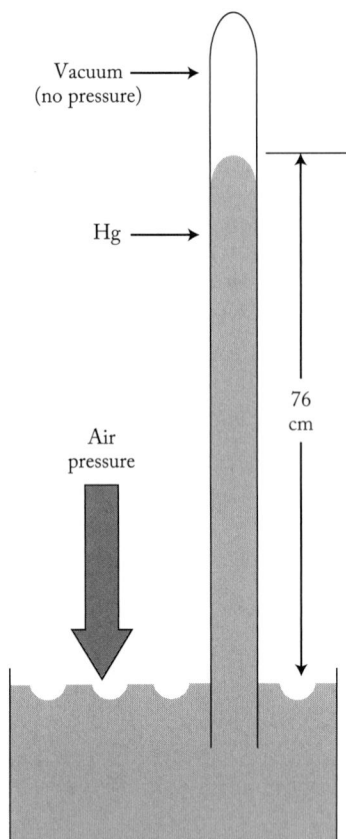

descent stops, forming a column of mercury inside the tube that is about 76 centimeters high. What is in the tube *above* the mercury column? A bit of mercury vapor, but no air. The space above the mercury in the tube is nearly a **vacuum**.

What happens if the top of the tube is snapped off? The mercury inside the tube behaves the same as a straw full of liquid when you take your finger off the top. The mercury in the tube falls quickly until it reaches the same level as the mercury in the beaker; however, as long as the tube remains sealed and is taller than 76 centimeters, the top of the mercury in the tube will remain about 76 centimeters above the top of the mercury in the beaker.

Why? The device created in this experiment is a mercury **barometer**. It measures the **pressure** of the air outside the tube. The pressure of the column of mercury *inside* the tube, pressing down on the top of the pool of mercury in the beaker, is *balanced* by the pressure of a 20-mile-high column of air, the atmospheric pressure, pressing down on the pool of mercury *outside* the tube.

If the air pressure outside the mercury column increases, the mercury is pushed higher in the tube. If the surrounding atmospheric pressure is lowered, the mercury level in the tube falls. When a weather forecast states that "the barometric pressure is 30.04 inches and falling," it is describing the height of the mercury column in a barometer (76 centimeters is about 30 inches). In meteorology, a high-pressure system is usually associated with fair weather. Falling barometers and low-pressure systems are often associated with clouds, storms, and precipitation.

Measuring Pressure

Pressure is measured in a variety of units. A gas pressure of exactly 760 **millimeters of mercury (mm Hg)** as measured by a barometer, also known as 760 **torr**, is defined as exactly 1 **atmosphere (atm)** of pressure. Normal atmospheric pressure at sea level on a fair-weather day is about 1 atmosphere. The SI unit for pressure is termed the **pascal** (abbreviated as Pa). Meteorology defines 100,000 pascals as 1 **bar**. The metric pressure units are summarized as follows:

Pressure Units		
1 atmosphere (atm)	$\equiv$ 760 mm Hg $\equiv$ 760 torr	(Exact definitions.)
	= 101 kilopascals (kPa)	(Not exact.)
	= 1.01 bars	

Any two of those measures can be used in a conversion factor for pressure units.

>   **Example:** 1 atm = 101 kPa

In this text, to learn gas calculations you will need to know the relationship for three of the pressure units. Apply this rule:

> If a problem involves *more* than one pressure unit, write in your DATA:
>
> 1 atm $\equiv$ 760 torr $\equiv$ 101 kPa

PRACTICE   A

1. Practice until you can write the "more-than-one pressure-unit equality" from memory. Use the equality recalled from memory in the calculations that follow.

2. The lowest atmospheric pressure at sea level in the Western Hemisphere was recorded in 2015 during Hurricane Patricia: a pressure of 656 torr. What is this pressure in atmospheres?

3. Solve without a calculator:

   a. A pressure of 2.00 atmospheres is how many kilopascals?

   b. A pressure of 76.0 torr is how many kilopascals?

Gas Temperature

To measure **temperature**, by the definitions of the *Celsius* (°C) scale (at 1 atm pressure):

- The *melting* and *freezing* temperature of *water* is **0°C**
- The *boiling* temperature of water is **100°C**

Temperature is defined as the **average kinetic energy** of particles. **Kinetic energy (KE)** is energy of **motion**, calculated by the equation

$$\text{Energy of motion} = \tfrac{1}{2}(\text{mass})(\text{velocity})^2, \text{ or } \mathbf{KE} = \tfrac{1}{2}\mathbf{m}\mathbf{v}^2$$

Particles cannot have zero mass, but they can have zero velocity: They can (in theory) stop moving. If the particles in a sample had zero velocity, kinetic energy would be zero and the temperature would be −273.15°C, which is defined as **absolute zero**. Absolute zero is the *bottom* of the temperature scale. Nothing can be colder than absolute zero.

A scale that defines **0** as absolute zero is termed an **absolute temperature scale**. In the SI system, absolute temperature is measured by the **Kelvin scale**, and the unit that measures absolute temperature is termed the **kelvin**. The abbreviation for kelvins is a capital **K** without a degree symbol.

Absolute zero ≡ 0 K = −273.15°C

The kelvin is the *same size* as a *degree Celsius*, but **0 K** is absolute zero.

To convert between Kelvin and Celsius temperatures, write the equation

K = °C + 273   (use 273.15 if additional precision is needed)

and solve for the symbol WANTED.

To check your answer: The temperature in kelvins is always 273 degrees *higher* than the temperature in degrees Celsius.

▷ TRY IT

Q. A temperature of 252 kelvins is how many degrees Celsius?

Answer:

$\boxed{K = °C + 273}$ Solve for °C: **°C** = K − 273 = 252 K − 273 = **−21°C**

S.F.: When using $\boxed{K = °C + 273}$ you are adding or subtracting, and answers are rounded to the highest *place* with doubt. All of the numbers shown have doubt in the ones place.

In scientific equations, the *symbol* for a temperature that must be measured by an absolute scale is a capital *T*. A lowercase *t* is often used as a symbol for temperature in degrees *Celsius*.

PRACTICE   B

1. Design flashcards for the shaded rules in this lesson (if you have different colors of cards, this chapter may be a good place to change to a new color). Practice until they are initially in memory. Then complete the problems below without looking back at the flashcards or rules.

2. A temperature of 25°C is how many kelvins? (Solve without a calculator.)

3. Complete the following chart (assume all pressures are 1 atmosphere).

	In Kelvins	In Degrees Celsius
Absolute Zero	_____	_____
Water Boils	_____	_____
Nitrogen Boils	_____	−196°C
Table Salt Melts	1074 K	_____
Water Freezes	_____	_____

4. In a problem, you calculate a temperature for a gas of −310°C. What do you notice about your answer?

Lesson 16.2   Gases at STP

In experiments with gases, it is preferred to measure gas *volumes* at **standard temperature and pressure**, abbreviated **STP**. For gas laws, **standard temperature** is the temperature of an ice–water mixture: **zero** degrees Celsius (273 K).

In gas measurements, there are two widely used but competing standards for **standard pressure**: Some textbooks use exactly 1 atmosphere, others use exactly 100 kilopascals. Because 1 atm = 101 kPa, these values are very close. In this text, we will define standard pressure as 1 atmosphere.

The following relationships must be committed to memory.

For Gas Law Calculations

- Standard temperature ≡ 0°C = +273.15 K
- Standard pressure ≡ 1 atm ≡ 760 torr = 101 kPa

Molar Gas Volume at STP

Gases have remarkably similar behavior. An important example involves gas volumes: If two samples of gases have the same number of *gas molecules* at the *same* temperature and pressure, the two samples will (usually) have the *same volume*. For gases exhibiting "ideal" behavior, this is true even if the gases have different molar masses, have different molecular formulas, or are a mixture of different gases.

In addition, if we know the temperature, pressure, and number of moles of gas particles in a sample, its volume is predictable.

> **Example:** For one mole of any ideal gas or any mixture of ideal gases, the volume at STP is 22.4 liters.

The volume of a mole of gas at STP (the **molar volume**) provides us with a conversion for conditions at STP.

The STP Prompt

In calculations, if a gas is stated to be at STP (or at a T and P that are the same as STP), write in the DATA:

1 mol gas = 22.4 L gas at STP

Note that in the above equality, "at STP" is attached only to the gas *volume*. Because gas volumes vary with temperature and pressure, *all* gas volumes *must* have pressure and temperature conditions stated if the volume is to be a measure of the number of particles in the sample.

All gas volumes must be labeled with a T and P if the T and P are known.

Calculations for Gases at STP

In gas calculations, the STP prompt is often used in conjunction with the grams and Avogadro prompts. You will also need to recall that if a problem asks for

- molar mass, you WANT grams *per* 1 mole;
- density, you WANT one mass unit (such as g or kg) over one volume unit (such as L, dm³, or mL).

These prompts and rules will allow you to solve most calculations for gases at STP by chaining conversions.

━━━━▶ **TRY IT**

Q. 2.0×10^{23} molecules of NO_2 gas would occupy how many liters at STP?

Answer:

WANTED: ? L NO_2 gas at STP =

DATA: 2.0×10^{23} molecules of NO_2 gas

6.02×10^{23} molecules = 1 mol   (Avogadro prompt.)

1 mol any gas = 22.4 L any gas at STP   (STP prompt.)

SOLVE: ? L $NO_2(g)$ STP = 2.0×10^{23} molec. NO_2 $\cdot$ $\dfrac{1 \text{ mol}}{6.02 \times 10^{23} \text{ molec.}}$

$\cdot \dfrac{22.4 \text{ L gas STP}}{1 \text{ mol gas}}$ = **7.4 L $NO_2(g)$ at STP**

Note the difference between these calculations and stoichiometry. The above problem involved only *one* substance. If stoichiometry steps are needed, your WANTED and DATA will have *two* substances involved in a chemical *reaction*.

Working the Examples

An effective technique in learning to solve problems in the physical sciences and math is to *work the examples* in the textbook. To do so:

- *Cover* the answer, read the question, and try to SOLVE.
- If you need help, peek at the answer, and then try the question again.

Apply that method to this problem.

━━━━▶ **TRY IT**

Q. Determine the density of O_2 gas at STP in grams per milliliter.

Answer:

WANTED: ? $\dfrac{\text{g } O_2}{\text{mL gas STP}}$ (Write complete labels.)

DATA: 32.0 g O_2 = 1 mol O_2 (*Grams* prompt in WANTED unit.)

1 mol = 22.4 L gas at STP   (STP in WANTED unit.)

SOLVE: (To review solving for a ratio, see Lesson 10.2.)

$\dfrac{? \text{ g } O_2}{\text{mL gas STP}} = \dfrac{32.0 \textbf{ g } O_2}{1 \text{ mol } O_2} \cdot \dfrac{1 \text{ mol gas}}{22.4 \text{ L gas STP}} \cdot \dfrac{10^{-3} \text{ L}}{1 \text{ mL}} = \dfrac{\textbf{1.43} \times \textbf{10}^{-3} \textbf{ g } O_2}{\textbf{mL gas STP}}$

(Those conversions, right-side up, may be in any order.)

Note that STP was attached to each gas *volume* unit (mL and L). Note also that:

- The conversions for *molar mass* (the grams prompt) and *particles per mole* (the Avogadro prompt) are valid whether a substance is a gas, liquid, or solid.

- The *STP* prompt is only true for *gases at STP*.

PRACTICE

1. Design flashcards for the shaded rules in this lesson. Combine them with the cards from Lesson 16.1 and run all of the cards to achieve quick recall. Then try the problems below.

2. Write values for standard pressure in three different units.

3. Write values for standard temperature for gases in two different units.

4. Calculate the density of SO_2 gas in g/L at STP.

5. Calculate the number of molecules in 1.12 L of CO_2 gas at STP.

6. The density of a gas at STP is 0.00205 g • mL^{-1}. What is its molar mass?

7. Calculate the volume in milliliters of 15.2 g of F_2 gas at 273 K and standard pressure.

8. If 0.0700 mol of a gas has a volume of 1,760 mL, what is the volume in liters of 1 mol of the gas under the same temperature and pressure conditions?

Lesson 16.3   Cancellation of Complex Units

In scientific calculations, some derived quantities and constants have **complex units**: units that are *reciprocal* or have *more* than one unit in the numerator or denominator. Let's review the mathematics of these fractions.

Why Not to Write "*A/B/C*"

The notation for complex fractions must be written with special care. Why?

TRY IT

On this page, write answers to the following. Do not use a calculator.

Q1. $\dfrac{8}{4}$ divided by 2 =   **Q2.** 8 divided by $\dfrac{4}{2}$ =   **Q3.** 8/4/2 =

Answers:

1. 2 divided by 2 = **1**   2. 8 divided by 2 = **4**

3. Could be **1** or **4**, depending on which is the fraction: 8/4 or 4/2.

The format *A/B/C* for numbers or units is ambiguous unless you know, from the context or prior steps, which is the fraction.

Let's try question 3 again, this time with the fraction identified by parentheses.

Use the rule from mathematics: Perform the operation inside parentheses *first*.

Q4. $(8/4)/2 =$ **Q5.** $8/(4/2) =$

STOP

Answers:

   4. $2/2 = $ **1**   5. $8/2 = $ **4**

With parentheses, the problem is easy, but for 8/4/2 without parentheses, you cannot be sure of the answer. For this reason, writing numbers or units in a format $A/B/C$ should be avoided.

In these lessons, to distinguish a numerator from a denominator, we will either use a heavier <u>division line</u> or group fractions in parentheses.

Multiplying and Dividing Fractions

By definition, the **reciprocal** of X is $1/X$. Also by definition:

$$1 \text{ divided by } (1/X) = \text{ the } \textit{reciprocal} \text{ of } (1/X) = \frac{1}{\frac{1}{X}} = X$$

Recall that $1/X$ can be written as X^{-1}. For exponentials, when you take an exponential to a power, you multiply the exponents. Using exponents, the rule for the reciprocal of a fraction can be explained as:

$$1 \text{ divided by } (1/X) = (X^{-1})^{-1} = (X^{+1}) = X$$

The rule $1/(1/X) = X$ is an example of this general rule:

Complex Fraction Rule 1

To simplify the *reciprocal* of a *fraction*, invert the fraction.

Q. Remove the parentheses and simplify: $1/(B/C) =$

STOP

Answer:

$1/(B/C)$ is a reciprocal of a fraction: 1 over a fraction. To simplify, invert the fraction (flip the *fraction* over).

In symbols: $\dfrac{1}{\frac{B}{C}} = \dfrac{C}{B}$

Using exponents: $(B/C)^{-1} = (B \cdot C^{-1})^{-1} = B^{-1} \cdot C$

Calculations in chemistry may involve fractions in the numerator, denominator, or both. To handle these cases systematically, use the following rule.

Complex Fraction Rule 2

When a term has *more than one* fraction line (either a _____ or a /), *separate* the terms that are *fractions*. To do so, apply these steps in order:

  a. If a term has a fraction in the *denominator*, separate the terms into (1/fraction in the denominator) multiplied by the remaining terms.

  b. If there is a fraction in the *numerator*, separate and multiply that fraction by the other terms in the fraction.

  c. Then simplify: Invert any reciprocal *fractions*, cancel units that cancel, and multiply the terms.

Let's learn these rules with some examples.

▶ **TRY IT**

Solve the questions below in your notebook.

Q1. Remove the parentheses and simplify: $A/(B/C)$. Apply rule 2a, then 2c.

STOP

Answer:

$$A/(B/C) = \frac{A}{\frac{B}{C}} = A \cdot \frac{1}{\frac{B}{C}} = A \cdot \frac{C}{B} = \frac{A \cdot C}{B}$$

$$\text{(rule 2a)} \qquad \text{(rule 2c)}$$

Because there is a fraction in the denominator, separate that fraction into a *reciprocal* times the other terms, then invert the reciprocal fraction and multiply.

You can also solve using exponents: $A \cdot (B \cdot C^{-1})^{-1} = A \cdot B^{-1} \cdot C$

Q2. Simplify $(A/B)/C$ by applying rule 2b.

STOP

Answer:

$$(A/B)/C = \frac{\frac{A}{B}}{C} = \frac{A}{B} \cdot \frac{1}{C} = \frac{A}{B \cdot C}$$

$$\text{(rule 2b)}$$
$$\text{(rewrite}$$
$$\text{for clarity)}$$

Note how this answer differs from the question 1 answer.

Q3. Simplify $(A/B)/(C \cdot (D/E)) =$

STOP

Answer:

$$(A/B)/(C \cdot (D/E)) = \frac{\dfrac{A}{B}}{C \cdot \dfrac{D}{E}} = \frac{A}{B} \cdot \frac{1}{\dfrac{D}{C}\dfrac{}{E}} = \frac{A}{B} \cdot \frac{1}{C} \cdot \frac{1}{\dfrac{D}{E}} = \frac{A \cdot E}{B \cdot C \cdot D}$$

(rewrite   (rule 2a)   (rule 2b)   (rule 2c)
for clarity)

To summarize rule 2:

> If a term has more than one fraction line, separate the fractions.

During problem solving, when *you* are *writing* terms with more than one fraction line, you will need to develop a systematic way to distinguish fractions in the numerator and denominator (such as thick lines or parentheses).

PRACTICE   **A**

1. Design, make, and practice flashcards for rules 1 and 2 in the subsection above.

2. Working in your notebook, simplify these.

    a. $X/(Y/Z) =$ b. $(D/E)^{-1} =$ c. $\dfrac{\text{meters}}{\dfrac{\text{meters}}{\text{second}}} =$

    d. $\dfrac{(\text{meters/second})}{\text{second}} =$ e. $\text{meters/second/second} =$

Dividing Complex Numbers and Units

In some derived quantities and constants, the units are complex fractions. Examples we will soon encounter include:

- The *gas constant* $= \boldsymbol{R} = 0.0821 \ \dfrac{\text{atm} \cdot \text{L}}{\text{mol} \cdot \text{K}}$

- The *specific heat capacity* of water $= 4.184 \ \text{J/g} \cdot \text{K}$

Joule (J) is a unit that measures energy.

In those units, the *dot* between two units is a notation meaning that the two units are *multiplied together* in either the numerator or denominator.

Examples:

- In 4.184 **J/g • K**, grams and kelvins are both in the *denominator*.

$$4.184 \text{ J/g} \cdot \text{K} = 4.184 \text{ J}/(\text{g} \cdot \text{K}) = 4.184 \text{ J} \cdot \text{g}^{-1} \cdot \text{K}^{-1}$$

- The gas constant

$$0.0821 \frac{\text{atm} \cdot \text{L}}{\text{mol} \cdot \text{K}}$$

can be written as 0.0821 atm • L/mol • K

When multiplying and dividing terms with complex units, you must do the math for both numbers and units. Numbers and units in fractions follow the rules of algebra reviewed in the subsection above.

▶ TRY IT

Q. Applying fraction rules 1 and 2, solve for a number and its unit.

$$\frac{360 \text{ J}}{18.0 \text{ K} \cdot 0.50 \dfrac{\text{J}}{\text{g} \cdot \text{K}}} =$$

🛑 **STOP**

Answer:

Note how the units are rearranged in each step.

$$\frac{360 \text{ J}}{18.0 \text{ K} \cdot 0.50 \dfrac{\text{J}}{\text{g} \cdot \text{K}}} = \frac{360 \text{ J}}{18.0 \text{ K}} \cdot \frac{1}{0.50 \dfrac{\text{J}}{\text{g} \cdot \text{K}}} = \frac{360 \text{ J}}{18.0 \text{ K}} \cdot \frac{\textbf{g} \cdot \textbf{K}}{0.50 \text{ J}} = \textbf{40. g}$$

Applying rule 2a: If there is a unit that has a fraction in the denominator, *separate* the fraction into *1/fraction in the denominator* multiplied by the other terms. Then, invert the reciprocal, cancel units that cancel, and multiply.

Mark the unit cancellation in the final step above.

Cancellation Shortcuts

Complex fractions can be simplified by first canceling *within* the numerator and denominator, and then canceling *between* the numerator and denominator. It must be remembered, however, that *canceling* a number or unit does *not* get rid of it: It replaces the number or unit with a 1.

⟶ TRY IT

Q. Try solving the numbers without a calculator. For the units, first cancel units *in* the denominator, then *between* the numerator and denominator:

$$\frac{4{,}800 \text{ J}}{12.0 \text{ g} \cdot 2.0 \dfrac{\text{J}}{\text{g} \cdot \text{K}}} =$$

STOP

Answer:

$$\frac{4{,}800 \text{ J}}{12 \text{ g} \cdot 2.0 \dfrac{\text{J}}{\text{g} \cdot \text{K}}} = \frac{4{,}800 \text{ J}}{12 \text{ g} \cdot 2.0 \dfrac{\text{J}}{\text{g} \cdot \text{K}}} = \frac{4{,}800}{24} \cdot \frac{1}{\dfrac{1}{\text{K}}} = 2.0 \times 10^2 \text{ K}$$

In the third term, the **1 terms** are important. *1/K* and *1/(1/K)* are not the same.

When in doubt about how to apply this type of unit cancellation, skip the cancellation shortcuts and the use systematic rules 1–2c.

PRACTICE   **B**

Apply the rules from memory to simplify these fractions.

1. $\dfrac{\text{calories}}{\text{calorie} \cdot \text{gram}}\Big/ {}^{\circ}\text{C} =$

   $$\frac{\text{calories}}{\dfrac{\text{calorie} \cdot \text{gram}}{{}^{\circ}\text{C}}} =$$

2. $$\frac{\text{atm}}{(\text{mol}) \cdot \dfrac{\text{atm} \cdot \text{L}}{\text{mol} \cdot \text{K}}} =$$

3. $$\frac{(\text{mol}) \cdot \dfrac{\text{atm} \cdot \text{L}}{\text{mol} \cdot \text{K}} \cdot (\text{K})}{\text{L}} =$$

Lesson 16.4   The Ideal Gas Law and Equations

If a gas is at STP, many calculations can be solved by conversions, but in most experiments and problems, gases are not at STP. In those cases, you will need gas *equations* to SOLVE.

The gas equation used most often is the **ideal gas law** (often remembered as *piv-nert*). For all "ideal" gases,

$$PV = nRT$$

where

- P and V are pressure and volume in any metric units,
- T is temperature in *kelvins*,
- n represents the number of *moles* of gas, and
- R is a number with units called the **gas constant**.

In this equation, P, V, T, and n are **variables**: Their values depend on the conditions in the problem. R is a number with units attached: For a given set of units, the number is always the same.

For any gas, if you know *any three* of the values for the four variables in the ideal gas law, and you know a value and units for R (which is usually supplied in a problem), you can use the ideal gas law to predict the value of the fourth variable.

Nonideal Behavior

The ideal gas law is derived mathematically based on assumptions, including that gas molecules do not attract when they collide, and that they do not condense to form liquids or solids. These assumptions do not hold under all conditions, and real gases can display "nonideal" behavior. But for most gases under typical gas conditions, the ideal gas law is a good *approximation* in predicting behavior. For gas calculations, assume that the *ideal* gas law will apply *unless* nonideal behavior is specified.

Solving Problems Requiring Equations

Calculations in chemistry can generally be put into three categories: those that can be solved with conversions, those that require equations, and those requiring both. We will use gas laws to illustrate a *system* for solving calculations that require equations. This system will be especially helpful if you take additional courses in chemistry or other sciences.

Let's start with a problem in which we know the correct equation to apply. This example can be solved in several ways, but we will solve with a method that has the advantage of working with more-difficult problems. Please try the steps described below.

▶ TRY IT

Q. A value for the gas constant R is usually supplied when solving gas problems; however, R can also be calculated based on values you know for a gas at STP.

> **Example:** One mole of any gas occupies 22.4 L at a pressure of 1 atmosphere and a temperature of 273 K.

Using the example DATA and the ideal gas law, calculate a value for R.

To SOLVE, complete the steps below in your notebook.

1. As always, begin by writing WANTED and the unit you are looking for. When solving for an *equation*, you might instead need to write the *symbol* you are looking for. If possible, write the symbol and unit WANTED.

> **Example:** WANTED: $? P =$ or   WANTED: $? P$ in atm $=$

2. *If* you know you need an equation to solve the problem *and* you know which equation you will need, *write* the equation. Put a box around it to say: This is important to remember.

   This problem says to use the ideal gas law. Write it below the WANTED line.

3. Next, make a DATA *table* that includes *each symbol* in the equation. For this problem, write:

   DATA: $P =$

   $V =$

   $n =$

   $R =$

   $T =$

4. Put a **?** after the *symbol* WANTED in the problem.

5. Read the problem again, and write each number and its unit after a symbol. Use *units* to match the symbol to the DATA (1 *mol* goes after *n*, 273 K goes after *T*, etc.).

Using the data from the example, in your notebook, complete steps 1–5.

Answer:

At this point, your paper should look like this:

   WANTED: **? R =**   (Write the unit and/or *symbol* WANTED.)

   $$\boxed{PV = nRT}$$

   DATA: $P = 1$ atm

   $V = 22.4$ L

   $n = 1$ mol

   $R = ?$

   $T = 273$ K

6. Write SOLVE. Then, using algebra, solve the fundamental, memorized equation for the *symbol* WANTED.

   SOLVE: **? = R =**

In your notebook, complete step 6.

$$? = R = \frac{PV}{nT}$$

7. Substitute DATA for the symbols and do the math. Cancel units that cancel, but write the units that do not cancel after the number that you calculate for the answer.

In your notebook, complete step 7 to finish the problem.

   SOLVE: $? = R = \dfrac{PV}{nT} = \dfrac{(1\ \text{atm})(22.4\ \text{L})}{(1\ \text{mol})(273\ \text{K})} = \mathbf{0.0821\ \dfrac{atm \cdot L}{mol \cdot K}}$

Be certain that your answer includes the *units* after the numbers.

When solving by applying equations, two key rules are:

Memorize fundamental equations in *one* format. Use algebra to SOLVE for each symbol WANTED.

Example: Memorize the fundamental equation: $PV = nRT$. When you need to SOLVE for one of the symbols, use algebra:

$n = PV/RT$ and $P = nRT/V$ but *don't* memorize the variations.

Do not plug in numbers until *after* you have solved for the WANTED symbol using symbols.

Symbols move more quickly (and with fewer mistakes) than numbers with their units. Let's summarize.

Steps for Solving with an Equation

1. Write the symbol and/or unit WANTED.
2. Write the memorized equation. Put a box around it.
3. Make a DATA table. On each line, put one symbol from the equation.
4. Write a **?** after the WANTED symbol. Write the unit WANTED if it is known.
5. Based on units, after each symbol write the matching DATA in the problem.
6. SOLVE your memorized equation for the WANTED **symbol** *before* substituting numbers and units.
7. Put both numbers *and* units into the equation when you SOLVE. Include in your answer all units that do not cancel.

PRACTICE   A

1. Write one or more possible metric *units* WANTED when you are asked to find:

   a. Molarity   b. Molar mass

   c. Volume   d. Mass

2. After each of these gas measurements, write the symbol from $PV = nRT$ that represents the dimension the unit is measuring.

   a. 0.50 mol   b. 202 kPa   c. 11.2 dm^3

   d. 373 K   e. 38 torr

(continued)

3. In this space, solve $PV = nRT$ in symbols for:

    a. $T =$ b. $V =$ c. $R =$

4. If 1 mol of a gas has a volume of 22.4 L but a pressure of 2.3 atm, what must the temperature be? Use $R = 0.0821$ atm • L/mol • K and the ideal gas law. As needed, look back at this lesson.

Ideal Gas Calculations and Temperature

In some cases, an equation will require certain units. As one example, scientific equations that include a capital T require temperature to be measured using an *absolute* temperature scale (which in the metric system means *kelvins*).

If an equation has a capital T for temperature:

- A DATA temperature that is not in kelvins must be converted to kelvins.
- If degrees Celsius is WANTED, kelvins must be found first.

▬▬▬✏▶ TRY IT

Q. What will be the volume of 1 mol of an ideal gas at a temperature of 25°C and a pressure of 0.500 atm? (Use the ideal gas law and $R = 0.0821$ atm • L/mol • K)

Solve in your notebook by applying the "Steps for Solving with an Equation." Try to apply the steps from memory, but look back if needed.

STOP

Your paper should include:

    WANTED: $? V =$ (Write the unit and/or *symbol* WANTED.)

                $\boxed{PV = nRT}$

    DATA: $P = 0.500$ atm

               $V = ?$

               $n = 1$ mol

               $R = 0.0821$ atm • L/mol • K

               $T = 25°C$ Must use **K**. $\boxed{K = °C + 273}$ $= 25 + 273 = \textbf{298 K}$

SOLVE for the WANTED unit in symbols, then plug in DATA with units.

STOP

    SOLVE: $PV = nRT$

$$? = V = \frac{nRT}{P} = nRT \cdot \frac{1}{P} = (1 \text{ mol})(0.0821 \frac{\text{atm} \cdot \text{L}}{\text{mol} \cdot \text{K}})(298 \text{ K}) \cdot \frac{1}{0.500 \text{ atm}} = \textbf{48.9 L}$$

:
:
:
:
:
:
:
:

Mark the unit cancellation in the line above *or* on your paper. Check that the unit of the answer is appropriate for the *symbol* WANTED.

Because a volume is WANTED, the volume unit *liters* obtained by unit cancellation is an indication that the calculation has been done correctly. If the unit after cancellation is not appropriate for the *symbol* WANTED, a mistake has been made in your work. The answer unit *must* be appropriate for the symbol WANTED.

· ▶

Note in the Try It that after the "SOLVE for *V* in symbols" step, the *symbol* on the bottom was *separated* from those on top before numbers were substituted. A suggested step is:

> If an equation solved in symbols has both a term with complex units and a symbol in the denominator, move the *denominator symbol(s)* to a *reciprocal*.

This step is not mandatory. You may do the math in any way you choose. However, in a fraction that includes symbols with complex units (such as *R*), converting "top-over-bottom symbols" to "top multiplied by 1/bottom symbols" will simplify unit cancellation.

PRACTICE   B

1. Design flashcards and/or "step summaries" for Lesson 16.4. Practice recall of your cards and lists.

2. Write one or more possible metric *units* WANTED when you are asked to find these quantities. For any part, if you cannot answer quickly, design a flashcard.

   a. Density   b. Speed or velocity

   c. Gas pressure   d. Temperature

3. A gas has a volume of 7.20 L at 2.50 atm pressure and 50.°C. How many moles are in the sample? Use $R = 0.0821$ atm · L/mol · K and the ideal gas law.

SUMMARY

In this chapter, we added the STP prompt to the grams, Avogadro, stoichiometry, and M prompts. What do they all have in common?

- Each prompt relates *moles* to other units. The way to make sense of most relationships in chemistry is to *count* the particles. Moles is the unit for counting an amount of particles that is visible.
- Each prompt adds a conversion to your DATA. Conversions quickly relate entities, dimensions, and units without having to recall or solve equations.
- This chapter included additional relationships to write automatically when a problem involves two pressure or two temperature units.

More complex relationships require equations, but as we will see in our next chapter, these prompts and cues will help in preparing DATA to simplify equation use.

REVIEW QUIZ

Use $R = 0.0821$ atm • L/mol • K

1. Mercury freezes at 234 K. What is this temperature in degrees Celsius?

2. A pressure of 50.5 kPa is how many atmospheres?

3. If 250. mL of a gas at STP weighs 0.313 g, what is the molar mass of the gas?

4. A sample of 3.00 moles of an ideal gas at 0°C has a volume of 11.2 liters. Using the ideal gas law, calculate the pressure exerted by the gas.

5. 2.80 liters of O_2 gas at STP has a mass of how many grams?

6. If 0.250 mole of helium gas has a volume of 15.3 liters at a pressure of 0.500 atmosphere, what will be its temperature in kelvins and in degrees Celsius?

ANSWERS

Lesson 16.1

Practice A   Some solutions are partial. Your work should include WANTED, DATA, and SOLVE.

2. WANTED: ? atm =
   DATA: 656 torr
             1 atm ≡ 760 torr = 101 kPa   (More than one P unit.)
   SOLVE: ? atm = 656 torr • $\dfrac{1\ atm}{760\ torr}$ = **0.863 atm**

3a. 202 kPa   3b. SOLVE: ? kPa = ^{1}76.0 torr • $\dfrac{101\ kPa}{_{10}760\ torr}$ = **10.1 kPa**

Practice B   2. $\boxed{K = °C + 273}$ = 25°C + 273 = 298 K

3.

	In Kelvins	In Degrees Celsius
Absolute Zero	**0 K**	**−273°C**
Water Boils	**373 K**	**100.°C**
Nitrogen Boils	**77 K**	**−196°C**
Table Salt Melts	1074 K	**801°C**
Water Freezes	**273 K**	**0°C**

4. A temperature of −310°C cannot be valid: −310°C is below absolute zero (−273°C). There cannot be a temperature colder than absolute zero.

Lesson 16.2

Practice   2. Standard pressure ≡ 1 atm ≡ 760 torr = 101 kPa

3. Standard temperature ≡ 0°C = 273 K

4. ? $\dfrac{g\ SO_2\ gas}{L\ SO_2\ gas\ at\ STP}$ = 64.1 g SO_2 • $\dfrac{1\ mol\ gas}{22.4\ L\ gas\ STP}$ = **2.86** $\dfrac{g\ SO_2(g)}{L\ SO_2\ gas\ at\ STP}$

5. ? molecules $CO_2(g)$ = 1.12 L $CO_2(g)$ STP • $\dfrac{1\ mol\ gas}{22.4\ L\ gas\ STP}$ • $\dfrac{6.02 \times 10^{23}\ molecules}{1\ mol}$

= **3.01 × 10²² molecules CO₂**

6. $? \dfrac{\text{g}}{\text{mol}} = \dfrac{0.00205 \text{ g gas}}{1 \text{ mL gas at STP}} \cdot \dfrac{1 \text{ mL}}{10^{-3} \text{ L}} \cdot \dfrac{22.4 \text{ L any gas at STP}}{1 \text{ mol gas}} = \textbf{45.9 } \dfrac{\textbf{g}}{\textbf{mol}}$

Reminders

- Attach temperature and pressure conditions, if known, to gas *volumes*.
- In the interest of readability, most unit cancellations in these answers are not marked. However, in your work, always ~~mark~~ your unit cancellations.

---

7. $? \text{ mL F}_2(g) \text{ STP} = 15.2 \text{ g F}_2(g) \cdot \dfrac{1 \text{ mol F}_2}{38.0 \text{ g F}_2} \cdot \dfrac{22.4 \text{ L gas STP}}{1 \text{ mol gas}} \cdot \dfrac{1 \text{ mL}}{10^{-3} \text{ L}} = \textbf{8.96} \times \textbf{10}^3 \textbf{ mL F}_2\textbf{(g) STP}$

8. WANTED: $\dfrac{? \text{ L gas at given P and T}}{1 \text{ mol}} =$

    Strategy: This problem does not specify STP, but you can SOLVE for the WANTED unit. Compare units: You WANT *liters* and *moles*. You are *given milliliters* and *moles*.

    DATA: 0.0700 mol gas = 1,760 mL gas at given T and P

    SOLVE: (Conversions may be in any order.)

$$\dfrac{? \text{ L gas at give } T \text{ and } P}{1 \textbf{ mol}} = \dfrac{1,760 \text{ mL gas at given } T \text{ and } P}{0.0700 \textbf{ mol}} \cdot \dfrac{10^{-3} \text{ L}}{1 \text{ mL}} = \textbf{25.1 L gas} \text{ at given } T \text{ and } P \over \textbf{mol}$$

Lesson 16.3

Practice A 2a. $X/(Y/Z) = X \cdot \dfrac{1}{\dfrac{Y}{Z}} = \dfrac{\textbf{X} \cdot \textbf{Z}}{\textbf{Y}}$

          2b. $(D/E)^{-1} = \dfrac{1}{\dfrac{D}{E}} = \dfrac{\textbf{E}}{\textbf{D}}$ or $(\textbf{D}^{-1} \cdot \textbf{E})$

      2c. $\dfrac{\text{meters}}{\dfrac{\text{meters}}{\text{second}}} = \text{meters} \cdot \dfrac{1}{\dfrac{\text{meters}}{\text{second}}} = \text{meters} \cdot \dfrac{\text{second}}{\text{meters}} = \textbf{seconds}$

      2d. $\dfrac{(\text{meters/second})}{\text{second}} = \dfrac{\text{meters}}{\text{second}} \cdot \dfrac{1}{\text{second}} = \dfrac{\textbf{meters}}{\textbf{second}^2}$

      2e. meters/second/second cannot be evaluated.

Practice B 1. $\dfrac{\text{calories}}{\dfrac{\text{calorie} \cdot \text{gram}}{°\text{C}}} = \dfrac{1}{\dfrac{\text{calorie}}{°\text{C}}} \cdot \dfrac{\text{calories}}{\text{gram}} = \dfrac{°\text{C}}{\text{calorie}} \cdot \dfrac{\text{calories}}{\text{gram}} = \dfrac{°\textbf{C}}{\textbf{gram}}$

      2. $\dfrac{\text{atm}}{(\text{mol}) \cdot \dfrac{\text{atm} \cdot \text{L}}{\text{mol} \cdot \text{K}}} = \text{atm} \cdot \dfrac{1}{\text{mol}} \cdot \dfrac{1}{\dfrac{\text{atm} \cdot \text{L}}{\text{mol} \cdot \text{K}}} = \dfrac{\text{atm} \cdot \text{mol} \cdot \text{K}}{\text{mol}} \cdot \dfrac{1}{\text{atm} \cdot \text{L}} = \dfrac{\textbf{K}}{\textbf{L}}$

      3. $(\text{mol}) \cdot \dfrac{\dfrac{\text{atm} \cdot \text{L}}{\text{mol} \cdot \text{K}}}{\text{L}} \cdot (\text{K}) = (\text{mol}) \cdot \dfrac{(\text{atm} \cdot \text{L})}{\text{mol} \cdot \text{K}} \cdot (\text{K}) \cdot \dfrac{1}{\text{L}} = \textbf{atm}$

Lesson 16.4

Practice A 1a. Molarity: $\dfrac{\textbf{mol}}{\textbf{1 L solution}}$ 1b. Molar mass: $\dfrac{\textbf{g}}{\textbf{1 mol}}$ 1c. Volume: **L, mL, dm³, cm³**   1d. Mass: **kg, g, mg**

      2a. 0.50 mol *n*   2b. 202 kPa *P*   2c. 11.2 dm³ *V*   2d. 373 K *T*   2e. 38 torr *P*

      3a. $T = \dfrac{PV}{nR}$ 3b. $V = \dfrac{nRT}{P}$ 3c. $R = \dfrac{PV}{nT}$

If you cannot solve this algebra correctly *every* time, find a friend or tutor who can help you learn the algebra needed for this chapter.

4. WANTED: **? T in kelvins =**   (T must be in kelvins.)

$$\boxed{PV = nRT}$$

DATA: $P = 2.3$ atm $V = 22.4$ L $n = 1.00$ mol

$R = 0.0821 \dfrac{\text{atm} \cdot \text{L}}{\text{mol} \cdot \text{K}}$

$T = ? K$

SOLVE: $PV = nRT$

$$? = T = \frac{PV}{nR} = \frac{(2.3\text{ atm})(22.4\text{ L})}{(1\text{ mol})\,0.0821\,\frac{(\text{atm} \cdot \text{L})}{\textbf{mol} \cdot \textbf{K}}} = \frac{(2.3\text{ atm})(22.4\text{ L})}{1\text{ mol}} \cdot \frac{\textbf{mol} \cdot \textbf{K}}{0.0821\text{ atm} \cdot \text{L}} = \textbf{630 K}$$

A Kelvin temperature is WANTED and is the unit after cancellation.

For 1 mol of gas at 22.4 L volume to cause a pressure of 2.3 atm (rather than 1 atm, as at STP), the gas must be very hot, with molecules moving much faster on average than at standard temperature (273 K).

Practice B   2a. Density $\dfrac{\text{any } \textbf{mass} \text{ unit (such as kg or g)}}{\text{any } \textbf{volume} \text{ unit (such as L, mL)}}$ 2b. Speed $\dfrac{\text{any } \textbf{distance} \text{ unit (such as cm, mi.)}}{\text{any } \textbf{time} \text{ unit (such as s, hr)}}$

2c. Gas pressure: **atm, torr, kPa**   2d. Temperature: **°C or K**

3. WANTED: **? n in mol =**   (Write the *symbol* and unit WANTED if known.)

$$\boxed{PV = nRT}$$

DATA: $P = 2.50$ atm $V = 7.20$ L $n = ?$ mol = WANTED

$R = 0.0821 \dfrac{\text{atm} \cdot \text{L}}{\text{mol} \cdot \text{K}}$

$T = 25°C$ Must use $\boxed{K = °C + 273} = 50. + 273 = \textbf{323 K}$

SOLVE: In symbols first.

$$? = n = \frac{PV}{RT} = \frac{(2.50\text{ atm})(7.20\text{ L})\,1}{(323\text{ K})(0.0821\,\frac{\textbf{atm} \cdot \textbf{L}}{\textbf{mol} \cdot \textbf{K}})} = \frac{(2.50\text{ atm})(7.20\text{ L})}{323\text{ K}} \cdot \frac{\textbf{mol} \cdot \textbf{K}}{0.0821\,\textbf{atm} \cdot \textbf{L}} = \textbf{0.679 mol}$$

Review Quiz

Partial solutions are shown. Your calculations must include listing DATA.

1. −39°C   2. 0.500 atm   3. If STP is mentioned, write the STP prompt and try to SOLVE with conversions.

SOLVE: (The conversions below may be in any order.)

$$? \ \frac{\textbf{g}}{\text{mol}} = \frac{0.313\ \textbf{g}\text{ gas}}{250.\text{ mL gas at STP}} \cdot \frac{1\text{ mL}}{10^{-3}\text{ L}} \cdot \frac{22.4\text{ L any gas at STP}}{1\text{ mol gas}} = \textbf{28.0}\ \frac{\textbf{g}}{\textbf{mol}}$$

4. WANTED: **? P =**   (Write the symbol *and/or* unit WANTED.)

$$\boxed{PV = nRT}$$

DATA: $P = ? =$ WANTED $V = 11.2$ L $n = 3.00$ mol

$R = 0.0821$ atm $\cdot$ L/mol $\cdot$ K

$T = 0°C$ Must use K. $\boxed{K = °C + 273} = 0 + 273 = \textbf{273 K}$

SOLVE: $? = P = \dfrac{nRT}{V} = nRT \cdot \dfrac{1}{V} = (3.00\text{ mol})(0.0821\,\frac{\text{atm} \cdot \text{L}}{\text{mol} \cdot \text{K}})\,(273\text{ K}) \cdot \dfrac{1}{11.2\text{ L}} = \textbf{6.00 atm}$

5. ? g O_2 gas $= 2.80$ L O_2 gas STP $\cdot \dfrac{1\text{ mol gas}}{22.4\text{ L gas STP}} \cdot \dfrac{32.0\text{ g } O_2}{1\text{ mol } O_2} = \textbf{4.00 g } O_2$

6. $T = \dfrac{PV}{nR} = \dfrac{(0.500\text{ atm})(15.3\text{ L})}{(0.250\text{ mol})} \cdot \dfrac{\text{mol} \cdot \text{K}}{0.0821\text{ atm} \cdot \text{L}} = \textbf{373 K}$

Solving for °C: $\boxed{K = °C + 273}$ °C = K − 273 = 373 K − 273 = **100.°C**

17

Gas Laws II: Choosing Equations

Lesson 17.1   Choosing Consistent Units

When you solve most equations in science, the numbers you substitute into equations must have units attached, and the units must cancel to result in the unit WANTED. In order for units to cancel, the units must be **consistent**. This means

> For each quantity in an equation, you must choose *one* unit to measure the quantity, then convert all DATA for that quantity to the chosen unit.

The ideal gas law ($PV = nRT$) involves four quantities: pressure, volume, particle counts, and temperature. Moles is our unit for particle counts, and temperature must be converted to kelvins. For pressure and volume, however, different units are possible. To have *consistent* units, you must choose

- *one* pressure unit (such as atm, kPa, or torr) and
- *one* volume unit (such as mL or L)

and convert all DATA to the units you chose.

Which units should you choose? A rule that simplifies problem solving is:

> Convert DATA to match the *units* of the most *complex* unit in the DATA.

In the ideal gas law, each of the four variables has *one* unit attached, but the constant R has *four* units attached. The easiest consistent units to choose will be the units attached to R.

Let's learn to apply these rules with an example.

━━━━━━━━━━━━━━▶ **TRY IT**

Q. Find the volume of 1 mol of an ideal gas at 25°C and a pressure of 380. torr. Use the ideal gas law and $R = 0.0821$ atm • L/mol • K

Apply the two shaded rules in this lesson and the "Steps for Solving with an Equation" in Lesson 16.4 (try to recall those steps from memory).

Answer:

WANTED: $? V =$ (Write the *symbol* and/or unit WANTED.)

$$\boxed{PV = nRT}$$

For your initial DATA table, list the DATA as written in the problem.

DATA: $P = 380.$ torr

$V = ? =$ WANTED

$n = 1$ mol

$R = 0.0821$ atm • L/mol • K

$T = 25°C$

Applying the rules above, temperature must be converted to kelvins, and the DATA for pressure is best converted to *atmospheres*: the pressure unit in the complex unit for R.

Make those changes in your DATA table.

DATA: $P = 380.$ torr • $\dfrac{1 \text{ atm}}{760 \text{ torr}}$ = **0.500 atm**

$V = ? =$ WANTED

$n = 1$ mol

$R = 0.0821$ atm • L/mol • K

$T = 25°C$ Use $\boxed{K = °C + 273}$ = 25 + 273 = **298 K**

All DATA *units* now match the units of the constant R. SOLVE for the WANTED unit in symbols, then plug in the DATA (with units).

SOLVE: $PV = nRT$

$$? = V = \frac{nRT}{P} = nRT \cdot \frac{1}{P} = (1 \text{ mol})\left(0.0821 \frac{\text{atm} \cdot \text{L}}{\text{mol} \cdot \text{K}}\right)(298 \text{ K}) \cdot \frac{1}{0.500 \text{ atm}} = \textbf{48.9 L}$$

Mark the unit cancellation in the line above *or* on your paper. Check that the unit of the answer is appropriate for the *symbol* WANTED.

Choosing an *R*

Consistent units are also important when you must *choose* an *R* value to use in a problem.

R is a number with units. When a constant includes units for several quantities, often different units can be used to measure one or more quantities, and different units change the *number* in *R*. The following are three equivalent values for *R*.

$$R = 0.0821 \text{ atm} \cdot \text{L/mol} \cdot \text{K}$$
$$= 8.31 \text{ kPa} \cdot \text{L/mol} \cdot \text{K}$$
$$= 62.4 \text{ torr} \cdot \text{L/mol} \cdot \text{K}$$

The number in each *R* depends on the pressure units; however, just as a speedometer shows that 55 miles/hour is the same as 88 kilometers/hour, the different values for *R* do not change the relationship between the quantities that *R* measures.

In gas calculations, you generally will not be asked to memorize values for *R*, but you will often need to *select* an *R* from a list of *R* values. To choose:

Pick the *R* with *units* that most closely match the other DATA units.

By doing so, the number of conversions to make DATA consistent will be minimized.

> **TRY IT**

Q. An ideal gas at 293 K and 202 kPa has a volume of 301 mL. How many moles are in the sample? (Use the ideal gas law and *one* of the three *R* values given above.)

Answer:

WANTED: **?** *n* in mol =   (Write the *symbol* and/or unit WANTED.)

$$\boxed{PV = nRT}$$

If you need a fundamental equation, write it first. The more often you write it, the better it will be remembered.

DATA: *P* = 202 kPa

V = 301 mL

n = ? **mol** = WANTED   (If the unit WANTED is known, include it.)

R = ?

T = 293 K

Which *R* value should you choose? None of the *R* values supplied use milliliters. Although *R* = 8.31 **kPa** · L/mol · K is *not* an exact match with the DATA units, it uses kilopascals, so it is the closest. Add that *R* to the DATA table.

One more change is needed. The units for P, T, and n match the chosen R units, but our best choice for R uses liters, while the supplied V is in milliliters. Each unit measuring volume must be the same or the units will not cancel and the equation will not work. Because we don't see an R value that uses milliliters, what's the best option?

STOP

Convert the supplied DATA to *liters* to match the volume unit in the chosen R. Do so in your DATA table, then complete the problem.

STOP

DATA: $P = 202$ kPa

$V = 301$ mL = **0.301 L**   (By inspection; see Lesson 12.1.)

$n = ?$ mol = WANTED

$R = \textbf{8.31 kPa} \cdot$ L/mol $\cdot$ K

$T = 293$ K

Are the units now consistent?

STOP

Yes. Solve $PV = nRT$ for the WANTED symbol *then* plug in numbers and units.

STOP

$$? = \textbf{\textit{n}} = \frac{PV}{RT} = \frac{(202\ \text{kPa})(0.301\ \text{L})}{(293\ \text{K})\ 8.31\ \dfrac{\text{kPa} \cdot \text{L}}{\text{mol} \cdot \text{K}}} = \frac{(202\ \text{kPa})(0.301\ \text{L})}{(293\ \text{K})} \cdot \frac{\textbf{mol} \cdot \textbf{K}}{8.31\ \text{kPa} \cdot \text{L}}$$

$$= \textbf{0.0250 mol}$$

Confirm that the units *cancel* to give an answer unit that fits the WANTED symbol. They must. Unit cancellation is an essential check on your work.

PRACTICE

1. Design flashcards for the shaded rules in this lesson. Practice recall of your cards from Chapter 16 and this lesson.

2. Write one or more possible metric *units* WANTED when you are asked to find these quantities. For any part, if you cannot answer quickly, design a flashcard.

    a. density

    b. speed or velocity

    c. gas pressure

    d. temperature

3. If 0.0500 mol of an ideal gas at 0°C has a volume of 560. mL, what will be its pressure in torr? Use one of the following values:

    $R = 0.0821$ atm $\cdot$ L/mol $\cdot$ K $= 8.31$ kPa $\cdot$ L/mol $\cdot$ K $= 62.4$ torr $\cdot$ L/mol $\cdot$ K

Lesson 17.2   The Combined Equation

Many gas law calculations involve the special case in which P, V, and/or T are changed, but the container is sealed so that the *moles* of gas in the sample does *not* change. If the number of gas moles in a sample is held constant, we can rewrite $PV = nRT$ as

$$\frac{PV}{T} = nR = (constant \text{ moles})(gas\ constant\ R) = (\text{a new } constant) \qquad (17.1)$$

When two constants are multiplied, the result is a new constant. Equation 17.1 means that if P, V, and/or T are changed while the moles of gas are held constant, the ratio "P times V over T" will keep the same value.

Another way to express this relationship: As long as the number of particles of gas does not change,

- if you have an *initial* set of conditions P_1, V_1, and T_1,
- and you change to a *new* set of conditions P_2, V_2, and T_2,

then the ratio PV/T must stay the same. Expressed in the elegant and efficient shorthand that is algebra:

$$nR = \text{a constant} = \frac{P_1 V_1}{T_1} = \frac{P_2 V_2}{T_2} \qquad (\text{if } n \text{ is held constant}) \qquad (17.2)$$

Equation 17.2 means that if five of the six variables among P_1, V_1, T_1, P_2, V_2, and T_2 are known, the sixth variable may be found using algebra, *without* knowing n or needing R. This relationship among the initial and final variables is emphasized by writing:

$$\frac{P_1 V_1}{T_1} = \frac{P_2 V_2}{T_2} \qquad (\text{if moles of gas are held constant}) \qquad (17.3)$$

We will call equation 17.3 the **combined equation** because it combines three historic gas laws. You may also see it referred to as the **two-point equation** because it is based on *initial* and *final* conditions. The combined equation is often memorized by repeated recitation of "*P* one *V* one over *T* one equals *P* two *V* two over *T* two."

When a problem says a sample of gas is *sealed* or *trapped* or has *constant moles*, and the sample has a change in *P, V,* and/or *T*, try the *combined equation* to SOLVE.

PRACTICE   A

In your notebook, solve the combined equation in *symbols* for

1. T_2 2. P_1 3. V_2

To solve with the combined equation (and with other equations), apply the "Steps for Solving with an Equation" in Lesson 16.4. To start:

- Write the symbol and/or unit WANTED.
- *Write* and box the fundamental, memorized equation.
- Make a DATA table that lists the *symbols* in the equation.

In a problem that requires the combined equation, write

$$\boxed{\frac{P_1V_1}{T_1} = \frac{P_2V_2}{T_2}}$$

and this DATA table:

DATA:	$P_1 =$	$P_2 =$
	$V_1 =$	$V_2 =$
	$T_1 =$	$T_2 =$

Then,

- Put the numbers and their units from the problem into the table. For the combined equation, write DATA at *initial* conditions in the *first* column and at final conditions in the second column.
- Label the symbol WANTED with a "?"; write the *units* WANTED if they are specified.
- SOLVE the fundamental equation in *symbols* for the WANTED symbol using algebra.
- Then substitute numbers *and* units and SOLVE. Apply unit cancellation as well as number math. Include units that do *not* cancel with your answer. Make sure that the answer unit fits the WANTED symbol.

TRY IT

Q. A sample of gas trapped by a moveable piston has a volume of 15.0 L at 273 K and 1.0 atm pressure. The pressure is increased to 2.5 atm while the temperature is increased to 373 K. What is the new volume of the gas?

WANTED: $? V\,new = V_2 =$ (Write the *symbol* and/or unit and a *label* WANTED.)

For a trapped sample of gas with a *change* to a *new* P, V, and/or T, try the combined equation.

$$\boxed{\frac{P_1V_1}{T_1} = \frac{P_2V_2}{T_2}}$$

DATA:	$P_1 = 1.0\ \text{atm}$	$P_2 = 2.5\ \text{atm}$
	$V_1 = 15.0\ \text{L}$	$V_2 = ?$
	$T_1 = 273\ \text{K}$	$T_2 = 373\ \text{K}$

SOLVE for the WANTED symbol, *then* substitute the DATA.

SOLVE: $? = V_2 = \dfrac{P_1V_1T_2}{P_2T_1} = \dfrac{(1.0\ \text{atm})(15.0\ \text{L})(373\ \text{K})}{(2.5\ \text{atm})(273\ \text{K})} = \mathbf{8.2\ L}$

The answer unit *liters* fits the *volume* dimension WANTED.

PRACTICE   **B**

1. In your notebook, solve the combined equation in symbols for

   a. T_1 b. P_2

2. A gas cylinder has a volume of 2.50 L. The pressure inside the tank is 100. atm at a temperature of 293 K. When a piston is withdrawn to expand the cylinder volume to 50.0 L, the pressure falls to 2.00 atm. What will be the new temperature in kelvins and in degrees Celsius?

The Combined Law and Consistent Units

In scientific equations, a capital T means *absolute* temperature. Because the combined equation is derived from $PV = nRT$, *both* equations *require* that:

- Temperature DATA must be converted to kelvins.
- If a temperature *not* in kelvins is WANTED, kelvins must be found first.

For both P and V, DATA must be converted to a *consistent* unit to SOLVE. In *combined* law calculations, because there is no R value with complex units that we are required to match, it does not matter which V or P units you choose.

- Volume may be in mL or L.
- Pressure may be expressed in kPa, atm, torr, or other pressure units.

However, *you* must choose *one unit* for P and for V, and all DATA for that quantity must be converted to that unit. It usually simplifies calculations if you convert all DATA to the unit WANTED if it is specified, but any units will work in an equation as long as the units are consistent. If non-WANTED units are found, WANTED units can be calculated at the final step.

Keeping consistent units in mind, try this problem.

▶ **TRY IT**

Q. A sample of gas has a volume of 250. mL under 4.5 atm pressure at 27°C. How many liters would the same moles of gas occupy at 50.5 kPa and standard temperature?

STOP

Answer:

WANTED: **?** $V\,new = V_2 =$ (Write the *symbol* and/or unit and a *label* WANTED.)

When one set of gas conditions is changed to *new* conditions, but *moles* are held *constant*, try the combined equation.

$$\boxed{\dfrac{P_1V_1}{T_1} = \dfrac{P_2V_2}{T_2}}$$

DATA: $P_1 = 4.5$ atm $P_2 = 50.5$ kPa $= \mathbf{0.500\ atm}$

               $V_1 = 250.\,\text{mL} = \mathbf{0.250\ L}$ $V_2 = ?$ L $=$ WANTED

               $T_1 = ?$ K $= 27°C + 273 = \mathbf{300.\ K}$ T_2 in K $=$ std. $T = \mathbf{273\ K}$

These unit conversions are used in the table:

T_1 *must* be in kelvins. $\boxed{\text{K} = {}^{\circ}\text{C} + 273}$ = 27°C + 273 = **300. K**

Because the DATA include two pressure units, write:

$$\boxed{1 \text{ atm} \equiv 760 \text{ torr} = 101 \text{ kPa}}$$

For P_1, choose one of the units in the DATA: kPa or atm. Either choice will result in the same answer. If you choose atm,

$$P_2 = ? \text{ atm} = 50.5 \text{ kPa} \cdot \frac{1 \text{ atm}}{101 \text{ kPa}} = \textbf{0.500 atm}$$

(If you choose kPa, P_1 = 455 kPa)

Because liters is WANTED but the initial volume is in milliliters, you must choose a consistent volume unit. It saves time to convert DATA to the WANTED unit if it is specified. The WANTED unit in this problem is liters.

$$V_1 = ? \text{ L} = 250. \text{ mL} = \textbf{0.250 L} \qquad \text{(By inspection; see Lesson 12.1.)}$$

SOLVE: $? = V_2 = \dfrac{P_1 V_1 T_2}{P_2 T_1} = \dfrac{(4.5 \text{ atm})(0.250 \text{ L})(273 \text{ K})}{(0.500 \text{ atm})(300. \text{ K})} = \textbf{2.0 L}$

PRACTICE   C

1. Make and practice flashcards for the rules in this lesson.

2. A sealed sample of hydrogen gas occupies 500. mL at 20.°C and 150. kPa. What would be the temperature in degrees Celsius if the volume of the container is increased to 2.00 L and the pressure is decreased to 0.550 atm?

Lesson 17.3   Simplifying Conditions

The combined equation includes a capital T, but *if* the *temperature* does *not change* in a problem using the equation, converting to kelvins is not necessary. Why?

If $T_1 = T_2$, then

$$\frac{P_1 V_1}{\cancel{T_1}} = \frac{P_2 V_2}{\cancel{T_2}}$$

and the combined equation simplifies to $\boxed{P_1 V_1 = P_2 V_2}$. This means that *if* the "initial and final" temperatures are the same, T is not needed to SOLVE, so conversion to kelvins is not required.

When using the *combined* equation, if *any* two symbols have the same value in a problem, those symbols can be cancelled because they are the same on both sides. Keep that in mind as you complete this question in your notebook.

▶ **TRY IT**

Q. A sample of chlorine gas has a volume of 22.4 L at 27°C and standard pressure. What will be the pressure in torr if the moles of gas and the temperature do not change but the volume is compressed to 16.8 L?

Answer:

WANTED: ? *P new* in **torr** = P_2 in **torr** =   (Write the *symbol* and/or *unit* and *label*.)

$$\frac{P_1V_1}{T_1} = \frac{P_2V_2}{T_2}$$

DATA: P_1 = std. *P*; use 760 **torr** to match → P_2 = ? in **torr**

V_1 = 22.4 L V_2 = 16.8 L

T_1 = 27°C = T_2

SOLVE: $\dfrac{P_1V_1}{\cancel{T_1}} = \dfrac{P_2V_2}{\cancel{T_2}}$; use $P_1V_1 = P_2V_2$

? = P_2 in **torr** = $\dfrac{P_1V_1}{V_2}$ = $\dfrac{(760\ \text{torr})(22.4\ \text{L})}{(16.8\ \text{L})}$ = **1,010 torr**

PRACTICE

1. Label each of these quantities with a *symbol* used in gas law equations.

   a. 122 K   b. 202 kPa   c. 190 torr   d. 30°C   e. 5.6 L

2. A sample of neon gas in a sealed metal (rigid) container is at 30.°C and 380. torr. What would be the pressure of the gas in kilopascals at standard temperature?

Lesson 17.4   Choosing the Right Equation

As a part of most physical science courses, you will likely be required to memorize at least some equations. When this is the case, a helpful strategy is to make a numbered list of the equations and to practice writing the list from memory. Then, at the beginning of each assignment or examination, write the numbered list from memory.

How can you tell whether to solve a problem using conversions or equations? And if you have multiple equations to choose from, how do you decide *which* equation to apply *when*?

The following system will work in both chemistry and other science courses. When you are not sure *which* equation to use, or *whether* to solve by equations or conversions, try these steps.

1. Write the WANTED *unit* and/or *symbol*.

2. List the DATA with units, substance formulas, and descriptive labels. Add any prompts.

3. Analyze whether the problem will require *conversions* or an *equation*.

   • Try conversions first. Conversions often work if *most* of the DATA can be listed as *equalities*.

   • If most DATA is in *single* units *or* the DATA includes *complex* units, you will likely need an equation. Recent lessons will suggest equations that may be needed.

   • Watch for hints at the need for a specific equation. For example, if R is mentioned in a gas problem, you may need $PV = nRT$

4. If conversions don't work but you are not sure *which* equation to apply:

   a. *Label* each item in your WANTED and DATA with a *symbol* based on its units.

      Examples:

      • 25 **kPa** would be labeled with a P for pressure

      • 293 **K** is assigned a T

      • L or mL or cm^3 are labeled with a V

   b. *Compare* the symbols listed in the WANTED and DATA to your written, memorized list of equations for the topic. Find the equation that uses *those* symbols. Write that equation.

   c. If no equations match the symbols exactly, see whether the problem's DATA can be converted to fit the symbols for a known equation. For example, degrees Celsius can be converted to kelvins, and grams can be converted to moles if you know the substance formula.

To summarize:

To Determine How to Solve a Problem

1. List the DATA. If it is mostly single units and equalities, try conversions. If not, try an equation.

2. When you are not sure *which* equation to use, label the DATA with symbols, then choose an equation that includes those symbols.

3. If no equation has matching symbols, see if the problem's quantities can be *converted* to quantities in a known equation.

PRACTICE

The problems in the Review Quiz for this chapter are a mix of gas calculations from Chapters 16 and 17. Use the strategies suggested in this lesson to choose how to solve each problem.

SUMMARY

How is solving with equations different from solving with conversions?

1. To choose the correct equation from several possibilities, when you look at a *unit* you must be able to quickly say: That's a pressure. That's a volume. That's a molarity. Because the number of metric units is limited, with flashcard practice this "dimension labeling" is simply a matter of practice.
2. To solve with equations, you need *fluency* in algebra. If you wish to major in the sciences but you cannot SOLVE for each symbol in an equation "with automaticity," set aside time for review and practice in an algebra textbook or tutorial (paper or online).
3. Apply the system introduced in Chapters 16 and 17 to organize your DATA.

With these skills and strategies, you will be able to solve equation problems across the sciences by applying the same steps.

REVIEW QUIZ

Use $R = 0.0821$ atm • L/mol • K $= 8.31$ kPa • L/mol • K $= 62.4$ torr • L/mol • K and the strategies in Lesson 17.4.

1. At standard temperature and a pressure of 190. torr, a gas has a volume of 14.4 liters. How many moles of gas are in the sample?

2. Find the density of oxygen gas (O_2) in grams per liter at STP.

3. A gas in a sealed but flexible balloon has a volume of 6.20 L at 30.°C and standard pressure. What will be its volume at $-10.$°C and 740. torr?

4. How many gas molecules will there be, per milliliter, for all ideal gases at STP?

5. A gas in a sealed glass bottle has a pressure of 112 kPa at 25°C. If the temperature is increased to 100.°C, what will be the pressure?

6. If 0.200 mol of UF_6 gas has a volume of 4.48 L and a pressure of 202.6 kPa, what is its temperature in degrees Celsius?

ANSWERS

Lesson 17.1

Practice   2a. Density: any **mass** unit (such as kg or g)   2b. Speed: any **distance** unit (such as cm, mi.)
                              any **volume** unit (such as L, mL)   any **time** unit (such as s, hr)

2c. Gas pressure: **atm, torr, kPa**   2d. Temperature: **°C** or **K**

3. WANTED: **? P** in **torr** =   (Write the *symbol* and/or *unit* WANTED.)

$$PV = nRT$$

DATA: $P = $ **? torr** = WANTED

$R = 62.4 \dfrac{\text{torr} \cdot \text{L}}{\text{mol} \cdot \text{K}}$ (Choose the R value that uses torr.)

$n = 0.0500$ mol

V must be in L to match R: 560. mL = **0.560 L**   (See Lesson 12.1.)

T must be in K: 0°C = **273 K**

SOLVE: In symbols first.

$$? = P = \frac{nRT}{V} = nRT \cdot \frac{1}{V} = (0.0500\ \text{mol})(62.4\ \frac{\text{torr} \cdot \text{L}}{\text{mol} \cdot \text{L}})(273\ \text{K}) \cdot \frac{1}{0.560\ \text{L}} = \textbf{1,520 torr}$$

Lesson 17.2

Practice A   1. $T_2 = \dfrac{P_2 V_2 T_1}{P_1 V_1}$ 2. $P_1 = \dfrac{P_2 V_2 T_1}{T_2 V_1}$ 3. $V_2 = \dfrac{P_1 V_1 T_2}{T_1 P_2}$

As a check, note the *patterns* above. The *P*s and *V*s, if multiplied together, have the same subscript, but if a *T* is multiplied with them, it will have the *opposite* subscript. In the fractions, there is always one more term on the top than on the bottom.

Practice B   1a. $T_1 = \dfrac{P_1 V_1 T_2}{P_2 V_2}$ 1b. $P_2 = \dfrac{P_1 V_1 T_2}{T_1 V_2}$

2. WANTED: ? *T* in **K** and *t* in °**C**   (Write the *symbol* and/or *unit* WANTED.)

If the moles of gas particles do not change, and one set of conditions is changed to new conditions, use

$$\boxed{\dfrac{P_1 V_1}{T_1} = \dfrac{P_2 V_2}{T_2}}$$

DATA: $P_1 = 100.\ \text{atm}$ $P_2 = 2.00\ \text{atm}$

$V_1 = 2.50\ \text{L}$ $V_2 = 50.0\ \text{L}$

$T_1 = 293\ \text{K}$ $T_2 = ?$ $t_2 = ?$

SOLVE: (In gas calculations, to find Celsius, SOLVE for kelvins first.)

$$? = T_2 = \frac{P_2 V_2 T_1}{P_1 V_1} = \frac{(2.00\ \text{atm})(50.0\ \text{L})(293\ \text{K})}{(100.\ \text{atm})(2.50\ \text{L})} = \textbf{117 K}$$

To find Celsius: $\boxed{\text{K} = {}^\circ\text{C} + 273}$ °C = K − 273 = 117 K − 273 = **−156°C**

Practice C   2. WANTED: ? *T* in **K** and *t* in °**C**   (Write the *symbol* and/or *unit* WANTED.)

For constant moles changing to new conditions, use

$$\boxed{\dfrac{P_1 V_1}{T_1} = \dfrac{P_2 V_2}{T_2}}$$

DATA: $P_1 = 150.\ \text{kPa}$ $P_2 = 0.550\ \text{atm} = \textbf{55.6 kPa}$

$V_1 = 500.\ \textbf{mL}$ $V_2 = 2.00\ \text{L} = \textbf{2,000 mL}$

$T_1 = 20.°\text{C} + 273 = 293\ \text{K}$ $T_2 = ?\ \textbf{K}$ $t_2 = ?\ °\textbf{C}$

Unit conversions for the DATA:

For T_1, K must be used. $\boxed{\text{K} = {}^\circ\text{C} + 273} = 20.°\text{C} + 273 = 293\ \text{K}$

For DATA with two *P* units, write: $\boxed{1\ \text{atm} = 760\ \text{torr} = 101\ \text{kPa}}$

For P_1, kPa *or* atm could be used as units. If you choose kPa,

$$?\ \text{kPa} = 0.550\ \text{atm} \cdot \frac{101\ \text{kPa}}{1\ \text{atm}} = 55.6\ \text{kPa}$$

SOLVE: (In gas problems, SOLVE in kelvins first.)

$$? = T_2 = \frac{P_2 V_2 T_1}{P_1 V_1} = \frac{(55.6\ \text{kPa})(2,000\ \text{mL})(293\ \text{K})}{(150.\ \text{kPa})(500.\ \text{mL})} = \textbf{434 K}$$

To find Celsius, *first* write the memorized equation, *then* SOLVE.

$\boxed{\text{K} = {}^\circ\text{C} + 273}$ °C = K − 273 = 434 K − 273 = **+161°C**

Different consistent units may result in slightly different answers due to rounding.

Lesson 17.3

Practice   1a. 122 K   *T*   1b. 202 kPa   *P*   1c. 190 torr   *P*   1d. 30°C   *t*   1e. 5.6 L   *V*

2. Hint: A sealed metal container will have a constant volume. $V_1 = V_2$

WANTED: ? *P new* in kPa = P_2 in **kPa** =

For trapped gas changing to new conditions, try

$$\boxed{\dfrac{P_1 V_1}{T_1} = \dfrac{P_2 V_2}{T_2}}$$

DATA: $P_1 = 380.\ \text{torr} = \textbf{50.5 kPa}$ $P_2 = ?\ \textbf{kPa}$

$V_1 = V_2$

$T_1 = 30.°\text{C} + 273 = \textbf{303 K}$ $T_2 = \textbf{273 K}$

For P_1, because kPa is the unit WANTED in P_2, convert to kPa.

$$?\ \text{kPa} = 380\ \text{torr} \cdot \frac{101\ \text{kPa}}{760\ \text{torr}} = \textbf{50.5 kPa}$$

SOLVE: $P_2 = \dfrac{P_1 \cancel{V_1} T_2}{T_1 \cancel{V_2}} = \dfrac{(50.5\ \text{kPa})(273\ \text{K})}{(303\ \text{K})} = \textbf{45.5 kPa}$

Review Quiz

Some problems may have multiple ways to solve. Any method may be used that results in the correct answer.

1. WANTED: ? mol =   *n*

   DATA: 190. torr   *P*   Standard temperature = 0°C = 273 K   *T*

   14.4 L   *V*

   Strategy: Based on the symbols above, which equation fits the DATA?

   $\boxed{PV = nRT}$ Which *R* value fits the DATA? The value using torr.

   $R = \textbf{62.4 torr} \cdot \text{L/mol} \cdot \text{K}$

   SOLVE: $PV = nRT$ in symbols first.

   $$? = n = \frac{PV}{RT} = \frac{(190.\ \text{torr})(14.4\ \text{L})\ 1}{(273\ \text{K})(62.4\ \frac{torr \cdot L}{mol \cdot K})} = \frac{(190\ \text{torr})(14.4\ \text{L})}{273\ \text{K}} \cdot \frac{mol \cdot K}{62.4\ torr \cdot L} = \textbf{0.161 mol}$$

   Is the answer unit appropriate for the *symbol* WANTED?

2. WANTED: $? \ \dfrac{\text{g } O_2 \text{ gas}}{\text{L } O_2 \text{ gas at STP}} =$

   DATA: 32.0 g O_2 = 1 mol O_2 (Grams prompt.)

   1 mol any gas = 22.4 L any gas at STP   (STP prompt.)

   Strategy: Analyze your DATA. Equalities lend themselves to conversions.

   SOLVE: $? \ \dfrac{\text{g } O_2 \text{ gas}}{\text{L } O_2 \text{ gas at STP}} = \dfrac{32.0 \text{ g } O_2}{1 \text{ mol } O_2} \cdot \dfrac{1 \text{ mol gas}}{22.4 \text{ L gas at STP}} = \textbf{1.43} \ \dfrac{\textbf{g } O_2\textbf{(g)}}{\textbf{L } O_2 \textbf{ at STP}}$

3. WANTED: ? *V* at end = V_2

   DATA: 6.20 L initial V_1

   30.°C + 273 = 303 K initial T_1

   Std. *P* = 760 torr initial, using the *P* units in the problem P_1

   −10.°C + 273 = 263 K final T_2

   740. torr final P_2

Strategy: The WANTED and DATA symbols match:

$$\boxed{\frac{P_1V_1}{T_1} = \frac{P_2V_2}{T_2}}$$

SOLVE: $\ ? = V_2 = \dfrac{P_1V_1T_2}{P_2T_1} = \dfrac{(760 \text{ torr})\,(6.20 \text{ L})\,(263 \text{ K})}{(740.\text{ torr})\,(303 \text{ K})} = \mathbf{5.53\,L}$

4. WANTED: $\ ? \ \dfrac{\text{molecules gas}}{\text{mL gas at STP}} =$

   DATA: 6.02×10^{23} molecules = 1 mol $\quad$ (10^{xx} = Avogadro prompt.)

   1 mol any gas = 22.4 L any gas at STP $\quad$ (STP prompt.)

   Strategy: All the DATA is in equalities. Try conversions.

   SOLVE: $\ ? \ \dfrac{\text{molecules gas}}{\text{mL gas at STP}} = \dfrac{6.02 \times 10^{23} \text{ molecules}}{1 \text{ mol}} \cdot \dfrac{1 \text{ mol gas}}{22.4 \text{ L gas STP}} \cdot \dfrac{10^{-3} \text{ L}}{1 \text{ mL}}$

   $$= \mathbf{2.69 \times 10^{19}} \ \dfrac{\textbf{molecules}}{\textbf{mL gas at STP}}$$

5. Hint: In a sealed glass container, V_1 equals V_2.

   WANTED: $\ ? \ P =$ $\quad$ Problem uses kPa units and wants P at **end** = P_2 =

   DATA: $P = 112$ kPa = P at start = P_1

   $t = 25°C$ initial $\quad$ K = 25°C + 273 = **298 K** = T_1

   $t = 100.°C$ final $\quad$ K = 100.°C + 273 = **373 K** = T_2

   $V_1 = V_2$

   Strategy: For a change in conditions with no change in the moles of gas, try the combined formula.

   SOLVE: $\dfrac{P_1\cancel{V_1}}{T_1} = \dfrac{P_2\cancel{V_2}}{T_2}$; $\dfrac{P_1}{T_1} = \dfrac{P_2}{T_2}$ $\quad ? = P_2 = \dfrac{P_1T_2}{T_1} = \dfrac{112 \text{ kPa} \cdot 373 \text{ K}}{298 \text{ K}} = \mathbf{140.\ kPa}$

6. WANTED: $\ ? \ °C =$ $\quad$ In gas calculations, find K first. $\quad T$

   DATA: 0.200 mol UF_6 $\quad n \quad$ 4.48 L gas $\quad V \quad$ 202.6 **kPa** $\quad P$

   The symbols are T, n, V, and P. Use which equation?

   $\boxed{PV = nRT}$ Because the P unit is kPa, choose the R with kPa.

   $R = 8.31$ **kPa** $\cdot$ L/mol $\cdot$ K

   SOLVE: $T = \dfrac{PV}{nR} = \dfrac{(202.6 \text{ kPa})(4.48 \text{ L})}{(0.200 \text{ mol})\left(8.31 \dfrac{\text{kPa} \cdot \text{L}}{\text{mol} \cdot \text{K}}\right)} = \dfrac{(202.6 \text{ kPa})(4.48 \text{ L})}{(0.200 \text{ mol})} \cdot \dfrac{\text{mol} \cdot \text{K}}{8.31 \text{ kPa} \cdot \text{L}} = \mathbf{546\ K}$

To find Celsius: $\boxed{\text{K} = °C + 273}$; $°C = K - 273 = 546 \text{ K} - 273 = \mathbf{+273°C}$

18

Phases and Energy

Lesson 18.1   Phases

In our environment on Earth, pure substances can exist in *three* **phases: solid, liquid,** or **gas** (the terms *states* and *phases*, in this context, have the same meaning).

Molecular compounds have relatively strong bonds that hold atoms together within the molecule. In addition, there are relatively weaker forces of attraction between different molecules that can cause them to "stick together" when they collide at relatively low speed. These weak attractions are a primary factor in explaining the phases and phase changes for molecules.

In the solid phase, the attractions between molecules hold the molecules in a crystal structure where, for most substances, they are as close as they can get. The molecules vibrate, but otherwise their motion within the crystal is very limited.

In their liquid phase, molecules gain some freedom: They can vibrate, rotate, and move past each other. However, in both the solid and liquid phases, molecules have minimal space between them, and as a result, solids and liquids do not compress (or compress only very slightly) when pressure is applied.

In the gas phase, molecules are separated by a considerable distance. Gases *can be compressed* because the *empty space* between the molecules can be reduced. If a gas is highly compressed or if its temperature is lowered, the relatively weak attractions between molecules become a larger factor, and all gases *condense* into a liquid or a solid at some point as pressure is increased and/or temperature is decreased.

Among the three phases, there are six types of **phase changes**:

- **Solid/liquid changes:** Solids **melt** (or **fuse**) to become liquids; liquids **freeze** (or **solidify**) to become solids.
- **Liquid/gas changes:** Liquids **boil** or **evaporate** to form gases; gases **condense** to become liquids.
- **Solid/gas changes:** Solids undergo **sublimation** to become gases. In the reverse process, gases can undergo **deposition** to form solids.

  Sublimation is a phase change less commonly encountered at room temperature and pressure, but you may be familiar with dry ice (solid carbon dioxide) or moth crystals (*para*-dichlorobenzene). At room temperature, these solids do not pass through a liquid phase as they convert from their solid phase to their gas phase. *Deposition* can be observed when water vapor forms ice crystals on a car windshield at temperatures below 0° Celsius.

For a pure substance, the temperature at which it melts (its **melting point**) will *equal* the temperature at which it solidifies (its **freezing point**).

> For a pure substance: melting point ≡ freezing point.

Under atmospheric pressures at or near standard pressure, for pure substances, *melting occurs at a characteristic temperature that can be used to identify the substance.* However, even small amounts of impurity in a substance will weaken its crystal structure and cause it to melt and freeze at a lower temperature.

PRACTICE A

Answer these questions before going on to the next section. Practice until you can answer the questions from memory.

1. Name three phases or states of matter. Name six different types of phase changes.

2. Which phases of matter can be significantly compressed in volume? Why?

3. Which has a higher temperature:
   a. The melting point or the freezing point of a pure substance?
   b. The melting point of a substance that is pure or of one that has impurities?

Vapor Pressure and Boiling Points

Nearly every liquid and most solid substances have a measurable tendency for their particles to become gas particles. At the surface of a liquid or solid, vibrating particles can break free and become part of the vapor above the liquid or solid. The *evaporation* of liquids at temperatures below their boiling points is one example of this tendency.

The molecules of a gas above its solid or liquid phase exert a gas pressure called the **vapor pressure** of the substance. The vapor pressure increases with increasing temperature.

A liquid will **boil** at any temperature at which its vapor pressure equals the atmospheric pressure above it.

Boiling points are characteristic temperatures that can be used to identify a substance. A **normal** boiling point is recorded at *standard* pressure (1 atmosphere). Boiling points must be recorded at a known pressure because liquids boil at a temperature that is highly dependent on the surrounding atmospheric pressure.

Examples:

- Water boils at 100°C *if* the pressure above the water is 101 kPa = 760 torr = 1 atm, a pressure that is about the average atmospheric pressure on a fair weather day at sea level.

- Water boils at about 95°C under the lower atmospheric pressure typically found in locations 1 mile above sea level (such as Denver, Colorado).

- In a *pressure cooker*, water boils at a higher temperature than at room pressure, and foods cook more quickly because the surrounding water is at a higher temperature.

Boiling points are affected by relatively small changes in the surrounding air pressure, such as those caused by altitude changes. In contrast, melting points are changed substantially only by larger changes in pressure.

Boiling is *not* the same as evaporating. Evaporation is a surface phenomenon, and measurable evaporation will occur from all liquids (and many solids) at any temperature. A liquid *boils* only when gas bubbles can form throughout the liquid and not just at its edges.

PRACTICE B

1. Design flashcards for the shaded vocabulary and rules to this point in this lesson. Practice with the cards before completing the questions below.

2. By definition, when does a liquid boil?

3. At what temperature does water boil at an atmospheric pressure of 101 kPa?

4. At approximately what temperature does water boil in a city that is 1 mile above sea level? What explains the difference from the boiling temperature at sea level?

5. Why does it take longer for an egg to "hard boil" in a "high-altitude" location than at sea level?

Lesson 18.2   Potential and Kinetic Energy

Energy

Chemistry is primarily concerned with matter and energy. Matter has mass and can be described in terms of particles such as protons, neutrons, and electrons. Energy has no mass. Energy can be divided into categories including mechanical energy, heat, electromagnetic energy (including X-rays, light, and radio waves), and electrochemical energy (such as the energy stored in batteries).

A fundamental principle of chemistry is the **law of conservation of energy**: Energy can neither be created nor destroyed. During chemical or physical processes, however, energy can *transfer* between substances, and energy can change its *form*.

Mechanical energy is the sum of two types of energy:

- **Kinetic energy** (KE) is defined as energy of *motion*.
- **Potential energy** (PE) is defined as *stored* energy.

Kinetic Energy

The kinetic energy of an object is calculated by the equation:

$$KE = 1/2(\text{mass})(\text{velocity})^2 = (1/2)mv^2$$

This equation means that

- If particle B has twice the mass of particle A but is moving at the same speed, particle B has *twice* as much kinetic energy.
- If particle C has the same mass as particle A but is moving twice as *fast* as particle A, it has *four times* as much kinetic energy.

Temperature is a measure of the *average* kinetic energy of particles. When the temperature of particles goes up, their average kinetic energy has increased. For this to occur, because the particles of a substance cannot change their mass, when the temperature in a sample increases, the particles must, on average, *move faster*.

PRACTICE **A**

Answer, and be able to answer from memory, these questions.

1. Define kinetic energy in words and by using symbols in an equation.

2. If car B is traveling at the same speed as car A, but car B has three times as much mass, compare the kinetic energy of car B to that of car A.

3. If car C has the same mass as car A, but car C is moving three times as fast, compare the kinetic energy of car C to that of car A.

4. Define temperature.

Potential Energy

Potential energy is stored energy. Lifting an object against gravity is one way to store energy in an object. If the object returns to its former lower position, it must release that added energy.

> **Example:** To raise a hammer, you must add energy. The energy is *stored* in the raised hammer as energy of position. If the hammer falls down to its original position, it must release the energy used to raise it. It can do so by creating *heat* where it hits. The hammer can also do *work*, such as driving nails. Heat, work, and energy of position are simply different forms of energy.

In chemical processes, forms of energy that can be stored and released include heat energy (such as the energy stored in plants during photosynthesis) and electrical energy (as is stored in batteries). Compressing a gas can also store energy. When a compressed gas moves a piston against resistance, the potential energy stored in the gas is converted to work, another form of energy.

Phases and Temperature

At a given pressure, the state of a substance is predictable based on its temperature. A visible amount of substance can either be in one state (solid, liquid, or gas) or in two states during a phase change. (The special case of "three states at once" occurs, but is beyond the scope of this text.)

For a substance at a given pressure, phase *changes* occur at characteristic temperatures. During a phase change, two phases will be present, and the particles in both phases will have the same temperature.

For a single substance, if two phases are mixed and both phases are present *after mixing*:

- The temperature of the particles will have *adjusted* to become the same in both phases.
- A mixture of the *solid* and *liquid* phases of a substance will always adjust to the temperature that is its *melting point*.
- A mixture of the *liquid* and *gas* phases in a boiling substance will always adjust to the temperature that is its *boiling point* at that pressure.

> **Examples:** For the substance H_2O, the melting and freezing temperature at standard pressure (1 atm) is always 0°C. In a mixture of crushed ice and water, the temperature in both phases will be 0°C for as long as both ice and water are present.
>
> For boiling water at 1 atm pressure, the temperature of both the water and the steam rising from the water will be 100°C.

One implication of this behavior is that when a substance is not at the temperature of a phase change, it will be in *one* state (solid, liquid, or gas), and you can predict from the pressure and temperature what state that will be.

> **Example:** At standard pressure, pure H_2O below 0°C is ice, above 100°C is steam, and between those two temperatures is liquid water.

Water has been used as an example because its behavior is familiar, but these rules for phases and phase changes are true (under most conditions) for every stable substance. The temperatures of the phase changes for every substance (at various pressures) are characteristic values that can be found in tables of scientific data.

PRACTICE B

1. Complete the flashcards for all of the shaded vocabulary and rules so far in this chapter, plus any summaries or examples you found helpful. Make a "practice until perfect" run through your cards before completing the questions below.

2. Define potential energy.

3. At standard pressure, what will be the state (phase) of H_2O at

     a. 20°C?   b. 105°C?   c. −12°C?

4. At standard pressure, the element mercury (Hg) melts at 234 K and boils at 630 K. What will be its state at 0°C?

Lesson 18.3   Phase Changes and Energy

Bonds and Energy

Changes in the potential energy of a chemical system can be the result of *chemical reactions* or *phase changes*.

Molecules and ions are held together by the attractions (bonds) arising from the protons and electrons within those particles. Energy always must be *added* to break chemical bonds, or to change a solid substance into a liquid and/or a gas. This added energy is needed to separate the attracting particles. If separated particles return to the bonding condition that existed before they were separated, the same amount of energy that was added and stored during their separation must be *released* from the particles into their environment.

In a chemical reaction, bonds between atoms break and new bonds form. As a result of a reaction, there is nearly always a characteristic *net* change in the energy stored in the substances. Energy must be added to break a bond, but *more* or *less* energy will be released when a different bond forms. This means that in a chemical reaction, energy is nearly always stored or released to some extent.

Potential Energy and Phases

For a molecular substance, during a phase change the bonds between the *atoms* in a molecule do not change, and the formula for the substance therefore does not change. As a molecular substance changes phase from *solid* to *liquid* and/or a *gas*, the energy added is the energy needed to separate the attracting *molecules*.

- During each phase change going from the solid to the gas phase, a characteristic amount of energy must be *stored* in each substance particle.
- During each phase change going from the gas to the liquid and/or solid phase, the same amount of energy stored during a phase change must *leave* the substance particles.

As a result, the *solid* phase of a substance will always have *less* stored (potential) energy than its *liquid* phase, which will always have less potential energy than its *gas* phase.

For a given substance: $PE_{solid} < PE_{liquid} < PE_{gas}$

Another way to say this rule is:

For a substance, its *gas* phase has the *most* stored (potential) energy and its *solid* phase has the *least*.

Example: In a glass of crushed ice and water placed in a room at "room temperature" (typically ~22°C), the liquid water and the ice will both be at 0°C. Because heat flows from warmer to cooler particles, heat from the warmer room will flow through the glass and into the mixture, and this heat melts the ice gradually. As long as ice remains in the glass, the ice and the liquid will remain at 0°C. At the point that *all* of the ice becomes melted, the glass contains all liquid water at 0°C. The heat from the warmer room will continue to flow into the colder water, and because only *one* phase is now present, the temperature will relatively quickly rise from 0°C toward room temperature.

A mixture of ice and water is a good *constant temperature bath* or *cold pack*. It will stay at 0°C for as long as *both* ice and liquid water are present.

PRACTICE A

1. On a gas stove, heat from a flame will enter water in a pot and cause the water to boil, but the water and steam during boiling both stay at 100°C. During boiling, where does the heat that enters the water from the much hotter flame go? What work does the heat from the flame do?

2. A small amount of boiling water at 100°C is added to a large quantity of an ice–water mixture in an insulated container. After stirring for 1 minute, the temperature is stable, and both ice and water remain.

   a. What is the temperature of the water in the cup?

   b. What is the temperature of the ice?

Energy and Phases

When energy is added to a substance (such as by heating) or removed from a substance (such as by cooling), whether its *kinetic* or its *potential* energy changes depends on whether the substance is present in one phase or two.

- When a substance is present in only one phase (all solid, all liquid, or all gas), adding or removing energy (such as by heating or cooling) changes the average *kinetic* energy of its particles (their temperature) but does *not* change the *potential* energy stored in the particles.

  While only one phase is present, adding or removing energy changes KE (and therefore temperature) but not PE.

- During a phase change (such as melting or boiling), *two* phases are present. If energy is added to or removed from a substance during a phase change, the *potential* energy stored in the substance changes, but the average *kinetic* energy of its particles does *not* change, and its temperature therefore stays constant.

  While two phases are present, adding or removing energy changes PE, but not KE (and not temperature).

  Example: For an ice cube that has just started to melt on a table in a room at room temperature, both the water on the surface of the ice and the ice itself are at 0°C. Minutes later, when the ice cube is almost finished melting, the ice and its puddle of water are both still at 0°C, though quite a bit of heat from the warmer room has entered the ice as it melts. That heat did not increase kinetic energy. Instead, during the phase change, the heat that entered the ice is stored as potential energy in the molecules of ice that melted and are now liquid.

PRACTICE B

1. Complete the flashcards for the shaded vocabulary and rules in this lesson. Make a "practice until perfect" run through all of the cards so far in this chapter before completing the questions below.

2. If substantial energy is added to a substance, and it remains the same substance but its temperature does not change, what does this tell you about the substance?

3. *Warm* water is added to an ice–water mixture in an insulated cup. After stirring for 1 minute, the temperature is stable, and both ice and water remain.

   a. During that minute, what physical changes occurred in the mixture?

   b. During that minute, what type of energy did the warm water lose? Where did this energy go? What type of energy did the warm water's lost energy become?

4. At standard pressure, small cubes of ice are removed from a freezer and placed in a teakettle. The kettle is placed on a lit gas stove and heated until 30 seconds after all of the water has boiled away. The graph that follows charts the change in the temperature of the H_2O molecules as they change from ice to water to steam.

THE TEAKETTLE PROBLEM

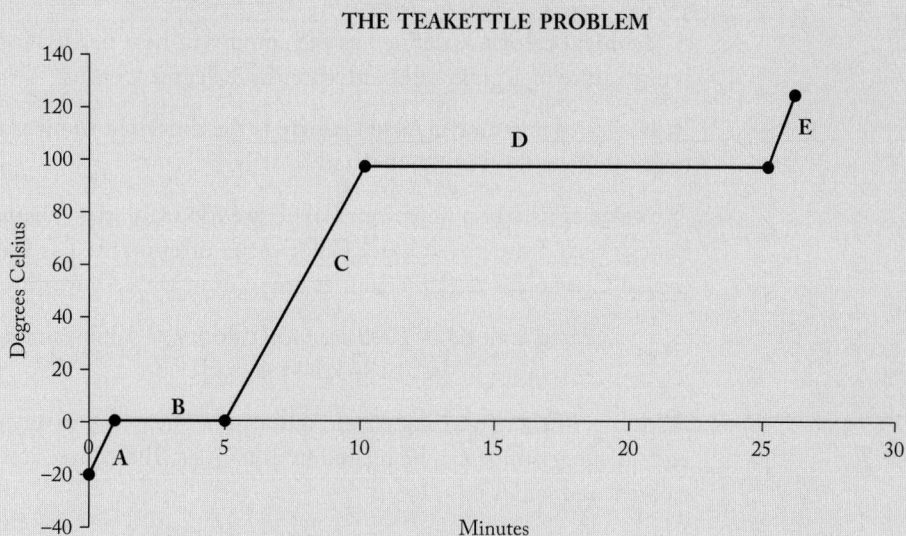

Using the graph and your knowledge in memory of phases and phase changes, answer these questions in your notebook.

a. How many phase changes occur from the beginning to the end of the process?

b. How many phases will have been present by the time the above process is completed?

c. Which segment of the graph represents water boiling to steam?

d. How can a change in the kinetic energy of the H_2O be recognized during the process?

e. In the graph, how can a change in the potential energy of the system be recognized?

f. In which lettered segments of the graph does potential energy remain constant?

g. In which segments of the graph does average kinetic energy remain constant?

h. Which portions of the graph show energy from the flame being converted into potential energy?

Lesson 18.4   Specific Heat Capacity

Before starting Lesson 18.4, if you have not already done so, complete Lesson 16.3 on cancellation of complex units.

Units That Measure Energy

In chemistry, energy is usually measured in joules or calories.

1. The **joule** (abbreviated **J**) is the SI unit measuring *energy*. A joule is defined as the amount of energy needed to accelerate 1 kilogram by 1 m/s^2 over a distance of 1 meter.

2. The **calorie** (abbreviated **cal**) is another metric unit used to measure energy.

   A **chemical calorie** is defined as the amount of heat needed to raise the temperature of 1 gram of liquid water by 1 degree Celsius.

3. In studies of nutrition, a **food Calorie** is the unit used to measure the heat released when food is metabolized in cells.

   The calories listed on nutritional labels are food Calories (chemical kilocalories). In textbooks, *food* Calories are often written as **C**alories with a capital **C**, whereas *chemical* **c**alories are written with a lowercase **c**.

$$1 \, food \, \text{Calorie} \equiv 1000 \, chemical \, \text{calories} \equiv 1 \text{ kilocalorie (kcal)}$$
$$= 4.184 \text{ kilojoules (kJ)}$$

4. All forms of energy are equivalent and can be measured in any energy units, and all energy units can be related by equalities. The conversion between calories and joules is

$$1 \text{ calorie} \equiv 4.184 \text{ joules (exactly)}$$

   In chemistry, the joule is the unit most often used to measure energy.

Change in Temperature

5. In science, the symbol Δ **(delta)** means *change in*. In heat equations, the symbol Δt (read "delta t") means *change in temperature*. In symbols, this definition is

$$\Delta t \equiv t_{\text{final}} - t_{\text{initial}}$$

   By this definition, a change in temperature is positive when temperature increases and negative when it decreases.

6. A *change* in temperature will be the same *number* of degrees when measured in the Celsius *or* Kelvin scales.

$$\Delta t \text{ measured in K} \equiv \Delta t \text{ measured in °C. In symbols, } \Delta t - \Delta T$$

   Why is this the case? The Kelvin and Celsius scales are defined to have the same *size* degree.

   Example: If, in Celsius, $\Delta t = 25°C - 0°C = \textbf{25}°C$, the same temperature measurements recorded in kelvins result in the *same number* for the *change*: $\Delta t = 298 \text{ K} - 273 \text{ K} = \textbf{25 K}$.

One implication of this definition for temperature units is:

For a Δt, the *units* **K** and **°C** and **degree** are all equivalent.

Specific Heat Capacity

7. **Heat** (symbol q) can be defined as a form of energy that *transfers* due to a difference in temperature. Heat transfers from warmer to colder matter.

8. The **specific heat capacity** (symbol lowercase c) of a substance is the amount of heat required to raise 1 *gram* of a substance in a single phase by 1 *degree* Celsius or kelvin. (Some textbooks use different terms and symbols for this quantity.)

9. For a substance in a single phase, when specific heat capacity is used to calculate a change in heat energy, we will use this **heat equation**:

$$q = c \cdot m \cdot \Delta t$$

The equation means:

The *heat energy* gained or lost by a substance =

(its specific heat capacity) $\times$ (its *mass*) $\times$ (its *change* in temperature)

10. In this text, a lowercase t is used in the heat equation to indicate that when using this equation, a conversion to kelvins is not required. Why? In a Δt, the *number* of degrees will be the same for both Kelvin and Celsius scale measurements (see point 6).

In equations, a capital T generally means a temperature in kelvins, but a lowercase t means that measurements may be recorded in either degrees Celsius or kelvins.

PRACTICE **A**

This lesson has introduced several new terms and equations, plus precise definitions for some familiar terms such as calorie and heat. Design flashcards for the shaded definitions, rules, and equations above, practice until each can be precisely recalled from memory, then continue with this lesson. If you continue to practice your cards over the next few days, you will be able to quickly master the calculations in this chapter.

Heat versus Temperature

11. Note the difference between *heat* and *temperature*.

- Temperature is an intensive property: When you measure a temperature, the value does not depend on the amount of matter being measured.
- Heat is an extensive property. When calculating the heat transferred in a process, what is being heated, the amount being heated, and how much heat is being transferred are all important.

  Example: On a gas stove, to bring water at room temperature to the point that it begins to boil, you must supply more heat for a larger amount of water than for a smaller amount.

Units and Signs

The equation relating *heat* (*q*) and *specific heat capacity* (*c*) is different in some respects from the equations for gas relationships. Let's take a look at the similarities and differences.

12. A value for heat (*q*) can be either positive or negative. In chemistry, the sign of heat is determined from the perspective of the chemical particles involved in a process. If chemical particles lose energy, *q* is either labeled as "heat lost" or is given a negative sign, matching the sign of the Δt for the chemical particles.

13. Like the gas constant (*R*), specific heat capacity (*c*) has a numeric value that will vary depending on its units.

   Example: The specific heat capacity of liquid water is

   $$c_{water} = \textbf{4.184 joules}/\text{gram} \cdot K = \textbf{1 calorie}/\text{gram} \cdot K$$

   One difference, however, is that *R* is the same constant for all ideal gases, but the value of *c* will vary for each substance and phase.

14. Recall that the *dot* between gram and K means that the units are multiplied together either in the numerator or the denominator. These three terms are equivalent:

   $$4.184 \text{ joules}/\textbf{gram} \cdot \textbf{K} = 4.184 \, \frac{J}{\textbf{g} \cdot \textbf{K}} = 4.184 \, J \cdot g^{-1} \cdot K^{-1}$$

   Specific heat capacity may also be measured in joules/*kilogram* • degree. You will learn below how to convert DATA to the consistent units needed to solve calculations.

15. Because for any Δt measurement, the word **degree** and the symbols °**C** and **K** are all *equivalent*:

   - A *c* value of "4.184 J/g • **degree**" is the same as "4.184 J/g • **K**"
   - In calculations based on Δt values, the *units* **degree** and **K** and °**C** can all *cancel*.

   Note how the units cancel in these terms:

   $$\frac{\text{joules} \cdot °C}{K} = \frac{\text{joules} \cdot °\cancel{C}}{\cancel{K}} = \textbf{joules} \quad \textit{and} \quad \frac{\text{calories} \cdot K}{\text{degree}} = \frac{\text{calories} \cdot \cancel{K}}{\cancel{\text{degree}}} = \textbf{calories}$$

PRACTICE B

Assume that the temperature units below are all based on measurements of a Δt. Simplify these terms using unit cancellation, then write the final unit. For a review of the rules for unit cancellation, see Lesson 16.3.

1. $\dfrac{J}{g \cdot K} \cdot g \cdot °C =$

2. $\dfrac{J}{\frac{J}{g \cdot K} \cdot g} =$

3. $\dfrac{\text{cal}}{\dfrac{\text{cal}}{\text{g} \cdot \text{degree}} \cdot {}^\circ\text{C}} =$

4. $\dfrac{\text{J}}{\text{g} \cdot {}^\circ\text{C}} =$

5. Name two metric units that can be used to measure q.

6. From memory, for each of the following symbols, state in one to three words what each symbol represents, then define in words what those one- to three-word terms mean.

    a. q b. c c. Δt

Lesson 18.5   Heat Calculations

Values for Specific Heat Capacity

In each phase, substances have a *characteristic* (precise and predictable) specific heat capacity (c). Some values for specific heat capacity are in the table below.

Substance	Specific Heat Capacity Values in J/g $\cdot$ K
H_2O liquid	4.184
H_2O solid	2.09
Cu solid	0.385
Fe solid	0.444

As in the case of H_2O in the table, each *phase* of each substance has a different c value.

Values for specific heat capacity apply only while a substance is in a single phase. A different measure for calculating heat changes is needed when a substance is *changing* phase.

To solve calculations that include specific heat capacity, we will apply:

The c Prompt

In a problem, if you see the term "specific heat capacity" or its symbol c, below the WANTED unit write the *heat equation* that uses c:

$$q = c \cdot m \cdot \Delta t$$

Solving Problems That Require Equations

To solve heat equation calculations, we will follow the steps that we applied to ideal gas equations in Lesson 16.4. Let us start with an example.

▶ **TRY IT**

Q. When 832 joules of heat is added to a sample of solid copper, the temperature rises from 15.0°C to 33.0°C. Based on the specific heat capacity in the table presented earlier, how many grams of copper were in the sample?

To solve, complete the following steps in your notebook.

1. As always, begin by writing "WANTED ?" and the unit and/or the equation symbol you are looking for. Include a chemical formula and/or other label describing what you are looking for.

2. This problem mentions "specific heat capacity." That's the *c prompt*. On the next line down, write the memorized equation that includes *c*. Put a box around the equation (and any equation that you use repeatedly).

3. Below the equation, make a DATA table that lists *each symbol* in the equation.

4. After each symbol, based on *units*, write the number and unit supplied in the problem that measures the quantity that the symbol represents.

5. After the symbol of the quantity you are *looking for*, write a **?** and the unit and label WANTED.

Do those steps, then check below.

STOP

Answer:

WANTED: $?\ g\ Cu =$

$$\boxed{q = c \cdot m \cdot \Delta t}$$

DATA: $q = 832\ J$

$c = 0.385\ J/g \cdot K$

$m = ?\ \mathbf{g\ Cu} = \text{WANTED}$

$\Delta t = t_{final} - t_{initial} = 33.0°C - 15.0°C = +\mathbf{18.0°C}$

6. Using algebra, SOLVE the fundamental memorized equation for the *symbol* WANTED. Do not plug in numbers until you have solved for the WANTED symbol.

STOP

SOLVE: $? = m = \dfrac{q}{c \cdot \Delta t}$

7. Substitute numbers and units and do the math. Cancel units that cancel, but include units that do not cancel after the number calculated.

$$? = m = \frac{q}{c \cdot \Delta t} = \frac{832\,\text{J}}{0.385\ \dfrac{\text{J}}{\text{g} \cdot \text{K}} \cdot 18.0°\text{C}} = \frac{832\,\text{J}}{18.0°\text{C}} \cdot \frac{1}{0.385\ \dfrac{\text{J}}{\text{g} \cdot \text{K}}}$$

$$= \frac{832\,\text{J}}{18.0°\text{C}} \cdot \frac{\text{g} \cdot \text{K}}{0.385\,\text{J}} = \mathbf{120.\ g\ Cu}$$

Mark the unit cancellation in the last line above. Make sure the answer unit makes sense for the symbol WANTED. Note that the term in the denominator that has a complex unit was separated and then simplified using the rules for reciprocals.

The above problem involved finding *grams* of Cu, but you did not need the molar mass to solve. This is because specific heat capacity is one of the rare quantities in chemistry that is defined based on grams rather than moles. Other heat problems *will* involve moles. Use this rule:

In calculations using $q = c \cdot m \cdot \Delta t$,

- If you see *moles* of a substance, write its molar mass in your DATA.
- If you don't see *moles*, you likely *won't* need the molar mass.

Finally, recall that when using equations, the rule is: Memorize equations in *one* format, then use algebra to solve for the symbol WANTED.

Example: Memorize $\boxed{q = c \cdot m \cdot \Delta t}$

Don't memorize: $m = \dfrac{q}{c \cdot \Delta t}$ and $c = \dfrac{q}{m \cdot \Delta t}$ and $\Delta t = \dfrac{q}{c \cdot m}$

PRACTICE

Add the rules from this lesson to your flashcards. Run your new flashcards and then all of the flashcards for this chapter before beginning the problems below.

1. In your notebook, without looking back at this lesson, write the equation that includes q and c.

2. In your notebook, using the symbols in the question 1 equation, solve in symbols for:

   a. $\Delta t =$ b. $m =$ c. $c =$

3. When 681 J of heat are added to 240. g of a solid substance, the temperature of the solid rises by 22.0 degrees. What is the specific heat capacity of the solid?

4. If 361 J of heat energy is added to a 32.5 g sample of solid iron (Fe) at 20.0°C, find the final temperature of the sample in °C using the value for c from the table presented earlier.

Lesson 18.6   Consistent Units in Heat Calculations

A substance that often supplies or absorbs heat in a chemical process is *liquid water*. In heat calculations that involve liquid water, use:

> **The *c* Water Prompt**
>
> If a problem mentions *energy* or *heat* or *joules—and* **liquid water**, below the WANTED unit or symbol write $q = c \cdot m \cdot \Delta t$.
>
> In the DATA table, write: $c = c_{water} = $ **4.184 J/g $\cdot$ K**

Using this prompt, problems involving heat and water can be solved in the same manner as the specific heat calculations in the previous lesson.

Though the common name for H_2O is water, in problems dealing with energy it is important to distinguish between *ice*, *water*, and *steam*. These three phases for H_2O have different values for *c*. In heat problems, however, unless ice or steam is specified, you should assume that *water* means *liquid* water.

In calculations involving liquid water, assume that 1 **milliliter** *liquid* H_2O = 1.00 **gram** *liquid* H_2O unless a more precise density is noted.

Consistent Units

To solve using an equation, each unit must match the requirements of the equation.

> **Example:** The equation for specific heat capacity requires *mass* (usually grams, occasionally kilograms). If DATA is supplied in *moles*, you must convert to mass units (grams or kilograms) before you SOLVE.

In addition, when solving equations, units must be *consistent*.

> **Example:** In an equation involving *mass*, grams *or* kilograms may be used, but not both. You must choose a mass unit, and then convert the other masses to that unit.

Which unit should you choose? In most cases, you may solve in any consistent units, but most problems will solve more quickly if you convert DATA to the units in the most *complex* unit in the DATA.

> **Example:** If a heat calculation includes a unit of "J/g $\cdot$ K," convert the units in the DATA to joules and grams.

The best time to convert to consistent units is *early* in a problem. The best place to convert to consistent units is in the DATA table.

━━━▷ TRY IT

Keeping those points in mind, try this example in your notebook. If you get stuck, peek at a bit of the answer, then try again.

Q. 16.0 mol of water at 25.0°C is supplied with 28.0 kJ of heat from a Bunsen burner. If all of the heat is absorbed by the water, what will be the water's final temperature in °C?

Complete the problem as far as placing the data into the DATA table, then stop.

STOP

Answer:

When you see joules and liquid water, use the *c water prompt*.

WANTED: $t_{final} = 25.0°C + \Delta t$

$$\boxed{q = c \cdot m \cdot \Delta t}$$

DATA: $q = 28.0$ kJ

$c = 4.184$ **J/g • K**

$m = 16.0$ mol H_2O

$\Delta t = ?$ (But t_{final} is WANTED.)

18.0 g $H_2O = 1$ mol H_2O (The data includes moles.)

The equation requires the substance *mass*. The problem does not supply a mass, but it does supply moles that you can convert to mass. In some cases, data in the problem will not fit the units required for the symbols in the equation. When that happens, write the data after the symbol that has units that the problem data can be converted to. Then, in the data table, convert the units supplied to the units required by the equation.

In a calculation using the heat equation, metric temperature units are equivalent, but the units that measure energy and mass must be consistent. Because the units supplied in this problem (kJ and mol) do *not* match the units of *c* (J and g), you must convert to consistent units. The easiest way to do so is to convert to the units of the most complex unit in the problem, which in this case is the unit for *c*. Convert the DATA to J and grams in your DATA table.

STOP

DATA: $q = 28.0$ kJ $= \mathbf{28.0 \times 10^3 \, J}$ (kilo means $\times 10^3$)

$c = 4.184$ J/g • K

$m = 16.0$ mol $H_2O \cdot \dfrac{18.0 \text{ g } H_2O}{1 \text{ mol } H_2O} = \mathbf{288 \, g \, H_2O}$

$\Delta t = ?$ But t_{final} WANTED $= 25.0°C + \Delta t$

All of the known DATA now has *consistent* units. Solve for the WANTED symbol.

STOP

SOLVE: $? = \Delta t = \dfrac{q}{c \cdot m} = \dfrac{28.0 \times 10^3 \text{ J}}{4.184 \dfrac{J}{g \cdot K} \cdot 288 \text{ g}}$

$$= 28{,}000 \text{ J} \cdot \dfrac{g \cdot K}{4.184 \text{ J}} \cdot \dfrac{1}{288 \text{ g}} = \mathbf{23.2 \text{ K increase}} = \Delta t$$

STOP

Mark the unit cancellation at the last step. Done? Always check the WANTED unit.

Because heat is being added to water, the temperature will increase. The final temperature is $25.0°C + \mathbf{23.2°C}$ or K $= \mathbf{48.2°C}$

Other Heat Equations

Some textbooks in chemistry, physics, or engineering use a variation of the heat equation that measures mass in SI units (kg) or uses moles in place of mass. If this is the case, the units of *c* will change, but otherwise all of the rules and steps above will apply. Our heat equation rules will work for a *c* in any units, as long as the units in the DATA are converted as needed to be *consistent*.

PRACTICE

1. Design a flashcard for the *c* water prompt and practice its recall, run the cards for this chapter, then complete the following problems.

2. Convert these to the units WANTED. Try to do so by inspection (in your head).

   a. ? J = 0.25 kJ = _____ = _____ = _____
      In: exponential notation   scientific notation   fixed notation

   b. 75 mL $H_2O(\ell)$ = _____ grams H_2O in fixed notation.

   c. ? g $H_2O(\ell)$ = 8.9 L $H_2O(\ell)$ in fixed and scientific notation.

3. A 36.0 mL sample of water is heated by 15.0°C. How many joules of energy are supplied?

4. A 15.0 mol sample of liquid water loses 6.70 kJ of heat. At the end of the process, the water temperature is 18.2°C. What was the original water temperature in °C?

SUMMARY

This chapter requires effort, repeated over several days, to memorize new vocabulary. Yet studies show that most students entering college can give reasonably accurate definitions for about 60,000 dictionary words, few of which required flashcards to learn. Why is learning chemistry more difficult?

Scientists who study learning explain that human children are born with a powerful instinctive drive to learn the language that they hear spoken around them. While sleeping, a child's brain can be observed (with new technologies) to be "playing back" unfamiliar words heard frequently and analyzing how to say them, what they mean, and how to use them in a sentence.

At about age 12, our ability to learn a new language diminishes, and in our native language, learning vocabulary for a topic that is *unfamiliar* requires effort. The good news is: Even after childhood, we learn new vocabulary with relative ease for *familiar* topics. For fields in which you have a well-developed conceptual framework, your brain can more easily determine where new words and meanings "fit." By study, as you increase your expertise, new knowledge in the field will enter your memory more rapidly.

REVIEW QUIZ

1. When does adding energy to a substance cause its temperature to rise?

2. For a given substance, which phase has the lowest amount of stored energy?

3. In a kitchen where the atmospheric pressure is standard pressure, water is placed in a teakettle and heated to boiling on a gas stove. After 5 minutes of boiling, about half of the water in the kettle has boiled away.

   a. What is the temperature of the liquid water in the kettle?

   b. What is the temperature of the steam above the water in the kettle?

   c. During the 5 minutes of boiling, has the kinetic energy of the water changed?

d. Has the potential energy of the molecules that are still liquid water changed?

e. Has the potential energy of the molecules that were converted from water to steam changed?

f. Where has the energy gone that was supplied by the gas stove in those 5 minutes? What kind of energy has it become?

4. In symbols, solve the heat equation for specific heat capacity.

5. How much heat (in joules) would be required to raise 4.50 mol of ice from −20.0°C to the temperature at which it begins to melt? (*c* for ice = 2.09 J/g • degree)

6. If 904 joules of heat are supplied to 0.500 mole of liquid water at 18.0°C, what is the water's final temperature?

ANSWERS

Lesson 18.1

Practice A   1. Phases: solid, liquid, gas. Changes: melting, freezing, boiling, condensing, sublimation, deposition.

   2. Only the gas phase. Only gases have substantial empty space between particles.

   3a. Melting point ≡ freezing point

   3b. A pure substance melts at a higher temperature than the same substance with impurities.

Practice B   2. When its vapor pressure equals the atmospheric pressure above it.

   3. 101 kPa is standard pressure: the pressure at which water boils at 100°C.

   4. Approximately 95°C. At high altitude, atmospheric pressure is lower, and therefore the water's vapor pressure will equal atmospheric pressure at a lower temperature.

   5. At high altitude, the water boils at a lower temperature, and at a lower temperature the changes needed to "cook" food take longer to occur.

Lesson 18.2

Practice A   1. Kinetic energy is energy of motion. KE = 1/2(mass)(velocity)^2 = $(1/2)mv^2$

   2. Car B is traveling with **three** times as much kinetic energy.

   3. Car C is traveling with **nine** times as much kinetic energy.

   4. Temperature is a measure of the average kinetic energy of chemical particles.

Practice B   2. Stored energy.

   3a. Liquid (between its melting and boiling temperatures).

   3b. Gas.   3c. Solid (the H_2O is below its melting point).

   4. 0°C is 273 K, which is between the melting and freezing temperatures, so the mercury will be in its liquid state.

Lesson 18.3

Practice A   1. The heat from the flame is stored in the molecules of water that change phase and become steam. The energy from the flame overcomes the force of attraction between the molecules in the liquid phase that is necessary for them to exist in their gas phase.

   2. If both ice and water are present, the temperature must be 0°C in both the water and the ice.

Practice B   2. Two phases must be present. If adding heat does not change temperature, the substance is undergoing a phase change.

3a. The warm water cooled to 0°C, and in the process the warm water melted some ice.

3b. The kinetic energy lost by the warm water in cooling to 0°C became *potential* energy that is stored in the molecules that were previously ice but were melted by the warm water.

4a. Two (melting and boiling)   4b. Three (solid, liquid, and gas)   4c. **D**

4d. The temperature **changes**.

4e. Where the graph has a "plateau" region, heat is being added from the stove for several minutes but the temperature (average kinetic energy) remains constant, so potential energy in the H_2O must be increasing.

4f. **A, C, and E**—when the kinetic energy is changing.

4g. **B and D**—temperature stays constant.

4h. **B and D**—during the two phase changes.

Lesson 18.4

Practice B   You may use other methods of unit cancellation that arrive at the same answers.

1. $\dfrac{\text{J}}{\cancel{\text{g}} \cdot \cancel{\text{K}}} \cdot \cancel{\text{g}} \cdot \cancel{°\text{C}} = \textbf{J}$

2. $\dfrac{\cancel{\text{J}}}{\dfrac{\cancel{\text{J}}}{\text{g} \cdot \text{K}} \cdot \text{g}} = \dfrac{1}{\dfrac{1}{\text{K}}} = \textbf{K}$

3. $\dfrac{\cancel{\text{cal}}}{\dfrac{\cancel{\text{cal}}}{\text{g} \cdot \cancel{\text{degree}}} \cdot \cancel{°\text{C}}} = \dfrac{1}{\dfrac{1}{\text{g}}} = \textbf{g}$

4. $\dfrac{\text{J}}{\text{g} \cdot °\text{C}} = \dfrac{\textbf{J}}{\textbf{g} \cdot °\textbf{C}}$ (In problem 4, nothing cancels.)

5. Joules and calories.

6a. q is the symbol for *heat:* energy that transfers due to a temperature difference.

6b. c is specific heat capacity: the amount of heat required to raise the temperature of 1 gram of a substance in a single phase by 1 kelvin or 1 degree Celsius.

6c. Δt is the symbol for "change in temperature," which is the difference between a final and an initial temperature.

Lesson 18.5

Practice   1. $q = c \cdot m \cdot \Delta t$

2a. $\Delta t = \dfrac{q}{m \cdot c}$ 2b. $m = \dfrac{q}{\Delta t \cdot c}$ 2c. $c = \dfrac{q}{\Delta t \cdot m}$

3. WANTED: $? = c$ (Write the unit and/or symbol WANTED.)

$\boxed{q = c \cdot m \cdot \Delta t}$ (If the problem mentions c, write the equation that includes c.)

DATA: $q = + 681\,J$

$c = ? =$ WANTED

$m = 240.\,g$ solid

$\Delta t = +22.0°C$

SOLVE: $\boxed{q = c \cdot m \cdot \Delta t}$, and c is WANTED.

$$? = c = \frac{q}{m \cdot \Delta t} = \frac{681\,J}{240.\,g \cdot 22.0°C} = \mathbf{0.129}\ \frac{\mathbf{J}}{\mathbf{g \cdot degree}}$$

4. WANTED: $? = t_{final} =$

Because $\boxed{\Delta t \equiv t_{final} - t_{initial}}$, $t_{final} = t_{initial} + \Delta t$

$\boxed{q = c \cdot m \cdot \Delta t}$

DATA: $q = +361\,J$

$c = 0.444\,J/g \cdot K$ for Fe

$m = 32.5\,g$ Fe

$\Delta t = ?$ WANTED $= t_{final} = 20.0°C + \Delta t$

SOLVE: $\boxed{q = c \cdot m \cdot \Delta t}$, and Δt is WANTED.

$$? = \Delta t = \frac{q}{c \cdot m} = \frac{361\,J}{0.444\,\dfrac{J}{g \cdot K} \cdot 32.5\,g} = +25.0\,K = \Delta t \qquad \text{Done?}$$

WANTED $= t_{final} = t_{initial} + \Delta t$

Final t = 20.0°C + Δt = 20.0°C + 25.0°C or K = 45.0°C

Lesson 18.6

Practice   2a. $?\,J = 0.25\,kJ = \mathbf{0.25 \times 10^3\,J = 2.5 \times 10^2\,J = 250\,J}$

2b. $?\,g\,H_2O(\ell) \approx 75\,mL\,H_2O(\ell) = \mathbf{75\,g\,H_2O(\ell)}$ $(1.00\,g$ liquid water $\approx 1\,mL)$

2c. $?\,g\,H_2O(\ell) = 8.9\,L\,H_2O(\ell) = 8{,}900\,mL\,H_2O(\ell) = \mathbf{8{,}900\,g\,H_2O(\ell)} = \mathbf{8.9 \times 10^3\,g\,H_2O(\ell)}$

3. WANTED: $?\,\mathbf{J}$ of heat

$\boxed{q = c \cdot m \cdot \Delta t}$

DATA: $q = ?\,J =$ WANTED

$c = 4.184\,J/g \cdot K$ for liquid water   (The c *water* prompt.)

$m = 36.0\,mL = 36.0\,g$ for liquid water

$\Delta t = ?\,K$ or $°C = 15.0°C$

(mL H_2O are converted to grams to match the c unit.)

SOLVE: $? = q = c \cdot m \cdot \Delta t = 4.184\,\dfrac{J}{g \cdot K} \cdot 36.0\,g \cdot 15.0°C = \mathbf{2{,}260\,J}$

4. WANTED: $t_{initial}$ Because $\boxed{\Delta t \equiv t_{final} - t_{initial}}$, $t_{initial} = t_{final} - \Delta t$

$\boxed{q = c \cdot m \cdot \Delta t}$

DATA: $q = -6.70 \text{ kJ} = -6.70 \times 10^3 \text{ J}$ (Convert to units in c.)

$c = 4.184 \text{ J/g} \cdot \text{K}$

$m = 15.0 \text{ mol } H_2O \cdot \dfrac{18.0 \text{ g } H_2O}{1 \text{ mol } H_2O} = \textbf{270. g } H_2O$

$\Delta t =$ ($t_{initial} =$ WANTED)

SOLVE: $? = \Delta t = \dfrac{q}{c \cdot m} = \dfrac{-6.70 \times 10^3 \text{ J}}{4.184 \dfrac{\text{J}}{\text{g} \cdot \text{K}} \cdot 270. \text{ g}} = -5.93 \text{ K} = \Delta t$

WANTED $= t_{initial} = t_{final} - \Delta t = 18.2°C - (-5.93°C \text{ or K}) = \textbf{24.1°C}$

(Because the water is losing heat, its initial temperature must be warmer than the final temperature.)

Review Quiz

1. When the substance is in a single phase and not undergoing a phase change.

2. The solid phase.

3a. 100°C   3b. 100°C   3c. No. The temperature has not changed.

3d. No. The potential energy does not change until the phase change.

3e. Yes. The molecules that have gone from liquid to gas phase gained PE.

3f. The energy from the burning stove has not changed the kinetic energy inside the teakettle, but the steam that left the boiling liquid had more potential energy in its molecules than they had in their liquid state.

4. $c = \dfrac{q}{\Delta t \cdot m}$

5. WANTED: q in joules

(Strategy: Note that the c value for ice is not the same as for liquid water. Because the problem mentions c, write the equation that uses c.)

$$\boxed{q = c \cdot m \cdot \Delta t}$$

DATA: $q = ? \text{ J} =$ WANTED

$c = 2.09 \text{ J/g} \cdot \text{K}$

$m = 4.50 \text{ mol } H_2O \cdot \dfrac{18.0 \text{ g } H_2O}{1 \text{ mol } H_2O} = \textbf{81.0 g } H_2O$

(Ice begins to melt when it reaches 0.0°C.)

$\Delta t = ? \text{ K or } °C = -20.0°C \text{ to } 0.0°C = +20.0°C$

SOLVE: $? = q = c \cdot m \cdot \Delta t = 2.09 \dfrac{\text{J}}{\text{g} \cdot \text{K}} \cdot 81.0 \text{ g} \cdot 20.0°C = \textbf{3,390 J}$ must be added.

6. $? = \Delta t = \dfrac{q}{c \cdot m} = \dfrac{904 \text{ J}}{4.184 \dfrac{\text{J}}{\text{g} \cdot \text{K}} \cdot 9.00 \text{ g}} = +24.0 \text{ K} = \Delta t$ Done?

WANTED: $t_{final} = t_{initial} + \Delta t = 18.0°C + 24.0°C \text{ or K} = \textbf{42.0°C}$ final temp.

19

Bonding

For this chapter, you will need a kit for building molecular models. These can be purchased at a college bookstore or online. For most model sets, sharing the cost to obtain half or one-third of a set will provide sufficient parts to build the molecules in these lessons. Take steps to acquire your models as soon as possible.

Introduction

Bonds are forces that hold atoms together. The nature of the chemical bond is a question at the heart of chemistry, but the answer is not completely understood. An explanation of bonding must take into account protons and electron pairs, electrical attraction and repulsion, and electrons that can behave as both particles and as waves. A theory that successfully explains all of the aspects of bonding does not yet exist.

However, with a combination of three simplified models—Lewis structures, VSEPR, and electronegativity—we can predict the composition, shape, and many chemical behaviors of a significant percentage of the substances within and around us.

Lesson 19.1   Lewis Structures

In ionic bonding, charged particles (ions) are held together by the electrical attraction of opposite charges. In covalent bonding, shared electrons form a bond that holds two atoms together. Covalent bonds may be single bonds with one pair of electrons, double bonds with two pairs, or triple bonds with three pairs.

If all of the atoms in an electrically neutral particle are bonded covalently, the particle is termed a molecule. The forces holding atoms together inside a molecule are strong compared to the forces between the molecules. As one result, compared to ionic compounds, many covalently bonded molecules can be pulled apart from each other at relatively low temperatures. A multiatom substance that is a gas or liquid at room temperature is always molecular: All of its bonds are covalent.

Valence Electrons

The electrons around a nucleus are found at different energy levels and at different average distances from the nucleus. The electrons of an atom that participate in bonding are termed its **valence electrons**. These are generally the electrons with relatively high energy and high average distance from the nucleus.

An atom can have at most eight valence electrons. For atoms in the eight "main groups" (tall columns) of the periodic table, the *number* of valence electrons in a neutral atom is the same as the *number* of the tall column in which the atom is located.

Examples:

- First-column neutral atoms have *one* valence electron.
- Neutral atoms in the carbon family (the fourth tall column) have four valence electrons.
- Noble gas atoms have eight valence electrons (except helium, which has two).

Lewis Structures

Lewis structures (also called **Lewis diagrams** or **electron dot diagrams**) can be drawn to represent the bonds and the nonbonding valence electrons in chemical particles.

Example: The Lewis structure of H_2 is **H $\colon$ H**

The shared valence electrons represented by the two dots compose the bond that holds the atoms together in the molecule H_2

Dot Diagrams for Neutral Atoms

To draw the Lewis structure for a multiatomic particle, begin by drawing the dot diagrams for its atoms. Follow these steps.

1. Write the symbol for the atom.
2. Determine the number of *valence* electrons in the atom.
3. Assume that each atom symbol has *four* sides. To represent the valence electrons, draw dots around the atom symbol. Put *one* electron on each of the four sides of the symbol *before* you start to *pair* electrons. The four sides are equivalent: You may place the electrons on any side.

   Examples:

   Boron has three valence electrons and is drawn as $\cdot \overset{\displaystyle \cdot}{B} \cdot$

   Nitrogen has five valence electrons and is drawn as $\cdot \overset{\displaystyle \cdot}{\underset{\displaystyle \cdot\cdot}{N}} \cdot$

   The dot diagram for a neutral boron atom has three **unpaired** electrons. Nitrogen has three unpaired and one **pair** of valence electrons.

PRACTICE   A

Use a periodic table as needed.

1. How many valence electrons are in these neutral atoms?

    a. Silicon   b. Phosphorus   c. Bromine   d. Sulfur

2. In the space below, draw the Lewis diagram for each atom in problem 1.

    a.   b.   c.   d.

The Octet Rule

To predict the bonding in multiatomic particles, we combine the dot diagrams for atoms by applying:

The Octet Rule

An atom has maximum stability when it is surrounded by *eight* valence electrons. (Hydrogen, an exception, is most stable with a **duet** of *two*.)

Combinations that result in all atoms being surrounded by eight valence electrons (two for H) tend to be *stable*. Stable combinations are likely to be found in nature and formed in chemical reactions.

> **Example:** The Lewis structure for the molecule hydrogen fluoride is

$$\text{H}\!:\!\overset{\displaystyle ..}{\underset{\displaystyle ..}{\text{F}}}\!:$$

> The two electrons between H and F form a *bond*. The F atom is also surrounded by three *lone pairs* of electrons.

Shared electrons are a **bonding pair**. A particle with more than one atom must have at least one bond. Electron pairs that are *not* between two atoms are termed **lone pairs** or **unshared pairs** or **nonbonding pairs** (all of those terms are equivalent).

A species that does not have a *satisfied octet* may exist, but it is likely to be unstable: It will tend to be a very *reactive* species.

Using Lewis Diagrams to Predict Bonding

Depending on the type of problem, there are two methods for drawing dot diagrams.

Method 1 This method predicts the molecular formula that is likely for cases in which the formula for a combination of one or two atoms is *not* supplied, but the molecule is known to have all single bonds. The method 1 steps are:

1. Draw the dot diagram for each different neutral atom.

2. Combine the diagrams of the atoms so that the *unpaired* electrons *pair* and are shared between two atoms. Stop when each symbol is surrounded by *eight* valence electrons (two for H).

━━━▶ TRY IT

Q. In your notebook, apply the method 1 steps to draw the Lewis structure for a stable molecule that contains only chlorine atoms with single bonds.

STOP

Answer:

To make a stable molecule, slide two chlorines together so that their unpaired electrons pair. *Each* chlorine is now surrounded by eight valence electrons. The *octet rule* is satisfied for both atoms. The two shared electrons form a *bond* between the two chlorines. The Cl_2 molecule has one *bond*, and each chlorine atom has three *lone pairs* of valence electrons.

$$:\!\ddot{C}l\!\cdot \;+\; \cdot\!\ddot{C}l\!: \;\longrightarrow\; :\!\ddot{C}l\!:\!\ddot{C}l\!: \;=\; Cl\!-\!Cl \;=\; Cl_2$$

(two *atom* dot diagrams)　　(*molecule* dot diagram)　　(structural and molecular formulas)

Method 1 is quite simplified, but it does predict bonding in many cases in which the molecular formula (and therefore the total number of valence electrons) is not known.

PRACTICE B

Use a periodic table. The atoms that form covalent bonds will most often be the nonmetals found toward the top right of the periodic table, *plus* hydrogen.

1. Using method 1, draw a Lewis diagram and then a structural formula for the predicted stable molecule formed by a combination of

    a. Fluorine and bromine atoms　　b. Hydrogen and chlorine atoms

2. For each of the molecules in problem 1, list the number of covalent bonds and the total number of lone pairs of electrons.

    Bonds: 1a. _____　1b. _____　　Lone pairs: 1a. _____　1b. _____

Method 2 If the formula for a molecule is *supplied*, a model that better predicts the nature of bonds is to combine the valence electrons for all of the atoms without regard to which atom contributes the electrons. Apply these steps:

1. *Count* the *total* number of valence electrons in the neutral atoms of the molecule.

2. Determine the **central atom**. In a simple molecule with more than two atoms, the *central atom* has the *most unpaired* electrons in the dot diagram for its *atom*. This central atom will also be the atom that is in a column that is closest to group 4A (the carbon family) in the periodic table.

3. Arrange the valence electrons around the central atom to satisfy the octet/duet rule without regard to which atom contributes which electrons.

> ▭▭▭▭▭➤ **TRY IT**

Q. Using method 2, draw the Lewis diagram for a water molecule, H_2O.

🛑

Answer:

1. Each of the two hydrogens has one valence electron and the one oxygen has six, so the molecule has eight valence electrons. The central atom is oxygen because the O atom dot diagram has two unpaired electrons, while H has only one.

2. The Lewis structure for water can be written in a variety of equivalent ways.

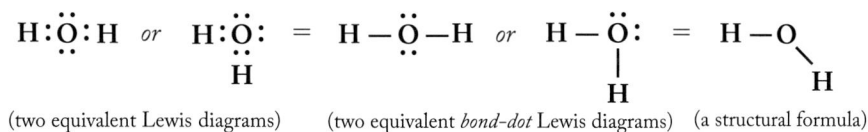

$$H:\overset{..}{\underset{..}{O}}:H \quad or \quad H:\overset{..}{\underset{H}{\underset{..}{O}}}: \quad = \quad H-\overset{..}{\underset{..}{O}}-H \quad or \quad H-\overset{..}{\underset{H}{\underset{..}{O}}}: \quad = \quad H-\overset{}{\underset{H}{O}}$$

(two equivalent Lewis diagrams)   (two equivalent *bond-dot* Lewis diagrams)   (a structural formula)

The first two structures above represent the *electron dot* form of a Lewis diagram showing all of the valence electrons. Because all four sides of an atom symbol are equivalent, the 90° and 180° drawings of the two Lewis diagrams are *equivalent*: both represent the same molecule. The electrons *between* adjacent atoms are the two *bonding pairs*. In addition, the oxygen has two *lone pairs*.

The third and fourth structures show an alternate type of Lewis structure: Each bonding pair is a line to represent the bond, but the lone pairs are represented by dots.

The last formula is a structural formula. A structural formula provides information about the location of the atoms in a molecule but often does not show the location of the lone pairs.

All of the diagrams show that in water there are two bonds and the oxygen atom is in the middle. The Lewis structure predicts that this is a stable combination because by sharing electrons, all of the atoms are surrounded by the number of valence electrons needed for stability.

To predict where the atoms and bonds will be, use method 1 if you are asked to *predict* a stable formula for a combination of one or two atoms and method 2 if the formula of a molecule is known.

Summary: Number of Bonds and Lone Pairs

Many of the frequently encountered molecules in both first-year and organic chemistry consist of hydrogen plus the second- and third-row nonmetals. In general, neutral atoms in the second row of the periodic table have the characteristics described below when they bond covalently. These patterns apply in many (but not all) cases for atoms below the second row. Learn this table so that given the terms in the first *column*, you can fill in the blanks.

Lewis Structure Predictions for Neutral Single-Bonded Central Atoms

Second-Row Symbol	Li	Be		B	C	N	O	F	Ne
Valence Electrons	1	2		3	4	5	6	7	8 or 0
Bonds	(Primarily ionic bonds)			3	4	3	2	1	0
Lone Pairs				0	0	1	2	3	4

In predicting formulas for molecules, it is helpful to remember: "Carbon bonds four times, nitrogen three times, oxygen twice, and hydrogen and halogens once."

PRACTICE C

1. Make and practice flashcards for the shaded rules in this lesson. In addition, in your notebook, copy the labels in the four rows of the first column of the summary table above, then practice filling in the rest of the table from memory.

2. Using method 2 and a periodic table, draw a Lewis diagram and then a structural formula for these.

   a. CH_4 b. PCl_3

3. For each molecule in problem 2, list the number of covalent bonds and the total number of lone pairs of electrons.

   Bonds: 2a. _____ 2b. _____   Lone pairs: 2a. _____ 2b. _____

4. By the rules for Lewis diagrams, how many bonds will be found around these neutral central atoms?

   a. Sulfur   b. Iodine   c. Silicon   d. Nitrogen

Lesson 19.2   # Molecular Shapes and Bond Angles

VSEPR

A key factor in the behavior of molecules is their shape: how the atoms and electrons are arranged in three-dimensional space.

The shapes and bond angles of most molecular compounds can be predicted with reasonable accuracy based on **valence shell electron-pair repulsion** theory (**VSEPR**). VSEPR assumes that all valence electron pairs (lone pairs and bonds) repel each other, and they will separate by the maximum possible angle around the nucleus.

Predicting the Shape of a Covalent Molecule

The components of the geometry of a chemical particle include:

- The shape and bond angles of its *electron pairs*.
- The shape and bond angles of the *atoms* in a molecule.

For particles containing one or two kinds of atoms and all single bonds, our goal is to be able to predict the shape and bond angles for both the electron pairs and the atoms. Central atoms in the same column of the periodic table will generally have the same electronic and molecular geometries. Let's examine an example from each column.

1. **Column 1 atoms:** The only atom in column 1 that shares its one valence electron is hydrogen, and the result is one single bond. The shape of the electron pairs around hydrogen is said to be **linear**. Because it takes three points to determine an angle, but hydrogen bonds only once, there is no "bond angle" around hydrogen.

   Example:

   $$H:H = H—H = H_2$$

   Each H bonds once. The molecule has a *linear* shape with *no* bond angles.

Dot diagrams can be drawn for the atoms in column 1 below hydrogen, but in compounds, alkali metal atoms lose their valence electron to form a positive ion, rather than sharing their valence electron in a covalent bond. VSEPR rules to determine shape apply to molecular, but not ionic, compounds.

2. **Column 2 atoms:** If an atom in column 2 *shares* its two valence electrons, it will be surrounded by two electron pairs, and both will be bonds. The two pairs will separate as much as possible, resulting in a *linear* shape around the atom for both the two bonds and the three atoms. For the two bonds and the three atoms, the **angle** around the central atom that allows the two pairs to get as far apart as possible is **180°**.

   All molecules vibrate, and as they do, their bond angles often change, but the angle predicted by VSEPR will be the *average* angle between two bonds.

   Example:

   $$:\ddot{C}l:Be:\ddot{C}l: = Cl—Be—Cl = BeCl_2$$

   The Be in $BeCl_2$ is surrounded by two electron pairs, and both are bonds. The shape of the electron pairs around the central atom and the shape of the molecule is linear with **180°** bond angles.

   Note that $BeCl_2$ is *electron deficient*: Around Be it violates the octet rule. $BeCl_2$ does form, but as an electron-deficient molecule it has some unusual properties.

In compounds, atoms below Be in column 2 tend to lose, rather than share, their two valence electrons, so that dot diagrams do not accurately predict the shapes and bond angles.

3. **Column 3 atoms:** If an atom in tall column 3 shares its three valence electrons, it will be surrounded by three electron pairs that are single bonds. The shape that allows the pairs to get as far apart as possible is termed **trigonal planar**. The three bonds are in a plane (flat) with 120° angles. Both the electron pairs and the molecule have a **trigonal planar** geometry and **120°** bond angles.

Example:

$$:\ddot{Br}:B:\ddot{Br}: = \begin{array}{c} Br \diagdown \diagup Br \\ B \\ | \end{array} 120° = BBr_3$$

In BBr_3, the shape of both the electron pairs and the molecule is trigonal planar, and all bond angles are 120°. Like $BeCl_2$, BBr_3 does form, but it violates the octet rule and has some unusual properties.

Atoms below boron in tall column 3 are metals, but in some cases (such as $AlCl_3$) they do form bonds that have covalent (shared electron) characteristics.

4. **Four pairs:** Because of the octet rule, *most* stable atoms in substances are surrounded by *four* electron pairs. The three-dimensional shape that allows four pairs to get as far apart as possible is termed **tetrahedral**. The angles of a tetrahedron are **109.47°**, which we will abbreviate as *about* **109° (~ 109°)**.

You will need molecular models for the examples below.

a. **Column 4 atoms**: For a single-bonded central atom in the *carbon* family (group 4A, tall column 4), all of its four valence electrons are shared in bonds. The shape of electron pairs and the molecule is said to be *tetrahedral* with ~109° angles.

Example:

$$H:\ddot{C}:H = \begin{array}{c} H \\ | \\ C \end{array} \sim 109° = CH_4$$

A three-dimensional tetrahedron is difficult to represent on two-dimensional paper. In the diagram above, the dotted line represents a bond going behind the plane of the paper, and the wedge represents a bond coming out of the paper. A 3-D model will assist in working with this important geometry.

TRY IT

Build a CH_4 molecule with your molecular models. Place the assembled model on a flat surface, then flip it so that it rests on three different points. Flip it again. Note the high symmetry of a three-dimensional tetrahedron: The shape of the molecule is the same no matter which three atoms the model sits upon.

In Lewis diagrams, we treat four sides around an atom symbol as equivalent because most nonmetal atoms are surrounded by four electron pairs. Those pairs repel into a tetrahedral shape, and the four sides of a tetrahedron are equivalent.

b. **Column 5 atoms:** A single-bonded central atom in the nitrogen family has five valence electrons that are most often arranged as *three bonds* and *one lone pair*.

The four electron pairs around the central atom assume the tetrahedral shape that places them as far apart as possible. The result is an *electronic* geometry that is tetrahedral, with angles between the electron pairs (bonds and the lone pair) of ~109°.

However, the lone pairs, though they contribute to determining the shape of bonds, are not considered when *naming* the *molecular* shape. The name of a molecular shape is based only on the positions of the bonds and atoms.

In the case of one lone pair and three bonds around a central atom, the four *atoms* are in the shape of a low pyramid, with the central atom above the plane of the three atoms to which it bonds. Because the pyramid rests on three points, the shape of the atoms is a **trigonal pyramid,** and the *molecular* geometry is termed **trigonal pyramidal**. But because the position of the bonds is determined by the total number of electron pairs, including the lone pair, the *angles* of the bonds are tetrahedral: ~**109°**.

Example:

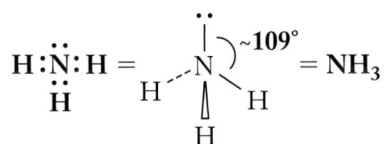

TRY IT

Build this NH_3 molecule using your molecular models. Starting from CH_4, change one bond to a lone pair (take off an atom, but leave the stick or spring). The four electron pairs remain repelled into a tetrahedral shape. Then take off the lone pair to look at just the shape and angles of the *bonds* and *atoms*. With the central nitrogen atom on top, check that the atoms form a low *pyramid* with tetrahedral (~109°) angles.

---

c. **Column 6 atoms:** A single-bonded neutral atom in the *oxygen* family (group 6A) is surrounded by *two bonds* and *two lone pairs*. The four electron pairs result in an electronic geometry that is tetrahedral with ~109° angles.

The two lone pairs count in deciding the shape, but they do not count when naming the shape of the molecule. The molecular geometry is said to be **bent** with ~109° angles.

Example:

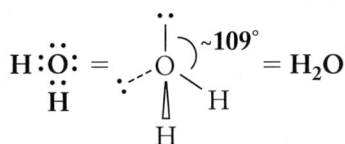

► TRY IT

Build the water molecule in the example above. Place two bonds and two lone pairs around the central atom. Then switch the position of one bond and one lone pair. Did this switch create a different molecule?

Answer:

No. Because of the symmetry of a tetrahedron, all four electron pairs around the central atom are in equivalent positions. The *same molecule* results no matter where the two bonds and two lone pairs are attached.

Now remove the two lone pairs. The molecular shape of H_2O is *bent*, and the angle formed by its two bonds is ~109°.

---

d. **Column 7 atoms:** Halogen atoms are surrounded by three lone pairs and one single bond. The electron pair geometry is tetrahedral, but because there is only one bond, the shape around the halogen atom is *linear*, and there is *no* bond angle.

► TRY IT

Build an HCl molecule. Place one bond and three lone pairs around the chlorine atom.

Switch the position of the bond to hydrogen. Note that this creates an equivalent molecule.

$$\text{H}:\ddot{\text{C}}\text{l}: \ = \ \ = \text{HCl}$$

The electron pairs around the chlorine are tetrahedral, but the H and Cl atoms form a line. The shape of the HCl molecule is linear with no bond angle.

Lone-Pair Repulsion

We can increase the accuracy of VSEPR bond-angle predictions with the following rule:

Lone pairs repel slightly more than bonds repel. The lone pairs need more room.

Because bonding pairs are more localized along the axis between the atomic nuclei, lone pairs tend to occupy slightly more space than bonds. This means that a lone pair repels other pairs slightly more than a bond repels other pairs. The angle between *bonds* is therefore slightly *smaller* than the other angles.

Build the water molecule. Include pieces for the two lone pairs. Push the lone pairs apart a bit. Squeeze the bonds together a bit.

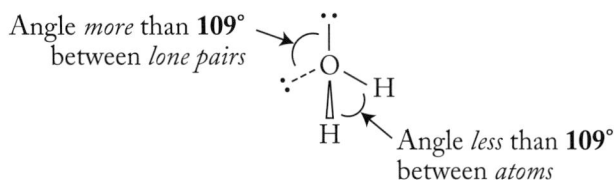

Angle *more* than **109°** between *lone pairs*

Angle *less* than **109°** between *atoms*

The general model predicts that a water molecule is bent with bond angles of ~109°, but because water has two lone pairs, they repel each other slightly more than the bonds repel each other. The actual average bond angle in water is **104.5°**, slightly less than the tetrahedral 109° angle. This angle in water is a typical value for a central atom surrounded by two lone pairs and two bonds.

For single-bonded central atoms, when will bond angles be *less* than 109° (<109°)? Only when the molecule has *one* or *two* lone pairs around the central atom. That will be the case when central atoms are in the *nitrogen* or *oxygen* family. Let's summarize this as:

The "Lone-Pair Scrunch" Rule

In molecules with all single bonds, around central atoms in the *nitrogen* or *oxygen* family the bond angles are predicted to be 103°–107° (slightly less than 109°).

Around single-bonded *carbon* family atoms, because there are no lone pairs, there is no lone-pair effect on the angles. The bond angles are tetrahedral: ~109°.

VSEPR Scope and Limitations

VSEPR rules to determine the shape around central atoms apply only to *molecular* compounds: those in which in most cases all of the atoms are nonmetals. There are only 12 nonmetals that bond, but because nonmetals tend to form strong covalent bonds with each other and often do so for multiple bonds in several directions, the nonmetals combine to form literally millions of different known molecules.

In some cases, for nonmetals in rows 3–6 of the periodic table, models for bonding that are more complex than VSEPR are needed to accurately predict molecular shapes and angles.

Summary: Shapes, Angles, and Families

For atoms in the second row of the periodic table, and with some exceptions for rows below the second row, VSEPR predicts that central atoms in neutral compounds will generally have the characteristics listed in the table below. Learn this table so that, given the terms in the first *column*, you can fill in the blanks, but you should also be able to explain these results based on the rules for VSEPR.

VSEPR Predictions for Neutral Single-Bonded Central Atoms

Second-Row Symbol	Li	Be	B	C	N	O	F	Ne
Valence Electrons	1	2	3	4	5	6	7	8
Bonds	(Primarily ionic bonds)		3	4	3	2	1	0
Lone Pairs			0	0	1	2	3	4
Electronic Geometry			Trigonal planar	Tetrahedral	Tetrahedral	Tetrahedral	Tetrahedral	
Molecular Geometry			Trigonal planar	Tetrahedral	Trigonal pyramidal	Bent	Linear	No bonds
Bond Angles			120°	109°	<109°	<109°	None	

Note that for single-bonded central atoms:

- In tall columns 3–4, the *electronic* shape and bond angles are the same as the *molecular* shape and bond angles.

- In tall columns 5–7, the *electronic* geometry is always tetrahedral, but the *molecular* geometry is not.

PRACTICE

1. Make and practice flashcards for the new vocabulary and rules in this lesson.

2. (Use a periodic table, plus models if needed.) For a molecule in which the central atom is surrounded by two bonds and two lone pairs:

   a. What is the geometry of the electron pairs?

   b. What is the shape of the molecule?

   c. What is the bond angle?

3. In your notebook, make a copy of the table below. Practice until, from memory, for the row 2 atoms in main groups 3 to 8 only, you can fill in the table with the VSEPR model predictions for the geometry of covalent compounds.

VSEPR Predictions for Neutral Single-Bonded Central Atoms

Family	-	-	B	C	N	O	F	Ne
Electronic Geometry	-	-						
Molecular Geometry	-	-						
Bond Angles	-	-						

4. Complete this table based on VSEPR predictions for neutral, single-bonded atoms.

Molecule	NF$_3$	SiH$_4$	AlCl$_3$	SI$_2$
Lewis Structure				
Electronic Geometry				
Molecular Shape				
Bond Angles				

5. Which molecule in problem 4 would likely be the least stable and most reactive? Why?

Lesson 19.3   Electronegativity

What decides if a particular bond will be ionic or covalent? Our "rule of thumb" has been that if a bond is between a metal and a nonmetal atom, it will likely be ionic, but between two nonmetals it will likely be covalent. A more precise view is that most bonds have a mixture of ionic and covalent character. This latter model for bonding will result in more accurate predictions of the properties of bonds and substances.

The **electronegativity model** predicts how strongly each atom attracts the electrons in a bond. The **electronegativity scale** assigns each atom an **electronegativity value (EN)**. Fluorine (EN = 4.0) is the strongest electron attractor. Cesium and francium (EN = 0.7) are the weakest electron attractors.

The following table lists the electronegativity values for atoms according to their position in the periodic table. These numbers are termed the *Pauling values*, named after the American chemist Linus Pauling, who first proposed the electronegativity scale. Other models assign slightly different EN values.

To speed your work, the electronegativity values in **bold** in the table below should be memorized. Note that the *second*-row numbers start at **1.0** and increase by **0.5** for each atom to the right. The values for hydrogen (2.1) and chlorine (3.0) are also encountered frequently and should be committed to memory.

Electronegativity Values

Row 1	**2.1**							
Row 2	**1.0**	**1.5**		**2.0**	**2.5**	**3.0**	**3.5**	**4.0**
Row 3	0.9	1.2		1.5	1.8	2.1	2.5	**3.0**
Row 4	0.8	1.0	1.3–1.9	1.6	1.8	2.0	2.4	2.8
Row 5	0.8	1.0	1.2–2.2	1.7	1.8	1.9	2.1	2.5
Row 6	0.7	0.9	1.0–2.4	1.8	1.9	1.9	2.0	2.2
Row 7	0.7	0.9						

In the table, note:

- The strongest electron attractors are toward the top right.
- Only four atoms have EN values of 3.0 and above: N, O, F, and Cl.
- Values generally (but not always) increase toward the top right corner of the periodic table: to the right across a row and up a column.

To predict *bond* behavior, the electronegativity model divides bonds into three types: ionic, polar covalent, and nonpolar covalent.

Ionic Bonds

In these lessons, our rule will be:

> If the *difference* in the electronegativities of two bonded atoms is
> - greater than 1.7, the bond is *ionic*;
> - 1.7 or less, the bond has primarily *covalent* character.

Textbooks use varying values, but 1.7 is a typical "dividing line" between ionic and covalent bonds.

An ionic bond can be thought of as a bond in which the difference in electron attraction is so strong that the more electronegative atom *removes* valence electrons from the other atom to form ions.

Polar versus Nonpolar Bonds

Covalent bonds can be divided into two types: **polar** and **nonpolar**.

> In a covalent bond with low or no electronegativity difference, the *bond* is said to be **nonpolar**.

Examples: These bonds are *nonpolar*.

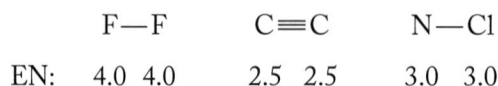

	F—F	C≡C	N—Cl
EN:	4.0 4.0	2.5 2.5	3.0 3.0

If two atoms in a bond have the same electronegativity values, the bond is nonpolar, and the bond electrons will be found on average at an equal distance between the two nuclei.

Whether bonds are single, double, or triple bonds (as in C≡C above) does not affect the electronegativity difference or bond polarity.

As the difference in the electronegativity of two bonded atoms increases, the bond becomes more *polar*. The electrons are still shared, but on average they are found closer to the more electron-attracting atom.

This uneven electron sharing creates a **dipole**: an uneven distribution of electric charge. The more electronegative atom can be described as having a *partial* negative charge, and the weaker electron attractor has a *partial* positive charge.

Dipoles may be represented using two types of notation. In math and science, a δ (a lowercase Greek *delta*) is often used as a symbol meaning **partial**. A polar bond can be labeled with two deltas. The atom that is the stronger electron attractor (with the higher electronegativity value) is labeled δ− (pronounced *delta minus*) to indicate a *partial* negative charge on the atom, and the weaker electron attractor is labeled δ+ (*delta plus*) to represent a partial positive charge.

Examples:

δ+ **C—O** δ−   δ+ **N—F** δ−   δ− **O—H** δ+

Another way to represent the dipole in a bond is to use an arrow in place of (or in addition to) the bond. The arrow points toward the atom that is the stronger electron attractor: toward the side of the bond where on average the electrons are more likely to be found.

Examples:

C → O   N → F   O ← H

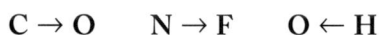

Use these rules:

To Label a Bond as Nonpolar, Polar, or Ionic

If the difference in electronegativity between two atoms is

- **0 to 0.4**, the bond is termed **nonpolar** covalent;
- **0.5 to 1.7**, the bond is **polar** covalent;
- **above 1.7**, the bond is *ionic.*

The breakpoints at 0.5 and 1.7 are typical, but some textbooks use slightly different values.

A more accurate view would be: As the difference in electronegativity in bonds rises from zero to as high as 3.3, the character of the bond *gradually* changes from nonpolar to polar to ionic. However, putting bonds into these three simplified categories will help in predicting many substance behaviors.

1. Make and practice flashcards for the new vocabulary and the shaded rules in this lesson. In addition, given a copy of the periodic table, practice until you can write the electronegativity values for the second-row atoms, plus H and Cl.

2. Draw the four bonds below on a line in your notebook. Leave space between each atom pair, plus a blank line above and room for several blank lines below. Then complete questions a–e. If needed, check answers after each part.

   C—H   N—Cl   C—F   O—B

   a. Using a periodic table, above each atom write its electronegativity value from memory.

   b. For bonds with a dipole, to the left and right of each atom, label the atom as $\delta+$ or $\delta-$.

   c. On the next line down, calculate the electronegativity difference (without a calculator).

   d. On the next line down, label the bond as nonpolar, polar, or ionic.

   e. On the next line down, write the bond as an arrow in the direction of the dipole.

Lesson 19.4	**Predicting the Polarity of Substances**

- What determines whether a substance will be a solid, liquid, or gas at room temperature?
- Why do table salt and sugar dissolve substantially in water but most substances do not?
- Can we predict formulas for new pharmaceuticals that will relieve pain and cure disease?

The answers to these practical and important questions are often found by investigating the shapes and polarities of substances.

Rules for Substance Polarity

In the previous lesson, *bonds* were classified as ionic, polar, and nonpolar. *Substances* can also be classified as having *ionic*, *polar*, and *nonpolar* character.

- An ionic substance has a complete charge separation: Its particles are ions.
- A polar molecule has partial charge separation: a net dipole.
- A nonpolar molecule has no charge separation: no dipole.

   Examples: NaCl behaves as an ionic compound, H_2O is polar, and CCl_4 is nonpolar.

A wide range of substance properties, including melting point, boiling point, and solubility in solvents, can be explained in part by whether the substance is ionic, polar, or nonpolar. These classifications can be predicted by applying the following:

Rules to Predict Substance Polarity

1. A substance with just *one* ionic bond will usually behave as an ionic compound even if it also has many nonpolar bonds.

2. A substance will be a *nonpolar* molecule if

   - it has *all* nonpolar bonds, *or*
   - it has polar bonds, but the dipoles cancel due to symmetry.

3. A substance with all covalent bonds will be a *polar* molecule if

   - it has polar bonds, *and*
   - the dipoles do *not* cancel due to molecular symmetry.

TRY IT

Q. Based on the rules above, label these substances as *ionic, polar,* or *nonpolar.* Use a periodic table without EN values (these atoms have values you should know).

   a. Cl_2 b. LiCl

Answers:

   a. In Cl_2, the shape must be **Cl—Cl**. The difference in EN values between the two atoms is zero. When the EN difference is 0 to 0.4, the bond is nonpolar. The *molecule* is **nonpolar** because its one bond has no charge separation.

   b. Li has a 1.0 EN and Cl has a 3.0 EN. The difference of 2.0 is above 1.7, so the *bond* is *ionic.* If one bond (or more) in a substance is ionic, the *substance* will have **ionic** behavior. The more electronegative atom will have the negative charge. In this case, the result is Li^+ ion and Cl^- ion.

Polarity in Two-Dimensional Molecules

For more-complex cases, the following *steps* to apply the rules will provide reasonably good predictions for whether a substance will have ionic, polar, or nonpolar character.

Steps to Determine Substance Polarity

1. Draw the Lewis diagram for the substance.

2. Draw the atoms, bonds, and shape of the substance. Label each atom with its electronegativity (EN) value.

3. Based on the EN *differences*, label each type of *bond* as nonpolar, polar, or ionic.

4. If the substance has *one* or more ionic bonds, predict that it will have *ionic* behavior.

5. If *all* of the bonds are *nonpolar*, predict the *substance* is *nonpolar*.

6. If steps 4 and 5 do not apply, one or more of the bonds must be polar. To determine the molecular polarity:

   a. On the drawing of the particle shape, draw arrows representing the dipoles. Use geometry and symmetry to see whether the dipoles cancel. If needed, make a 3-D model.

   b. If the dipoles *cancel*, the molecule is *nonpolar*. If the dipoles do *not* cancel, there is a net dipole, and the molecule is *polar*.

Dipole Cancellation

Step 6 above requires *adding bond dipoles*. Dipoles add using *vector addition*. You may have had practice adding vectors in math or physics classes. In analyzing the polarity of molecules, a key rule is:

Equal but opposite vectors (and dipoles) *cancel.*

Let's learn the method by example.

> ► **TRY IT**

Q. Label these substances as *ionic, polar,* or *nonpolar*. Use a periodic table without EN values.

   a. O=C=O (linear shape)   b. HCl

Answer:

a. CO_2 has two double bonds. An EN difference has the same impact on polarity whether the bond is single, double, or triple. The carbon EN is 2.5, and oxygen's EN is 3.5. The EN difference is 1.0, so both *bonds* are *polar*.

   When bonds are polar, the symmetry test must be applied to see whether the dipoles cancel. If we add the dipole arrows to the molecular shape, the dipole diagram is $O \leftarrow C \rightarrow O$. When dipoles are *equal* but in *opposite directions*, they *cancel* due to symmetry. The $O \leftarrow C \rightarrow O$ has polar bonds but is a **nonpolar** molecule because the dipoles *cancel*.

b. H has a 2.1 EN and Cl a 3.0 EN. The difference is 0.9, which is in the range of 0.5 to 1.7, so the bond is polar. The shape for this molecule is H—Cl. Because Cl is more electronegative, the dipole points toward Cl: $H \rightarrow Cl$. Because there are no other bonds to cancel this dipole, the *molecule* has a dipole and is **polar**.

   The Cl atom has lone pairs, but in this model for predicting polarity, we assume that bonds, not lone pairs, contain the dipoles.

PRACTICE **A**

Use a periodic table that does not include electronegativity values (you should know these from their table position). Based on VSEPR and electronegativity, predict whether these compounds will be *ionic*, *polar*, or *nonpolar*. If needed, check answers after each part.

1. BeH_2 2. LiF   3. $\begin{matrix} H \\ \diagdown \\ C=O \\ \diagup \\ H \end{matrix}$ (flat shape, 120° angles)   4. BCl_3

Polarity in 3-D Molecules

In evaluating molecular polarity, all of our examples so far have been two-dimensional shapes that can be drawn on paper or screen. For compounds that are three-dimensional, a model should be made and the dipoles analyzed.

▶ TRY IT

Complete the parts below one at a time, checking your answers after each part.

Q. Using the "Steps to Determine Substance Polarity," label these compounds as *ionic*, *polar*, or *nonpolar*. Use a periodic table without electronegativity values. Make the molecular models for each compound.

a. H_2O b. CCl_4 c. NH_3

STOP

Answers:

To evaluate molecular polarity, first evaluate bond polarity. *If* the bonds are *polar*, evaluate symmetry to see whether the dipoles cancel.

a. H_2O: H has an EN of 2.1, and O has an EN of 3.5. The difference of 1.4 makes each bond polar. If one or more bonds are polar, evaluate the symmetry.

Water: $:\!\ddot{O}\!:\!H =$ 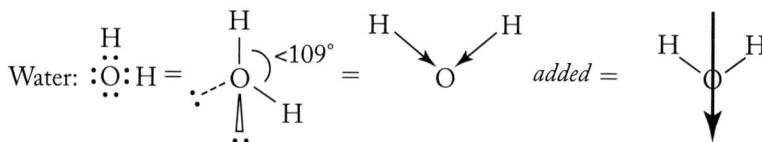   *added* =

The bonds have a bent shape and both dipoles point toward oxygen, so they do *not* cancel. The bonds are polar *and* the water *molecule* is **polar**. The net dipole in water can be represented by an arrow (above) or using the δ notation below.

$$\delta+ \ \begin{matrix} H \\ \diagdown \\ \diagup \\ H \end{matrix}\!O \ \ \delta-$$

The H side of water is δ+ and the O side is δ−. The net dipole means that water is *polar*.

The polarity of water is central to understanding the behavior of aqueous solutions that are important in chemistry and biology.

b. **CCl₄:** C is EN 2.5 and Cl is EN 3.0, so the C—Cl bond is weakly polar, and the dipoles point toward Cl.

CCl₄ is *tetrahedral* with four *equal* bond dipoles. Turning the model so that two bonds are up and two down, the top and the bottom two dipoles cancel side to side. The resultants are two dipoles, one pointing up and the other down. These two resultant dipoles also cancel, because they are equal but in opposite directions.

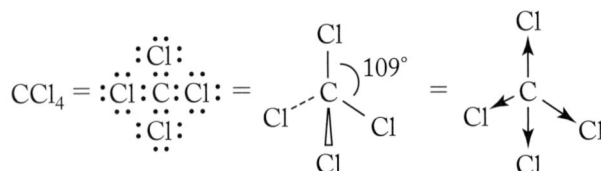

When a central atom is surrounded by four tetrahedral bonds to the same kind of atom, the molecule is always nonpolar: Any dipoles will cancel due to symmetry.

c. **NH₃:** First evaluate *bond* polarity. The EN difference of 0.9 means that the bonds are polar, with the dipoles pointing toward N.

To check for dipole cancellation, make the tetrahedral model, then take off the lone pair to focus on the bonds that determine the polarity. If the model is placed so that the central N is up, all of the dipoles point upward from the H's toward N. The dipoles are equal but not *opposite*: they do not cancel. NH₃ is **polar**.

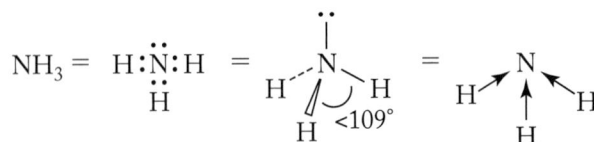

In trigonal pyramidal and bent molecules, dipoles in the same direction, pointing either to or from a central atom, result in polar molecules.

Some general rules for dipole addition are:

* When a central atom in tall columns 3 or 4 is surrounded by single bonds to the same atom, any dipoles will cancel due to symmetry: The molecule will always be nonpolar.

* If a central atom is in tall columns 5 or 6, dipoles in the same direction, pointing either to or from a central atom, result in polar molecules.

But rather than memorize these rules, if bonds are polar it is best to draw or make the model to see if the dipoles cancel.

Exceptions

The above rules classify substances as nonpolar, polar, and ionic. In reality, polarity is not that simple.

- The polarity of substances can be measured numerically to obtain a **dipole moment,** and those measurements show a continuum of values from totally nonpolar to highly ionic.
- There are factors that affect polarity in addition to the electronegativity and atom geometry considered in our model above.
- Our rules classify molecules as nonpolar that in reality are slightly polar.

That said, our simplified model is a *starting point* for the prediction of properties based on polarity for a variety of substances.

PRACTICE B

1. Design flashcards for the shaded rules in this lesson.

2. For these problems, use a periodic table *and* a table of electronegativity values. Based on VSEPR and electronegativity, predict whether these substances will be *ionic*, *polar*, or *nonpolar*.

   a. SF_2 b. PCl_3 c. SiH_4 d. LiCl

SUMMARY

A strategy for study that cognitive scientists say will speed your rate of learning is to:

- Break new knowledge into small elements.
- Move those elements into long-term memory.
- Practice using this new knowledge in a variety of distinctive contexts.

One suggested explanation for why this method works involves the powerful drive in humans to learn to speak. Even without schooling, all hearing children learn the language that is spoken around them. A child's brain instinctively responds to the stimulus of speech by storing the sound of a word in a brain cell (neuron). As the word is heard repeatedly, the brain constructs linkages between that neuron and other cells that describe the word's meaning and the rules for its use. Learning that is not as powerfully instinctive as "learning to speak during childhood" requires mental effort, but a study strategy of "move an element into memory, and then explore its application" aligns with how the brain is structured to store, understand, and apply the words we speak.

REVIEW QUIZ

You may use a periodic table and molecular models, but no calculator and no table of electronegativity values.

1. Answer in scientific notation without a calculator.
   a. Multiply: $(6.0 \times 10^{23})(9.0 \times 10^{-5}) =$

   b. $\dfrac{1.0 \times 10^{-14}}{4.0 \times 10^{-4}}$

2. Fill in the blanks.

Molecule	PH$_3$	H$_2$	HBr
Lewis Structure			
Electronic Geometry			
Molecular Geometry			
Bond Angles			
Bond Polarity			
Molecular Polarity			

3. Fill in the blanks.

Molecule	OF$_2$	SI$_2$	BI$_3$
Lewis Structure			
Electron Geometry			
Name of Shape and Sketch of Shape			
Bond Angles			
Bond Polarity			
Molecular Polarity			

ANSWERS

Lesson 19.1

Practice A   1a. Silicon: **4**   1b. Phosphorus: **5**   1c. Bromine: **7**   1d. Sulfur: **6**

2. It does not matter which sides have the paired or unpaired electrons: The four sides are equivalent.

2a. ·Ṡi·   2b. ·P̈·   2c. :B̈r·   2d. :S̈·

Practice B   1a. :F̈:B̈r: = F—Br   1b. H:C̈l: = H—Cl

2. Bonds: 1a. **1**   1b. **1**   Lone pairs: 1a. **6**   1b. **3**

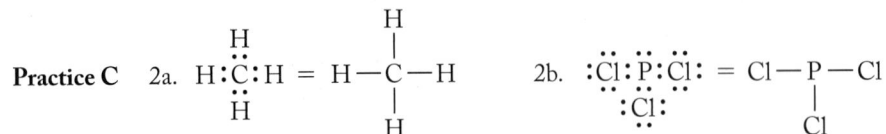

Practice C   2a.
$$
\begin{array}{c}
\text{H:C:H} \\
\end{array}
= \text{H—C—H}
$$

2a. H:C̈:H = H—C—H
(with H above and below)

2b. :C̈l:P̈:C̈l: = Cl—P—Cl
:C̈l: Cl
(with Cl below)

3. Bonds: 2a. **4**   2b. **3**   Lone pairs: 2a. **0**   2b. **10**

4a. Sulfur: **2**   4b. Iodine: **1**   4c. Silicon: **4**   4d. Nitrogen: **3**

Lesson 19.2

Practice   2a. The four electron pairs repel into a tetrahedral shape.   2b. The three atoms are in a bent shape.

2c. The bond angle is slightly *less* than 109°.

3.

Family	-	-	B	C	N	O	F	Ne
Electronic Geometry	-	-	Trigonal planar	Tetrahedral	Tetrahedral	Tetrahedral	Tetrahedral	
Molecular Geometry	-	-	Trigonal planar	Tetrahedral	Trigonal pyramidal	Bent	Linear	No bonds
Bond Angles	-	-	120°	~109°	<109°	<109°	None	

4.

Molecule	NF_3	SiH_4	$AlCl_3$	SI_2
Lewis Structure	:F:N:F: ̈ :F:	H H:Si:H H	:Cl:Al:Cl: :Cl:	:I:S:I: (also could be drawn at 90°)
Electronic Geometry	Tetrahedral	Tetrahedral	Trigonal planar	Tetrahedral
Molecular Shape	Trigonal pyramidal	Tetrahedral	Trigonal planar	Bent
Bond Angles	<109°	~109°	120°	<109°

5. $AlCl_3$ would be predicted to be the least stable and most reactive because **Al** has an unsatisfied octet.

Lesson 19.3

Practice   2a.   **2.5 2.1**   **3.0 3.0**   **2.5 4.0**   **3.5 2.0**

2b. δ– **C—H** δ+   **N—Cl**   δ+ **C—F** δ–   δ– **O—B** δ+

2c. **0.4**   **0**   **1.5**   **1.5**

2d. **nonpolar or slightly polar**   nonpolar   polar   polar

2e. **C ← H**   **N—Cl (no dipole)**   **C → F**   **O ← B**

Lesson 19.4

Practice A   All of the molecules in Practice A are *two-dimensional*: Their shape can be drawn on paper.

1. **BeH_2:** First assign EN values to categorize the bonds. 2.1 H−1.5 Be = **0.6** > 0.4 = *polar bonds. If* bonds are polar, draw the Lewis diagram and shape to see whether the dipoles cancel. The central atom **Be** is predicted by VSEPR to have a **linear** shape and **180°** bond angles. Add the dipoles to the diagram. Because they are equal and in opposite directions, the dipoles cancel. The VSEPR prediction is that the molecule is **nonpolar**.

EN: 2.1 1.5 2.1

H:Be:H   H—Be—H   H ← Be → H = **nonpolar** molecule

2. **LiF**: First assign EN values to categorize the bonds. 4.0 F⁻ 1.0 Li = **3.0** > 1.7 = an *ionic bond*. If one bond is ionic, the substance is **ionic**.

3. **H₂C═O**: Assign EN values to categorize the bonds.

   2.1 H − 2.5 C = **0.4** difference, which means a slight dipole toward C.

   3.5 O − 2.5 C = **1.0** difference, which means a stronger dipole toward O.

   Electronegativity differences apply in the same way to single and double bonds.

   If one or more bonds are polar, draw the Lewis diagram, sketch the shape, add the dipoles, and see whether the dipoles cancel. Because this molecule is flat, it can be analyzed on paper.

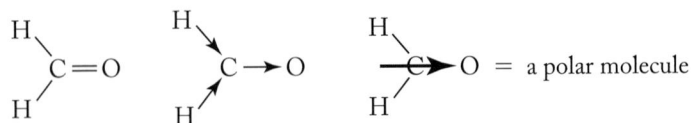

   Because these dipoles are all in the same direction, they do not cancel. The molecule is **polar**.

4. **BCl₃**: Assign EN values to categorize the bonds. 3.0 Cl − 2.0 B = **1.0** = **polar bonds**.

   If bonds are polar, draw the Lewis diagram and shape to see whether the dipoles cancel.

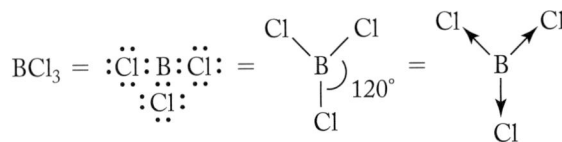

   The central atom **B** is predicted by VSEPR to have a **trigonal planar** shape for its bonds and **120°** bond angles. Adding the dipoles by vector addition, they are equal and in opposite directions, so they cancel. VSEPR predicts that the molecule is **nonpolar**. Three bonds that are the same in a trigonal planar shape will always result in a nonpolar molecule. Any dipole will cancel due to symmetry.

Practice B   2a. **SF₂**: Assign EN values to categorize the bonds. 4.0 F − 2.5 S = **1.5** = a **polar bond**.

   If bonds are polar, draw the Lewis diagram and shape, make the model if needed, then analyze the vectors and see whether the dipoles cancel. The dot diagram predicts a bent molecular shape and slightly less than 109° bond angles. Because the two dipoles are slightly less than 109° apart rather than 180°, they are equal but *not* opposite. Adding the two dipoles by vector addition gives a net resultant dipole. The *molecule* is predicted to be:

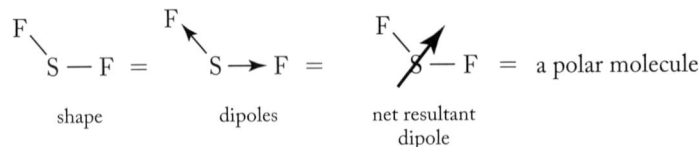

2b. **PCl₃**: First label the bonds using their electronegativities. 3.0 Cl − 2.1 P = **0.9** = **polar bonds**. Do the dipoles cancel? Based on the dot diagram and 3-D model, the shape of PCl₃ is *trigonal pyramidal* with slightly less than **109°** bond angles.

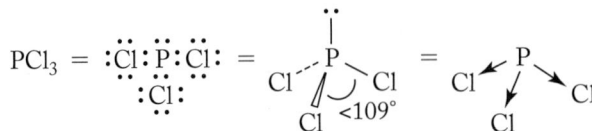

   The three *bond* dipoles all point *down* from P: They do not cancel. The molecule is **polar**.

2c. **SiH_4:** For a central atom in the carbon family, whenever four of the same atoms are attached, any dipoles will cancel due to the symmetry of a tetrahedron. SiH_4 is a **nonpolar** molecule.

2d. **LiCl:** Lithium has an EN of 1.0, and chlorine has an EN of 3.0. Because the EN *difference* is 2.0, the substance is **ionic**. The rules for using VSEPR to determine shape and angles apply only to molecular compounds (in which all bonds are covalent).

Review Quiz

1a. $(6.0 \times 10^{23})(9.0 \times 10^{-5}) = 54 \times 10^{18} = \mathbf{5.4 \times 10^{19}}$

1b. $\dfrac{1.0 \times 10^{-14}}{4.0 \times 10^{-4}} = 0.25 \times 10^{-10} = \mathbf{2.5 \times 10^{-11}}$

2.

Molecule	PH_3	H_2	HBr
Lewis Structure	H:P̈:H H	H:H	H:B̈r:
Electronic Geometry	Tetrahedral	Linear	Tetrahedral
Molecular Geometry	Trigonal pyramidal	Linear	Linear
Bond Angles	<109°	No angle	No angle
Bond Polarity	Nonpolar	Nonpolar	Polar
Molecular Polarity	Nonpolar	Nonpolar	Polar

3.

Molecule	OF_2	SI_2	BI_3
Lewis Structure	:F̈:Ö:F̈: (also could be drawn at 90°)	:Ï:S̈:Ï: (also could be drawn at 90°)	:Ï:B:Ï: :Ï:
Shape of Electron Pairs	Tetrahedral	Tetrahedral	Trigonal planar
Name of Shape and Sketch of Shape	Bent F—O⟋F	Bent I—S⟋I	Trigonal planar I | B I⟋ ⟍I
Bond Angles	<109°	<109°	120°
Bond Polarity	4.0 – 3.5 = 0.5 = **Polar**	2.5 – 2.5 = 0 = **Nonpolar**	2.5 – 2.0 = 0.5 = **Polar**
Molecular Polarity	**Polar** (bent with polar bonds)	**Nonpolar** (nonpolar bonds)	**Nonpolar** (the dipoles cancel)

20

Introduction to Equilibrium

Lesson 20.1   Reversible Reactions

Reaction Rates

For most chemical reactions to take place, two reactant particles must collide. If temperature is increased, a reaction nearly always occurs at a faster rate. Why? Reactant particles at higher temperatures, on average, are traveling faster. This means that they have higher average kinetic energy when they collide and are more likely to change in some way (react).

Reactions also proceed faster when the reacting particles are more concentrated, because they collide more often.

Reactions That Go to Equilibrium

Chemical reactions can be divided into three types.

1. **Reactions that go nearly 100% to completion.** Burning gasoline to form carbon dioxide and water is one such reaction. Once it begins, the reaction continues until one of the reactants (the gasoline or oxygen) is essentially used up.

2. **Reactions that don't go.** Trying to convert carbon dioxide and water into gasoline and oxygen gas is theoretically possible, but in practice, reversing the burning of gasoline is very difficult to accomplish in the laboratory.

3. **Reactions that are *reversible* and go *partially* to completion.** Some reactions go from products back to reactants with relative ease. As a reversible reaction proceeds, the reactants are gradually used up and the *forward* reaction slows down. As product concentrations increase, they more frequently collide and re-form the reactants. This process continues until both the forward and reverse reactions are occuring at the same rate. At that point, the system is said to be at **equilibrium**.

For equilibrium to exist, both of these conditions must be met:

- All reactants and products must be present in at least small quantities.
- The reaction must be in a *closed* system: No particles or energy can be entering or leaving the reaction vessel.

For a reaction at equilibrium, no change seems to be occurring, but this appearance is deceiving. A system at equilibrium is **dynamic**: the forward and reverse reactions continue. However, because the rate of the forward reaction is equal to the rate of the reverse reaction, there is no apparent change.

To summarize:

For a Reaction at Equilibrium

- The forward and reverse reactions are proceeding at the same rate.
- No *net* change is occurring.
- The system is closed, and all reactants and products are present.

PRACTICE

1. State two ways to increase the rate of a reaction.

2. What is true for the reaction rates in a reaction that has reached equilibrium?

3. What must be true for a reaction vessel to be "closed"?

Lesson 20.2 | Le Châtelier's Principle

In these lessons, we will use a two-way arrow ($\leftrightarrows$) to indicate that a reaction is reversible.

Equilibrium is important because many reactions are reversible in practice. For those reactions, we want to be able to:

- Predict what will happen when a system at equilibrium is *disrupted*.
- *Shift* an equilibrium to make as much of a desired substance as possible.

When a reaction mixture at equilibrium is subjected to a change, the direction that a reversible reaction will shift can be predicted by:

Le Châtelier's Principle

If a system at equilibrium is subjected to a change, processes occur that tend to counteract that change.

Changes in Concentration

To predict shifts in equilibrium due to changes in *concentration*, it is helpful to restate Le Châtelier's principle to focus on concentration.

Predicting Equilibrium Shifts Due to Concentration Change

For a reversible reaction at equilibrium, write the balanced equation using a *two-way* ($\leftrightarrows$) arrow, then apply the following rules.

- *Increasing* a [substance] that appears on one side of an equilibrium equation shifts an equilibrium toward the *other* side. The other substance concentrations on the *same* side as the increased [substance] are *decreased*, and the substance concentrations on the *other* side are *increased*.
- *Decreasing* a [substance] that appears on one side of an equilibrium equation shifts the equilibrium *toward* that side. The other [substances] on the *same* side are *increased*, and the [substances] on the *other* side are *decreased*.

▶ **TRY IT**

Refer to the statements of Le Châtelier's principle above to answer these questions.

Q. Chromate ions react with acids to form dichromate ions in this reversible reaction.

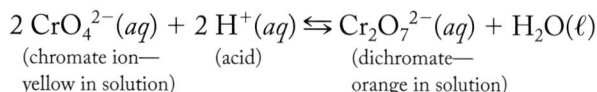

$$2\,CrO_4{}^{2-}(aq) + 2\,H^+(aq) \leftrightarrows Cr_2O_7{}^{2-}(aq) + H_2O(\ell)$$

(chromate ion—   (acid)   (dichromate—
yellow in solution)   orange in solution)

For the above reaction, if acid (H^+) is added to a yellow chromate ion solution at equilibrium:

a. In which direction will the equilibrium shift (left or right)? _____.

b. The [$CrO_4{}^{2-}$] will (increase or decrease?) _____.

c. The [dichromate ion] will (increase or decrease?) _____.

d. What color change will tend to occur? _____.

STOP

Answers:

a. The equilibrium will shift to the **right**. Increasing the concentration of the H^+ found on the left side shifts the equilibrium toward the right side.

b. The $[CrO_4^{2-}]$ will **decrease.** Increasing the concentration of a substance that appears on one side decreases the concentration of the other substances on that side.

c. The [dichromate ion] will **increase.** Increasing the $[H^+]$ that appears on the left increases the concentration of the substances on the right.

d. Because adding acid decreases the [chromate] and increases the [dichromate], the solution color shifts **from yellow toward orange.**

To explain these shifts, we examine what is happening in the reaction at the molecular level.

- If $[H^+]$ is increased, there are more frequent collisions between the H^+ and the chromate ions. Though the percentage of collisions that result in a reaction stays the same if the temperature remains constant, more frequent collisions means a higher rate of forward reaction. Increasing the $[H^+]$ causes the rates of the forward and reverse reactions, equal at equilibrium, to be thrown out of balance. The increased forward reaction uses up chromate and forms more dichromate ions.

- As more dichromate forms, its collisions with water increase, and the speed of the reverse reaction increases. A new balance is reached, but only after some of the yellow chromate has been used up and more orange dichromate has formed.

▶ TRY IT

Q. For the same chromate–dichromate reaction, if a strong *base* is added to the orange solution that results after adding acid:

a. The [acid] in the solution will (increase or decrease?) _____.

   [If you are unsure of your part (a) answer, check the answer before continuing.]

b. In which direction will the equilibrium shift (left or right)? _____.

c. The $[CrO_4^{2-}]$ will (increase or decrease?) _____.

d. The $[Cr_2O_7^{2-}]$ will (increase or decrease?) _____.

e. What color change will occur? _____.

STOP

Answers:

a. Bases neutralize acids (see Lesson 14.2). This will **decrease** the $[H^+]$.

b. Decreasing the $[H^+]$, which is a reactant on the **left** side, will shift the equilibrium *toward* that side.

c. Shifting to the left means that the $[CrO_4^{2-}]$ will **increase.**

d. Decreasing the concentration of a term on the left shifts the equilibrium to the left and decreases the concentration of terms on the right. The $[Cr_2O_7^{2-}]$ will **decrease.**

e. Adding base decreases the [acid]. The equilibrium shifts toward the side with the acid term, increasing the [yellow chromate] and decreasing the [orange dichromate]. The color shifts **from orange toward yellow.**

Based on what is happening at the molecular level, these shifts are logical. When $[H^+]$ is decreased, there will be a lower rate of collisions between the acid and the chromate ions, so that the reverse reaction is now faster than the slowed forward reaction. The reverse reaction uses up orange dichromate and forms yellow chromate until a new balanced equilibrium is reached.

PRACTICE A

1. Design flashcards that state Le Châtelier's principle in one or more ways that will help in solving problems like those in the Try Its above. Practice your cards, then complete the following problems, applying the rules from memory. Check your answers as you go.

2. The Haber process converts nitrogen to ammonia by this reversible reaction:

$$N_2(g) + 3\,H_2(g) \leftrightarrows 2\,NH_3(g)$$

For a mixture at equilibrium, if the $[H_2]$ is increased:

a. Equilibrium will shift to the (left or right?) _____.

b. $[N_2]$ will (increase or decrease?) _____.

c. $[NH_3]$ will (increase or decrease?) _____.

3. For the same reaction at equilibrium, if the $[N_2]$ is decreased:

a. Equilibrium will shift to the (left or right?) _____.

b. $[H_2]$ will (increase or decrease?) _____.   c. $[NH_3]$ will _____.

Energy Changes

For a system at equilibrium, if the reaction equation includes an energy term, changes in energy produce shifts in equilibrium that can be predicted by Le Châtelier's principle in a manner similar to changes in concentration.

- Adding energy shifts the equilibrium away from the side with an energy term.
- Removing energy shifts the equilibrium toward the side with an energy term.

One way to add energy to a system is to heat it. Energy can be removed by cooling a system.

> **Example:** The melting of ice and the freezing of water can be described by this reaction:
>
> $$H_2O(s) + 6.03\ kJ/mol \leftrightarrows H_2O(\ell)$$
>
> Adding heat to a mixture of ice and water drives the equilibrium to the right: Some of the ice melts. Cooling the system (such as by placing it in a freezer) removes energy and shifts the equilibrium to the left: Liquid water changes to ice.
>
> Le Châtelier's principle predicts changes only while *both* ice and liquid water remain in the system. To be at equilibrium, all of the reactants and products must be present.

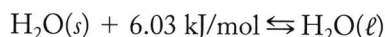

> **TRY IT**

Q1. For a system at equilibrium described by

$$N_2(g) + O_2(g) + 90.0 \text{ kJ} \rightleftharpoons 2 \text{ NO}(g)$$

If the reaction vessel is heated:

    a. The equilibrium will shift to the (left or right?) ——————————.

    b. $[O_2]$ will (increase or decrease?) ——————————.

    c. $[NO]$ will ——————————.

STOP

Answers:

    a. The equilibrium shifts to the **right**. Energy is a term on the left, so as energy is added, the system uses up energy, shifting the equilibrium to the right.

    b. $[O_2]$ will **decrease**. If heat is needed to cause a reaction on one side, adding heat will cause the rate of that reaction to increase, using up other substances on that side.

    c. $[NO]$ will **increase**. Increasing energy (a term on the left) causes more of substances on the right to form.

Q2. For the system in Q1 above, if the $[O_2]$ is decreased:

    a. Equilibrium will shift to the ——————————.

    b. $[N_2]$ will ——————————.   c. The temperature will ——————————.

STOP

Answers:

    a. The equilibrium shifts to the **left**.   b. $[N_2]$ will **increase**.

    c. Shifting the equilibrium to the left increases the amount of terms shown on the left, including energy. In this system, the increased heat will be manifested as an increase in average kinetic energy, and the temperature will **increase**.

PRACTICE   **B**

1. Design flashcards that summarize the energy rules above.

2. In a mixture at equilibrium for this synthesis of methanol:

$$CO(g) + 2 H_2(g) \rightleftharpoons CH_3OH(g) + \text{energy}$$

    a. If the temperature of the system is decreased:

        i. Equilibrium will shift to the (left or right?) ——————————.

        ii. $[CH_3OH]$ will (increase or decrease?) ——————————.

b. If the [H_2] is increased:

    i. Equilibrium will shift to the _____.

    ii. [CH_3OH] will _____.

    iii. The temperature in the vessel will (increase or decrease?) _____.

3. In a closed system at equilibrium, for the reaction

$$PCl_3(g) + Cl_2(g) \leftrightarrows PCl_5(g) + energy$$

  a. If the temperature is increased:

    i. Equilibrium shifts to the _____.   ii. [PCl_5] will _____.

  b. If the [Cl_2] is increased:

    i. Equilibrium shifts to the _____.   ii. [PCl_3] will _____.

    iii. The temperature in the vessel will (increase or decrease?) _____.

Concentration Changes: Special Cases

There are two special rules for concentration changes.

1. Adding or removing a *solid* or a *liquid* reacting particle does not shift an equilibrium.

   A solid has a constant concentration determined by its density. Adding or removing a *solid* in the reaction equation does not change the concentration of the solid, and therefore does not shift the equilibrium.

   Example: If solid table salt (NaCl) is added to a glass of water, initially it all dissolves, but if enough salt is added and stirred, the solution becomes **saturated:** At a given temperature it reaches the limit of [salt] that dissolves. A mixture of a saturated solution and its solid is a solubility equilibrium, represented in this case by

   $$NaCl(s) \leftrightarrows NaCl(aq)$$

   Once equilibrium is reached, adding more salt crystals increases the *amount* of solid salt on the bottom but does not increase the *concentration* (density) of the solid salt or the *dissolved* salt concentration. Adding more *solid* does not shift the equilibrium concentrations.

   A substance in its *liquid* state also has a constant concentration at a given temperature. Adding or removing a pure liquid from a system at equilibrium will not change the liquid's concentration and will not shift the equilibrium.

2. If the *solvent* for the reaction is a term in the equation, using up or forming solvent in the reaction does not substantially shift the equilibrium.

By definition, a solvent is a substance present in very high concentration compared to the other substances present in a solution. As a result, the [solvent] will remain very close to constant even if it is used up or formed by a reaction occurring in the solvent.

Catalysts: No Shift

A **catalyst** is a substance that is not used up in a reaction, but when added to a reaction mixture increases the rate of the reaction. Adding a catalyst will *not shift* an equilibrium, but adding a catalyst will cause a reaction to reach equilibrium more quickly.

Driving a Reversible Reaction

Given a reversible reaction at equilibrium, if the system is opened and a reactant or product is allowed to escape (such as letting a gas product escape from a reaction in a solution), the reversible reaction is no longer in a closed system at equilibrium. In this case, the reaction and the mixture composition will shift toward the side that contains the escaping particle. This type of shift can be used to drive a reversible reaction toward a side with products that are wanted.

PRACTICE **C**

1. Design and practice flashcards that cover the rules in this lesson, then complete the problems below.

2. For this reaction at equilibrium:

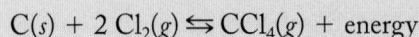

$$C(s) + 2\,Cl_2(g) \leftrightarrows CCl_4(g) + energy$$

   a. If the [Cl_2] is decreased:

      i. Equilibrium will shift to the (left or right?) _____.

      ii. [CCl_4] will (increase or decrease?) _____.   iii. [C] will _____.

      iv. The temperature in the reaction vessel will _____.

   b. If a catalyst is added, equilibrium will _____.

3. For the reaction at equilibrium:

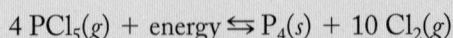

$$4\,PCl_5(g) + energy \leftrightarrows P_4(s) + 10\,Cl_2(g)$$

   If the temperature in the reaction vessel is decreased:

   a. Equilibrium will shift to the _____.

   b. [Cl_2] will _____.   c. [PCl_5] will _____.

   d. [P_4] will _____.

Lesson 20.3   Equilibrium Constants

At a fixed temperature, for the general reversible reaction

$$a\text{A} + b\text{B} \rightleftharpoons c\text{C} + d\text{D}$$

the **law of mass action** states that for a reaction at equilibrium, the ratio

$$\frac{[\text{C}]^c[\text{D}]^d}{[\text{A}]^a[\text{B}]^b} = \frac{\text{product of the [products]}}{\text{product of the [reactants]}}$$

is *constant*. This ratio is called the **equilibrium constant**, which is given the symbol K.

Example: For the reaction

$$\text{S}_2(g) + 2\,\text{H}_2(g) \rightleftharpoons 2\,\text{H}_2\text{S}(g)$$

the equilibrium constant **expression** *is*

$$K = \frac{[\text{H}_2\text{S}]^2}{[\text{S}_2][\text{H}_2]^2}$$

and the equilibrium constant **value** is 1.1×10^7 at 973 K.

The equilibrium constant **equation** is a combination of the K expression, the K value, and the temperature at which the value is recorded:

$$K = \frac{[\text{H}_2\text{S}]^2}{[\text{S}_2][\text{H}_2]^2} = 1.1 \times 10^7 \text{ at 973 K}$$

The equilibrium constant *expression* is the K ratio that shows the *symbols* for substance concentrations and their powers.

- The concentrations of the particles in the products are multiplied in the numerator.
- The concentrations of particles in the reactants are multiplied in the denominator.
- The *coefficient* of each substance in the balanced equation becomes the *power* of its concentration in the equilibrium constant expression.

As temperature changes, K values change, but the K expression stays the same.

Why *K* Is Important

One of the goals of science is to predict the results of change. Many processes important in chemical and biological systems involve reactions at equilibrium that can be subjected to change. Using Le Châtelier's principle, we can predict the *direction* in which a change will shift a system at equilibrium.

In addition, according to the law of mass action, if a system at equilibrium is shifted by changing the concentration of one of the reacting particles, the concentrations of other particles will change to keep the same K ratio. If we know the new concentration, by using K equations we can often precisely predict, at the new equilibrium, what the concentrations of the other particles will be.

PRACTICE   A

1. Define an equilibrium constant expression:

   a. In words.   b. Using an equation.

2. In an equilibrium constant expression, which terms are written in the denominator?

3. What is the difference between an equilibrium constant expression and an equilibrium constant equation?

4. In a K equation, if the temperature of a system at equilibrium is changed:

   a. What changes?   b. What does not change?

K Notation

For problems involving equilibrium constants, note these conventions for the symbols and abbreviations used in textbooks.

- The abbreviation for kelvins is K. The equilibrium constant is an italicized K.

- If a concentration is written in a K expression, it must be a concentration measured at equilibrium, written as $[X]_{\text{at equilibrium}}$ or $[X]_{\text{at eq}}$. As a simplification, the label *at equilibrium* is often left off, but you should assume that all concentrations in a K expression are measured *at equilibrium*.

Because K expressions can be written using units other than concentration, a K based on molar concentrations is often given the symbol K_c, but if no subscript after a K is written, assume that a K value and expression is a K_c based on molarity.

Equilibrium Constant Expressions

To write the K expression for a reaction, all that is needed is a balanced equation.

TRY IT

Q. For the Haber process reaction

$$N_2(g) + 3\,H_2(g) \leftrightharpoons 2\,NH_3(g)$$

write the equilibrium constant expression.

STOP

Answer:

$$K = \frac{[NH_3]^2}{[N_2][H_2]^3}$$

PRACTICE   B

Write the equilibrium constant expression for these reactions.

1. $CH_4(g) + 2\,O_2(g) \leftrightharpoons CO_2(g) + 2\,H_2O(g)$

2. $2\,C_4H_{10}(g) + 13\,O_2(g) \leftrightharpoons 8\,CO_2(g) + 10\,H_2O(g)$

Concentrations That Are Constant

The correlation between a balanced equation and its equilibrium constant expression is simple, but there is one important exception. By convention, if the concentration of a particle involved in a reaction is essentially constant, that constant value is included in the *value* of the equilibrium constant, and the number **1** is substituted for the concentration term in the K expression. The concentrations that are assigned a value of 1 in a K expression are the same as those that do not shift an equilibrium when applying Le Châtelier's principle.

To summarize:

- Only concentrations that can substantially *change* are included in K expressions.

- Solids, pure liquids, and solvents have concentrations that do not change significantly during reactions.

- In place of terms for the concentration of solids, pure liquids, and solvents, a 1 is substituted in the K expression.

> TRY IT

Q. Write the equilibrium constant expression for this reaction.

$$CaCl_2(s) \leftrightharpoons Ca^{2+}(aq) + 2\,Cl^-(aq)$$

Answer:

$$K = [Ca^{2+}]\,[Cl^-]^2$$

The balanced equation shows that the ions on the right are dissolved in water. The concentration of substances dissolved in a solvent can vary, and terms for concentrations that can vary are included in a K expression.

Because the reactant on the left is a solid, its concentration is constant, and a 1 is substituted for [a solid] in a K expression. When a 1 is in a denominator of a fraction, it is omitted as understood.

A 1 is omitted if it is alone in the denominator, but mathematically a 1 cannot be omitted if it is *alone* in the numerator (on top).

━━━━━━━━━━▶ TRY IT

Q. Write the equilibrium constant expression for this reaction.

$$BaO(s) + CO_2(g) \leftrightarrows BaCO_3(s)$$

🛑 **Answer:**

$$K = \frac{1}{[CO_2]}$$

Both sides of the equation have solids, and in the place of [solid], we substitute a value of **1**. Because the concentration of a gas can vary, the term for [gas] must be included in the K expression. Mathematically, a **1** must be written only if it is by itself in the numerator.

Substance States

The *phase* or *state* (gas or aqueous) of a substance must be known in the terms of a K expression. However, if the balanced equation that includes phases is written to accompany the K expression, the states can be omitted as understood in the K expression terms.

Water follows the rules for K equations, but the use of (*aq*) to represent the phase of particles in a solution makes water a somewhat special case.

- The "substitute 1 for constant concentration terms" rule means that a $[H_2O]$ term will be represented by a 1 in a K expression when the water is a solid (ice), a pure liquid, or a solvent. Water is the solvent when any of the other particles in a reaction are in the (*aq*) state.

- If the water is a reactant or product in its gas phase (as vapor or steam), its concentration can *vary* because all gases are compressible. A term for $[H_2O(g)]$ must therefore be included in the K expression.

To summarize, for water, the rule is:

> In K expressions; $[H_2O(g)]$ is included, but $[H_2O(s)]$ and $[H_2O(\ell)]$ are replaced by a **1**.

Flashcards

For this lesson, you need the following items firmly in memory: this lesson's shaded rules and vocabulary, plus the answers to the questions in Practice A. Design and practice flashcards, then complete the problem set below.

PRACTICE **C**

Write the equilibrium constant expression for each of these reactions.

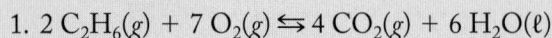

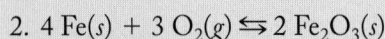

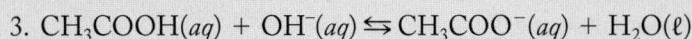

1. $2 C_2H_6(g) + 7 O_2(g) \leftrightarrows 4 CO_2(g) + 6 H_2O(\ell)$

2. $4 Fe(s) + 3 O_2(g) \leftrightarrows 2 Fe_2O_3(s)$

3. $CH_3COOH(aq) + OH^-(aq) \leftrightarrows CH_3COO^-(aq) + H_2O(\ell)$

Lesson 20.4   Equilibrium Constant Values

Lesson 20.4   Equilibrium Constant Values

K Values and the Favored Side

K values are derived from a fraction:

$$K = \frac{\text{product of the [products]}}{\text{product of the [reactants]}}$$

If an equilibrium *favors the products*, the concentrations in the numerator of the fraction will be relatively large compared to the values in the denominator, and the resulting value for K will be a positive number that is greater than 1, usually written in scientific notation with a positive power of 10.

If an equilibrium favors the reactants, the concentrations on top will be relatively small and on the bottom will be relatively large. The resulting K value will be a number between 0 and 1, usually expressed in scientific notation with a power of 10 that is negative.

> **Examples:** For these reactions at 25°C:
>
> A. $2\,N_2(g) + O_2(g) \leftrightarrows 2\,NO(g)$ $K = 1 \times 10^{-35}$
>
> B. $Ag^+(aq) + 2\,NH_3(aq) \leftrightarrows Ag(NH_3)_2^+(aq)$ $K = 1.7 \times 10^7$
>
> C. $HSO_4^-(aq) \leftrightarrows H^+(aq) + SO_4^{2-}(aq)$ $K = 0.013$

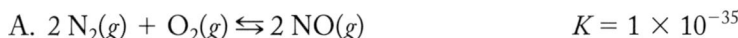

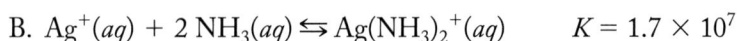

Reaction A has a K value that is much less than 1. At equilibrium, the substances on the left side (the reactants) will be favored.

Reaction B has a K value that is much greater than 1. At equilibrium, the substances on the right side (the products) will be favored.

Reaction C has a K value that is smaller than 1, favoring the left side, but compared to most K values, K is not far from 1. At equilibrium in reaction C, you would expect to find a more balanced mixture of reactants and products than in reaction A or B.

To summarize, for K values written in scientific notation:

- If K has a positive power of 10, equilibrium favors the products.
- If K has a negative power of 10, equilibrium favors the reactants.

PRACTICE   A

For the following reactions at 25°C:

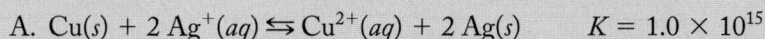

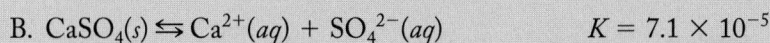

A. $Cu(s) + 2\,Ag^+(aq) \leftrightarrows Cu^{2+}(aq) + 2\,Ag(s)$ $K = 1.0 \times 10^{15}$

B. $CaSO_4(s) \leftrightarrows Ca^{2+}(aq) + SO_4^{2-}(aq)$ $K = 7.1 \times 10^{-5}$

C. $AgI(s) \leftrightarrows Ag^+(aq) + I^-(aq)$ $K = 1.5 \times 10^{-16}$

1. Write the K expression for each reaction.

2. Which equilibrium most favors the substances on the right side of the equation?

3. Which reaction will form the lowest concentrations of product?

K Values for Reversed Reactions

Equilibrium is the result of a reversible reaction. Reversible reactions can be written in either direction. For this reason, for every *K* value, the direction of the reaction must be shown.

> **Example:** The conversion of nitrogen dioxide to dinitrogen tetroxide is reversible. If we designate this as the forward reaction:
>
> $$2\ NO_2(g) \leftrightarrows N_2O_4(g) \qquad K_{forward} = \frac{[N_2O_4]}{[NO_2]^2}$$
>
> The reverse reaction is
>
> $$N_2O_4(g) \leftrightarrows 2\ NO_2(g) \qquad K_{reverse} = \frac{[NO_2]^2}{[N_2O_4]}$$
>
> The two equilibrium constant expressions are different, but related:
>
> $$K_{reverse} = \frac{[NO_2]^2}{[N_2O_4]} = \frac{1}{\frac{[N_2O_4]}{[NO_2]^2}} = \frac{1}{K_{forward}}$$

The relationship derived in the example:

$$K_r = 1/K_f$$

will be true for all reactions. Stated in words:

> At a given temperature, if a reaction is written in the reverse direction, its *K* value is the reciprocal of the original *K*.

Reciprocals of Exponential Notation

Recall that:

- The reciprocal of $X = 1/X = X^{-1}$
- When you take the reciprocal of an exponential term, you change the sign of the exponent.

    Examples: $1/10^5 = 10^{-5}$ The reciprocal of 10^{-7} is 10^7

On a calculator, there are two ways to calculate the reciprocal of a number:

- divide **1** by the number, or
- use a reciprocal key, usually labeled as $\boxed{1/x}$ or $\boxed{x^{-1}}$.

However, you also need to be able to estimate a reciprocal using "pencil and paper" math as a check on your calculator use.

Example: Solved on a calculator, the reciprocal of 1.8×10^{-8} is

$$1/(1.8 \times 10^{-8}) = \mathbf{5.6 \times 10^7}$$

A mental math *estimate* of the reciprocal of 1.8×10^{-8} is

$$\frac{1}{2 \times 10^{-8}} = \frac{1}{2} \times \frac{1}{10^{-8}} = 0.5 \times 10^{+8} = \mathbf{5 \times 10^7}$$

The two answers are *close*—a check on calculator use.

TRY IT

Q1. In this space, using mental arithmetic, *estimate* the reciprocal of 4.6×10^{12}. Write the final answer in scientific notation.

STOP

Answer:

One possibility is:

$$\frac{1}{5 \times 10^{12}} = \frac{1}{5} \times \frac{1}{10^{12}} = 0.2 \times 10^{-12} = \mathbf{2 \times 10^{-13}}$$

To review exponential mental math, see Lessons 1.3−1.5.

Q2. Using a calculator, find the reciprocal of 4.6×10^{12}.

STOP

Answer:

$\mathbf{2.2 \times 10^{-13}}$. The estimate and calculator answer are close.

All calculator answers must be checked, either by a different key sequence or (preferably) by a process or estimate that includes mental math.

PRACTICE   B

1. Design flashcards for the rule on *K* values for reversed reactions.

2. Without a calculator, write the reciprocal, expressed as a power of 10, of:

   a. 10^{-14} b. 100   c. 0.001

   Report numeric answers for the remaining questions in scientific notation.

3. Without a calculator, find the reciprocal of 3.0×10^{-9}.

(continued)

4. Without a calculator, *estimate* the reciprocal of 3.7×10^6.

5. With a calculator, calculate the reciprocal of 3.7×10^6. Is the answer close to your question 4 estimate?

6. For the reaction $N_2(g) + 2\,H_2O(g) \leftrightarrows 2\,NO(g) + 2\,H_2(g)$, the equilibrium constant value is calculated to be 1.6×10^{-3}.

   a. Which side is favored at equilibrium: products or reactants?

   b. What is the value of K for the reverse reaction at the same temperature?

7. For the reaction at 25°C:

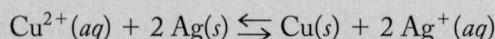

$$Cu^{2+}(aq) + 2\,Ag(s) \leftrightarrows Cu(s) + 2\,Ag^+(aq)$$

   a. Write the K expression.

   b. Write the K value for the reaction. Use the K data in Practice A of this lesson.

SUMMARY

Science tells us that when solving problems, there are two types of understanding. Experts in a field have "explicit understanding," meaning that they can both solve a problem and explain *why* they took the steps they did. But when using tools from related fields to solve problems within your field, your brain most often relies on "implicit understanding," which is a correct intuitive sense of what to do even if you cannot explain why.

The brain gains *implicit* understanding by practice and from exposure to concepts. Explicit understanding requires additional study and time. As one example, in learning to speak, your brain learns very complex rules of grammar and applies them effortlessly, even if you are unable to explain what many of the rules are. For useful work, knowing *how* to use tools such as math and language and computers is necessary; being able to explain *why* they work is less important if that is not the focus of your discipline.

Learning science requires a balance between implicit understanding of the many tools of math and science and the explicit understanding of topics in your major. Be patient! The explicit understanding of an expert develops gradually—over years of study and practice.

REVIEW QUIZ

For this quiz, you will not need a calculator or a periodic table.

1. Which of the following situations represents this reaction at equilibrium:

$$H_2O(s) + heat \leftrightarrows H_2O(\ell)$$

   a. A water-and-ice mixture in a well-insulated container.

   b. A cup of water.

   c. A melting cube of ice on a warm kitchen countertop.

2. For the reaction at equilibrium:

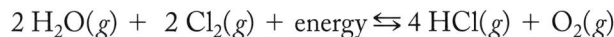

$$2\,H_2O(g)\; +\; 2\,Cl_2(g)\; +\; \text{energy} \leftrightarrows 4\,HCl(g)\; +\; O_2(g)$$

a. If the $[O_2]$ is increased:

    i. Equilibrium will shift to the (right or left?) _____.

    ii. [HCl] will (increase or decrease?)_____.

b. If the [HCl] is decreased:

    i. Equilibrium will shift to the _____.

    ii. $[O_2]$ will _____.

    iii. $[Cl_2]$ will _____.

    iv. The temperature in the reaction vessel will _____.

c. If the temperature in the reaction vessel is decreased:

    i. Equilibrium will shift to the _____.

    ii. $[O_2]$ will _____.

    iii. $[H_2O]$ will _____.

d. If a catalyst is added to the reaction mixture:

    i. Equilibrium will shift to the _____.

    ii. $[O_2]$ will _____.

3. Write the equilibrium constant expression for:

a. $Cl_2(g) + PCl_3(g) \leftrightarrows PCl_5(g)$

b. $Ca_3(PO_4)_2(s) \leftrightarrows 3\,Ca^{2+}(aq) + 2\,PO_4^{3-}(aq)$

4. Without a calculator, find the reciprocal of 4.0×10^{-12}.

5. For the following reactions at 25°C:

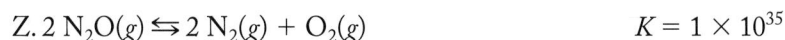

    X. $H_2O(\ell) \leftrightarrows H^+(aq) + OH^-(aq)$ $K = 1.0 \times 10^{-14}$

    Y. $CH_3COOH(aq) \leftrightarrows H^+(aq) + CH_3COO^-(aq)$ $K = 2 \times 10^{-5}$

    Z. $2\,N_2O(g) \leftrightarrows 2\,N_2(g) + O_2(g)$ $K = 1 \times 10^{35}$

a. Write the K expression for reactions X, Y, and Z.

b. Which equilibrium favors the products?

c. At the same temperature, what is the K value for reaction Y written in the reverse direction? (Answer without a calculator and in scientific notation.)

ANSWERS

Lesson 20.1

Practice

1. Increase the temperature or increase the concentration of one or more of the reactants.

2. The rate of the forward reaction must be equal to the rate of the reverse reaction.

3. The vessel must be sealed so that no particles enter or leave, and it must be insulated so that no energy enters or leaves.

Lesson 20.2

Practice A 2a. Equilibrium shifts **right**.   2b. $[N_2]$ will **decrease**.   2c. $[NH_3]$ will **increase**.

3a. Equilibrium shifts **left**.   3b. $[H_2]$ will **increase**.   3c. $[NH_3]$ will **decrease**.

Practice B 2a.i. Equilibrium shifts **right**.   2a.ii. $[CH_3OH]$ will **increase**.

2b.i. Equilibrium shifts **right**.   2b.ii. $[CH_3OH]$ will **increase**.   2b.iii. Temperature will **increase**.

3a.i. Equilibrium shifts **left**.   3a.ii. $[PCl_5]$ will **decrease**.

3b.i Equilibrium shifts **right**.   3b.ii. $[PCl_3]$ will **decrease**.   3b.iii. Temperature will **increase**.

Practice C 2a.i. Equilibrium shifts **left**.   2a.ii. $[CCl_4]$ will **decrease**.   2a.iii. $[C]$ will **not change** because C is solid.

2a.iv. Temperature will **decrease**.

2b. Catalysts do not shift the position of an equilibrium.

3a. **Left**.   3b. **Decrease**.   3c. **Increase**.   3d. [Solid] **will not change**.

Lesson 20.3

Practice A 1a. The product of the concentrations of the products divided by the product of the concentrations of the reactants.

1b. $K_{expression} = \dfrac{\text{product of the [products]}}{\text{product of the [reactants]}}$

or if $aA + bB \rightleftarrows cC + dD$, then $K_{expression} = \dfrac{[C]^c[D]^d}{[A]^a[B]^b}$

2. The product of the concentrations of the reactants.

3. The equation includes the K expression, the K numeric value, and the temperature.

4a. The K value changes.   4b. The K expression does not change.

Practice B   1. $K = \dfrac{[CO_2][H_2O]^2}{[CH_4][O_2]^2}$ 2. $K = \dfrac{[CO_2]^8[H_2O]^{10}}{[C_4H_{10}]^2[O_2]^{13}}$

Practice C   1. $K = \dfrac{[CO_2]^4}{[C_2H_6]^2[O_2]^7}$ 2. $K = \dfrac{1}{[O_2]^3}$ 3. $K = \dfrac{[CH_3COO^-]}{[CH_3COOH][OH^-]}$

Lesson 20.4

Practice A 1A. $K = \dfrac{[Cu^{2+}]}{[Ag^+]^2}$ 1B. $K = [Ca^{2+}][SO_4{}^{2-}]$ 1C. $K = [Ag^+][I^-]$

2. Reaction A, with the largest K value, most favors the right side (products).

3. Reaction C, with the K value much smaller than the others, will most favor the left side (reactants), and will form the smallest concentrations of products.

Practice B 2a. 10^{14} 2b. $100 = 10^2$, reciprocal $= 10^{-2}$ 2c. $0.001 = 10^{-3}$, reciprocal $= 10^3$

3. $\dfrac{1}{3.0 \times 10^{-9}} = \dfrac{1}{3} \times \dfrac{1}{10^{-9}} = 0.333 \times 10^{+9} = \mathbf{3.3 \times 10^8}$

4. One possible estimate: $\dfrac{1}{4 \times 10^6} = \dfrac{1}{4} \times \dfrac{1}{10^6} = 0.25 \times 10^{-6} = \mathbf{2.5 \times 10^{-7}}$

5. $1/(3.7 \times 10^6) = 2.703 \times 10^{-7} = \mathbf{2.7 \times 10^{-7}}$ (Close to the estimate.)

6a. Reactants, because the K value is much less than one.

6b. $K_{reverse} = \dfrac{1}{K_{forward}} = \dfrac{1}{1.6 \times 10^{-3}} = 0.625 \times 10^{+3} = \mathbf{6.3 \times 10^2}$

7a. $K_{expression} = \dfrac{[Ag^+]^2}{[Cu^{2+}]}$

7b. The equation is the reverse of reaction A in Practice A.

$$K_{reverse} = \dfrac{1}{K_{forward}} = \dfrac{1}{1.0 \times 10^{15}} = \mathbf{1.0 \times 10^{-15}}$$

Review Quiz

1. Only situation (a) is at equilibrium. In (b), not all of the reactants and products are present. In case (c), heat is entering the system, and at equilibrium the system must be closed to particles and heat entering or leaving.

2a.i. **Left.**   2a.ii. [HCl] will **decrease.**   2b.i. **Right.**   2b.ii. **Increase.**   2b.iii. **Decrease.**

2b.iv. Temperature will **decrease.**   2c.i. **Left.**   2c.ii. **Decrease.**   2c.iii. [H_2O *gas*] will **increase.**

2d.i. Equilibrium will **not shift.**   2d.ii. [O_2] will **not change.**

3a. $K = \dfrac{[PCl_5]}{[PCl_3][Cl_2]}$ 3b. $K = [Ca^{2+}]^3[PO_4^{3-}]^2$ 4. $0.25 \times 10^{12} = \mathbf{2.5 \times 10^{11}}$

5a.X. $K = [H^+][OH^-]$ 5a.Y. $K = \dfrac{[H^+][CH_3COO^-]}{[CH_3COOH]}$ 5a.Z. $K = \dfrac{[N_2]^2[O_2]}{[N_2O]^2}$

5b. Z favors products.   5c. $K_{reverse} = \dfrac{1}{2 \times 10^{-5}} = 0.5 \times 10^{+5} = \mathbf{5 \times 10^4}$

21

Equilibrium Calculations

Lesson 21.1   Powers and Roots of Exponential Notation

PRETEST   If you can solve these problems correctly, skip to Lesson 21.2. Express your final answers in scientific notation. Answers are at the end of the chapter.

1. Without a calculator, solve $(3.0 \times 10^3)^3 =$

2. Use a calculator. $(6.50 \times 10^{-4})^2 =$

3. Do not use a calculator. The cube root of $8.0 \times 10^{-24} =$

4. Use a calculator as needed: $(2.7 \times 10^{-14})^{1/3} =$

Taking Numbers to a Power

To solve calculations based on equilibrium constants, you will need to find powers and roots of exponential notation.

Many calculators have an x^2 key. To calculate both squares and higher powers, most calculators also have *power* functions labeled x^y or y^x or $(y)^x$ or $\wedge$.

To learn how to use the power keys, you may check your calculator manual online, experiment using simple examples for which you know the answers, or (preferably) do both.

TRY IT

Q1. Answer using mental math: $2^3 =$

STOP

Answer:

$2^3 = 2 \times 2 \times 2 = 8$

Now, with the name and model of your calculator, search online for its user manual. In the manual, find "taking a number to a power." (Online videos may be available as well.) Using the key sequence the online manual recommends, calculate 2^3 (and make sure the answer is 8). In this space, write the key sequence that solved this question correctly:

TRY IT

Q1. Using the calculator power keys, answer in fixed decimal notation. (To review notation, see Lesson 1.2.) In this lesson for purposes of significant figures, assume one- or two-digit numbers without a decimal are exact.

    a. $16^3 =$ b. $2.50^4 =$ c. $0.50^2 =$

STOP

Q2. Check your Q1 answers by using the calculator to multiply the number by itself for the number of times indicated by the power.

    a. $16^3 =$ b. $2.50^4 =$ c. $0.50^2 =$

STOP

Answers:

Both methods must result in:

    a. 4,096   b. 39.1   c. 0.25

Calculator operations should always be performed in two different ways as a check on your work.

Taking Exponential Notation to a Power

On a calculator, exponential notation can be taken to a power without entering the exponential term. Doing so is another way to check a calculator answer. Let's review the rules.

1. To take exponential terms to a power, multiply the exponents.

   Examples: $(10^3)^2 = 10^6$ $(10^5)^{-2} = 10^{-10}$

2. When taking exponential notation to a power, use number rules for numbers and exponential rules for exponents.

   Example: $(3.0 \times 10^4)^3 = (3.0)^3 \times (10^4)^3 = 27 \times 10^{12} = 2.7 \times 10^{13}$

➤ TRY IT

Without a calculator, write answers, then convert the answers to scientific notation.

Q1. $(5 \times 10^4)^2 =$

Q2. $(2 \times 10^{-3})^4 =$

STOP

Answers:

1. $(5)^2 \times (10^4)^2 = 25 \times 10^8 = $ **2.5×10^9**

2. $(2)^4 \times (10^{-3})^4 = 16 \times 10^{-12} = $ **1.6×10^{-11}**

For Q3 and Q4, use a calculator to take the significand to the power, use mental math to take the exponential to the power, and convert the final answer to scientific notation.

Q3. $(9.5 \times 10^6)^3 =$

Q4. $(3.4 \times 10^{-7})^4 =$

STOP

Answers:

3. $(9.5)^3 \times (10^6)^3 = 857 \times 10^{18} = 8.57 \times 10^{20} = $ **8.6×10^{20}**

4. $(3.4)^4 \times (10^{-7})^4 = 134 \times 10^{-28} = 1.34 \times 10^{-26} = $ **1.3×10^{-26}**

On a calculator, you can also enter *all* of the exponential notation and take it to a power. Search the online manual for "exponential notation to a power." Then, using the recommended key sequence, solve:

$(7.5 \times 10^{-2})^3 =$

Make sure the unrounded answer includes $4.2187... \times 10^{-4}$. You may want to adjust the calculator settings to display scientific notation (see the manual for assistance). Note how your calculator displays the exponent in the answer.

Record the key sequence that solved this problem correctly. To do so, *either* circle one of the following sequences that worked:

- A standard TI-type calculator *may* use: 7.5 ⌈E or EE⌉ 2 ⌈+/−⌉ ⌈y^x⌉ 3 ⌈=⌉
- A graphing calculator *might* use: 7.5 ⌈EE⌉ ⌈(−)⌉ 2 ⌈enter⌉ ⌈^⌉ 3 ⌈enter⌉
- An RPN calculator may use: 7.5 ⌈E or EE or EXP⌉ 2 ⌈+/−⌉ ⌈enter⌉ 3 ⌈y^x⌉

OR write in this space the key sequence that worked on *your* calculator:

► TRY IT

Solve each of the following problems by entering both the significand and exponential term into the calculator. Write the calculator answer including the first five reported digits in the significand, then round the final answer to proper significant figures.

Q1. $(2.1 \times 10^6)^4 =$

Q2. $(3.92 \times 10^{-2})^3 =$

Answers:

1. $(2.1 \times 10^6)^4 = 1.9448 \times 10^{25} = \mathbf{1.9 \times 10^{25}}$

2. $(3.92 \times 10^{-2})^3 = 6.0236 \times 10^{-5} = \mathbf{6.02 \times 10^{-5}}$

To check your calculator answers, solve Q1 and Q2 again. This time, use the calculator to take the significand to the power, use mental math to take the exponential to the power, combine the two parts, and convert to scientific notation.

Q3. $(2.1 \times 10^6)^4 =$

Q4. $(3.92 \times 10^{-2})^3 =$

Answers:

3. $= (2.1)^4 \times (10^6)^4 = 19.45 \times 10^{24} = 1.945 \times 10^{25} = \mathbf{1.9 \times 10^{25}}$

4. $= (3.92)^3 \times (10^{-2})^3 = 60.236 \times 10^{-6} = 6.0236 \times 10^{-5} = \mathbf{6.02 \times 10^{-5}}$

"Exponents by mental math" should result in the same answers as "all on the calculator."

PRACTICE   A

Report final answers in scientific notation.

1. Calculate using "significands on the calculator, but exponents by mental math."

    a. $(7.7 \times 10^4)^4 =$

    b. $(5.52 \times 10^{-2})^3 =$

2. Solve by a method that enters both the significand and exponential term into the calculator.

    a. $(7.7 \times 10^4)^4 =$

    b. $(5.52 \times 10^{-2})^3 =$

3. Did the answers using the two different methods agree?

Roots

Because 3 cubed is $3 \times 3 \times 3 = 27$, the **cube root** of 27 is 3.

Mathematically, taking the root of a quantity is the same as assigning the quantity the **reciprocal exponent** of the root. In symbols:

$$\sqrt[y]{x} \equiv x^{1/y}$$

Examples:

- The *cube root* of x can be written as $\sqrt[3]{x}$ or as $x^{1/3}$.
- $16^{1/4}$ and $\sqrt[4]{16}$ both mean the *fourth root* of 16.

<hr>

━━━━━━▶ TRY IT

Q. Calculate by mental math: The cube root of $8 = 8^{1/3} =$

STOP

Answer:

$8 = 2 \times 2 \times 2$, so $8^{1/3} = $ **2**

Now search online with the name and model of your calculator and "taking a root." Using the key sequence recommended by the manual, calculate the cube root of 8. Try to find *two* recommended key sequences that result in the answer **2**. Write the sequences below:

To take a cube root of 8: _____

and _____.

STOP

Answer:

The right answer is whatever works on your calculator, but circle any of the sequences below that work.

- On a standard TI calculator, try **8** (2nd *or* INV) (y^x) **3** (=) and/or **8** (y^x) (() **1** (÷) **3** ()) (=) and/or **8** (y^x) **0.33333333** (=)
- Graphing: Try **8** (^) (() **1** (÷) **3** ()) (enter) and/or **8** (^) **3** (1/x *or* x^{-1}) (enter)
- In RPN, try **8** (enter) **3** (1/x) (y^x) and/or **8** (enter) **0.33333** (y^x)

Try to follow the logic of two sequences that work on your calculator.

PRACTICE B

From the sequences written or circled above, choose two that make sense to you, then apply them to the following problems. In this practice set, assume these are exact numbers and answer in fixed decimal notation.

1. $\sqrt[4]{16}$ By method 1 =   By method 2 =

2. $125^{1/3}$ By method 1 =   By method 2 =

3. The cube root of 0.008   By method 1 =   By method 2 =

Roots of Exponential Notation

A calculator can find a root by entering the entire exponential notation term. Search your manual for "root of exponential notation," then calculate:

$(8.04 \times 10^{-5})^{1/3} =$

The unrounded answer should begin as $4.316\ldots \times 10^{-2}$. Write the key sequence:

If any of the key sequences below also work, circle them.

- On a standard TI: Try **8.04** ⎡EE⎤ **5** ⎡+/−⎤ ⎡2nd *or* INV⎤ ⎡y^x⎤ **3** ⎡=⎤ and try **8.04** ⎡E *or* EE *or* EXP⎤ **5** ⎡+/−⎤ ⎡y^x⎤ ⎡(⎤ 1 ⎡÷⎤ 3 ⎡)⎤ ⎡=⎤
- Graphing: Try **8.04** ⎡EE⎤ ⎡(−)⎤ **5** ⎡enter⎤ ⎡^⎤ ⎡(⎤ 1 ⎡÷⎤ 3 ⎡)⎤ ⎡enter⎤
- In RPN, try **8.04** ⎡E *or* EE *or* EXP⎤ **5** ⎡+/−⎤ ⎡enter⎤ 3 ⎡1/x⎤ ⎡y^x⎤

(All positive numbers have two even roots, positive and negative [±], but in this chapter, we will write all roots as positive values.)

PRACTICE   C

Solve these using *one* calculator method. Express answers in fixed decimal notation.

1. $\sqrt[5]{0.01024} =$

2. $(6.20 \times 10^4)^{1/8} =$

Roots of Divisible Powers of 10

To check roots of exponential notation, it is important to be able to take a root in *part* without a calculator.

- Roots of exponential terms (10^x) can be taken *without* a calculator if the power of 10, when multiplied by the reciprocal exponent, results in an *integer*.

Another way to say this:

- The root of 10^x can be found without the calculator if x is *evenly divisible* by **2** to find a *square* root, by **3** to find a *cube* root, etc.

To calculate the root of an evenly divisible power of 10, use these steps.

1. Write the root as a reciprocal exponent.

2. Apply the rule: To take an exponential term to a power, multiply the exponents.

   Example: The cube root of $10^{-9} = (10^{-9})^{1/3} = \mathbf{10^{-3}}$

To calculate the root of a value in exponential notation, if the exponent is *evenly divisible* by the root, take the root of the significand on the calculator if needed, then take the root of the exponential term by mental math.

▶ TRY IT

Q1. Without a calculator, solve: $(8.0 \times 10^{15})^{1/3} =$

Answer:

$$(8.0 \times 10^{15})^{1/3} = (8.0)^{1/3} \times (10^{15})^{1/3} = \mathbf{2.0 \times 10^5}$$

Q2. Use a calculator for the root of the significand, but solve the exponential root by mental arithmetic. Answer in scientific notation.

$$\sqrt[3]{68.4 \times 10^{-12}} =$$

Answer:

$$(68.4 \times 10^{-12})^{1/3} = (68.4)^{1/3} \times (10^{-12})^{1/3} = 4.089 \times 10^{-4} = \mathbf{4.09 \times 10^{-4}}$$

PRACTICE   D

1. Do not use a calculator. Write answers as 10 to a power.

   a. The square root of $10^{12} =$

   b. $\sqrt[4]{10^{-24}} =$

2. Use a calculator for the root of the significand, but solve the exponential root by mental arithmetic. Answer in scientific notation with proper significant figures.

   a. $(97.7 \times 10^{-35})^{1/5} =$

   b. $\sqrt[4]{8.35 \times 10^{-20}} =$

3. Which root is equivalent to $x^{0.125}$?

4. Without a calculator: $81^{0.25} =$

When a value in exponential notation does not have an evenly divisible power, the exponent can be made divisible by moving the decimal.

4. To take the root of a value in exponential notation with a power that is not evenly divisible by the root:

   a. Make the exponent *smaller* until the exponent times the reciprocal power results in a positive or negative *whole number*. Adjust the significand to keep the same numeric value.

   b. Take the root of the significand on the calculator, then take the root of the exponential term by mental math.

   Example: The fourth root of $3.11 \times 10^{10} =$

   $$(3.11 \times 10^{10})^{1/4} = (311 \times 10^{8})^{1/4} = (311)^{1/4} \times (10^{8})^{1/4} = \mathbf{4.20 \times 10^{2}}$$

TRY IT

Q. Use a calculator for the significand, but solve the exponential root by mental arithmetic. Answer in scientific notation.

$$(55.1 \times 10^{-11})^{1/3} =$$

Answer:

$$(551 \times 10^{-12})^{1/3} = (551)^{1/3} \times (10^{-12})^{1/3} = 8.198 \times 10^{-4} = \mathbf{8.20 \times 10^{-4}}$$

5. To take the root of a power of 10 (a 10^x alone) that is not evenly divisible without entering the exponential term into the calculator, place a "1 ×" in front of the exponential term, then follow step 4 above.

Example: $\sqrt{10^{-7}} = (\mathbf{1} \times 10^{-7})^{1/2} = (\mathbf{10} \times 10^{-8})^{1/2} = \mathbf{3.16 \times 10^{-4}}$

In these lessons, round roots of non-divisible powers of 10 to 3 *s.f.*

▶ TRY IT

Q1. In your notebook, without entering the exponent into the calculator, take the cube root of 10^{17}.

(STOP)

Answer:

$$(10^{17})^{1/3} = (\mathbf{1} \times 10^{17})^{1/3} = (\mathbf{100} \times 10^{15})^{1/3} = (100^{1/3} \times 10^5) = \mathbf{4.64 \times 10^5}$$

Q2. Now take the cube root of 10^{17} "all on the calculator." You *may* need to enter $\mathbf{1} \times 10^{17}$ to take the root. Compare your answer to the answer above.

If your method to check a calculator answer includes some mental math, you will have a better sense of what is a reasonable answer. In addition, knowing the mental math, the steps to take when using the calculator will be easier to remember because they make more sense.

PRACTICE   E

Solve in your notebook. Do as many as you need to feel confident, but be sure to do problems 3a and 3b.

1. Solve first "all on the calculator," and then "significands on the calculator but exponentials by mental math."

   a. $(6.0 \times 10^{23})^{1/3} =$ b. $\sqrt[4]{10^{-21}} =$ c. The cube root of $1.25 \times 10^{-7} =$

2. Use a calculator as needed, but solve each problem using two different methods.

   a. $(0.25)^3 =$ b. $\sqrt[5]{2.37 \times 10^{-16}} =$ c. $(4.7 \times 10^{-4})^3 =$ d. $(3.3 \times 10^{-3})^{1/9} =$

3. Solve in your notebook:

   a. If $\dfrac{[X]^2}{(0.020)} = 0.125$; find $[X]$ in fixed decimal notation.   b. $\dfrac{(0.040)^2(0.10)^6}{(0.020)^2} =$

Lesson 21.2   Equilibrium Constant Calculations

In theory, all reactions that occur are reversible, and all reactions go to equilibrium. In practice, many equilibria favor the products so much that the reaction is considered to go to completion: The initial amount of limiting reactant decides how much of each reactant is used up and how much of each product forms, and calculations can be solved by conversion stoichiometry.

When a reaction goes only *partially* to completion, there is no limiting reactant. In such a case, solving of how much of each reactant reacts and how much of each product forms requires the use of a K equation.

Calculations Based on *K* Equations

To organize our work for reactions that go to equilibrium, our rule will be:

If a calculation involves K and concentrations, write the **WRECK** steps, then solve for the symbol WANTED.

The steps are:

1. **W** (WANTED): Write the WANTED unit and/or symbol.
2. **R** (*R*eaction): Write a balanced equation for the *R*eaction.
3. **E** (*E*xtent): Write the *E*xtent, such as "(Goes partially)" or "(Goes ~ 100%)."
4. **C**: List the *C*oncentrations at equilibrium for each particle in the reaction.
5. **K**: Write the *K* equation.
6. SOLVE the *K* equation in symbols for the term that *includes* the WANTED symbol.
7. Substitute numbers *without units*, then solve for the symbol WANTED.
8. If finding a concentration, add the unit *mol/L* (M) to the answer.

▶ TRY IT

Apply the *WRECK* steps to the following problem. If you get stuck, read a bit of the answer until you are unstuck, then complete your work.

Q. For a reversible reaction

$$H_2(g) + I_2(g) \leftrightarrows 2\,HI(g)$$

concentrations at equilibrium are $[H_2] = 0.020$ M and $[I_2] = 0.32$ M, and $K = 25$. Find $[HI]$ at equilibrium.

Answer:

1. **W**: (WANTED): ? = $[HI]_{eq.}$ in mol/L
2. **R**: $H_2(g) + I_2(g) \leftrightarrows 2\,HI(g)$ (Goes partially.)
3. **E**: (Extent: If a reaction goes to equilibrium, it *goes partially* to completion.)
4. **C**: $[H_2]_{eq.} = 0.020$ M

    $[I_2]_{eq.} = 0.32$ M

    $[HI]_{eq.} = ?$ M
5. **K**: $K = \dfrac{[HI]^2}{[H_2][I_2]} = 25$ at equilibrium.

6. $[HI]^2 = K[H_2][I_2]$

7, 8. $[HI]^2 = (25)\,(0.020)\,(0.32) = 0.160$ (Carry extra *s.f.* until final step.)

$[HI]$ = square root of $[HI]^2 = ([HI]^2)^{1/2} = (0.160)^{1/2} = \textbf{0.40 mol/L}$

Significant figures in K calculations: All numbers in the original data had 2 *s.f.*, so the answer is rounded to 2 *s.f.* Because coefficients are exact, powers based on coefficients are exact and do not affect the number of digits reported.

To summarize:

> If a calculation involves K and concentrations:
>
> - Write the **WRECK** steps.
> - Solve the K equation for the term that includes the WANTED symbol.
> - Substitute numbers without units to SOLVE for the WANTED symbol.
> - If a concentration is WANTED, add M to the answer.

Units When Solving with *K* Equations

K equations include concentrations that have units, but K values listed as equal to the K expression are numbers without units. Where did the units go?

Similar to the dilution equation studied in Lesson 12.1, the K equation is a simplification derived from more-complex relationships. K expressions are actually based on a quantity called **activity** that does not have units. Concentration is related to activity but is easier to measure, so we use concentration as a practical way to solve calculations involving K. Units are omitted because concentrations and K values are based on activity.

For calculations based on "real" relationships in science, units must cancel, but because the K equation is a simplification, units may not cancel properly and unit cancellation will not catch mistakes. For now, remember:

> For the step at which a K equation is used:
>
> - K values and concentrations do not include units.
> - When solving for concentration, mol/L must be added to the answer.
> - Double-check your algebra and substitution of numbers for symbols.

PRACTICE

1. In your notebook, write an abbreviated version of the shaded summary lists in this lesson. Practice writing your summaries until you can do so from memory, then try the problems below.

2. Given the reaction

$$2A + B \rightleftharpoons 4C \qquad \text{(All gases.)}$$

if K = 0.020, and at equilibrium [A] = 0.050 M and [B] = 0.125 M, find [C].

3. For the reaction

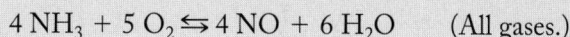

$$4\,NH_3 + 5\,O_2 \leftrightharpoons 4\,NO + 6\,H_2O \qquad \text{(All gases.)}$$

the equilibrium concentrations are [NH_3] = 0.050 M, [O_2] = 0.0020 M, [NO] = 0.50 M, and [H_2O] = 0.20 M. What is the value of K at this temperature?

Lesson 21.3   *RICE* Tables

So far in our K problems, the concentrations *at equilibrium* needed to solve K equations have been supplied. However, if concentrations are known for a mixture of *reactants initially* (with no products yet formed), as well as for any *one* reactant or product after the reaction has reached equilibrium, concentrations at equilibrium for all particles in the reaction (and a value for K) can be calculated.

To solve these equilibrium calculations, we will use a "chemistry accounting system" that we will call a *RICE* table.

- A *RICE* table has four rows.
- In column 1, the rows are labeled **R**eaction, **I**nitial, **C**hange, **E**quilibrium.
- Each reactant and product is placed in its own column.
- A double line separates reactants from products.

Let's illustrate this type of table with a problem.

► TRY IT

Q. The morning chemistry lab assistant is filling lab drawers. The initial inventory contains 95 burners, 220 racks, and 2,500 test tubes. Into each drawer is placed one Bunsen burner, two test-tube racks, and 20 test tubes. When the afternoon assistant arrives, she finds 60 racks in the inventory. How many drawers were filled? How many burners and test tubes remain in the inventory at the end of the process?

To solve, follow these steps. First, balance this "equation" for the process:

_____ burners + _____ racks + _____ test tubes → **1** drawer

Next, without a calculator, fill in all of the blanks and empty cells in this table.

Reaction/Process	_____ Burner	_____ Racks	_____ Test tubes	_____ Drawer
Initial Count				
Change (use + and −)				
At End/Equilibrium				

Answer:

Based on the initial data:

Reaction/Process	1 Burner	2 Racks	20 Test tubes	1 Drawer
Initial Count	95	220	2,500	0
Change (use + and −)		*−160*		
At End/Equilibrium		60		

Calculate the one *change* for which there is sufficient data. From that number, use the ratios of the process (in the **R** row) to complete the *Change* row. In that row, use a − *sign* for components used up and a + sign for those formed. Then calculate the amount of each component present at the end.

Adjust your work if needed and complete the table.

Reaction/Process	1 Burner	2 Racks	20 Test tubes	1 Drawer
Initial Count	95	220	2,500	0
Change (use + and −)	**−80**	−160	**−1,600**	**+80**
At End/Equilibrium	**15**	60	**900**	**80**

The bottom row answers the question. Note a key to the table:

> The *ratios* in row 1 (**R**eaction) must match the ratios in row 3 (**C**hange). The coefficients of the process determine the **C**hange ratios.

This same method can be used to find moles of chemical substances present at equilibrium.

TRY IT

Without a calculator, fill in the *RICE* table to solve this problem.

Q. In a Haber process experiment, initially 6.00 mol N_2, 1.00 mol H_2, and no moles of NH_3 are mixed. At equilibrium, 5.80 mol N_2 remains. How many moles of H_2 and NH_3 are present at equilibrium? The reaction is:

$$N_2 + 3\,H_2 \rightleftharpoons 2\,NH_3 \quad \text{(All gases.)}$$

Reaction	____ N_2	____ H_2	____ NH_3
Initial			
Change (use + and −)			
Equilibrium			

Answer:

Reaction	1 N$_2$	3 H$_2$	2 NH$_3$
Initial	**6.00 mol**	**1.00 mol**	**0 mol**
Change (use + and −)	*−0.20 mol*		
Equilibrium	**5.80 mol**		

- Calculate the *change* that you can.
- From the one known **C**hange and the ratios of reaction (the coefficients), calculate the other **C**hanges.
- Calculate the **E**quilibrium row.

Reaction	1 N$_2$	3 H$_2$	2 NH$_3$
Initial	6.00 mol	1.00 mol	0 mol
Change (use + and −)	−0.20 mol	**−0.60 mol**	**+0.40 mol**
Equilibrium	5.80 mol	**0.40 mol**	**0.40 mol**

PRACTICE

1. Design and practice flashcards for the shaded rules in this lesson.

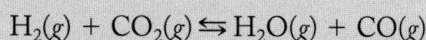

2. For the reaction

$$H_2(g) + CO_2(g) \leftrightarrows H_2O(g) + CO(g)$$

the initial mixture is composed of 2.00 mol H$_2$ and 1.00 mol CO$_2$. At equilibrium, 0.30 mol CO gas are found. Calculate the moles of the other substances present at equilibrium.

Lesson 21.4   *RICE* Tables and *K* Calculations

Which *units* can be used in a *RICE* table?

1. *Moles* can always be used. *RICE* calculations are based on coefficients, and coefficients can always be read as moles.

2. *Concentrations* (in mol/L) can also be used in *RICE* tables *if* all of the moles in a problem are measured in the *same volume*.

   Why? Coefficients are mole ratios. However, if all of the moles are contained in the same volume, dividing each of the moles by that volume will not change the ratios. The mole and the mol/L *ratios* will be the same.

Unless otherwise noted, you should assume that in equilibrium calculations,

- a gas equilibrium is in a container of constant volume, and
- particles dissolved in a solution are in a constant volume.

In those cases, the units in *RICE* tables can be all moles *or* all moles per liter.

3. Numbers in *RICE* tables below row 1 must have units attached, and all units must be the same.

If the data in a *RICE* table is moles per liter or can be converted to moles per liter, the values in the bottom row of the table can be substituted into the *K* equation for the reaction.

When to Use *WRECK* and *RICE*

In which problems will you need a *RICE* table? The purpose of the table is to find concentrations at equilibrium that can be substituted into the *K* equation.

- For problems involving reactions that go to equilibrium, if all of the known and WANTED values are concentrations *at equilibrium*, you can solve using the *WRECK* steps without a *RICE* table.

- For reactions that go to equilibrium in which more than one of the concentrations *at equilibrium* is *not* known, you will usually need a *RICE* table to gather the concentrations needed for the *K* equation.

- Usually, if a calculation supplies the *initial* moles or concentrations plus *one* value at equilibrium, it will require a *RICE* table.

⬛ ✏️━━━━➤ **TRY IT**

For the reaction in a sealed container at a constant temperature:

$$H_2 + I_2 \leftrightharpoons 2\,HI \qquad \text{(All gases.)}$$

initial concentrations are 0.050 M H_2, 0.050 M I_2, and no HI. At equilibrium, $[H_2] = 0.010$ M.

Q1. Calculate the concentrations of all of the substances present at equilibrium.

🛑

Answer:

WANTED: Concentrations of H_2, I_2, and HI at equilibrium.

Strategy: When *initial* moles or mol/L of reactants and products are supplied and values *at equilibrium* are WANTED, solve with a *RICE* table.

Reaction	1 H_2	1 I_2	2 HI
Initial	0.050 M	0.050 M	0 M
Change (use + and −)	−0.040 M	−0.040 M	+0.080 M
At Equilibrium	0.010 M	0.010 M	0.080 M

In *RICE* tables, units must be stated and must all be the same. The bottom row answers Q1.

Q2. Calculate the value of *K* under the reaction conditions.

STOP

Answer:

For calculations involving concentrations and *K*, write the *WRECK* steps.

1. *W*ANTED: *K*

2. *R*eaction: $H_2 + I_2 \leftrightarrows 2\,HI$ (All gases.)

3. *E*xtent: Goes to equilibrium. Use a *K* equation to solve.

4. *C*oncentrations at equilibrium: See bottom row of *RICE* table.

5. *K*: $K = \dfrac{[HI]^2}{[H_2][I_2]}$ (All measured at equilibrium.)

6. SOLVE: $K = \dfrac{(0.080)^2}{(0.010)^2} = \dfrac{(8.0 \times 10^{-2})^2}{(10^{-2})^2} = \dfrac{64 \times 10^{-4}}{10^{-4}} = \mathbf{64}$ or $\mathbf{6.4 \times 10^1}$

At the point when concentrations are substituted into a *K* equation, the units are omitted, but until that step, the units must be included with all DATA except for *K* values. *K* does not have units.

PRACTICE

1. In your notebook, summarize the shaded rules above, convert the rules to flashcards and practice their recall, then complete the following problems.

2. For the unbalanced reaction equation:

$$SO_2(g) + O_2(g) \leftrightarrows SO_3(g)$$

in a sealed glass vessel with temperature held constant, the initial gas mixture contains $[SO_2] = 0.0100$ M, $[O_2] = 0.0030$ M, and no SO_3. At equilibrium, $[SO_3] = 0.0040$ M.

   a. Balance the equation using lowest-whole-number coefficients (then check your answer before continuing).

   b. Calculate the concentrations of all reactants and products at equilibrium.

   c. Calculate a value for the equilibrium constant under these conditions.

   d. Calculate *K* for the reverse reaction at the same temperature.

3. For this reversible reaction of gases in a sealed glass vessel at a constant temperature:

$$2\,HI \leftrightarrows H_2 + I_2$$

initially the vessel contains $[HI] = 0.220$ M but no H_2 or I_2. When the system reaches equilibrium, $[H_2] = 0.010$ M. Calculate *K*. (Try completing this problem without using a calculator.)

SUMMARY

One strategy for solving math and science problems is to apply a stepwise procedure (an algorithm) often identified with a phrase or mnemonic. In these lessons, our procedures have included Wanted/Data/Solve, *WDBB, REC, WRECK,* and *RICE.* Why do we focus on algorithms?

During the initial study of math and science, most problems are "well structured," meaning that they have clear "right answers." Scientists who study the brain agree that to solve a well-structured problem that has some complexity, a sequenced procedure is required as a way around limits in working memory. The procedure should be applicable in clearly defined situations and be easy to remember. Phrases and mnemonics can serve as "memory devices" that help in recalling the steps of procedures.

A part of learning science is building a library of facts, algorithms, situational cues, and memory devices in your long-term memory. With practice, procedures become automated and can effortlessly be applied to problems. In science, solving problems is the goal.

REVIEW QUIZ

1. Use a calculator as needed, but solve using two different methods.

   a. $(1.6 \times 10^{-11})^{1/4} =$

   b. $(2.0 \times 10^5)^6 =$

   c. $(10^{-6})^{1/5} =$

   d. The sixth root of $9.5 \times 10^{15} =$

2. (Try without a calculator.) Given the reaction

$$3\,A + B \leftrightarrows 2\,C \qquad \text{(All gases.)}$$

   if $K = 0.020$, and at equilibrium $[B] = 0.40$ M and $[C] = 0.0080$ M, find $[A]$.

3. For this system in a sealed container at equilibrium:

$$CO_2(g) + H_2(g) \leftrightarrows CO(g) + H_2O(\ell)$$

   gas concentrations are 0.020 M CO_2 and 0.045 M H_2. If the value of the equilibrium constant is 4.8, find $[CO]$ in the mixture.

4. (Try without a calculator.) At a fixed temperature, for the reaction:

$$2\,NO + Cl_2 \leftrightarrows 2\,NOCl \qquad \text{(All gases.)}$$

   0.100 M NO and 0.150 M Cl_2 are originally mixed in a sealed glass vessel. At equilibrium, $[NO] = 0.050$ M.

   a. Calculate the $[Cl_2]$ and $[NOCl]$ present at equilibrium.
   b. Calculate the value for K under the above conditions.

ANSWERS

Some answers are partial. Your work should show each of the steps in Try It examples.

Lesson 21.1

Pretest   1. 2.7×10^{10} 2. 4.23×10^{-7} 3. 2.0×10^{-8} 4. 3.0×10^{-5}

Practice A

1a and 2a. $(7.7 \times 10^4)^4 = 3{,}515 \times 10^{16} = $ **3.5×10^{19}**

1b and 2b. $(5.52 \times 10^{-2})^3 = 168.2 \times 10^{-6} = 1.682 \times 10^{-4} = $ **1.68×10^{-4}**

Practice B   1. 2   2. 5   3. 0.2

Practice C   1. $\sqrt[5]{0.01024} = $ **0.4000**   2. $(6.20 \times 10^4)^{1/8} = $ **3.97**

Practice D   1a. 10^6 1b. 10^{-6} 2a. $(97.7)^{1/5} \times (10^{-35})^{1/5} = $ **2.50×10^{-7}**

2b. $(8.35)^{1/4} \times (10^{-20})^{1/4} = $ **1.70×10^{-5}**   3. $x^{0.125} = x^{1/8} = $ the **eighth** root

4. $81^{0.25} = 81^{1/4} = (81^{1/2})^{1/2} = (9)^{1/2} = $ **3.0**

Practice E   Solve using any two methods you choose.

1a. $(6.0 \times 10^{23})^{1/3} = (600 \times 10^{21})^{1/3} = (600^{1/3} \times 10^7) = $ **8.4×10^7**

1b. $(10^{-21})^{1/4} = (1{,}000 \times 10^{-24})^{1/4} = (1{,}000)^{1/4} \times (10^{-24})^{1/4} = $ **5.62×10^{-6}**

1c. $(1.25 \times 10^{-7})^{1/3} = (125 \times 10^{-9})^{1/3} = (125^{1/3} \times 10^{-3}) = $ **5.00×10^{-3}**

2a. 0.016   2b. 7.50×10^{-4} 2c. 1.0×10^{-10} 2d. $0.53 = 5.3 \times 10^{-1}$

3a. $[X]^2 = (0.020)(0.125) = 2.5 \times 10^{-3}; [X] = (2.5 \times 10^{-3})^{1/2} = $ **0.050**

3b. $\dfrac{(0.040)^2 (0.10)^6}{(0.020)^2} = \dfrac{(1.6 \times 10^{-3})(1.0 \times 10^{-6})}{(4.0 \times 10^{-4})} = \dfrac{1.6}{4.0} \times \dfrac{10^{-9}}{10^{-4}} = 0.40 \times 10^{-5} = $ **4.0×10^{-6}**

Lesson 21.2

Practice   2. For calculations involving K and concentrations, write the *WRECK* steps.

WANTED: **[C] = ?**

*R*xn: $2A + B \leftrightarrows 4C$ (All gases.)

*E*xtent: Goes to equilibrium, use K.

*C*onc@Eq.:   (See list in problem. If DATA is clearly labeled in the problem, there is no need to rewrite it. If the labels on the DATA in a word problem are not clear, list the values and assign symbols before substituting into an equation.)

K: $K = \dfrac{[C]^4}{[A]^2 [B]}$

SOLVE: First solve K for the term that *includes* the WANTED symbol.

$[C]^4 = K \cdot [A]^2 \cdot [B] = (0.020)(0.050)^2 (0.125) = 6.25 \times 10^{-6}$

Then solve for the WANTED symbol.

$[C] = (6.25 \times 10^{-6})^{1/4} = (625 \times 10^{-8})^{1/4} = $ **5.0×10^{-2} M** = **0.050 M**

3. For calculations involving K and concentrations, write the *WRECK* steps.

WANTED: $? = K$

*R*xn: (See problem.)

*E*xtent: Goes to equilibrium, use K equation.

*C*onc@Eq.:   (See list in problem.)

K: $K = \dfrac{[NO]^4 [H_2O]^6}{[NH_3]^4 [O_2]^5}$ (Because this H_2O is a gas, it is included in the K equation.)

SOLVE: The equation as written solves for the WANTED symbol.

$$? = K = \frac{[0.50]^4[0.20]^6}{[0.050]^4[0.0020]^5} = \frac{(5.0 \times 10^{-1})^4(2.0 \times 10^{-1})^6}{(5.0 \times 10^{-2})^4(2.0 \times 10^{-3})^5} = \frac{(625 \times 10^{-4})(2.0)^6 \times 10^{-6}}{(625 \times 10^{-8})(2.0)^5 \times 10^{-15}}$$

$$= \frac{(2.0)^6 \times 10^{-10}}{(2.0)^5 \times 10^{-23}} = \mathbf{2.0 \times 10^{13}} \qquad (K \text{ values are written without units.})$$

The arithmetic may be done in any way that results in a correct answer.

Lesson 21.3

Practice   2. WANTED: Moles of H_2, CO_2, and H_2O at equilibrium.

Strategy: To find values at equilibrium when only one item of data at equilibrium is supplied, use a *RICE* table.

Initial data:

Reaction	1 H$_2$	1 CO$_2$	1 H$_2$O	1 CO
Initial	2.00 mol	1.00 mol	0	0
Change (+ and −)				+0.30 mol
At Equilibrium				0.30 mol

Calculate the change row based on coefficients, then values at equilibrium.

Reaction	1 H$_2$	1 CO$_2$	1 H$_2$O	1 CO
Initial	2.00 mol	1.00 mol	0 mol	0 mol
Change (+ and −)	−0.30 mol	−0.30 mol	+0.30 mol	+0.30 mol
At Equilibrium	1.70 mol	0.70 mol	0.30 mol	0.30 mol

Lesson 21.4

Practice   2a. $2 SO_2(g) + O_2(g) \rightleftharpoons 2 SO_3(g)$

2b. WANTED: Concentrations at equilibrium for all three substances, in mol/L.

DATA: Based on initial data, measurements at equilibrium are WANTED. Use a *RICE* table as needed.

Reaction	2 SO$_2$	1 O$_2$	2 SO$_3$
Initial	0.0100 M	0.0030 M	0 M
Change (+ and −)	−0.0040 M	−0.0020 M	+0.0040 M
At Equilibrium	0.0060 M	0.0010 M	0.0040 M

The bottom row shows concentrations at equilibrium.

2c. For K calculations, write the *WRECK* steps.

$$K: \qquad K = \frac{[SO_3]^2}{[SO_2]^2[O_2]}$$

Values substituted to find K must be mol/L at equilibrium—the values in the bottom row of the *RICE* table.

SOLVE: $$K = \frac{(4.0 \times 10^{-3})^2}{(6.0 \times 10^{-3})^2(1.0 \times 10^{-3})} = \frac{16 \times 10^{-6}}{36 \times 10^{-9}} = 0.444 \times 10^3 = \mathbf{440}$$

2d. $K_{reverse} = \dfrac{1}{K_{forward}} = \dfrac{1}{440} = \mathbf{2.3 \times 10^{-3}}$

3. For calculations involving concentrations and K, write the *WRECK* steps.

WANTED: $K = \dfrac{[H_2][I_2]}{[HI]^2}$

Concentrations: To find measurements at equilibrium when some supplied data is not, use a *RICE* table.

Reaction	2 HI	1 H$_2$	1 I$_2$
Initial	0.220 M	0 M	0 M
Change (+ and −)	−0.020 M	+0.010 M	+0.010 M
At Equilibrium	0.200 M	0.010 M	0.010 M

$K = \dfrac{[H_2][I_2]}{[HI]^2} = \dfrac{(1.0 \times 10^{-2})^2}{(2.0 \times 10^{-1})^2} = \dfrac{1.0 \times 10^{-4}}{4.0 \times 10^{-2}} = 0.25 \times 10^{-2} = \mathbf{2.5 \times 10^{-3}}$

Review Quiz

1a. $(1.6 \times 10^{-11})^{1/4} = (16 \times 10^{-12})^{1/4} = (16^{1/4}) \times (10^{-12})^{1/4} = \mathbf{2.0 \times 10^{-3}}$

1b. 6.4×10^{31} 1c. $(10^{-6})^{1/5} = (10,000 \times 10^{-10})^{1/5} = \mathbf{6.31 \times 10^{-2}}$

1d. 4.6×10^2

2. SOLVE: $[A]^3 = \dfrac{[C]^2}{K \cdot [B]} = \dfrac{(8 \times 10^{-3})^2}{(2.0 \times 10^{-2})(4.0 \times 10^{-1})} = \dfrac{64 \times 10^{-6}}{(8.0 \times 10^{-3})} = 8.0 \times 10^{-3}$

   $[A] = \mathbf{0.20\,M}$ or $\mathbf{2.0 \times 10^{-1}\,M}$

3. SOLVE: $[CO] = K \cdot [CO_2] \cdot [H_2] = (4.8)(0.020)(0.045) = \mathbf{4.3 \times 10^{-3}\,M\,CO}$

4a. WANTED: mol/L of reactants and products at equilibrium.

   DATA: To find moles or mol/L at equilibrium when some supplied data is not at equilibrium, use a *RICE* table.

Reaction	2 NO	1 Cl$_2$	2 NOCl
Initial	0.100 M	0.150 M	0 M
Change (+ and −)	−0.050 M	−0.025 mol	+0.050 M
At Equilibrium	0.050 M	0.125 M	0.050 M

4b. SOLVE: $K = \dfrac{(0.050)^2}{(0.050)^2(0.125)} = \dfrac{1}{0.125} = \mathbf{8.0}$

22

Acid–Base Fundamentals

Lesson 22.1   Acid–Base Math Review

The practice set below reviews math needed for acid–base concentration calculations. Try the *last* lettered *part* of each question. If you solve that problem correctly, go to the next question. If not, review the Lesson 1.1 rules for decimal equivalents and Lesson 1.4 rules for exponential calculations, then complete the other parts of the question.

PRACTICE

Solve problems 1–6 without a calculator.

1. Write answers as 10 to a power.

   a. $(10^{-8})(10^{+2}) =$

   b. $(10^{-3})(10^{-12}) =$

   c. $(x)(10^{-12}) = 10^{-14}; x =$

   d. $(10^{-3})(x) = 10^{-14}; x =$

   e. $\dfrac{10^{-14}}{10^{-5}} =$

   f. $\dfrac{10^{-14}}{10^{-11}} =$

(continued)

2. Convert to scientific notation.

    a. $324 \times 10^{+12} =$ b. $0.050 \times 10^{-11} =$

For the remaining problems in this Practice, convert final answers to scientific notation.

3. Solve on this page.

    a. $(2.0 \times 10^1)(3.0 \times 10^{-11}) =$

    b. $\dfrac{1.0 \times 10^{-14}}{2.0 \times 10^4} =$

4. Solve in your notebook.

    a. $(x)(2.0 \times 10^{-8}) = 10. \times 10^{-15};\ x =$

    b. $(2.5 \times 10^{-6})(x) = 10. \times 10^{-15};\ x =$

5. Convert fixed decimals to exponential notation, then solve without a calculator.

    a. $\dfrac{1.0 \times 10^{-14}}{0.040} =$

    b. $\dfrac{1.0 \times 10^{-14}}{0.0030} =$

6. Solve in your notebook.

    a. $(x)(0.20) = 1.0 \times 10^{-14};\ x =$

    b. $(0.0125)(x) = 1.0 \times 10^{-14};\ x =$

7. Use a calculator for the numbers but not for exponents.

    a. $\dfrac{1.0 \times 10^{-14}}{8.8 \times 10^{-4}} =$

    b. $\dfrac{1.0 \times 10^{-14}}{2.4} =$

8. Solve in your notebook. Use a calculator for all or part of the calculation.

    a. $(x)(6.7 \times 10^{-12}) = 1.0 \times 10^{-14};\ x =$

    b. $(1.25 \times 10^{-7})(x) = 1.0 \times 10^{-14};\ x =$

Lesson 22.2 K_w Calculations

For the acid–base calculations in Chapter 14, our goal was to find the amount needed to precisely neutralize an amount of acid or base. Beginning in this chapter, we return to acids and bases to ask additional questions important in both chemistry and biology. What is the nature of acidic and basic solutions before they react? Which particles and ions are present? How do strong acids such as hydrochloric acid differ from weak acids such as vinegar that we frequently consume as food?

Let's start with the molecule that has the highest concentration in aqueous solutions: H_2O.

The Ionization of Water

Recall that a water molecule has two bonds, two lone pairs, a bent shape, bond angles slightly less than 109°, and two dipoles that result in a net dipole.

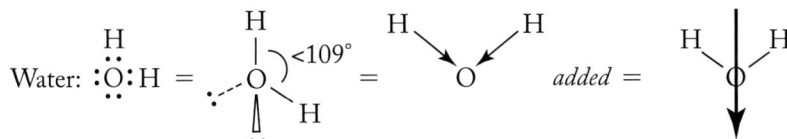

At temperatures above absolute zero, the bonds in H_2O bend and stretch. At room temperature, the liquid molecules also move and collide at high average speeds. In part as a result of these motions, for about one in 500 million molecules at room temperature, one of the bonds in liquid water is broken. The result is the formation of two ions, H^+ and OH^-.

When a bond that has a mix of ionic and covalent character separates to form ions, the process may be termed *ionization* (forming ions) or *dissociation*. The separation of water into ions is reversible and can be represented by:

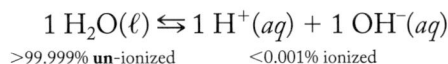

$$1\ H_2O(\ell) \leftrightarrows 1\ H^+(aq) + 1\ OH^-(aq)$$

>99.999% **un**-ionized   <0.001% ionized

In pure liquid water, the *number* of H^+ and OH^- ions must be equal, because they are formed in a 1:1 ratio. The *concentration* of the H^+ and OH^- ions must also be equal, because the moles of ions are in the same volume of water. At room temperature, these concentrations are very small: 1.0×10^{-7} M for each ion. The concentration of the un-ionized water molecules is very large in comparison: about 55 M. However, even at these low concentrations, small changes in the balance between H^+ and OH^- can have a substantial influence on reactions in aqueous solutions.

Water's Ionization Constant: K_w

For the reversible ionization of water:

$$H_2O(\ell) \leftrightarrows H^+(aq) + OH^-(aq)$$

TRY IT

Q. Write the K expression for the above reaction.

Answer:

Water is both a liquid and the solvent for the ions. As a result, in the K expression the water concentration term is represented by 1. Because a 1 in the denominator is omitted from K expressions (see Lesson 20.3), the equilibrium constant expression is written as

$$K = [H^+][OH^-]$$

At room temperature (25°C), the K value for the ionization of water is $\mathbf{1.0 \times 10^{-14}}$. This small number (0.000 000 000 000 0**10**) indicates that the reaction strongly favors the reactants: Very few water molecules separate into ions.

As temperature increases, the molecules of water collide with higher average energy, the bonds bend and stretch more vigorously, and the bonds break more often. Of importance in biology, at body temperature in most mammals (37°C) the $[H^+]$ in water is about $\mathbf{1.6 \times 10^{-7}}$ M. However, for acid–base calculations in chemistry, you should assume a temperature of 25°C unless otherwise noted.

Because the ionization of water is a part of many calculations, its K equation is given a special symbol: K_w. In aqueous solutions:

$$K_w = [H^+][OH^-] = 1.0 \times 10^{-14} \text{ at 25°C}$$

This relationship between $[H^+]$ and $[OH^-]$ is an inverse proportion. If substances are added to water that make one ion concentration increase, the other must decrease in the same proportion: If one ion concentration triples, the other ion concentration must become one third of its original value.

Acid–Base Terminology

In pure water, the concentration of H^+ and OH^- ions must be equal, but if acids or bases are dissolved into water, this balance is upset. By what are termed the **classical** definitions of acids and bases:

* An **acid** is a substance that ionizes in water and forms H^+ ions.
* A **base** is a substance that dissociates in water to form OH^- ions.

These are also called the **Arrhenius** definitions for acids and bases, after the Swedish chemist Svante Arrhenius who first proposed the existence of electrically charged particles (ions).

When an acid or base is added to water, if either $[H^+]$ or $[OH^-]$ is known, the concentration of the other can be solved using:

The K_w Prompt

See the symbols $[H^+]$ *and* $[OH^-]$ in a problem? *Write*:

$$K_w = \boxed{[H^+][OH^-] = 1.0 \times 10^{-14}}$$

and solve for the symbol WANTED.

Some problems will ask for the approximate $[H^+]$ and $[OH^-]$ in acidic or basic solutions. Those calculations can be completed by quick mental arithmetic.

▶ **TRY IT**

Q. In a solution with an $[OH^-]$ of about 10^{-2} M, what is the approximate $[H^+]$? (Answer as 10 to a power.)

STOP

Answer:

See $[H^+]$ and $[OH^-]$? Write $K_w = \boxed{[H^+][OH^-] = 1.0 \times 10^{-14}}$

STOP

SOLVE by substituting the $[OH^-]$ and using mental arithmetic:

$$[H^+](\sim 10^{-2}) = 10^{-14}; \; \mathbf{[H^+] \approx 10^{-12}\,M}$$

As always when solving K equations, units are omitted during calculations, but if a concentration is WANTED, add mol/L (M) to the answer.

In other problems, you will need to calculate [ions] more precisely.

TRY IT

Solve in your notebook using WANTED, DATA, and SOLVE, without a calculator.

Q. The $[H^+]$ in an aqueous solution is 5.0×10^{-3} M. Find the $[OH^-]$.

STOP

Answer:

WANTED: $[OH^-] = ?$

DATA: $[H^+] = 5.0 \times 10^{-3}$ M

$\boxed{[H^+][OH^-] = 1.0 \times 10^{-14}}$ (K_w prompt.)

SOLVE the equation for the WANTED *symbol, then* plug in the DATA.

$$? = [OH^-] = \frac{1.0 \times 10^{-14}}{[H^+]} = \frac{1.0 \times 10^{-14}}{5.0 \times 10^{-3}} = 0.20 \times 10^{-11} = \mathbf{2.0 \times 10^{-12}\,M}$$

K_w Check

Each time you solve using the K_w equation, do a check: *estimate* $[H^+]$ times $[OH^-]$. To do so, multiply the *answer* value times the number in the *position* in the denominator that is in *italics* above. Multiply the significands first, then the exponents. The result must equal $\mathbf{10 \times 10^{-15}}$ *or* $\mathbf{1.0 \times 10^{-14}}$. Try that for the Try It problem above. Does the answer check?

PRACTICE

Solve these in your notebook. Do not use a calculator.

1. Find the $[H^+]$ if the $[OH^-]$ is

     a. 10^{-11} mol/L   b. 0.010 molar   c. 5.0×10^{-11} M

2. Find the $[OH^-]$ if the $[H^+]$ is

     a. 10^{-9} mol/L   b. 3.0×10^{-5} M   c. 2.0 molar

Lesson 22.3	Strong Acid Solutions

By the classical (Arrhenius) definitions in chemistry:

- **Strong acids** are compounds that ionize essentially 100% in water to form H^+ ions.

- **Strong bases** dissociate essentially 100% in water to form OH^- ions.

- **Weak** acids or bases ionize only partially when dissolved in water.

An H^+ ion, with one proton and no electrons, is often referred to as a *proton*. Acids can be classified as **monoprotic** (containing *one* hydrogen atom that can ionize) or **polyprotic** (containing more than one acidic proton). Molecules can react as acids and bases in a variety of ways.

> **Example:** One type of acid–base reaction is the mixing of hydrogen chloride gas and ammonia gas to form solid ammonium chloride.

$$HCl(g) + NH_3(g) \rightarrow NH_4Cl(s)$$

However, *most* acid–base reactions are conducted in water, and in these lessons, if no state for an acidic or basic particle is shown, assume that the state is aqueous (*aq*).

Strong Acids

In chemistry, we often have a need for strong acid solutions. The strong acids employed most frequently are HCl, HNO_3, and H_2SO_4.

- HCl and HNO_3 are strong *monoprotic* acids. Both are highly soluble in water and ionize essentially 100% to release one H^+ ion.

  Other strong monoprotic acids include HBr, HI, $HClO_4$, and $HMnO_4$, but because these substances may undergo redox as well as acid–base reactions, they are used less often for reactions that require a strong acid.

- H_2SO_4 is a strong *diprotic* acid. When H_2SO_4 is dissolved in water, the first proton ionizes essentially 100%, but the second ionizes only partially.

The ionization of acids is more complex than the dissociation of most ionic compounds, in part because the bond of the H in acids has a balance of ionic and covalent character. Because of the mixed nature of bonds to hydrogen, we need three sets of rules to calculate ion concentrations in acidic solutions:

- One for monoprotic acids (such as HCl and HNO_3) that ionize completely

- One for weak acids, in which some but not all molecules ionize

- One for polyprotic acids (such as H_2SO_4) in which some H atoms ionize more easily than others

Let's start with solutions of the strong acids HCl and HNO_3.

Expressing Concentrations in Strong Acid Solutions

The concentration of an HCl or HNO_3 solution is usually expressed using the molecular formula, but the molecular formula does not represent the particles that are actually present in the solution.

As one example, molecules of the gas hydrogen chloride (HCl) readily dissolve in water to form a solution of *hydrochloric acid*. If 0.20 mole of HCl is dissolved per liter of solution, the solution concentration is written as "[HCl] = 0.20 M" which is termed the "[HCl] as mixed." However, there are very few molecules of HCl in an "HCl solution." As HCl dissolves in water, its molecules immediately separate to form ions:

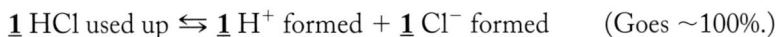

$$\underline{1}\text{ HCl used up} \leftrightarrows \underline{1}\text{ H}^+ \text{ formed} + \underline{1}\text{ Cl}^- \text{ formed} \qquad \text{(Goes } \sim100\%.\text{)}$$

This reaction is reversible, but the right side is so strongly favored at equilibrium that essentially all of the HCl is converted to ions. In a solution labeled [HCl] = 0.20 M, the concentrations from the acid are

$$[\text{HCl}] = 0\text{ M}; \qquad [\text{H}^+] = 0.20\text{ M}; \qquad [\text{Cl}^-] = 0.20\text{ M}$$

However, if a problem asks for the concentration of a compound that ionizes 100%, unless otherwise noted, you should assume it is asking for the [compound *as mixed*].

Calculating Concentrations in Strong Acid Solutions

The ionization of a strong acid parallels what happens to other ionic compounds that in water dissolve and separate into ions essentially 100% (see Lesson 12.2). Calculations for the dissociation of HCl and HNO_3 are simplified because the reaction ratios are 1 to 1 to 1.

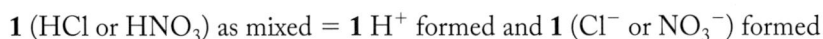

$$\mathbf{1}\text{ (HCl or HNO}_3\text{) as mixed} = \mathbf{1}\text{ H}^+ \text{ formed and } \mathbf{1}\text{ (Cl}^- \text{ or NO}_3{}^-\text{) formed}$$

Coefficients supply *mole* ratios. In a solution, because all of the particles are in the same constant volume, the coefficients are also mole *per liter* ratios. Based on the 1 to 1 to 1 ratios above, we can write

$$[\text{HCl or HNO}_3]\text{ as mixed} = [\text{H}^+]\text{ in solution} = [\text{Cl}^- \text{ or NO}_3{}^-]\text{ in solution}$$

Let's summarize this behavior with two rules:

Strong monoprotic acids ionize $\sim100\%$ to form H^+ and an anion.

and

See H^+ and HCl or HNO_3? Write:

$$[\text{HCl or HNO}_3]_{\text{as mixed}} = [\text{H}^+]_{\text{in soln.}} = [\text{Cl}^- \text{ or NO}_3{}^-]_{\text{in soln.}}$$

TRY IT

Q. In a 0.45 M HCl solution, write the

a. $[\text{HCl}]_{\text{as mixed}}$ 　　　b. $[\text{H}^+]_{\text{in soln.}}$ 　　　c. $[\text{Cl}^-]_{\text{in soln.}}$

Answers:

See H^+ and HCl or HNO_3? Write:

$$[\text{HCl}]_{\text{as mixed}} = [\text{H}^+]_{\text{in soln.}} = [\text{Cl}^-]_{\text{in soln.}} = \mathbf{0.45\ M} \text{ from the problem data}$$

In some problems, to find [ions], the [HCl] or [HNO_3] as mixed will need to be calculated first.

━━▶ TRY IT

Q. If 0.030 mol of HNO_3 is mixed with water to form 150 mL of solution, find

   a. $[HNO_3]_{\text{as mixed}}$ b. $[H^+]_{\text{in soln.}}$ c. $[NO_3^-]_{\text{in soln.}}$

STOP

Answers:

   a. WANTED: $? = \dfrac{\text{mol } HNO_3}{\text{L soln.}}$

      DATA: 0.030 mol HNO_3 = 150 mL soln.   (Two measures; same soln.)

         (When a ratio unit is WANTED, all DATA will be in equalities.)

      SOLVE: (To review molarity calculations, see Lesson 10.3.)

   $$? = \frac{\text{mol } HNO_3}{\text{L soln.}} = \frac{0.030 \text{ mol } HNO_3}{150 \text{ mL soln.}} \cdot \frac{1 \text{ mL}}{10^{-3} \text{ L}} = \mathbf{0.20\,M\ HNO_3}$$

   b, c. WANTED: $[H^+]_{\text{in soln.}}$ and $[NO_3^-]_{\text{in soln.}}$

      See H^+ and HNO_3? Write:

      $[HNO_3]_{\text{as mixed}} = [H^+]_{\text{in soln.}} = [NO_3^-]_{\text{in soln.}} = \mathbf{0.20\,M}$ from part (a).

PRACTICE A

1. In a 0.15 M solution of nitric acid, find the

   a. $[H^+]$ b. $[NO_3^-]$

2. 7.30 grams of hydrochloric acid is dissolved in water to make 250 milliliters of solution. In your notebook, find the

   a. Moles of HCl dissolved per liter   b. $[H^+]$ c. $[Cl^-]$

The HCl and HNO₃ *Quick Rule*

In most strong acid calculations, only $[H^+]$ is WANTED, and you can use this

   Quick rule: **[HCl] *or* [HNO_3] = [H⁺]**

Example: In a solution labeled 0.35 M HCl, **[H⁺] = 0.35 M**

H$^+$ and H$_3$O$^+$: Equivalent

In aqueous solutions, the proton released by an acid is nearly always found attached to a water molecule, forming a **hydronium ion** (H$_3$O$^+$). This reaction can be represented as

$$1\,H^+ + 1\,H_2O \rightleftharpoons 1\,H_3O^+$$

However, because the reaction is easily reversed, and the proton behaves in most respects as if it were a free particle, in most instances the symbols H$^+$ and H$_3$O$^+$ are considered to be equivalent. In calculations, apply this rule:

If you see **H$_3$O$^+$**, write: **H$_3$O$^+$ = H$^+$**

Why [Acid] Determines [H$^+$]

When a strong acid is dissolved in water, H$^+$ is contributed by *both* the acid and the water. However, if the strong acid is mixed in any significant concentration, the share of H$^+$ ions contributed by the water can be ignored.

Why? Consider a 0.20 M HCl solution. The strong acid contributes 0.20 mole of H$^+$ ions per liter to the water. Before the acid was added, the water contained only 10^{-7} mole of H$^+$ ions per liter. Note the uncertainty in those two amounts:

0.2$\underline{0}$ M H$^+$ from the acid, with doubt in the hundredths place

0.000000$\underline{1}$ M H$^+$ initially in the water, with doubt in a much lower place

This is one indication that any initial H$^+$ contribution from the water's ions is too small to be significant. In solutions of acids, because the acid ionization is the *dominant* reaction, the rule for calculations is

In an acid solution, use rules for acid ionization to find [H$^+$].

Let's summarize:

Rules for Solutions of Acids

1. See [H$^+$] and [OH$^-$]? Write $K_w = $ $\boxed{[H^+][OH^-] = 1.0 \times 10^{-14}}$

2. Strong monoprotic acids ionize ~100% to form H$^+$ and an anion.

3. See H$^+$ and HCl or HNO$_3$? Write:

$$[\text{HCl or HNO}_3]_{\text{as mixed}} = [H^+]_{\text{in soln.}} = [\text{Cl}^- \text{ or NO}_3^-]_{\text{in soln.}}$$

4. *Quick rule*: [HCl *or* HNO$_3$] = [H$^+$]

5. See H$_3$O$^+$? Write: H$_3$O$^+$ = H$^+$

6. In the solution of an acid, use rules for acid ionization to find [H$^+$].

PRACTICE **B**

1. In your notebook, write a brief, memorable summary of the preceding six points. Practice writing your summary until you can do so from memory.

2. In 0.25 M HNO_3:

   a. $[H^+] =$ b. $[H_3O^+] =$

3. In 2.0 M HCl:

   a. $[H_3O^+] =$ b. $[H^+] =$

Lesson 22.4   [OH⁻] in Acid Solutions

Adding acid to water increases the $[H^+]$ in the solution. This shifts the equilibrium for the "auto-ionization" reaction that is always taking place in water:

$$1\ H_2O(\ell) \leftrightharpoons 1\ H^+(aq) + 1\ OH^-(aq) \quad \text{(Goes slightly.)}$$

The *value* of the $[OH^-]$ after the shift can be calculated by the equilibrium constant for the water ionization reaction:

$$K_w = [H^+][OH^-] = 1.0 \times 10^{-14}$$

To solve for $[OH^-]$ in an *acidic* solution, let us add to rule 6 from the previous lesson as follows.

> 6. In the solution of an acid, use rules for acid ionization to find $[H^+]$, *then use K_w to find $[OH^-]$.*

TRY IT

Q. In a 0.40 M HCl solution, find:

  a. $[H^+]$ b. $[Cl^-]$ c. $[OH^-]$

Answers:

a, b. See H^+ and HCl or HNO_3? Write:

$$[HCl]_{\text{as mixed}} = [H^+]_{\text{in soln.}} = [Cl^-]_{\text{in soln.}} = \mathbf{0.40\ M}$$

  c. WANTED: $[OH^-] = ?$

Apply rule 6. *Part (a)* found $[H^+] = 0.40$ M

$$\boxed{[H^+][OH^-] = 1.0 \times 10^{-14}}\ \text{SOLVE for the symbol WANTED.}$$

$$? = [OH^-] = \frac{1.0 \times 10^{-14}}{[H^+]} = \frac{1.0 \times 10^{-14}}{0.40} = \boxed{2.5 \times 10^{-14}\ \text{M}}$$

K_w *check*: $[H^+] \times [OH^-]$ (circled above) must $\approx \mathbf{10 \times 10^{-15}}$ *or* $\mathbf{1.0 \times 10^{-14}}$.

PRACTICE

Work in your notebook.

1. From memory, write the rules for acidic solutions. Include the modified rule 6.

2. In a solution labeled 10^{-3} M HNO_3, answer with 10 to a power:

   a. $[H^+] =$ b. $[NO_3^-] =$

   c. $[H_3O^+] =$ d. $[OH^-] =$

3. In a solution labeled 0.020 M HCl:

   a. Write the balanced equation for the ionization of this acid.

   b. Which side will be favored in this ionization: products or reactants?

   c. What is the *mole-to-mole* ratio between HCl used up and H^+ formed?

   d. What is the *mol/L* ratio between HCl used up and H^+ formed?

   e. $[H^+] =$ f. $[Cl^-] =$

   g. $[OH^-] =$ h. $[H_3O^+] =$

Lesson 22.5   Strong Base Solutions

Strong bases, by the classical (Arrhenius) definitions, dissociate essentially 100% in water to form OH^- ions. The substances most frequently used to make strong base solutions are two water-soluble ionic solids: sodium hydroxide (NaOH) and potassium hydroxide (KOH). When mixed with water,

$$1\ NaOH(s) \rightarrow 1\ Na^+(aq) + 1\ OH^-(aq) \qquad \text{(Goes ~100\%.)}$$

$$1\ KOH(s) \rightarrow 1\ K^+(aq) + 1\ OH^-(aq) \qquad \text{(Goes ~100\%.)}$$

There are other types of strong bases, but in these lessons we will limit our attention to the strong bases that are *alkali metal hydroxides*.

[Ions] in Strong Base Solutions

As with strong acids, if a solution of a strong base is *labeled* [NaOH] = 0.15 M, this represents how the solution is *mixed*, but not the particles present in the solution. Because NaOH dissociates ~100%, what is actually present in "0.15 M NaOH" is **0** M NaOH, 0.15 M Na^+, and 0.15 M OH^-. However, if a problem asks for [NaOH] or [KOH], assume it is asking not for 0 M but for the concentration *as mixed*.

The rules for NaOH and KOH are similar to those for the monoprotic strong acids. Add the following rules to your list:

Rules for Solutions of Acids and Bases

7. Strong bases NaOH and KOH dissociate ~100% to form OH^- and Na^+ or K^+.

8. See OH^- and NaOH or KOH? Write:

$$[\text{NaOH or KOH}]_{\text{as mixed}} = [OH^-]_{\text{in soln.}} = [Na^+ \text{ or } K^+]_{\text{in soln.}}$$

9. *Quick rule*: $[\text{NaOH or KOH}] = [OH^-]$

10. In the solution of a base, use rules for base behavior to find $[OH^-]$, *then* K_w to find $[H^+]$.

▷ TRY IT

Q. In a solution labeled 0.0030 M NaOH:

   a. What three ions are present in the solution?

   b. What is the concentration of each ion?

Answers:

   a. NaOH dissociates to form Na^+ and OH^- ions. Water ionizes to form H^+ and OH^-. The three ions are **Na^+, OH^-, and H^+.**

   b. See OH^- and NaOH or KOH? Write:

$$[\text{NaOH or KOH}]_{\text{as mixed}} = \mathbf{[OH^-]_{\text{in soln.}}} = \mathbf{[Na^+ \text{ or } K^+]_{\text{in soln.}}} = \mathbf{0.0030\,M}$$

To find $[H^+]$ in the solution of a base, first use the base rules to find $[OH^-]$ (done), then use K_w to find $[H^+]$.

$$\boxed{[H^+][OH^-] = 1.0 \times 10^{-14}} \quad (K_w \text{ prompt.})$$

$$[H^+] = \frac{1.0 \times 10^{-14}}{[OH^-]} = \frac{1.0 \times 10^{-14}}{3.0 \times 10^{-3}} = 0.33 \times 10^{-11} = \mathbf{3.3 \times 10^{-12}\,M\,H^+}$$

K_w check: $3.0 \times 3.3 \approx \mathbf{10}$; $10^{-3} \times 10^{-12} = 10^{-15}$, combined $\approx \mathbf{10 \times 10^{-15}}$

Another way to check that your answers are reasonable is to compare the calculated $[H^+]$ and $[OH^-]$ values. In a solution of an acid, the $[H^+]$ will be the higher of the two concentrations. In the solution of a base, $[OH^-]$ will be higher than $[H^+]$.

PRACTICE

Work in your notebook.

1. Write a condensed version of rules 7–10. Practice until you can recall and write your summary from memory.

2. In a solution labeled 10^{-1} M NaOH:

    a. Write the balanced equation for dissociation.

    b. Which side is favored in the reaction: products or reactants?

    c. $[OH^-] =$ d. $[Na^+] =$ e. $[H^+] =$ f. $[H_3O^+] =$

3. In a solution labeled 8.0×10^{-3} M KOH, find the:

    a. $[OH^-]$ b. $[H_3O^+]$ c. $[K^+]$ d. $[H^+]$

SUMMARY

This chapter includes a "numbered summary list" of facts and problem-solving procedures. Summary lists are an effective way to learn information that is related and/or best remembered in a sequence.

Except for a few items of data, your brain can only reason with facts and procedures that you can recall from long-term memory. Recall depends on cues that activate associations. A part of building your mental library is to provide a "hierarchical structure": a kind of Dewey decimal system that provides a framework for what you know. During initial learning, when you see cues such as the formulas for the strong acids and bases, if you write the rules that apply, you help your memory to organize knowledge logically and efficiently. Gradually, after practice in a variety of contexts, cues in problems will activate associated elements automatically, without having to write the lists or rules.

REVIEW QUIZ

Work in your notebook.

1. Without a calculator: $(4.0 \times 10^{-9})(x) = 1.0 \times 10^{-14}$; $x =$

2. With a calculator: $\dfrac{1.0 \times 10^{-14}}{4.3 \times 10^{-4}} =$

3. From memory, write the ten rules for strong acid and strong base solutions.

4. In a solution labeled 10^{-2} M HCl, represent with 10 to a power:

    a. $[H^+]$ b. $[Cl^-]$

    c. $[H_3O^+]$ d. $[OH^-]$

5. In an aqueous solution, find $[H^+]$ if $[OH^-]$ is 0.036 M.

6. If 20.0 millimoles of OH^- ions are dissolved in 400. mL of solution, find:

    a. $[OH^-]$ b. $[H^+]$

7. If 5.00 millimoles of NaOH are dissolved in 0.250 liters of solution, find:

    a. $[OH^-]$ b. $[H^+]$ c. $[H_3O^+]$

ANSWERS

Answers and some partial solutions are shown. Your work should include all of the steps in the Try It examples.

Lesson 22.1

Practice   1a. 10^{-6} 1b. 10^{-15} 1c. 10^{-2} 1d. 10^{-11} 1e. 10^{-9} 1f. 10^{-3}

2a. 3.24×10^{14} 2b. 5.0×10^{-13} 3a. 6.0×10^{-10} 3b. 5.0×10^{-19}

4a. $x = \dfrac{10. \times 10^{-15}}{2.0 \times 10^{-8}} = 5.0 \times 10^{-15-(-8)} = 5.0 \times 10^{-15+8} = \mathbf{5.0 \times 10^{-7}}$

4b. 4.0×10^{-9} 5a. 2.5×10^{-13} 5b. 3.3×10^{-12} 6a. 5.0×10^{-14}

6b. $x = \dfrac{1.0 \times 10^{-14}}{1.25 \times 10^{-2}} = 0.80 \times 10^{-12} = \mathbf{8.0 \times 10^{-13}}$

7a. $\dfrac{1.0 \times 10^{-14}}{8.8 \times 10^{-4}} = 0.11 \times 10^{-10} = \mathbf{1.1 \times 10^{-11}}$

7b. 4.2×10^{-15} 8a. 1.5×10^{-3}

8b. $x = \dfrac{1.0 \times 10^{-14}}{1.25 \times 10^{-7}} = 0.80 \times 10^{-7} = \mathbf{8.0 \times 10^{-8}}$

Lesson 22.2

Practice   1a. $[H^+][OH^-] = 1.0 \times 10^{-14}$; $[H^+](10^{-11}) = 10^{-14}$; $\mathbf{[H^+] = 10^{-3}\,M}$

1b. WANTED: $[H^+] = ?$

DATA: $[OH^-] = 0.010\,M$

$\boxed{[H^+][OH^-] = 1.0 \times 10^{-14}}$ (K_w prompt. SOLVE for the symbol WANTED.)

$[H^+] = \dfrac{1 \times 10^{-14}}{0.010} = \dfrac{1 \times 10^{-14}}{1.0 \times 10^{-2}} = 1.0 \times 10^{-14-(-2)} = 1.0 \times 10^{-14+2} = \mathbf{1.0 \times 10^{-12}\,M}$

1c. SOLVE: $[H^+] = \dfrac{1.0 \times 10^{-14}}{[OH^-]} = \dfrac{1.0 \times 10^{-14}}{5.0 \times 10^{-11}} = 0.20 \times 10^{-3} = \mathbf{2.0 \times 10^{-4}\,M\,H^+}$

K_w check: Answer times bottom $= 2 \times 5 = \mathbf{10}$; $10^{-4} \times 10^{-11} = \mathbf{10^{-15}}$. Check.

2a. $[H^+][OH^-] = 1.0 \times 10^{-14}$; $(10^{-9})[OH^-] = 10^{-14}$; $\mathbf{[OH^-] = 10^{-5}\,M}$

2b. WANTED: $[OH^-] = ?$

DATA: $[H^+] = 3.0 \times 10^{-5}\,M$ $\boxed{[H^+][OH^-] = 1.0 \times 10^{-14}}$

SOLVE: $[OH^-] = \dfrac{1.0 \times 10^{-14}}{[H^+]} = \dfrac{1.0 \times 10^{-14}}{3.0 \times 10^{-5}} = 0.33 \times 10^{-9} = \mathbf{3.3 \times 10^{-10}\,M\,OH^-}$

K_w check: Answer times bottom $= 3.3 \times 3 = \mathbf{9.9}$; $10^{-10} \times 10^{-5} = \mathbf{10^{-15}}$. Close enough.

2c. SOLVE: $[OH^-] = \dfrac{1.0 \times 10^{-14}}{[H^+]} = \dfrac{1.0 \times 10^{-14}}{2.0} = 0.50 \times 10^{-14} = \mathbf{5.0 \times 10^{-15}\,M}$

K_w check: Answer times bottom $= 5 \times 2 = \mathbf{10}$; $10^{-15} \times 10^0 = \mathbf{10^{-15}}$. Check.

Lesson 22.3

Practice A

1a, 1b. See H^+ and HCl or HNO_3? Write:

$[\text{HCl or }\mathbf{HNO_3}]_{\text{as mixed}} = [H^+]_{\text{in soln.}} = [Cl^- \text{ or } NO_3^-]_{\text{in soln.}} = \mathbf{0.15\,M}$

2a. WANTED: $? \dfrac{\text{mol HCl}}{\text{L soln.}} = [\text{HCl}]_{\text{as mixed}}$

DATA: 7.30 g HCl = 250 mL soln.   (*Equivalent*: two measures; same soln.)

36.5 g HCl = 1 mol HCl   (*Grams prompt*.)

SOLVE: $? \dfrac{\text{mol HCl}}{\text{L soln.}} = \dfrac{7.30\ \text{g HCl}}{250\ \text{mL soln.}} \cdot \dfrac{1\ \text{mol HCl}}{36.5\ \text{g HCl}} \cdot \dfrac{1\ \text{mL}}{10^{-3}\ \text{L}} = \mathbf{0.80\ M\ HCl}$

2b, 2c. $[\mathbf{HCl}]_{\textbf{as mixed}} = [\mathbf{H^+}]_{\textbf{in soln.}} = [\mathbf{Cl^-}]_{\textbf{in soln.}} = \mathbf{0.80\ M}$

Practice B   2. Using the quick rule:

a. $[\text{H}^+] = \mathbf{0.25\ M}$ 2b. $[\text{H}_3\text{O}^+] = [\text{H}^+] = \mathbf{0.25\ M}$

3. First, apply the quick rule to 3b.

3b. $[\text{H}^+] = \mathbf{2.0\ M}$ 3a. $[\text{H}_3\text{O}^+] = [\text{H}^+] = \mathbf{2.0\ M}$

Lesson 22.4

Practice

2a, 2b. $[\text{HNO}_3]_{\text{as mixed}} = [\text{H}^+]_{\text{in soln.}} = [\text{NO}_3^-]_{\text{in soln.}} = \mathbf{10^{-3}\ M}$

2c. $[\mathbf{H_3O^+}] = [\mathbf{H^+}] = \mathbf{10^{-3}\ M}$

2d. In an acid solution, use the acid rules to find $[\text{H}^+]$, then K_w to find $[\text{OH}^-]$.

Because $[\text{H}^+] = 10^{-3}$ M and $[\text{H}^+][\text{OH}^-] = 1.0 \times 10^{-14}$, $[\mathbf{OH^-}] = \mathbf{10^{-11}\ M}$

3a, 3b. $1\ \text{HCl}(aq) \leftrightarrows 1\ \text{H}^+(aq) + 1\ \text{Cl}^-(aq)$ (Goes 100%.) **Products favored.**

3c. 1 mol HCl used up = 1 mol H^+ formed.

3d. Because all moles are in the same liters, mole and mol/L ratios are the same: 1 to 1.

3e, 3f. $[\text{HCl}]_{\text{as mixed}} = \mathbf{0.020\ M} = [\mathbf{H^+}]_{\textbf{in soln.}} = [\mathbf{Cl^-}]_{\textbf{in soln.}}$

3g. In an acid solution, use the acid rules to find $[\text{H}^+]$ (above), then K_w to find $[\text{OH}^-]$.

$$[\mathbf{OH^-}] = \dfrac{1.0 \times 10^{-14}}{[\text{H}^+]} = \dfrac{1.0 \times 10^{-14}}{2.0 \times 10^{-2}} = 0.50 \times 10^{-12} = \mathbf{5.0 \times 10^{-13}\ M}$$

K_w check: $[\text{H}^+][\text{OH}^-] = 2 \times 5 = \mathbf{10}$; $10^{-2} \times 10^{-13} = \mathbf{10^{-15}}$. Check.

3h. $[\text{H}_3\text{O}^+] = [\mathbf{H^+}] = \mathbf{0.020\ M}$ *or* $\mathbf{2.0 \times 10^{-2}\ M}$

Lesson 22.5

Practice   2a. $1\ \text{NaOH} \rightarrow 1\ \text{Na}^+ + 1\ \text{OH}^-$

2b. Strong bases dissociate 100%; products are favored.

2c, 2d. See OH^- and NaOH or KOH? Write:

$$[\text{NaOH or KOH}]_{\text{as mixed}} = [\text{OH}^-]_{\text{in soln.}} = [\text{Na}^+ \text{ or K}^+]_{\text{in soln.}} = \mathbf{10^{-1}\ M}$$

2e, 2f. To find $[\text{H}^+]$ in a solution of a base, first find $[\text{OH}^-]$, then $[\text{H}^+]$ using K_w.

$$[\text{H}^+] = \dfrac{1.0 \times 10^{-14}}{[\text{OH}^-]} = \dfrac{1.0 \times 10^{-14}}{10^{-1}} = \mathbf{10^{-13}\ M} = [\mathbf{H^+}] = [\mathbf{H_3O^+}]$$

3a, 3c. $[\text{NaOH or KOH}]_{\text{as mixed}} = [\text{OH}^-]_{\text{in soln.}} = [\text{Na}^+ \text{ or K}^+]_{\text{in soln.}} = \mathbf{8.0 \times 10^{-3}\ M}$

3b, 3d. $[\text{H}_3\text{O}^+] = [\mathbf{H^+}] = \dfrac{1.0 \times 10^{-14}}{[\text{OH}^-]} = \dfrac{1.0 \times 10^{-14}}{8.0 \times 10^{-3}} = 0.125 \times 10^{-11} = \mathbf{1.3 \times 10^{-12}\ M\ H^+}$

Review Quiz

You may do the arithmetic on calculations in any way you choose.

1. $x = \dfrac{1.0 \times 10^{-14}}{4.0 \times 10^{-9}} = 0.25 \times 10^{-14-(-9)} = 0.25 \times 10^{-5} = \mathbf{2.5 \times 10^{-6}}$

2. 2.3×10^{-11}

4a, 4b, 4c. $[\text{HCl}]_{\text{as mixed}} = [\text{H}^+] = [\text{Cl}^-] = [\text{H}_3\text{O}^+] = \mathbf{10^{-2}\,M}$

4d. Because $[\text{H}^+] = 10^{-2}\,M$ and $[\text{H}^+][\text{OH}^-] = 1.0 \times 10^{-14}$, $[\text{OH}^-] = \mathbf{10^{-12}\,M}$

5. SOLVE: $[\text{H}^+] = \dfrac{1.0 \times 10^{-14}}{[\text{OH}^-]} = \dfrac{1.0 \times 10^{-14}}{3.6 \times 10^{-2}} = 0.28 \times 10^{-12} = \mathbf{2.8 \times 10^{-13}\,M\,H^+}$

K_w check (estimate): $\sim 3 \times \sim 4 = \sim \mathbf{12}$; $10^{-13} \times 10^{-2} = \mathbf{10^{-15}}$. Close. Check.

6a. SOLVE: $\dfrac{?\ \text{mol OH}^-}{\text{L soln.}} = \dfrac{20.0\ \text{mmol OH}^-}{400.\ \text{mL soln.}} = \mathbf{0.0500\ \dfrac{mol\ OH^-}{L\ soln.}}$

Because a prefix is an abbreviation for an exponential, if the same prefix is on the top and bottom, it can cancel, just as two equal exponential terms can cancel.

6b. $[\text{H}^+] = \dfrac{1.0 \times 10^{-14}}{[\text{OH}^-]} = \dfrac{1.0 \times 10^{-14}}{5.00 \times 10^{-2}} = 0.20 \times 10^{-12} = \mathbf{2.0 \times 10^{-13}\,M\,H^+}$

7a. SOLVE: $\dfrac{?\ \text{mol NaOH}}{\text{L soln.}} = \dfrac{5.00\ \text{mmol NaOH}}{0.250\ \text{L soln.}} \cdot \dfrac{10^{-3}\ \text{mol}}{1\ \text{mmol}} = \mathbf{2.00 \times 10^{-2}\ \dfrac{mol}{L}} = [\text{NaOH}] = [\text{OH}^-]$

7b, 7c. $[\text{H}^+] = \dfrac{1.0 \times 10^{-14}}{[\text{OH}^-]} = \dfrac{1.0 \times 10^{-14}}{2.00 \times 10^{-2}} = 0.50 \times 10^{-12} = \mathbf{5.0 \times 10^{-13}\,M} = [\text{H}^+] = [\text{H}_3\text{O}^+]$

23

pH and Weak Acids

Lesson 23.1   Base 10 Logarithms

Numeric values can be expressed as fixed decimal numbers, in exponential notation, or as a base number to a power. When a value is expressed as a base to a power, the power can be either an integer or a number with decimals.

> **TRY IT**
>
> **Q.** Answer these without a calculator.
>
>   a. 10^2 = the number 100, and 10^3 = the number _____.
>
>   b. Without a calculator, *estimate* the value of $10^{2.5}$ = _____.
>
> **STOP**
>
> **Answers:**
>
>   a. 10^3 = **1000**   b. **100** = $10^2 < \mathbf{10^{2.5}} < 10^3$ = **1000**
>
> The answer should be somewhere very roughly halfway between 100 and 1000.

Now, find the value expressed in fixed decimal notation. Use the calculator you are permitted to use during examinations.

━━━━━▶ TRY IT

Q. $10^{2.5}$ = what fixed decimal number? _____.

Answer:

- On a "standard" calculator, you might try **2.5** $\boxed{10^x}$ *and/or* 10 $\boxed{y^x}$ **2.5** $\boxed{=}$ *and/or* **2.5** $\boxed{2^{nd} \text{ or INV}}$ $\boxed{\log}$.
- Graphing: Try **10** $\boxed{\wedge}$ **2.5** $\boxed{\text{enter}}$ RPN: Try **2.5** $\boxed{\text{enter}}$ $\boxed{10^x}$

Circle or write a sequence that results in **316** (which may be displayed on your calculator as **3.16 × 10²**).

For significant figures in calculations involving exponential terms, rules may differ for different topics and bases. For now, use this rule:

- If a value is an *integer* power of 10, assume it is an exact number.
- If an exponent has a decimal, round answers based on that value to three significant figures.

━━━━━▶ TRY IT

Q. Use your calculator to convert the following values to scientific notation.

    a. $10^{23.7798}$ =   b. $10^{-3.9}$ =

Answers:

    a. 6.02×10^{23} (Rounded to 3 *s.f.*)

    Note how your calculator *displayed* the exponent. You will need to translate the calculator display into scientific notation when writing answers.

    b. 1.26×10^{-4} Entering a negative exponent may require a $\boxed{+/-}$ or $\boxed{(-)}$ key.

PRACTICE   **A**

Convert the following to scientific notation with three significant figures.

    1. $10^{+16.5}$ =   2. $10^{-16.5}$ =   3. $10^{-0.7}$ =

Logarithm Definitions

In words, we will define a logarithm in two ways.

- A logarithm is an exponent.
- A logarithm answers this question: If a number is written as a base number to a power, what is the power?

A logarithm can be a power of any base.

Example: Because $2^4 = 16$, the *base 2 log* of 16 can be written $\log_2 16 = 4$

In science, *base 10* and *base e* are used most often. If no base is specified, you should assume that **log** means a *base 10* log. The symbol for a *base e* log (a *natural* log) is **ln**.

Base 10 Logs

The *log* function answers this question: If a value is written as **10** to a power, what is the power?

TRY IT

Using that rule, answer these without using a calculator.

Q. Write the log of

a. 10^2 b. 10,000   c. 0.001

STOP

Answers:

a. The log of 10^2 is **2**   b. $\log 10,000 = \log 10^4 = 4$ c. $\log 0.001 = \log 10^{-3} = -3$

Logs cannot be taken of negative numbers. For fixed decimal numbers greater than 1, log values will be positive. For positive numbers between 0 and 1, log values will be negative. Check those rules against the answers in the preceding Try It.

The *equation* defining a base 10 log is $\log 10^x = x$. It is also helpful to remember this example: The log of 100 is 2

When working with this type of problem, one way to check that you are doing calculator operations properly is to do a *simple* calculation, first in your head or on paper, and then using the calculator. The two answers should agree.

▶ TRY IT

Q1. Without using a calculator, write the log of

    a. 1000   b. 100,000   c. 0.01

Q2. The $\boxed{\text{log}}$ button finds a *base 10* log. Using a calculator, find the log of

    a. 1000   b. 100,000   c. 0.01

Answers:

1. Without a calculator:

    a. $\log(1000) = \log(10^3) = $ **3**

    b. The log of $100,000 = \log 10^5 = $ **5**

    c. The log of $0.01 = \log 10^{-2} = $ **−2**

2. For part (a), try 1000 $\boxed{\text{log}}$ or 1000 $\boxed{\text{enter}}$ $\boxed{\text{log}}$. Circle or write the sequence that works.

Your calculator and mental-arithmetic answers must agree.

Rounding Base 10 Calculations

To convey *approximate* uncertainty, in these lessons we will apply the following conventions to *base 10* log calculations.

- When a value is a whole number power of 10, write the log as an integer.

    Example: $x = 0.0001$ or 10^{-4}, $\log(x) = -4$

- When converting between a value in scientific notation and a log, round so that the number of digits in the *significand* of scientific notation equals the number of digits *after the decimal* in the log.

    Examples:

        $x = \mathbf{5} \times 10^{-6}; \log(x) = -5.\mathbf{3}$

        $x = \mathbf{5.1} \times 10^{-6}; \log(x) = -5.\mathbf{29}$

        $\log(x) = -11.6; x = \mathbf{3} \times 10^{-12}$

Checking Calculated Logarithms

To check a conversion between a log and a number, apply this rule:

Log check: The power of 10 in scientific notation must agree with the log value within ±1.

➡ TRY IT

Q. Use your calculator to solve these, then apply the rule above to check your answers.

a. $\text{Log}(7.4 \times 10^6) =$ (Log and exponent ±1?) _____

b. $\text{Log}(7.4 \times 10^{-6}) =$ (Log and exponent ±1?) _____

c. $\text{Log } 2{,}000 =$ (Log and exponent ±1?) _____

STOP

Answers:

a. $\text{Log}(7.4 \times 10^6) =$ **6.87** (±1? ✓)

b. $\text{Log}(7.4 \times 10^{-6}) =$ **−5.13** (±1? ✓)

c. $\text{Log } 2{,}000 = \log(2 \times 10^3) =$ **3.3** (±1? ✓)

Let's summarize the log rules so far.

Rules for Base 10 Logarithms

1. A *logarithm* is an exponent: the power of any base number to a power.

2. A base 10 logarithm answers the question: If a number is written as 10 to a power, what is the power?

3. A (log) button converts fixed decimal notation to *10 to a power* notation.

4. **Log 10^x = x**; the *log* of 100 is 2.

5. The log of a fixed decimal number greater than 1 is positive.

   The log of a fixed decimal number between 0 and 1 is negative.

6. *Log check*: The scientific-notation *exponent* and the log must agree within ±1.

PRACTICE B

1. In your notebook, write a memorable version of the six log rules. Practice writing your summary list until you can do so from memory.

2. Except as noted, convert these values to scientific notation.

a. $10^{-5.4} =$ (±1?) _____ b. $10^{-11.5} =$ (±1?) _____

c. $10^{-0.5} =$ (±1?) _____ d. $10^{-0.5} =$ (fixed decimal): _____

3. Report these base 10 log values in fixed decimal notation.

a. $\text{Log}(6.8 \times 10^{12}) =$ b. $\text{Log}(6.8 \times 10^{-12})$ (±1?) _____

c. $\text{Log } 4.6 =$ d. $\text{Log } 0.0020 =$

Converting from Logs to Numbers

Given the log of a value, we need to be able to convert to the value in both fixed decimal and scientific notation. This is called **taking the antilog** or taking the **inverse log**, but it is easier to remember what these terms mean (and what buttons to press) if you remember what a log is.

▶ **TRY IT**

🛑

Q. If the log of a number is 2, the number is _____.

Answer:

If the log is 2, the number is $10^2 = 100$: A log is an exponent of 10.

- ▶

The equation that converts a log value to its corresponding fixed decimal is

$$10^{\log x} = x$$

Recite and repeat to remember: "10 to the log x equals x."

▶ **TRY IT**

🛑

Without a calculator, apply the rules above.

Q. If these values are the logs of numbers, write the numbers in fixed decimal notation.

    a. 6   b. −1   c. 0

Answers:

    a. When the log is 6, the number is $10^6 = \mathbf{1{,}000{,}000}$

    b. When the log is −1, the number is $10^{-1} = \mathbf{0.1}$

    c. When log $= 0$, number $= 10^0 = \mathbf{1}$ (Any positive number to the power 0 equals 1.)

- ▶

Knowing the log, to find the number take the *antilog*: Write the log as a power of 10, then convert that exponential term to a number (or to scientific notation). On a calculator, try these steps.

- Input the **log**, then "take the antilog:" press ⟨INV⟩ ⟨LOG⟩ or ⟨2nd⟩ ⟨LOG⟩
- Input the **log value**, then press ⟨10^x⟩. A log is an exponent of 10.
- On some calculators the sequence is: 10, ⟨$x^\wedge y$⟩, **log value**, ⟨=⟩.

Circle or write, and note the logic, of the steps that work on your calculator.

▶ **TRY IT**

Using your key sequence, answer in scientific notation.

    a. If log $x = 8.7, x =$ (±1? ____)

    b. Log A $= -10.7$, A $=$ (±1? ____)

Note the same "is it reasonable?" quick check. The *log* and the *exponent* of the value in scientific notation must agree within ±1.

STOP

Answers:

    a. 5×10^8 (±1? ✓)   b. 2×10^{-11} (±1? ✓)

Add these to your summary list of "Rules for Base 10 Logarithms."

> 7. To go from the log to the number, take the antilog. Use $10^{\log x} = x$
>
> 8. Recite and repeat to remember: "10 to the log x equals x."

Logarithms and Measurements

In science, one use of logarithms is to report measurements that vary over a wide range as familiar fixed decimal numbers. The Richter scale for earthquakes, decibels for sound, magnitudes for star brightness, and pH scale for solutions are examples of scales based on base 10 logarithms.

On a logarithmic scale, each increase of 1 represents an actual numeric increase of 10. As one example, an earthquake of 8.0 on the Richter scale is 100 times stronger than a 6.0 Richter scale earthquake.

PRACTICE **C**

1. Practice writing the eight "Rules for Base 10 Logarithms" until you can do so from memory. You may want to design flashcards for these rules as well.

2. Convert these values to scientific notation.

    a. Log $x = 12.400$; $x =$ (±1?)

    b. Log $x = -5.90$; $x =$ (±1?)

    c. Log $x = -0.25$; $x =$

    d. $10^{-3.3} =$

    e. Log $x = 1.100$; $x =$

Lesson 23.2   The pH System

A **pH** is a number between −2 and 16 that indicates the acidity or basicity (bay-SIS-uh-tee) of an aqueous solution. The pH is a measure of the $[H^+]$ in a solution.

THE pH SCALE

```
---- |-------------------- |-------------------- |------
     0                     7                     14
```
Acidic   Neutral   Basic

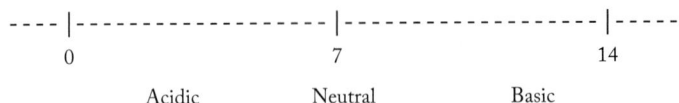

In these lessons, we will use the following simplified definitions of pH:

- pH is the negative log of the hydrogen ion concentration.
- In equation form, $pH \equiv -\log [H^+]$ *and* $[H^+] \equiv 10^{-pH}$

    Examples: In a solution, if $[H^+] = 10^{-2}$ M, the pH = **2**.

    If the pH = **11**, the $[H^+] = 1 \times 10^{-11}$ M.

It may help to remember pH as the "**power of H**:" the number located after the minus sign when $[H^+]$ is written as a negative power of 10.

In calculations that include pH, apply this rule:

The pH Prompt

See pH? Write: $\boxed{pH \equiv -\log [H^+]}$ and $\boxed{[H^+] \equiv 10^{-pH}}$

You will also need to apply from memory these rules for aqueous solutions at 25°C:

- Pure water has a pH of **7.0.**
- In a *pH-neutral* solution, $[H^+] = [OH^-] = 1.0 \times 10^{-7}$ M and pH = 7.0
- In *acidic* solutions, the pH is *less* than **7**. In *basic* solutions (also called **alkaline** solutions), the pH is *greater* than **7**.
- *A lower* pH means a *higher acidity.* A higher pH means a higher basicity.
- Because pH is a *negative* log, *increasing* $[H^+]$ by a factor of 10 *lowers* the pH value by one.

Units and Logarithms

Mathematically, a logarithmic value cannot have units, and logarithms cannot be taken of values with units attached. All precisely stated scientific relationships must obey these rules, but some of the equations we write in chemistry are simplified to speed problem solving. To simplify pH calculations, we apply this rule:

A pH value is not assigned a unit, but when a concentration is calculated based on a pH, the unit *mol/L* (M) must be added to the answer.

Integer pH

In aqueous solutions, if the value of $[H^+]$ *or* $[OH^-]$ *or* pH is known, the other two values can be calculated. When pH values are integers, finding pH does not require a calculator.

▶ **TRY IT**

Q. Apply the pH rules to these problems.

   a. In an aqueous solution, if $[H^+] = 10^{-1}$ M, find the pH.

   b. If $[H^+] = 0.001$ M, find the pH. Is the solution acidic or basic?

   c. If pH = 9, what is the $[OH^-]$? Is the solution acidic or basic?

🛑

Answers:

   a. $\boxed{pH \equiv -\log [H^+]}$ and $\boxed{[H^+] \equiv 10^{-pH}}$

   $[H^+] \equiv 10^{-pH} = 10^{-1}$ M; **pH = 1**

   b. $[H^+] \equiv 10^{-pH} = 0.001$ M $= 10^{-3}$ M; **pH = 3**

   Because the pH is less than 7, the solution is **acidic**.

   c. Because $[H^+] = 10^{-pH} = 10^{-9}$ M, $[OH^-] = 10^{-5}$ M based on K_w. A **basic** solution has a pH greater than 7.0. In a basic solution, $[OH^-]$ is always larger than $[H^+]$, as is the case in this aqueous solution.

PRACTICE **A**

1. In your notebook, design flashcards for the shaded rules so far in this lesson. Run the flashcards until perfect, and then complete the problems that follow.

   For these problems, write answers as integers or integer powers of 10, and solve using mental or "pencil-and-paper" arithmetic (not a calculator). In each part (c), circle the correct answer.

2. If $[H^+] = 10^{-4}$ M,

   a. $[OH^-] =$ b. pH =   c. Acidic or basic solution?

3. If pH = 8,

   a. $[H^+] =$ b. $[OH^-] =$ c. Acidic or basic solution?

4. If $[OH^-] = 10^{-3}$ M,

   a. $[H^+] =$ b. pH =   c. Acidic or alkaline solution?

(continued)

5. If $[OH^-] = 10^{-14}$ M,

    a. $[H^+] =$ b. pH =   c. Acidic or basic solution?

6 If $[H^+] = 10$ M,

    a. $[OH^-] =$ b. pH =   c. Acidic or basic solution?

From [H⁺] to Decimal pH

If the $[H^+]$ is not a whole-number power of 10, you can convert among $[H^+]$, $[OH^-]$, and pH using a calculator.

TRY IT

Q. For an aqueous solution, if $[H^+] = 5.0 \times 10^{-4}$ M, find the pH.

Answer:

See pH? Write: $\boxed{pH \equiv -\log [H^+]}$ and $\boxed{[H^+] \equiv 10^{-pH}}$

    WANTED: pH = ?   Use the equation that solves *for pH from* $[H^+]$.

$$pH \equiv -\log [H^+] = -\log(5.0 \times 10^{-4}) = -(-3.30) = \textbf{3.30}$$

To simplify calculator use when solving a "minus the log" term, solve stepwise: First take the log of the exponential notation, then apply the minus sign to change the sign of the answer. For any solution with an $[H^+]$ lower than 1.0 M, the pH must be *positive*. If the $[H^+]$ is greater than 1.0 M, which can occur in relatively concentrated solutions of strong acids, the pH will be a *negative* number.

Checking pH Calculations

Does the answer for the previous Try It above make sense? For $[H^+]$:

$$\textbf{1.0} \times 10^{-4} < \textbf{5.0} \times 10^{-4} < \textbf{10.} \times 10^{-4} \equiv \textbf{1.0} \times 10^{-3}$$

pH = **4.00**   pH = ?   pH = **3.00**

Based on the above, the pH should be between 4 and 3, which **3.30** is.

The estimation logic above can be used to *check* pH calculations by:

The pH Check

A decimal pH value rounded *up* to the next whole number must equal the number after the exponential minus sign when $[H^+]$ is written in scientific notation.

For the above Try It, pH = **3.30** rounds *up* to **4**. The [H$^+$] exponential term in scientific notation must therefore be 10^{-4}, and it is.

The check rule can also be used to do a quick estimate of pH from [H$^+$].

▶ TRY IT

Q. If [H$^+$] = 3.0 × 10^{-8} M, estimate the pH, and then find the precise pH.

Answer:

The pH must be a decimal number that rounds up to 8, so it must be **7.xx**

The precise answer is pH ≡ −log [H$^+$] = −log(3.0 × 10^{-8}) = **7.52**   Check!

PRACTICE B

1. Design a flashcard for the pH-check rule. Using the rule, *estimate* the pH for the questions that follow.

    a. Problem 2: pH ≈   b. Problem 3: pH ≈   c. Problem 5: pH ≈

For problems 2–6, solve in your notebook and use a calculator.

2. Find pH when [H$^+$] = 2 × 10^{-3} M.

3. If [H$_3$O$^+$] = 8.2 × 10^{-11} M, pH = ?

4. If [OH$^-$] = 2.0 × 10^{-4} M, find the pH.

5. Find the pH of a 0.040 M HCl solution.

6. If 0.012 gram of NaOH is dissolved to make 100. milliliters of solution, what is the solution pH?

From Decimal pH to [H$^+$]

To convert from decimal pH to [H$^+$], apply the pH prompt and the rules for antilogs.

▶ TRY IT

Q. If the pH of a solution is 9.70, what is the [H$^+$]?

Answer:

See pH? Write: pH ≡ −log [H$^+$] and [H$^+$] ≡ 10^{-pH}

The equation that finds $[H^+]$ from pH is $[H^+] \equiv 10^{-pH}$. $[H^+]$ therefore equals $\mathbf{10^{-9.70}}$ **M.** Mathematically that's correct, but in chemistry we express large or small values in scientific notation. Complete the steps of that conversion.

$$[H^+] = 10^{-9.70} \text{ M} = \mathbf{2.0 \times 10^{-10}} \text{ M}$$

When a concentration is calculated from a pH, the unit M must be added to the answer. Apply the *pH check* to the answer above.

The pH = 9.70 rounds up to the integer 10, so the $[H^+]$ exponential in scientific notation should be 10^{-10}, and it is.

PRACTICE C

Either create and practice a summary list for the pH rules or practice flashcards for the rules (or do both). Then solve the following problems in your notebook.

1. For pH = 5.5,

   a. $[H^+] =$ b. Does the pH check?

2. For pH = 8.20,

   a. $[H^+] =$ b. Does the pH check?   c. $[OH^-] =$

3. In an HCl solution of pH = 3.60, find:

   a. $[H^+]$ b. $[OH^-]$ c. $[Cl^-]$ d. $[HCl]_{as\ mixed}$

4. If pH = 1.7, $[H^+] =$

5. If pH = 12.5, $[H^+] =$

6. If pH = –0.50, $[H^+] =$

7. Which solution in problems 3–6 is the most acidic?

Lesson 23.3   Weak Acids and K_a Expressions

Strong acids ionize 100% to form H^+ ions when dissolved in water. Weak acids can be defined as substances that, when dissolved in water, form H^+ ions but do so only slightly. For strong acid ionization, equilibrium favors the products, but for weak acids, equilibrium favors the reactants.

Vinegar is a familiar weak acid solution: a mixture composed primarily of water and *acetic acid,* a weak acid that is soluble in water. The vinegar solution in oil-and-vinegar salad dressings is about one volume of acetic acid per 10 to 20 volumes of water.

In water, acetic acid ionizes slightly to produce H^+ ions. As the products form, the reverse reaction of the proton returning to the acetate ion also occurs, and a solution will quickly reach an *equilibrium* state where no further net change takes place.

The behavior of acetic acid in water can be represented by the equation:

$$1\ CH_3COOH \leftrightarrows 1\ H^+ + 1\ CH_3COO^- \qquad \text{(Goes ~1%.)}$$

In equations representing weak acid ionization, if no state is shown after a particle formula, assume the state is (*aq*).

[H$^+$] in Weak versus Strong Acid Solutions

A weak acid solution has fewer H$^+$ ions and has a *higher* pH (is less acidic) than a strong acid solution that has the same acid concentration.

> **Example:** In **0.10 M** HCl, a solution of a *strong* acid, the approximate concentrations and pH include
>
> [HCl not ionized] = **0 M**; **[H$^+$]** = **0.10 M** = 10^{-1} M; pH = **1**
>
> In **0.10 M** *acetic* acid, roughly 1 acetic acid particle per 100 is ionized. In this solution,
>
> [CH$_3$COOH] $\approx$ 0.099 M $\approx$ **0.10 M**; **[H$^+$]** $\approx$ **0.001 M** $\approx$ 10^{-3} M; pH $\approx$ **3**
>
> Compare the [H$^+$] in the two solutions. At the same mixed concentrations, the *hydrochloric* acid solution has about 100 times more protons than the *acetic* acid solution.

Vinegar can be safely used as a food ingredient because acetic acid solutions do not contain high concentrations of the reactive H$^+$ ions. However, even at low concentrations, the [H$^+$] in weak acid solutions can have a major influence on reactions.

Conjugate Bases

The ionization of a strong or weak acid can be represented by this general equation:

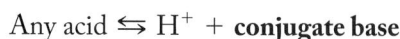

$$\text{Any acid} \leftrightarrows H^+ + \textbf{conjugate base}$$

> Conjugate base is the term for the particle remaining after a proton leaves an acid. The formula for the conjugate base is the acid formula with *one less H* atom and *one less positive charge*.

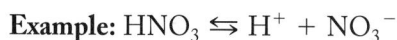

> **Example:** $HNO_3 \leftrightarrows H^+ + NO_3^-$
>
> For nitric acid, nitrate ion is the conjugate base.

An acid can be defined as any particle that can *lose a proton*. This means that acids can be positive or negative ions as well as electrically neutral particles.

▶ **TRY IT**

Q. Fill in the conjugate base for the ionization of these acids.

a. $HCl \leftrightharpoons H^+ +$

b. $NH_4^+ \leftrightharpoons H^+ +$

c. $H_2PO_4^- \leftrightharpoons H^+ +$

STOP

Answers:

a. $HCl \leftrightharpoons H^+ + \mathbf{Cl^-}$ b. $NH_4^+ \leftrightharpoons H^+ + \mathbf{NH_3}$ c. $H_2PO_4^- \leftrightharpoons H^+ + \mathbf{HPO_4^{2-}}$

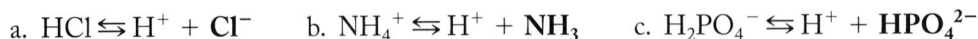

Representing Weak Acid Ionization

The ionization of a specific weak acid can be represented by three formats: general, specific ionization, and hydrolysis. For example, for the weak acid hydrofluoric acid in an aqueous solution, its acidic behavior can be represented by:

- A general reaction: 1 weak acid $\leftrightharpoons$ 1 H^+ + 1 conjugate base   (Goes ~1%.)
- As an *ionization*: $HF \leftrightharpoons H^+ + F^-$ (Goes ~1%.)
- As a *hydrolysis*: $HF + H_2O(\ell) \leftrightharpoons H_3O^+ + F^-$ (Goes ~1%.)

Recall that hydrolysis is a reaction with water, and H^+ and H_3O^+ are equivalent ways of representing the proton released by an acid. These three equations are *equivalent* ways of representing the same reaction.

▶ **TRY IT**

Q. Assume the first particle in each reaction is acting as an *acid*. Complete the reaction by writing formulas for the products. Circle the conjugate base.

a. $HCN(aq) \leftrightharpoons$

b. $NH_4^+ + H_2O(\ell) \leftrightharpoons$

STOP

Answers:

a. $HCN(aq) \leftrightharpoons H^+(aq) +$ $\boxed{CN^-(aq)}$

An acid ionizes to produce a proton (H^+) and the acid's conjugate base.

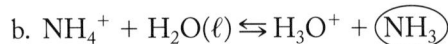

b. $NH_4^+ + H_2O(\ell) \leftrightharpoons H_3O^+ +$ $\boxed{NH_3}$

When a weak acid ionization is written in the hydrolysis format (losing a proton by donating the proton to water), one of the product particles is always the hydronium ion (H_3O^+) and the other is the conjugate base.

All reaction equations must be balanced for atoms and charge. Check the balancing in answers (a) and (b).

PRACTICE   A

Assuming the first particle is acting as a weak acid in an aqueous solution, write the formulas for the products. Circle the conjugate base.

1. $HI \rightleftharpoons$

2. $HCO_3^- \rightleftharpoons$

3. $HS^- + H_2O(\ell) \rightleftharpoons$

4. $HPO_4^{2-} + H_2O(\ell) \rightleftharpoons$

The K_a Expression

For any reaction in which one acid particle (strong or weak) ionizes to form one H^+ (or reacts with water to form one H_3O^+), the equilibrium constant K is given a special name, the **acid dissociation constant**, and a special symbol, K_a. By convention, acid ionization or hydrolysis is written with one un-ionized acid particle on the left and one H^+ (or H_3O^+) on the right.

For all acid ionization reactions:

$$Acid \rightleftharpoons H^+ + conjugate\ base \qquad K_a = acid\ dissociation\ constant$$

$$K_a \equiv \frac{[H^+]_{at\ eq.} \cdot [conjugate\ base]_{at\ eq.}}{[nondissociated\ acid]_{at\ eq.}}$$

▶ TRY IT

Q. Write the K expression for the weak acid behavior represented by:

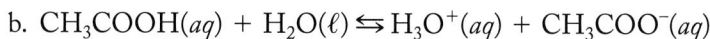

a. $CH_3COOH(aq) \rightleftharpoons H^+(aq) + CH_3COO^-(aq)$

b. $CH_3COOH(aq) + H_2O(\ell) \rightleftharpoons H_3O^+(aq) + CH_3COO^-(aq)$

Answers:

Because each reaction is an acid releasing one proton, the K is a K_a. For the above two reactions:

a. $K_a = \dfrac{[H^+]_{at\ eq.}[CH_3COO^-]_{at\ eq.}}{[CH_3COOH]_{at\ eq.}}$ b. $K_a = \dfrac{[H_3O^+][CH_3COO^-]}{[CH_3COOH]}$

A K_a expression must include $[H^+]^1$ *or* $[H_3O^+]^1$ on *top* and $[nondissociated\ acid]^1$ on the bottom.

For all K_a expressions:

- Because H^+ and H_3O^+ are equivalent, the two K_a expressions are equivalent, and one format may be substituted for the other.

- If the state of a particle is omitted, it is assumed to be aqueous.

- Because a K is based on concentrations at equilibrium, if a label after a concentration term is omitted, $[\ ]_{at\ equilibrium}$ is understood.

- The powers of all of the concentrations are **1** and are omitted as understood.

- For the hydrolysis reaction, $[H_2O]$ is represented by a 1 in the K_a expression.

Polyprotic Acids

A polyprotic acid can ionize to lose more than one proton, but each successive hydrogen ionizes to a lesser extent than the previous hydrogen. In K calculations, these successive ionizations are written separately, with the conjugate base of the first ionization becoming the weak acid in the second, etc. A K_a value always refers to the ionization of a weak acid to produce *one* H^+ or *one* H_3O^+.

PRACTICE B

Assuming the first particle is acting as an acid in an aqueous solution, complete the reaction by writing the formulas for the products, then write the K_a *expression* for each reaction.

1. $H_2CO_3 \leftrightarrows$

2. $HCO_3^- \leftrightarrows$

3. $H_2S + H_2O(\ell) \leftrightarrows$

4. $HS^- + H_2O(\ell) \leftrightarrows$

K_a Values

Each weak acid has a characteristic K_a *value* at a given temperature.

Sample K_a Values at 25°C

Hydrochloric acid (strong)	$HCl \leftrightarrows H^+ + Cl^-$	K_a = very large
Hydrofluoric acid (weak)	$HF \leftrightarrows H^+ + F^-$	$K_a = 6.8 \times 10^{-4}$
Acetic acid (weak)	$CH_3COOH \leftrightarrows H^+ + CH_3COO^-$	$K_a = 1.8 \times 10^{-5}$
Hydrocyanic acid (weak)	$HCN \leftrightarrows H^+ + CN^-$	$K_a = 6.2 \times 10^{-10}$

For strong acids, ionization strongly favors the products, and the K_a values for strong acids will be a number much greater than 1. Because strong acid ionization is considered to go to completion rather than to equilibrium, the K value is not needed for calculations.

Weak acids have K_a values between 1.0 and 10^{-16}. Written in scientific notation, each will have a negative power of 10. Comparing two weak acids, the weaker acid has a lower fraction of hydrogen atoms that ionize and a lower K_a value.

TRY IT

Q. In the preceding table of K_a values, which is the weakest acid?

Answer:

HCN, because it has the lowest K_a.

As the temperature of a weak acid solution rises, more bonds to H break, more protons form, and K_a values increase. However, unless otherwise noted, assume K_a calculations are based on reactions at 25°C (77°F).

For each H atom in a compound, a K_a value can be measured that represents the tendency of the H to ionize. However, if a K_a value is less than 10^{-16}, the H will react as an acid only with bases that are *very* strong: stronger than hydroxides. Under most circumstances, an H with a K_a of less than 10^{-16} is considered a nonreactive (nonacidic) hydrogen (see Lesson 14.1).

PRACTICE   C

1. In your notebook, make a summary list of the shaded rules in this lesson. Design a flashcard or two that defines or gives an example for each rule. Run the cards until each has been recalled correctly, then complete the problems below.

2. For the weak acid NH_4^+, the K_a value is 5.6×10^{-10}.

    a. Write the reaction for which this is the K value.

    b. Write the K_a expression for this reaction.

    c. What term does a K_a expression always have by itself in its denominator?

3. The ion HSO_3^- has a K_a value of 6.2×10^{-8}.

    a. Write the reaction for which this is the K value.

    b. Write the K_a expression for this reaction.

4. Which ion is the weaker acid: NH_4^+ or HSO_3^-?

Lesson 23.4 K_a Equations

For all *weak* acid (WA) ionizations, the general R and E (Reaction and Extent) can be written

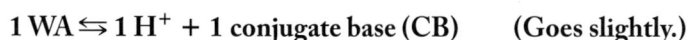

$$1\,WA \rightleftharpoons 1\,H^+ + 1 \text{ conjugate base (CB)} \qquad \textbf{(Goes slightly.)}$$

Q. Write the K expression for the general reaction above.

Answer:

$$K_a \equiv \frac{[H^+]_{at\ eq.} \cdot [CB]_{at\ eq.}}{[WA]_{at\ eq.}} \qquad \text{(Definition)}$$

Because this ionization forms H^+, the K is a K_a. The $[\ ]_{at\ equilibrium}$ or "at eq." subscript is often omitted as understood in K expressions, but in this case it will be helpful in the discussion that follows.

In Lesson 21.3, to solve K calculations we used a *RICE* table. However, compared to general K calculations, *RICE* tables for weak acid ionization are simplified because

- The weak acid is always a reactant.
- One reaction product is always $[H^+]$ or $[H_3O^+]$ and the other is the conjugate base.
- The coefficients are always **1** weak acid particle consumed *equals* **1** H^+ ion formed *and* **1** conjugate-base particle formed.

This means:

> For all weak acid ionization or hydrolysis reactions, the *RICE* table for the *general* reaction is the same.

Because all weak acid ionization calculations are similar, instead of writing a full *RICE* table we can write symbols for the concentrations at equilibrium. Those symbols will be the same in each problem.

Let's walk through the logic for applying the *WRECK* steps to weak acid ionization.

1. **Use x to represent small concentrations.** The coefficients in a weak acid ionization are all 1:

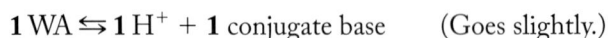

   $$1\,WA \rightleftharpoons 1\,H^+ + 1 \text{ conjugate base} \qquad \text{(Goes slightly.)}$$

   This means that the three following concentrations are equal and can all be represented by an x that represents a relatively small concentration compared to the original weak acid concentration:

   $$\text{Small } [WA]_{that\ ionizes} = [H^+]_{formed} = [CB]_{formed} = x = small$$

2. **Concentrations at equilibrium.** In a solution, the $[WA]$ at equilibrium, after ionization, is usually difficult to measure directly. However, the mol/L

of weak acid in a solution that we mix is usually a quantity that we know. Substituting x as defined above simplifies the math. By definition:

$$[\text{WA}]_{\text{at eq.}} \equiv [\text{WA}]_{\text{mixed}} - \text{small } [\text{WA}]_{\text{that ionizes}} \equiv [\text{WA}]_{\text{mixed}} - x$$

If we can solve for x, we solve for all of these quantities:

$$x = [\text{H}^+]_{\text{formed}} = [\text{CB}]_{\text{formed}} = [\text{WA}]_{\text{that } ionizes} \text{ and } [\text{WA}]_{\text{at eq.}} (\equiv [\text{WA}]_{\text{mixed}} - x)$$

▶ TRY IT

Q. Substitute a term for each concentration into the blanks in the **C** step below. Write a term that *includes x* and uses the symbols for measurable concentrations in the equalities above.

Reaction and Extent: 1 WA $\leftrightarrows$ 1 H$^+$ + 1 CB   (Goes slightly.)

$\land$ $\land$ $\land$

Conc. at eq.:   _____   ____ ____

Answer:

Reaction and Extent: 1 WA $\leftrightarrows$ 1 H$^+$ + 1 CB   (Goes slightly.)

$\land$ $\land$ $\land$

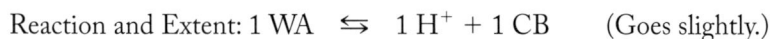

Conc. at eq.: $[\text{WA}]_{\text{mixed}} - x$ x x

The **C** row above will be the same for *all* weak acid ionization reactions.

3. **The K step.** For each of the three *concentration* terms in the K_a definition below, we can substitute a term from the **C** row that includes an x.

▶ TRY IT

Q. Substitute the terms from the **C** row to fill in the top and bottom of the *fraction* term below. Each term will include x.

$$K_a \equiv \frac{[\text{H}^+]_{\text{at eq.}} \cdot [\text{CB}]_{\text{at eq.}}}{[\text{WA}]_{\text{at eq.}}} \equiv \underline{\hspace{3cm}}$$

Answer:

$$K_a \equiv \frac{[\text{H}^+]_{\text{at eq.}} \cdot [\text{CB}]_{\text{at eq.}}}{[\text{WA}]_{\text{at eq.}}} \equiv \frac{x^2}{[\text{WA}]_{\text{mixed}} - x}$$

$\land$ $\land$

Definition based on R row   *Exact* based on C row

Both the *definition* and *exact* K_a equations are true at equilibrium, but the *exact* equation has an advantage. Our goal is usually to solve for x, and the exact equation can be solved for x if the K_a value and the [WA *as mixed*] are known (as they usually are).

However, because the exact equation has both x^2 and x terms, it is a quadratic equation. Quadratic equations can be converted to the general form

$$ax^2 + bx + c = 0$$

Solving a quadratic equation for x is not difficult but can be time consuming, and for most weak acid calculations, the following approximation may be used that solves for x more *quickly*.

4. **The K approximation.** In a weak acid solution, the value of x is small compared to the value of the [WA] as mixed. This allows us to simplify the measurement of the [WA] at equilibrium. The mathematical rule is:

> In approximations, you can *ignore small* numbers if they are *added to* or *subtracted from* much larger numbers.

► TRY IT

Q. Using the approximation rule, write the simplified, *approximate* term in the blank below.

$$[WA]_{\text{at equilibrium}} \equiv [WA]_{\text{mixed}} - x \approx \underline{\hspace{2cm}}$$

Answer:

$$[WA]_{\text{at equilibrium}} \equiv [WA]_{\text{mixed}} - x \approx \mathbf{[WA]_{mixed}}$$

When a small quantity is added to or subtracted from a larger quantity, the larger quantity remains approximately the same. In a weak acid solution, the weak acid concentration at equilibrium is not very different from the concentration as mixed, which is a value we usually know.

► TRY IT

Q. Write an approximation for the exact equation.

$$K_a \equiv \frac{[H^+]_{\text{eq.}} [CB]_{\text{eq.}}}{[WA]_{\text{at eq.}}} \equiv \frac{x \cdot x}{[WA]_{\text{mixed}} - x} \approx \boxed{\underline{\hspace{2cm}}}$$

$\qquad\qquad$ ∧ $\qquad\qquad\qquad$ ∧ $\qquad\qquad\qquad$ ∧

$\qquad\quad$ Definition $\qquad\qquad$ Exact $\qquad\qquad$ Approximation

Answer:

The K_a expression can be simplified to this approximation:

$$K_a \equiv \frac{[H^+]_{\text{eq.}} [CB]_{\text{eq.}}}{[WA]_{\text{at eq.}}} \equiv \frac{x^2}{[WA]_{\text{mixed}} - x} \approx \boxed{\frac{x^2}{[WA]_{\text{mixed}}}}$$

$\qquad\qquad$ ∧ $\qquad\qquad\qquad$ ∧ $\qquad\qquad\qquad$ ∧

$\qquad\quad$ Definition $\qquad\qquad$ Exact $\qquad\qquad$ Approximation

For K_a calculations, we will first solve the approximation equation, and in most cases the approximation will be an acceptable answer.

Weak Acid Ionization and Particle Concentrations

As a foundation for additional topics of importance in chemistry and other scientific disciplines, you will need to write the *WRECK* steps for general weak acid ionization from memory, including the definition, exact, and approximate forms of the K expression.

For Calculations Involving the Ionization of a Weak Acid

Write the *WRECK* steps as follows:

WANTED: Write the symbol for the term that is WANTED.

Rxn. and Extent: $1\,WA \;\rightleftharpoons\; 1\,H^+ \;+\; 1\,CB$ (Goes slightly.)

Conc. at eq.: $[WA]_{mixed} - x \qquad x \qquad x$

$$K_a \equiv \frac{[H^+]_{eq.}\,[CB]_{eq.}}{[WA]_{at\ eq.}} \equiv \frac{x^2}{[WA]_{mixed} - x} \approx \frac{x^2}{[WA]_{mixed}} \approx K_a$$

Definition   Exact   Approximation

Practice

1. For the ionization of a weak acid (WA) to form a proton and its conjugate base (CB):

   a. Write just the *REC* steps. In the *C* row, each term should include x.

   b. Write the K_a definition based on the *R* row reaction.

   c. Write the K_a exact equation using *C* row terms that include x.

   d. Write the K_a approximation equation.

2. For the ionization of the weak acid NH_4^+:

   a. Write the *R*eaction and its *E*xtent, using the particle formulas in the reaction.

   b. Below the *R*eaction and *E*xtent, write the *C*oncentrations at equilibrium, defined so that each term includes x.

   c. Write the K_a definition equation, using the particle formulas.

   d. Write the K_a exact equation using the terms defined in the **Concentrations at equilibrium.**

   e. Write the K_a approximation equation based on your answer to part (d).

Lesson 23.5   Math for K_a Calculations

Squares and Square Roots of Exponential Notation

K_a calculations involve both squaring and taking the square root of exponential notation. Rules for powers and roots were covered in Lesson 21.1. The following additional rules are specific to square roots.

1. A precise square root of exponential notation can be found without entering the exponent into the calculator. The steps are:

   a. Write the square root as the exponential notation taken to the 1/2 power.

   b. Make the *exponent even*. If the exponent is odd, make the *exponent* 10 times (one number) *smaller* and the *significand* 10 times *larger*.

   c. Take the square root of the *significand* on the calculator. Take the square root of the exponential term by multiplying the exponent by 1/2.

   Example:
   $$\sqrt{2.5 \times 10^{-7}} = (2.5 \times 10^{-7})^{1/2} = (25 \times 10^{-8})^{1/2} = 5.0 \times 10^{-4}$$

2. When solving for a square root, there are two results: one positive and one negative: $\sqrt{4} = \pm 2$ When solving science problems, usually only one root will make sense. The general rule is: Choose the one root that makes sense in the problem. In this chapter, assume the needed root is the *positive* root.

3. To estimate a square root, make the exponent even if needed, estimate the square root of the significand, and then attach the root of the exponent solved by mental arithmetic.

PRACTICE

Do the *last lettered part* of each numbered problem, then more *if* you need more practice. If you need additional review of the rules, see Lesson 21.1.

1. *Estimate* the positive square root of:

   a. 45   b. 95   c. 7

2. Use a calculator to take a square root, then compare your answers to problem 1.

   a. 45   b. 95   c. 7

For problems 3–7, convert final answers to scientific notation.

3. Solve without a calculator.

   a. $(7.0 \times 10^{-5})^2 =$

   b. $\sqrt{0.0064} =$

4. Solve Q3a and Q3b using the square and square root functions on your calculator.

    a. $(7.0 \times 10^{-5})^2 =$ b. $\sqrt{0.0064} =$

5. Use a calculator for the significand, but solve for the exponential term by inspection.

    a. $(2.56 \times 10^{-4})^{1/2} =$

    b. $\sqrt{1.44 \times 10^{-6}} =$

6. Estimate the square root. Need help? See rule 3 above.

    a. $(4.2 \times 10^{-5})^{1/2} =$

    b. $\sqrt{7.2 \times 10^{-3}} =$

7. Find an exact square root. Use a calculator as needed. Compare answers to Q6.

    a. $(4.2 \times 10^{-5})^{1/2} =$ b. $\sqrt{7.2 \times 10^{-3}} =$

Lesson 23.6 K_a Calculations

The *WRECK* Steps for Weak Acid Ionization

Weak acids ionize only slightly, going to equilibrium rather than to completion. Our general rule is:

> In calculations involving concentrations for a reaction that goes to equilibrium, write the *WRECK* steps and use the K equation to solve.

For weak acid ionization, our rule will be:

The K_a Prompt

In calculations involving K_a or concentrations in weak acid solutions, write the *WRECK* steps and use the K_a approximation to solve.

We will implement this rule by applying the following steps:

1. Write the equation for ionization of the specific weak acid.

2. Write the *WRECK* steps for *general* weak acid ionization.

(continued)

3. Make a DATA table using the symbols in the approximation equation; fill in the problem's specific numbers and particle formulas.

4. Solve the K_a approximation equation for the symbol WANTED.

K_a calculations follow the same conventions as other K calculations:

- Units are not attached to K values.
- Units are omitted when substituting numbers into a K expression.
- When a concentration is WANTED, the *unit* moles/liter (M) must be added to the answer.

▶ TRY IT

Q. In a 2.0 M acetic acid (CH_3COOH) solution, calculate [H^+]. Use $K_a = 1.8 \times 10^{-5}$. Try to solve this problem without a calculator.

STOP

Answer:

Specific reaction: $1\,CH_3COOH \leftrightharpoons 1\,H^+ + 1\,CH_3COO^-$

WANTED: $[H^+] = x = ?$

R and E: $1\,WA \quad\leftrightharpoons\quad 1\,H^+ + 1\,CB$ (Goes slightly.)

Conc. at eq.: $[WA]_{mixed} - x \qquad x \qquad x$

K: $K_a \equiv \dfrac{[H^+]_{eq.}[CB]_{eq.}}{[WA]_{at\,eq.}} \equiv \dfrac{x^2}{[WA]_{mixed}-x} \approx \boxed{\dfrac{x^2}{[WA]_{mixed}}} \approx K_a$

Definition   Exact   Approximation

DATA: $x = [H^+] = [CH_3COO^-] = ? = $ WANTED

$[WA]_{as\,mixed} = [CH_3COOH]_{as\,mixed} = 2.0\,M$

$K_a = 1.8 \times 10^{-5}$

SOLVE the approximation equation for the term that includes the symbol WANTED, then solve for the symbol WANTED.

$$x^2 = (K_a)([WA]_{mixed}) = (1.8 \times 10^{-5})(2.0) = 3.6 \times 10^{-5} = 36 \times 10^{-6}$$

$$x = 6.0 \times 10^{-3}\,M = [H^+]$$

If the value of x solved by the approximation is more than 5% of the [WA], the quadratic equation is generally solved to obtain a more precise answer, but in this introductory text we will restrict our attention to problems in which the approximation gives acceptable results.

PRACTICE   A

When in doubt, check your answers after each part.

1. Dissolved in water, hydrogen cyanide (HCN) forms a weak acid solution.

   a. Calculate the $[H^+]$ in a 0.50 M HCN solution ($K_a = 6.2 \times 10^{-10}$).

   b. Calculate the pH of the HCN solution.

2. At a nonstandard temperature, the $[F^-]$ in a 0.25 M hydrofluoric acid (HF) solution is measured to be 0.012 molar.

   a. Find the K_a value for HF at the temperature in this experiment. Solve first using the K_a approximation equation, and then solve with the exact K_a equation.

   b. Find the pH of the solution.

The K_a Quick Steps

It is important to be able to solve K_a calculations by writing out the methodical *WRECK* steps. For calculations involving weak acids mixed with other particles, we will need those methodical steps. However, once you master the *WRECK* steps, you may solve K_a calculations more quickly by starting at the *K* step: Write the approximation equation and its DATA table, then solve. The rule is:

The K_a Quick Steps

See K_a and $[H^+]$ or pH? Write and SOLVE the K_a approximation equation.

The steps are:

1. Write the general K_a approximation equation: $\boxed{K_a \approx \dfrac{x^2}{[WA]_{mixed}}}$

2. Make a DATA table using the equation symbols.

3. Solve the K_a approximation for the symbol WANTED.

TRY IT

Q. Calculate the $[H^+]$ in 0.40 M hypochlorous acid.

$$HOCl \leftrightarrows H^+ + OCl^- \qquad K_a = 3.5 \times 10^{-8}$$

STOP

Answer:

$$K_a \approx \frac{x^2}{[WA]_{mixed}}$$

DATA: $x = [H^+] = [OCl^-] = ? =$ WANTED

$[WA]_{mixed} = [HOCl]_{mixed} = 0.40$ M

$K_a = 3.5 \times 10^{-8}$

SOLVE: $x^2 = (K_a)([WA]_{mixed}) = (3.5 \times 10^{-8})(0.40) = 1.4 \times 10^{-8}$

$x = \mathbf{1.2 \times 10^{-4}\,M} = \mathbf{[H^+]}$

PRACTICE   B

1. Using the quick steps, calculate the pH of a 0.50 M formic acid (HCOOH) solution ($K_a = 1.8 \times 10^{-4}$).

2. A 0.25 molar solution of a monoprotic weak acid has a pH of 4.49.

   a. Calculate K_a b. Find the $[OH^-]$ in this solution.

SUMMARY

Research on learning shows that simply highlighting text is not an efficient way to study, but there are ways to use highlighting as a part of study that can help in building memory.

Try this: When you read a passage with unfamiliar content, underline or highlight words with meanings that you cannot recall or relationships that are not in memory, then convert the highlighted text to flashcards that you practice. In this way, highlighting can be an important first step in effective learning.

Similarly, simply re-reading a text or notes generally does not result in efficient learning. However, if you re-read a text *after* moving into memory the meaning of unfamiliar vocabulary, re-reading can assist in gaining an intuitive sense of when new information needs to be activated in memory. When reading about a new topic, stopping frequently to answer questions about content, and re-reading as needed to find answers, can also help to drive new facts, procedures, and concepts into memory. In learning science, building long-term memory is the goal.

REVIEW QUIZ

1. $Log(5 \times 10^{-5}) =$ 2. $Log\ x = -6.30; x =$

3. If pH = 7.22:

   a. $[H^+] =$ b. $[OH^-] =$

4. In a KOH solution of pH = 11.3, find:

   a. $[H^+]$ b. $[OH^-]$ c. $[K^+]$ d. $[KOH]_{as\ mixed}$

5. $(2.5 \times 10^{-5})^2 =$

6. Solve without a calculator: $(81 \times 10^{-16})^{1/2} =$

7. Benzoic acid ionizes to form the benzoate ion by this reaction.

$$C_6H_5COOH \rightleftharpoons H^+ + C_6H_5COO^-   K_a = 6.3 \times 10^{-5}$$

   In a 0.040 M benzoic acid solution:

   a. Calculate $[H^+]$ using the quick steps.   b. Find the pH.

8. In a 0.50 M solution of a weak acid, the concentration of the base conjugate is measured to be 4.7×10^{-3} M. Using the approximation equation, find the K_a under these conditions.

ANSWERS

Lesson 23.1

Practice A   1. $10^{+16.5} = \mathbf{3.16 \times 10^{16}}$ 2. $10^{-16.5} = \mathbf{3.16 \times 10^{-17}}$ 3. $10^{-0.7} = \mathbf{2.00 \times 10^{-1}}$

Practice B   2a. 4×10^{-6} (±1?) ✓   2b. 3×10^{-12} (±1?) ✓   2c. 3×10^{-1} (±1?) ✓   2d. 0.3

3a. 12.83   3b. -11.17 (±1?) ✓   3c. 0.66   3d. $Log\ 0.0020 = log\ (2.0 \times 10^{-3}) = \mathbf{-2.70}$

Practice C   2a. 2.51×10^{12} 2b. 1.3×10^{-6} 2c. 5.6×10^{-1}

2d. 5×10^{-4} 2e. 1.26×10^1

Lesson 23.2

Practice A   2a. 10^{-10} M   2b. 4   2c. Acidic

3a. 10^{-8} M   3b. 10^{-6} M   3c. Basic

4a. 10^{-11} M   4b. 11   4c. Alkaline

5a. 1.0 M   5b. pH = ?, $[H^+] = 1.0 = 10^0 = 10^{-pH} = 10^{-0}$; **pH = 0**

In the 1.0 M solution of a strong monoprotic acid, the pH is 0.

5c. **Acidic.** pH = 0, which is less than 7; $[H^+]$ is higher than $[OH^-]$.

6a. 10^{-15} M   6b. $[H^+] = 10^1 = 10^{-pH} = 10^{-(-1)}$; **pH = −1** (pH is negative when $[H^+] > 1.0$ M)

6c. Highly acidic.

Practice B   1a. Problem 2: pH = **2.x** (Rounds up to 3.)   1b. Problem 3: pH = **10.xx** (Rounds up to 11.)

1c. Problem 5: 0.040 M HCl = 4×10^{-2} M H$^+$; **pH = 1.xx**

2. See pH? Write $\boxed{pH \equiv -\log[H^+]}$ and $\boxed{[H^+] \equiv 10^{-pH}}$

When pH is WANTED, use pH $\equiv -\log[H^+]$

pH $= -\log(2 \times 10^{-3}) =$ **2.7**   Estimate was 2.x in answer (1a); they agree.

3. $[H_3O^+] = 8.2 \times 10^{-11} = [H^+]$

If pH is WANTED, use pH $\equiv -\log[H^+] = -\log(8.2 \times 10^{-11}) =$ **10.09**   Check.

4. See pH? Write $\boxed{pH \equiv -\log[H^+]}$ and $\boxed{[H^+] \equiv 10^{-pH}}$

WANT pH, need [H$^+$] to find pH but are given [OH$^-$].

See H$^+$ and OH$^-$? Write $K_w = [H^+][OH^-] = 1.0 \times 10^{-14}$. Use K_w to find [H$^+$]:

$$[H^+] = \frac{1.0 \times 10^{-14}}{[OH^-]} = \frac{1.0 \times 10^{-14}}{2.0 \times 10^{-4}} = 0.50 \times 10^{-10} = 5.0 \times 10^{-11} \text{ M H}^+$$

Then pH $\equiv -\log[H^+] = -\log(5.0 \times 10^{-11}) =$ **10.30**   (Rounds up to 11. Check.)

5. To find pH, need [H$^+$] first. Quick rule: [HCl] = 0.040 M = [H$^+$]

pH $\equiv -\log[H^+] = -\log(0.040) = -\log(4.0 \times 10^{-2}) =$ **1.40**   (Rounded up = 2. Check.)

6. To find pH, you need [H$^+$]. To find [H$^+$], you need [NaOH].

WANTED: $[NaOH]_{\text{as mixed}} = ?\ \dfrac{\text{mol NaOH}}{\text{L soln.}}$

SOLVE: $\dfrac{\text{mol NaOH}}{\text{L soln.}} = \dfrac{0.012 \text{ g NaOH}}{100.\text{ mL soln.}} \cdot \dfrac{1 \text{ mol NaOH}}{40.0 \text{ g NaOH}} \cdot \dfrac{1 \text{ mL}}{10^{-3} \text{ L}} =$ **0.0030 M** = **[NaOH]**

To find [H$^+$], use K_w. In NaOH solutions, [NaOH] = [OH$^-$]

$$[H^+] = \frac{1.0 \times 10^{-14}}{[OH^-]} = \frac{1.0 \times 10^{-14}}{3.0 \times 10^{-3}} = 0.33 \times 10^{-11} = 3.3 \times 10^{-12} \text{ M H}^+$$

pH $= -\log(3.3 \times 10^{-12}) =$ **11.48**   (Base solution, basic pH; 11.48 rounded up = 12. Check.)

Practice C   1a. See pH? Write $\boxed{pH \equiv -\log[H^+]}$ and $\boxed{[H^+] \equiv 10^{-pH}}$

To find [H$^+$] from pH, use: $\boxed{[H^+] \equiv 10^{-pH}} = 10^{-5.5} =$ **3 × 10^{-6} M**

1b. pH = **5.5** rounded up is 6; [H$^+$] exponent is 10^{-6} M. Check.

2a. $\boxed{[H^+] \equiv 10^{-pH}} = 10^{-8.20} =$ **6.3 × 10^{-9} M H$^+$**

2b. pH = **8.20** rounds up to 9, [H$^+$] should = ? × 10^{-9} M. Check.

2c. WANT [OH$^-$]. Know [H$^+$]. Use K_w: $[H^+][OH^-] = 1.0 \times 10^{-14}$

$? = [OH^-] = \dfrac{1.0 \times 10^{-14}}{[H^+]} = \dfrac{1.0 \times 10^{-14}}{6.3 \times 10^{-9}} = 0.16 \times 10^{-5} =$ **1.6 × 10^{-6} M OH$^-$**

3a. $[H^+] = ? = 10^{-pH} = 10^{-3.60} =$ **2.5 × 10^{-4} M**

3b. [OH$^-$] = ? Knowing [H$^+$], to find [OH$^-$], use K_w. [OH$^-$] = **4.0 × 10^{-11} M**

3c, 3d. $[HCl]_{\text{mixed}}$ based on part (a) = **2.5 × 10^{-4} M** = [Cl$^-$]

4. $\boxed{[H^+] \equiv 10^{-pH}} = 10^{-1.7} =$ **2 × 10^{-2} M**   5. $\boxed{[H^+] \equiv 10^{-pH}} = 10^{-12.5} =$ **3 × 10^{-13} M**

6. $\boxed{[H^+] \equiv 10^{-pH}}$ $= 10^{-(-0.50)} = 10^{0.50} =$ **3.2 M** or **3.2 × 10⁰ M**

   Solutions with an $[H^+]$ higher than 1.0 M have a negative pH.

7. The **problem 6** solution has both the lowest pH and the highest $[H^+]$. By either measure it is the most acidic.

Lesson 23.3

Practice A   1. $HI \leftrightharpoons H^+ + \boxed{I^-}$ 2. $HCO_3^- \leftrightharpoons H^+ + \boxed{CO_3^{2-}}$

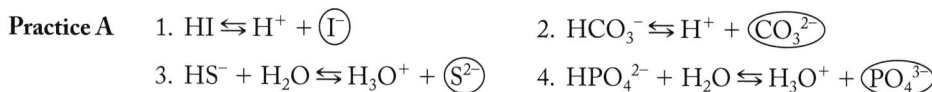

3. $HS^- + H_2O \leftrightharpoons H_3O^+ + \boxed{S^{2-}}$ 4. $HPO_4^{2-} + H_2O \leftrightharpoons H_3O^+ + \boxed{PO_4^{3-}}$

Practice B

In K expressions, all concentrations are assumed to be at equilibrium unless indicated otherwise.

1. $H_2CO_3 \leftrightharpoons H^+ + HCO_3^-$

$K_a = \dfrac{[H^+][HCO_3^-]}{[H_2CO_3]}$

2. $HCO_3^- \leftrightharpoons H^+ + CO_3^{2-}$

$K_a = \dfrac{[H^+][CO_3^{2-}]}{[HCO_3^-]}$

3. $H_2S + H_2O(\ell) \leftrightharpoons H_3O^+ + HS^-$

$K_a = \dfrac{[H_3O^+][HS^-]}{[H_2S]}$

4. $HS^- + H_2O(\ell) \leftrightharpoons H_3O^+ + S^{2-}$

$K_a = \dfrac{[H_3O^+][S^{2-}]}{[HS^-]}$

Practice C   2a. $NH_4^+ \leftrightharpoons H^+ + NH_3$

A K_a is a K for the reaction where an acid loses an H^+ ion. For a reaction with a K_a supplied, an H^+ or H_3O^+ ion will always be on the right side.

2b. $K_a = \dfrac{[H^+][NH_3]}{[NH_4^+]}$

2c. A K_a expression always has [Nondissociated Acid]$_{\text{at equilibrium}}$ alone in its denominator.

3a. $HSO_3^- \leftrightharpoons H^+ + SO_3^{2-}$

3b. $K_a = \dfrac{[H^+][SO_3^{2-}]}{[HSO_3^-]}$

4. The weaker acid has the lower K_a value: **NH₄⁺**

Lesson 23.4

Practice   1a. **R and E:** $1\ WA \leftrightharpoons 1\ H^+ + 1\ CB$ (Goes slightly.)

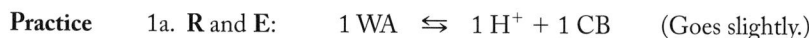

$\qquad\qquad$ **C**$_{\text{at eq.}}$: $[WA]_{\text{mixed}} - x \quad x \quad x$

1b, 1c, 1d. $K_a \equiv \dfrac{[H^+]_{\text{eq.}}[CB]_{\text{eq.}}}{[WA]_{\text{eq.}}} \equiv \dfrac{x^2}{[WA]_{\text{mixed}} - x} \approx \dfrac{x^2}{[WA]_{\text{mixed}}}$

$\qquad\qquad\qquad$ Definition $\qquad\qquad$ Exact $\qquad\qquad$ Approximation

2a. **R** and **E**: $1\,NH_4^+ \leftrightarrows 1\,H^+ + 1\,NH_3$ (Goes slightly.)

2b. **C**$_{at\,eq.}$: $[NH_4^+]_{mixed} - x$ x x

2c. $K_a \equiv \dfrac{[H^+]_{eq.}\,[NH_3]_{eq.}}{[NH_4^+]_{at\,eq.}}$ 2d. $K_a \equiv \dfrac{x \cdot x}{[NH_4^+]_{mixed} - x}$

2e. $K_a \approx \dfrac{x^2}{[NH_4^+]_{as\,mixed}}$

Lesson 23.5

Practice   1a. $45: 6^2 = 36$ and $7^2 = 49$; estimate $\approx$ **6.7**

1b. $95: 9^2 = 81$ and $10^2 = 100$; estimate $\approx$ **9.7**

1c. $7: 2^2 = 4$ and $3^2 = 9$; estimate $\approx$ **2.7**

2a. 6.71   2b. 9.75   2c. 2.65

3a and 3b. Answers should be close to the 4a and 4b answers.

4a. 4.9×10^{-9}

4b. $(64 \times 10^{-4})^{1/2} =$ **8.0×10^{-2}**

5a. 1.60×10^{-2}

5b. 1.20×10^{-3}

6a and 6b. Answers should be close to the 7a and 7b answers.

7a. 6.5×10^{-3}

7b. 8.5×10^{-2}

Lesson 23.6

Practice A   1a. See K_a and $[H^+]$ or pH? Write the *WRECK* steps, solve the approximation.

Specific reaction: $1\,HCN \leftrightarrows 1\,H^+ + 1\,CN^-$ (Goes slightly.)

WANTED: $[H^+] = ? = x$

R and **E**: $1\,WA \leftrightarrows 1\,H^+ + 1\,CB$ (Goes slightly.)

Conc. at eq.: $[WA]_{mixed} - x$ x x

$K_a \equiv \dfrac{[H^+]_{eq.}\,[CB]_{eq.}}{[WA]_{at\,eq.}} \equiv \dfrac{x^2}{[WA]_{mixed} - x} \approx \dfrac{x^2}{[WA]_{mixed}} \approx K_a$

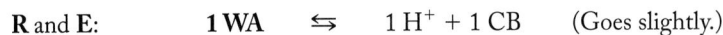

  Definition   Exact   Approximation

Make a data table using the approximation symbols.

DATA: $x = [H^+] = ? =$ WANTED

$[WA]_{mixed} = 0.50$ M HCN

$K_a = 6.2 \times 10^{-10}$

SOLVE: $x^2 \approx (K_a)\,([WA]_{mixed}) = (6.2 \times 10^{-10})\,(0.50) = 3.1 \times 10^{-10}$

$[H^+] = x = $ (estimate: $1-2 \times 10^{-5} = $ **$1.76 \times 10^{-5}\,M\,H^+$**

(This answer carries an extra *s.f.* until the final step.)

1b. See pH? Write $\boxed{pH \equiv -\log\,[H^+]}$ and $\boxed{[H^+] \equiv 10^{-pH}}$

WANTED: pH $= -\log(1.76 \times 10^{-5}) = $ **4.75**   (4.75 rounded up = 5: Check.)

2a. See K_a and $[H^+]$ or pH? Write the *WRECK* steps, solve the approximation.

Specific reaction: $1\,HF \leftrightarrows 1\,H^+ + 1\,F^-$ (Goes slightly.)

WANTED: $K_a = ?$

R and *E*:   **$1\,WA$** $\leftrightarrows$ $1H^+ + 1\,CB$ (Goes slightly.)
                            $\wedge$ $\wedge$ $\wedge$

*C*onc. at eq.: $[WA]_{mixed} - x$ x x

K: $K_a \equiv \dfrac{[H^+]_{eq.}\,[CB]_{eq.}}{[WA]_{at\,eq.}} \equiv \dfrac{x^2}{[WA]_{mixed} - x} \approx \dfrac{x^2}{[WA]_{mixed}} \approx K_a$
                $\wedge$ $\wedge$ $\wedge$
            Definition   Exact   Approximation

DATA: $x = [F^-] = [H^+] = 0.012\,M = 1.2 \times 10^{-2}\,M$

$[WA]_{mixed} = 0.25\,M\,HF$

$K_a = ? = $ WANTED

The K_a approximation as written solves for the WANTED symbol:

$K_a \approx \dfrac{x^2}{[WA]_{mixed}} \approx \dfrac{(1.2 \times 10^{-2})^2}{0.25} \approx \dfrac{1.44 \times 10^{-4}}{0.25} \approx $ **5.8×10^{-4}**

Because this problem supplies x, you can easily solve the exact equation:

$K_a = \dfrac{x^2}{[WA]_{mixed} - x} = \dfrac{(1.2 \times 10^{-2})^2}{0.25 - 0.012} = \dfrac{1.44 \times 10^{-4}}{0.238} = $ **6.1×10^{-4}**

A K value is not assigned units. Note that the approximation and exact answers are close. Because K values have significant uncertainty, if $[H^+]$ is less than 5% of $[WA]$, solving with the relatively quick approximation is considered "close enough."

2b. WANTED: pH $\boxed{pH \equiv -\log\,[H^+]}$ and $\boxed{[H^+] \equiv 10^{-pH}}$

$[F^-] = [H^+] = 0.012\,M$; pH $= -\log\,[H^+] = -\log(1.2 \times 10^{-2}) = 1.? = $ **1.92**

Practice B   1. WANTED: pH $\boxed{pH \equiv -\log\,[H^+]}$ and $\boxed{[H^+] \equiv 10^{-pH}}$

Before you can calculate pH, you need $[H^+]$. See K_a? Try the quick steps.

$K_a \approx \dfrac{x^2}{[WA]_{mixed}}$

DATA: $x = [H^+] = ? = $ WANTED

$[WA]_{mixed} = [HCOOH]_{mixed} = 0.50\,M$

$K_a = 1.8 \times 10^{-4}$

SOLVE: $x^2 = (K_a)\,([WA]_{mixed}) = (1.8 \times 10^{-4})(0.50) = 0.90 \times 10^{-4} = 90 \times 10^{-6}$

$x = [H^+] = 9.5 \times 10^{-3}$ M.   But always look back at what was WANTED.

SOLVE: $\mathbf{pH} = -\log[H^+] = -\log(9.5 \times 10^{-3}) = 2.? = \mathbf{2.02}$

2a. See pH? Write $\boxed{pH \equiv -\log[H^+]}$ and $\boxed{[H^+] \equiv 10^{-pH}}$

WANTED: $K_a = ?$ Apply the quick steps.

$$\boxed{K_a \approx \frac{x^2}{[WA]_{mixed}}}$$

DATA: $x = [H^+] = ?$ but we can find $[H^+]$ from the pH.

$[WA]_{mixed} = 0.25$ M WA

$K_a = ? =$ WANTED

$[H^+] = 10^{-pH} = 10^{-4.49} = 3.2 \times 10^{-5}$ M $= x$ (Check: 4.49 rounded up = 5.)

SOLVE: $K_a = \dfrac{x^2}{[WA]_{mixed}} = \dfrac{(3.2 \times 10^{-5})^2}{0.25} = \dfrac{10.2 \times 10^{-10}}{0.25} = 41.0 \times 10^{-10} = \mathbf{4.1 \times 10^{-9}}$

2b. WANTED: $[OH^-] = ?$ In an acid solution, find $[OH^-]$ using K_w.

$[OH^-] = \dfrac{1.0 \times 10^{-14}}{3.2 \times 10^{-5}} = 0.31 \times 10^{-9} = \mathbf{3.1 \times 10^{-10}}$

Review Quiz

1. -4.3 2. 5.0×10^{-7} 3. $[H^+] = 10^{-7.22} = \mathbf{6.0 \times 10^{-8}}$ M, $[OH^-] = \mathbf{1.7 \times 10^{-7}}$ M

4a. $[H^+] = \mathbf{5.0 \times 10^{-12}}$ M   4b, 4c, 4d. $[OH^-] = \mathbf{2.0 \times 10^{-3}}$ M $= [K^+] = [KOH]_{as\ mixed}$

5. 6.3×10^{-10} 6. 9.0×10^{-8}

7a. $x^2 = (6.3 \times 10^{-5})(0.040) = 2.5 \times 10^{-6}$; $x = \mathbf{1.6 \times 10^{-3}}$ M $= [H^+]$ 7b. pH $= \mathbf{2.80}$

8. $K_a = \dfrac{x^2}{[WA]_{mixed}} = \dfrac{(4.7 \times 10^{-3})^2}{0.50} = \dfrac{22.1 \times 10^{-6}}{0.50} = 44.2 \times 10^{-6} = \mathbf{4.4 \times 10^{-5}} = K_a$

24

Nuclear Chemistry

Lesson 24.1	Isotopes and Radioactive Decay

In nature on our planet, most atoms have nuclei that have not changed in composition in the past 5 billion years, but a small percentage of nuclei (less than one in 10,000) is radioactive. A radioactive isotope over time will *decay* to form a nucleus with a different number of protons and neutrons. Some radioactive nuclides decay on average in seconds, while others decay on average after several billion years.

Nuclei that are radioactive but are not found in nature can be manufactured in nuclear reactors, and these nuclei are of special importance in chemistry and medicine. Stable and radioactive isotopes of an atom have the same chemical behavior, but when a radioactive isotope decays, it emits radiation that can reveal its location. Using chemical reactions, substances can be synthesized that have a higher than average percentage of radioactive nuclei. During chemical and biological processes, where these "tagged" atoms go and how they react can be measured. The *radioactive dyes* used in medical imaging are an example of this technology.

Representing Nuclides

Recall that

- Each different combination of protons and neutrons is termed a *nuclide*.

- Each nuclide has a *mass number*, which is the *sum* of its number of protons and neutrons.

- Nuclides with the same number of protons but differing numbers of neutrons are termed *isotopes*.

A nuclide can be identified in three ways: by its number of protons and number of neutrons, by a nuclide symbol that includes its mass number, or by its isotope name.

Examples: The three isotopes of hydrogen can be identified as follows:

- One proton + no neutrons *or* as ^{1}H, named hydrogen-1
- One proton + one neutron *or* as ^{2}H, named hydrogen-2
- One proton + two neutrons *or* as ^{3}H, named hydrogen-3

In the special case of hydrogen, its isotopes are also given the nonsystematic names protium, deuterium, and tritium, respectively.

Nuclide symbols may also be written with the nuclear charge below the mass number. This is called *A–Z notation*, illustrated for a tritium nucleus below. *A* is the symbol for mass number and *Z* represents nuclear charge.

$$^3_1\text{H}$$

Z values are not required to identify an atom: The nuclear charge of an atom is its number of protons, which is identified by its symbol. However, for nuclear particles that do not contain protons, termed *subatomic* particles, the nuclear charge and the number of protons are not the same, and showing *Z* values may be required.

PRACTICE A

1. The mass number of a nucleus is determined by its number of _____.

2. Isotopes have the same number of _____ but different numbers of _____.

3. Write the nuclide (isotope) symbol for a single proton using *A–Z* notation.

4. Consulting a periodic table, fill in each column below.

Protons	Neutrons	Atomic Number	Mass Number	Nuclide Symbol	Isotope Name
2	2				
	118	79			
82			206		
				^{242}Pu	
					uranium-235

Alpha and Beta Decay

A nucleus that is radioactive has a tendency to *decay*: to gradually expel or capture particles until a stable neutron and proton combination is achieved. There are more than a dozen types of nuclear processes that are termed decay, but the two types that occur most frequently for radioactive nuclei are **alpha (α) decay** and **beta (β) decay**.

When a particle composed of two protons and two neutrons is ejected from a nucleus, it is termed an **alpha particle**. Because an alpha particle has the same structure as a helium-4 nucleus, it is given the same isotopic symbol.

$$\text{Alpha particle} = \alpha \text{ particle} = {}_{2}^{4}\text{He}$$

The ejection of an alpha particle from a nucleus is termed *alpha decay*.

> **Example:** The isotope U-238 undergoes alpha decay, a nuclear reaction that can be represented as

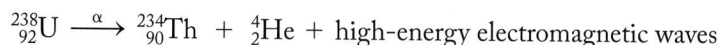

$${}_{92}^{238}\text{U} \xrightarrow{\alpha} {}_{90}^{234}\text{Th} + {}_{2}^{4}\text{He} + \text{high-energy electromagnetic waves}$$

In **beta decay**, a neutron in a nucleus decays into a proton and an electron, and the electron is expelled from the nucleus at high speed. When an electron is formed in this manner, it is termed a **beta particle**. Because an electron has no protons and no neutrons, its mass number is zero. Because it has a negative charge, when an electron is formed in the nucleus its *nuclear charge* is −1. A beta particle can be represented in these ways:

$$\text{Beta particle} = \beta \text{ particle} = {}_{-1}^{0}\beta \text{ or } {}_{-1}^{0}\text{e}$$

In beta decay, the number of neutrons in a nucleus *decreases* by one, but the number of protons increases by one, so the mass number of the decaying isotope stays the same.

> **Example:** The beta decay of lead-210 can be written as

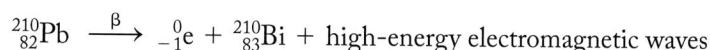

$${}_{82}^{210}\text{Pb} \xrightarrow{\beta} {}_{-1}^{0}\text{e} + {}_{83}^{210}\text{Bi} + \text{high-energy electromagnetic waves}$$

PRACTICE B

1. Write two ways to represent a beta particle in *A–Z* notation.

2. Name the type of decay represented by these nuclear reactions.

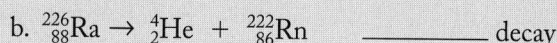

    a. ${}_{6}^{14}\text{C} \rightarrow {}_{-1}^{0}\text{e} + {}_{7}^{14}\text{N}$ _____ decay

    b. ${}_{88}^{226}\text{Ra} \rightarrow {}_{2}^{4}\text{He} + {}_{86}^{222}\text{Rn}$ _____ decay

| Lesson 24.2 | **Radioactive Half-Life** |

Fractions and Percentages: Review

In calculations involving radioactive half-life, you will need to use mental arithmetic to convert quickly between a fraction and its percentage. To refresh your memory on fractions and percentages, review Lesson 12.3 at this time, repeat working the Try It problems in Lesson 12.3, and then complete the practice below. If additional review is needed, complete additional practice problems in Lesson 12.3.

PRACTICE A

Do not use a calculator.

1. 1/8 has what decimal equivalent and what percentage?

2. 16% has what decimal equivalent?

3. 0.24% has what decimal equivalent?

4. 7.4/10,000 has what decimal equivalent and percentage?

5. After decay, if 6 parts remain out of an original 30:

   a. What fraction remains?   b. What percentage remains?

Half-Life

The **half-life** (symbol $t_{1/2}$) of a reactant is the time required for half of the substance to be used up in a reaction. Each radioactive nuclide has a characteristic half-life: In any sample, the time in which half of the nuclides decay is *constant*. It is not possible to predict when any one nucleus will decay, but in any sample of more than a few hundred of a given nuclide, we can accurately predict how long it will take for any percentage of the nuclei to decay.

The rate of *chemical* reactions varies with changes in temperature, but radioactive decay is a nuclear reaction, and the rate of decay does not vary measurably even over temperature ranges of 1000 K.

Half-Life Calculations for Simple Multiples

In calculations involving half-life, the two variables will generally be the average *time* in which a given nuclide in a sample decays and the *percentage* of those nuclei that remain. If the time period for decay is equal to either the half-life or a simple multiple of the half-life, calculations can be completed by mental arithmetic.

• If a sample of a given nucleus has decayed for a time equal to *one* half-life, half of the original nuclei have *decayed* and half *remain*.

- After *two* half-lives (double the time of the half-life), the number of nuclei remaining is half of the half that remained after the first half-life: half of 1/2 = 1/4 (25%) of the original nuclei remain, and 75% have decayed.

- At triple the half-life, 1/2 of 1/4 = 1/8 (12.5%) of the original nuclei remain.

In decay calculations, you must distinguish between quantities *decayed* and *remaining*.

- 100% − percentage decayed = percentage remaining
- 1.000 − fraction decayed = fraction remaining

Examples: If 15% remains, 85% has decayed. If the fraction decayed is 0.35, the fraction remaining is 0.65.

▶ **TRY IT**

Q. Fluorine-18, a radioactive isotope used in nuclear medicine, has a half-life of 1.8 hours. How long will it take for 87.5% of the ^{18}F nuclei in a sample to decay?

Answer:

If 87.5% has decayed, 12.5% remains. After two half-lives, 25% remains, and after *three* half-lives, 25% × 1/2 = 12.5% remains.

$$3 \text{ half-lives} \times 1.8 \text{ hr/half-life} = \mathbf{5.4\,hr}$$

In decay calculations for "simple multiple" cases, given the headings in the table below you will need to be able to fill in the rest of the table from memory. This should not be difficult: In the two middle columns, each number is half of the value of the number above it.

Whole Number Half-Life Values

For Radioactive Nuclei, at Time =	Fraction Remaining	Percentage Remaining	Percentage Decayed
0	1	100%	0%
One half-life	1/2	50%	50%
Two half-lives	1/4	25%	75%
Three half-lives	1/8	12.5%	87.5%

For half-life calculations that are not easy multiples, we will use the logic of this chart to make estimates to check answers.

PRACTICE   B

1. Write the table above until, given the headings, you can fill in the four rows from memory.

2. If 90.% of a sample has decayed, what fraction remains?

3. If the fraction of a sample that has decayed is 0.40, what percentage remains?

4. The nucleus of plutonium-239 undergoes radioactive decay with a half-life of 24,400 years. In a sample containing ^{239}Pu:

   a. After how many years will 25% of the original ^{239}Pu nuclei remain?

   b. After how many half-lives will the count of ^{239}Pu isotopes be 1/16th of its original number?

   c. What percentage of the ^{239}Pu has decayed after exactly four half-lives?

Lesson 24.3   **Natural Logarithms**

Natural logarithms have a central role in explaining rates of radioactive decay and other calculations in the sciences. The rules for natural logs parallel those for base 10 logs, but because our number system is based on 10, it is easier to learn the logic of base 10 logs first.

If you have not completed Lesson 23.1 on base 10 logarithms, do so now. If you have previously completed that lesson, review the eight "Rules for Base 10 Logarithms" in Lesson 23.1. In either case, complete Practice A below before continuing with this lesson.

PRACTICE   A

Practice with the calculator you will use on tests.

1. Without a calculator, answer each of these with an integer.

   a. $\text{Log}(10^{14}) =$

   b. $\text{Log}(10) =$

   c. $\text{Log}(1) =$

2. Use a calculator to answer these questions.

   a. $10^{3.2} =$ (in scientific notation): $(\pm 1?)$ _____

   b. $10^{-12.30} =$ $(\pm 1?)$ _____

   c. $10^{-0.200} =$ (fixed notation):   (scientific notation): $(\pm 1?)$ _____

d. $Log(2 \times 10^{14}) =$ $(\pm 1?)$ _____

e. $Log(2.0 \times 10^{-14}) =$ f. $Log(5.00) =$

g. $Log\ X = 4.7; X =$ h. $Log\ A = -8.2; A =$

i. $Log(0.0050) =$ j. $Log\ D = -0.50; D =$

k. $10^{-9.5} =$ l. $Log(3.0 \times 10^{-5}) =$

The Number *e*

In mathematical and scientific equations, the lowercase **e** is an abbreviation for a number: **2.7182818. . . .** The rounded value $\boxed{e = 2.718}$ must be recallable from memory.

In science, **e** is found in many equations that predict natural phenomena. In these equations, *e* is the base for values expressed in the form e^x, and *e* is termed the **natural exponential**. To solve calculations involving radioactive half-life, we will use both the natural exponential *e* and the natural log function **ln**.

Calculating with the Natural Exponential

> **TRY IT**
>
> **Q1.** We know that e^1 equals what number?
>
> **Answer:**
>
> **2.718. . . .** Write or circle a key sequence on your calculator that converts e^1 to a number.
> You might try **1** $\boxed{e^x}$ or **1** $\boxed{enter}$ $\boxed{e^x}$.
>
> **Q2.** Use your key sequence to answer parts (a) to (d).
>
> During **e** and **ln** calculations in these lessons, for answers that are not integers, round fixed decimals and significands to three significant figures.
>
> a. $e^2 =$ b. $e^{2.5} =$
>
> c. $e^{-1} =$ d. $e^{-2.5} =$
>
> **Answers:**
>
> a. $e^2 =$ **7.39**   b. $e^{2.5} =$ **12.2**
>
> Recall that entering a negative number may require a $\boxed{+/-}$ or $\boxed{(-)}$ key.
>
> c. $e^{-1} (= 1/e = 1/2.718) =$ **0.368**   d. $e^{-2.5} (= 1/e^{2.5}) =$ **0.0821**

Some calculators use an **E** at the right side of the answer screen to show the power of **10** for numbers in scientific notation. This **E** is *not* the same as the symbol *e* for the natural exponential.

Calculating Natural Logs

The **ln** function (the **natural log**) answers this question: If a number is written as *e* to a power, what is the power?

Just as by definition, $\boxed{\log 10^x \equiv x}$, the natural log definition is $\boxed{\ln e^x \equiv x}$. Note the similarities in the two definitions.

▶ TRY IT

Q1. Use the natural log definition to complete these without a calculator.

    a. $\ln e^0 =$ b. $\ln e^1 =$ c. $\ln e^{-4} =$

Answers:

    a. $\ln e^0 = \mathbf{0}$ b. $\ln e^1 = \mathbf{1}$ c. $\ln e^{-4} = \mathbf{-4}$

Q2. Use mental math: $\ln(2.718)$ should equal about _____.

    $\ln(2.718) \approx \ln e \approx \ln e^1 = \mathbf{1}$

Q3. Use your calculator for the same calculation: $\ln(2.718) =$ _____.

The calculator answer should be close to the mental arithmetic answer.

Write the key sequence that works to solve **ln(2.718)** above. The same steps should take the natural log of any positive number.

▶ TRY IT

Q. Using your calculator: $\ln(314) =$

Answer:

$\ln(314) = \mathbf{5.75}$

Significant Figures and Natural Log Calculations

In base *e* calculations, numbers are often exact, such as $\ln(1/2)$, but the corresponding numbers are rounded. In addition, ways to express uncertainty in base *e* are debated as to their statistical validity. In these lessons, in base *e* calculations we will report non-integer results with 3 significant figures and allow for rounding in calculated ln values.

PRACTICE B

Round your answers to three significant figures.

1. $e^2 =$

2. $e^{-4.7} =$

3. $e^{-11} =$

4. $\ln(42) =$

5. $\ln(1/2) =$

6. $\ln(5.00 \times 10^{-4}) =$

7. $\ln(10^{-4}) =$

Converting ln Values to Numbers

A base **10** logarithm can be defined by $\boxed{10^{\log x} \equiv x}$. A similar base **e** definition is $\boxed{e^{\ln x} \equiv x}$. Using this base **e** definition above, some calculations involving ln and **e** can be solved by mental math.

▶ **TRY IT**

Q. $e^{\ln(11)} =$

Answer:

$e^{\ln(11)} = \mathbf{11}$

The equation $\boxed{e^{\ln x} \equiv x}$ also means that you can convert an ln value to its corresponding number by writing the ln value as a power of e.

▶ **TRY IT**

Q. If $\ln(X) = 1$, the fixed decimal number X (use "paper-and-pencil" math) is _____.

Answer:

If $\ln(X) = 1$, $X = e^{\ln X} = e^1 = \mathbf{2.718.\ldots}$

Knowing that answer, do the same ln-to-number conversion on your calculator by taking the antilog.

▶ **TRY IT**

Q. If $\ln(X) = 1$, the number X obtained using the calculator is _____.

Answer:

Input **1** then either $\boxed{\text{INV or 2nd}}$ $\boxed{\ln}$ or $\boxed{e^x}$

Write or circle the key sequence that converted ln = 1 to the number **2.718.\ldots**

Use your key sequence to convert the following ln values to numbers. Add an exponent to e, then fill in the remaining blanks.

➤ TRY IT

Q1. If $\ln(X) = 6, X = e$ _____ = (number): _____ = (scientific notation): _____

Answer:

If $\ln(X) = 6, X = e^6$ = (fixed decimal): **403** = (scientific notation): **4.03 × 10^2**

Q2. Apply the same steps to these.

   a. If $\ln(X) = -4.5$; $X = e$ _____ = (fixed): _____ = (scientific notation): _____

   b. $\ln(X) = 57.2$; $X = e$ _____ = (scientific notation): _____

   c. $\ln(X) = 0.0300$; $X = e$ _____ = (fixed decimal): _____

Answers:

   a. If $\ln(X) = -4.5, X = e^{-4.5}$ = **0.0111** = **1.11 × 10^{-2}**

   b. $\ln(X) = 57.2$; $X = e^{57.2}$ = **6.94 × 10^{24}**

   c. $\ln(X) = 0.0300$; $X = e^{0.0300}$ = **1.03**

If you get lost on a natural log calculation, a good strategy is to solve a similar and simple base 10 mental computation, and then apply the same logic to the natural log case. Simple base 10 calculations can be solved in your head, and the formulas and steps for base 10 and base e calculations are parallel.

Converting between Base 10 and Natural Logs

A general rule for the logarithms of any base **b** is

$$\text{Log}_b(x) = \ln(x)/\ln(b) \tag{24.1}$$

For base 10 logs, this equation becomes

$$\text{Log}_{10}(x) = \ln(x)/\ln(10) = \ln(x)/2.303 \tag{24.2}$$

For use in checking ln calculations, it is helpful to remember the relationship between base e and base 10 as

$$\ln(x) = 2.303 \log(x) \tag{24.3}$$

or as "The *natural* log of a number is always its base 10 log multiplied by about 2.3."

> To check an ln calculation: Estimate the base 10 log of a value, then multiply by about 2.3.

PRACTICE **C**

Answer questions 1–4 without a calculator.

1. The log of 100 = _____, so the ln of 100 ≈ _____

2. If ln X = 23, log X ≈ _____ and X ≈ 10 to which power? _____

3. If ln X = 0.23, log X ≈ _____

4. If ln X = –4.6, log X ≈ _____ and X ≈ 10 to which power? _____

Let's summarize.

Rules for Natural Logarithms

1. The symbol e is an abbreviation for a number with special properties:
   e = **2.718. . . .**

2. The **ln** (natural log) function answers this question: If a number is written as e to a power, what is the power?

3. Knowing the ln, to find the number take the *antilog*. On most calculators, input the ln value, then (INV) (ln) or (e^x). An ln is an exponent of e.

4. When you encounter e and **ln** in calculations, write the following "log and 10" chart:

   $$\log 10^x = x \qquad \text{and} \qquad \ln e^x = x$$
   $$10^{\log x} = x \qquad \text{and} \qquad e^{\ln x} = x$$

   Note the patterns. Note the logic: A log is an exponent.

5. A base e log is the base 10 log multiplied by 2.303: $\boxed{\ln(x) = 2.303 \log(x)}$

   Use this relationship to estimate and check calculated ln values.

PRACTICE **D**

1. Practice writing the five "Rules for Natural Logarithms" until you can do so from memory. You may want to design flashcards for the rules as well.

2. Try every other lettered problem first. Complete the remaining problems for additional practice or pretest review. In e and ln calculations, round non-integer answers to 3 *s.f.*

   a. $e^{5.2}$ =   b. $e^{-1.7}$ =

   c. $e^{-20.75}$ =   d. ln(1066) =

   e. ln(1/4) =   f. ln(3 × 10^8) =

g. $\ln(14.92 \times 10^{-6}) =$ h. $\ln e^{6.2} =$

i. $e^{\ln(42)} =$

j. If $\ln X = -6.8, X = e$ $=$ (number in scientific notation):

k. If $\ln D = 7.4822, D =$

l. If $\log A = -9, A =$

m. If $\log x = 13.7, x =$

n. $\text{Log } B = -13.7; B =$

o. $10^{-11.7} =$

p. $\ln B = -13.7; B =$

q. $e^{-11.7} =$

r. $\ln(0.050) =$

s. $e^{-0.693} =$

3. Solve these problems in your notebook.

   a. If $\log(x) = 5.0, \ln(x) \approx$

   b. If $\ln(x) = 34.5, \log(x) \approx$

   c. If $\ln(x) = -(0.075 \text{ day}^{-1})(4.0 \text{ days}), x = ?$

4. Given this equation: $\ln(A) = (-0.0173 \text{ s}^{-1})(t)$

   a. If $t = 20.0$ seconds, $(A) = ?$ b. If $A = 0.500, t = ?$

5. Given this equation: $\ln(A) = (-0.0241 \text{ yr}^{-1})(t)$

   a. If $A = 0.00250, t = ?$ b. If $t = 28.8$ years, $A = ?$

Lesson 24.4 Radioactive Half-Life Calculations

Rate Constants for Radioactive Decay

We can solve half-life calculations that do not involve simple multiples by using **rate equations**. Each radioactive nucleus has a **rate constant** (k) for decay that is characteristic: a value that is constant. Different radioactive nuclides have different values for k. The values for k have $time^{-1}$ units, such as s^{-1} or yr^{-1}.

The equation that predicts the decay rate for radioactive nuclei can be written as

$$\ln\left(\frac{N_t}{N_0}\right) = -kt \tag{24.4}$$

In equation 24.4, N_t/N_0 is the *fraction* measuring the number of radioactive nuclei *remaining* at time $= t$ compared to time $t = 0$. Equation 24.4 can be abbreviated as

$$\ln(\textit{fraction remaining}) = -kt \qquad\qquad (24.5)$$

Example: After two half-lives, half of half (25%) of the original radioactive nuclei in a sample will remain, and **0.25** is the *fraction* remaining.

Because the nuclei are decaying, the value of N at time $= t$ will be *less* than it was at time $= 0$, and the value of the fraction must be less than 1. This means that in decay calculations, the *fraction remaining* must have a value between 0 and 1.00 (such as 0.55 or 0.020).

Decay calculations usually involve fractions or percentages of isotopes, and in those cases the equation stated in terms of *fraction remaining* is more convenient to use.

TRY IT

Apply the equation that includes *fraction remaining* to this problem.

Q. If a radioactive isotope with a short half-life has a rate constant for decay of $+0.00500$ s^{-1}, what fraction of this isotope remains after 45 seconds?

Answer:

WANTED: Fraction remaining $=$

Equation: $\boxed{\ln(\text{fraction remaining}) = -kt}$

Make a DATA table listing each term in the equation. To solve for the fraction remaining, you will need to find **ln**(fraction remaining) first.

DATA: $\ln(\text{fraction remaining}) = ?$

$k = 0.00500$ s^{-1}

$t = 45$ s

SOLVE: $\ln(\text{fraction remaining}) = -kt = -(0.00500\ s^{-1})\,(45\ s) = \mathbf{-0.225}$

Note how the units multiplied: $(s^{-1})(s) = s^{-1+1} = s^0 = \mathbf{1}$

From $\ln(\text{fraction}) = -0.225$, how can you calculate the fraction?

$$\text{Fraction remaining} = e^{\ln(\text{fraction})} = e^{-0.225} = \mathbf{0.80}$$

As one check on a decay calculation, recall that a *fraction remaining* must have a decimal equivalent value between 0 and 1, and 0.80 does.

Solving with Percentages

To use the equation with the term *fraction remaining*, you must calculate using fractions and not percentages. This means:

- If a percentage is WANTED, you will need to solve for the *fraction* first.
- If a percentage is *given*, you must convert to its decimal-equivalent fraction to substitute into the equation.

You will also need to distinguish between quantities *decayed* and *remaining*.

Example: If 45% remains, 55% has decayed.

PRACTICE **A**

Commit to memory the equation above that includes *fraction remaining*, then solve the following problems.

1. For a radioactive sample, −1.386 is the value for ln(fraction remaining).

   a. What fraction of the sample remains?   b. What percentage has decayed?

2. For the decay of a radioactive nucleus, if the rate constant is $k = 0.04606$ hr^{-1}, what percentage remains after 50.0 hr?

3. The plutonium-238 used in nuclear power supplies for interplanetary spacecraft has a half-life of 87.7 years: At that time, 50.0% of the original nuclei remain. Calculate the rate constant for the decay of this isotope.

Half-Life Calculations Beyond Simple Multiples

For a radioactive nucleus, after a time equal to one half-life (symbol $t_{1/2}$), half of a sample has decayed and half remains. Substituting into equation 24.5,

$$\ln(\text{fraction remaining}) = -kt$$

at a time equal to one half-life, we can write

$$\ln(1/2) = -kt_{1/2} \tag{24.6}$$

TRY IT

Q. The rate constant for the decay of the tritium isotope of hydrogen is 0.0562 yr^{-1}. Using equation 24.6, calculate the half-life of tritium.

Answer:

WANTED: $t_{1/2}$ for tritium

DATA: 0.0562 yr$^{-1} = k$

Equation: $\boxed{\ln(1/2) = -kt_{1/2}}$

SOLVE the equation for the WANTED variable in symbols, then substitute the DATA.

$$t_{1/2} = \frac{\ln(1/2)}{-k} = \frac{\ln(0.500)}{-k} = \frac{-0.693}{-0.0562 \text{ yr}^{-1}} = \textbf{12.3 yr}$$

In equation 24.6, the rate constant (k) and half-life ($t_{1/2}$) are variables: Their numeric values will differ for different radioactive nuclei. The term $\ln(1/2)$ is a constant that can be converted to a numeric value.

$$\ln(1/2) = \ln(0.500) = -0.693$$

Substituting the value for the constant into equation 24.6 results in:

$$-0.693 = -kt_{1/2} \tag{24.7}$$

and solving equation 24.7 for half life:

$$t_{1/2} = \frac{-(-0.693)}{k} = \frac{0.693}{k} \tag{24.8}$$

For radioactive half-life calculations, we need an equation that includes half-life ($t_{1/2}$). Equations 24.6, 24.7, and 24.8 include half-life, but they are all equivalent, and it is best if we memorize just one definition and solve for other formats using algebra. The best definitions to memorize are those that are both based on fundamentals and easy to remember. In these lessons, we will follow this rule:

The Radioactive-Decay Prompt

If a decay calculation includes a *half-life* and/or a *fraction* or *percentage*, and the answer cannot be calculated using whole-number multiples, write in the DATA

$$\ln(\text{fraction remaining}) = -kt \qquad \text{and} \qquad \ln(1/2) = -kt_{1/2}$$

Note that the second equation is simply a special case of the first: When the fraction remaining is 1/2, the time is equal to the half-life.

Most decay calculations can be solved using one of those two equations or both. If the WANTED and/or DATA in a problem include both a time of decay (t) and a half-life ($t_{1/2}$), both equations will be needed. In that case, one way to solve is to find the value for the variable k that is common to both equations. Apply this rule:

In a radioactive-decay problem, if the WANTED and/or DATA include both t and $t_{1/2}$:

- Find k using the equation that can be solved for k with the DATA provided.
- Substitute that value for k into the equation that includes the symbol WANTED.

───────────────→ **TRY IT**

Q. Iodine-131, an isotope used to treat thyroid disorders, has a half-life of 8.1 days. What percentage of these isotopes remains after 48 hours?

STOP

Answer:

WANTED: % ^{131}I remaining

Strategy: See radioactive half-life and fraction or percentage, with no simple multiples of half life? Write:

$$\boxed{\ln(\text{fraction remaining}) = -kt}\quad\text{and}\quad\boxed{\ln(1/2) = -kt_{1/2}}$$

DATA: 8.1 **days** = half-life = $t_{1/2}$

48 hr = 2.0 **days** = t

(Convert equation DATA to *consistent* units. Which unit you choose will not affect the answer, but here, converting to days is easier.)

If needed, adjust your work and solve from here.

STOP

To find % remaining, find fraction remaining first. In the equation that includes fraction remaining, we know t but not k, but the second equation contains only two *variables*, and we know $t_{1/2}$. We can solve for k then substitute k into the first equation to find **ln**(fraction remaining), then the fraction remaining, then the % remaining.

STOP

$$k = \frac{-\ln(1/2)}{t_{1/2}} = \frac{-(-0.693)}{t_{1/2}} = \frac{+0.693}{8.1\ \text{days}} = 0.0856\ \text{day}^{-1}$$

$$\ln(\text{fraction remaining}) = -kt = -(0.0856\ \text{day}^{-1})(2.0\ \text{days}) = \mathbf{-0.171}$$

Knowing ln(fraction remaining), how can the fraction remaining be found?

$$\boxed{\text{Fraction remaining} = e^{\ln(\text{fraction remaining})}}$$

Finish from here.

STOP

$$\text{Fraction} = e^{\ln(\text{fraction})} = e^{-0.171} = 0.843 = \mathbf{84\%}\ \text{of I-131}\ \textit{remains}\ \text{after 2.0 days}$$

▶

PRACTICE   B

Add the equations of the decay prompt to your flashcards and practice the cards for this chapter, then try the problems below.

1. Strontium-90 is a radioactive nuclide found in **fallout**: dust particles in the cloud produced by the atmospheric testing of nuclear weapons. In chemical and biological systems, strontium behaves much like

calcium. If dairy cattle consume crops exposed to dust or rain containing fallout, dairy products containing calcium will also contain ^{90}Sr. Similar to calcium, ^{90}Sr will be deposited in the bones of dairy product consumers, increasing a risk of cancer. In part for this reason, most (but not all) nations conducting nuclear tests signed a 1963 treaty that banned atmospheric testing. Strontium-90 undergoes beta decay with a half-life of 28.8 yr. What percentage of an original count of ^{90}Sr isotopes in bones will remain after 40.0 yr?

    a. Estimate the answer.   b. Calculate the answer.

2. The element polonium was first isolated by Dr. Marie Sklodowska Curie and named for her native country, Poland. If 20.0% of ^{210}Po remains in a sample after 321 days of alpha decay:

    a. Estimate the half-life of ^{210}Po.   b. Calculate a precise half-life of ^{210}Po.

3. In a sample of radon-222, 10.0% remains after 12.6 days of alpha decay.

    a. What is the composition of the radon-222 nucleus?

    b. Estimate the half-life for ^{222}Rn.

    c. Calculate a precise half-life of ^{222}Rn. Compare your answer to your part (b) estimate.

SUMMARY

Base 10 and base e logarithms are a part of calculations in additional topics in chemistry such as kinetics (the study of rates of reaction), buffer solutions (mixtures of acids and bases), thermodynamics (including the impact of temperature, heat, and mechanical work on chemical processes), and electrochemistry (including batteries for cell phones and automobiles). The good news is: Once you have mastered the algebra of logarithm calculations for pH and radioactive decay, you are ready for the math of additional topics you will encounter across the sciences.

REVIEW QUIZ

Solve questions 1–5 without a calculator.

1. A carbon-14 nucleus contains how many protons and neutrons?

2. Draw the symbol for an alpha particle in A–Z notation.

3. Name the type of decay represented by these nuclear reactions.

    a. $^{90}_{38}Sr \rightarrow {}^{0}_{-1}e + {}^{90}_{39}Y$ _____ decay

    b. $^{222}_{86}Rn \rightarrow {}^{4}_{2}He + {}^{218}_{84}Po$ _____ decay

4. After decay, if 7 parts remain out of an original 56:

    a. What fraction has decayed?   b. What percentage remains?

5. If a sample initially has exactly 1,600 radioactive particles, how many remain after 3 half lives?

Solve questions 6–10 with a calculator.

6. If $\log X = 6.6$, $X =$

7. $\ln(9 \times 10^5) =$

8. If $\ln X = -12.5$, $X =$

9. In bananas, a dietary source of potassium, about 1 potassium atom per 10,000 is radioactive potassium-40. The rate constant for the decay of ^{40}K is 5.54×10^{-10} yr^{-1}. Calculate the ^{40}K half-life in both scientific and fixed-decimal notation.

10. The half-life of carbon-14 is 5,730 yr. What fraction of the original carbon-14 in a sample has decayed after 1,650 yr? Estimate, then calculate.

ANSWERS

Lesson 24.1

Practice A   1. Protons + neutrons

2. Same number of **protons**, different number of **neutrons**.

3. $_{1}^{1}H$

4.

Protons	Neutrons	Atomic Number	Mass Number	Nuclide Symbol	Isotope Name
2	2	2	4	^{4}He	helium-4
79	118	79	197	^{197}Au	gold-197
82	124	82	206	^{206}Pb	lead-206
94	148	94	242	^{242}Pu	plutonium-242
92	143	92	235	^{235}U	uranium-235

In writing the nuclide or isotope symbols for atoms, a nuclear charge below the mass number is optional.

Practice B   1. β particle = $_{-1}^{0}β$ or $_{-1}^{0}e$ 2a. Beta decay   2b. Alpha decay

Lesson 24.2

Practice A   1. Decimal equivalent = **0.125**;   percentage = **12.5%**

2. 0.16   3. 0.0024 = 2.4×10^{-3}

4. Decimal equivalent = **0.00074** = 7.4×10^{-4}; percentage = **0.074%** = 7.4×10^{-2}%

5. Fraction = $\dfrac{\text{part}}{\text{total}}$ = $\dfrac{6 \text{ remain}}{30}$ = $\dfrac{1}{5}$ remains = **0.20 remains = 20% remains**

Practice B   2. If 90.% has decayed, 10.% remains: fraction remaining = **0.10**

3. Decayed = 40.%; remaining = **60.%**

4a. Half of half (25%) remains after two half-lives.
Two half-lives = 2 × 24,400 yr = **48,800 yr**

4b. Half remains after one half-life, 1/4th after two, 1/8th after three, 1/16th after **four** half-lives.

4c. 1/16th remains; 1/16 = 0.0625 = 6.25% remains, so **93.75% has decayed**.

Lesson 24.3

Practice A   1a. 14   1b. $\text{Log}(10) = \log(10^1) = \mathbf{1}$ 1c. $\text{Log}(1) = \log(10^0) = \mathbf{0}$

2a. 2×10^3 2b. 5.0×10^{-13} 2c. $0.631 = 6.31 \times 10^{-1}$

2d. 14.3   2e. -13.7 2f. 0.699   2g. $X = \mathbf{5.01 \times 10^4}$

2h. $A = \mathbf{6 \times 10^{-9}}$ 2i. -2.30 2j. $D = \mathbf{0.32}$ 2k. 3×10^{-10}

2l. -4.52

Practice B   1. 7.39

2. 9.10×10^{-3}

3. 1.67×10^{-5}

4. 3.74

5. $\ln(1/2) = \ln(0.5) = \mathbf{-0.693}$

6. -7.60

7. $\ln(10^{-4}) = \ln(1 \times 10^{-4}) = \mathbf{-9.21}$

Practice C   1. The log of $100 = \log 10^2 = \mathbf{2}$, so the ln of $100 \approx 2.3 \times 2 \approx \mathbf{4.6}$

2. If $\ln X = 23$, $\log X \approx \mathbf{10}$ and $X \approx 10$ to which power? **10**

3. If $\ln X = 0.23$, $\log X \approx \mathbf{0.10}$

4. If $\ln X = -4.6$ $\log X \approx \mathbf{-2.0}$ and $X \approx 10$ to which power? **−2**

Practice D   2a. 181   2b. 0.183   2c. 9.74×10^{-10} 2d. 6.97

2e. $\ln(1/4) = \ln(0.25) = \mathbf{-1.39}$ 2f. 19.5   2g. -11.1 2h. 6.20

2i. 42.0   2j. $X = e^{-6.8} = \mathbf{1.11 \times 10^{-3}}$ 2k. $D = \mathbf{1,780}$

2l. $A = \mathbf{1 \times 10^{-9}}$ 2m. $x = \mathbf{5 \times 10^{13}}$ 2n. $B = \mathbf{2 \times 10^{-14}}$

2o. 2×10^{-12} 2p. $B = \mathbf{1.12 \times 10^{-6}}$ 2q. 8.29×10^{-6}

2r. -3.00 2s. 0.500

3a. $\ln(x) = 2.303 \log(x)$; $\ln(x) = (2.303)(5.00) = \mathbf{11.5}$

3b. $\ln(x) = 2.303 \log(x)$; $\log(x) = 34.5/2.303 = \mathbf{15.0}$

3c. $\ln(x) = -0.300(\text{day}^{-1})(\text{days}) = -0.300 \text{ day}^0 = -0.300(1) = -0.300$; $x = e^{\ln(x)} = e^{-0.300} = \mathbf{0.741}$

4a. To find A, first solve for ln(A).

$$\mathbf{\ln(A)} = (-0.0173 \text{ s}^{-1})(20.0 \text{ s}) = -0.346; \mathbf{A} = e^{\ln(A)} = \mathbf{0.708}$$

b. $t = \dfrac{\ln(A)}{-0.0173 \text{ s}^{-1}} = \dfrac{\ln(0.500)}{-0.0173 \text{ s}^{-1}} = \dfrac{-0.693}{-0.0173 \text{ s}^{-1}} = \mathbf{40.1 \text{ s}}$

Answer units: $1/\text{s}^{-1} = (\text{s}^{-1})^{-1} = \text{s}$

5a. $t = \dfrac{\ln(A)}{-0.0241/\text{yr}} = \dfrac{\ln(0.0025)}{-0.0241/\text{yr}} = \dfrac{-5.99}{-0.0241/\text{yr}} = \textbf{249 yr}$

b. $\ln(A) = (-0.0241/\text{yr})(28.8\ \text{yr}) = -0.694;\ A = e^{\ln(A)} = e^{-0.694} = \textbf{0.500}$

Lesson 24.4

Practice A 1a. WANTED: Fraction of sample remaining

         DATA: $\ln(\text{fraction remaining}) = -1.386$

         SOLVE: $\boxed{\text{Fraction remaining} = e^{\ln\ (\text{fraction remaining})}} = e^{-1.386} = \textbf{0.250}$

   1b. If the fraction remaining is 0.250, 25.0% remains, and the percentage decayed is **75.0%**.

    2. WANTED: % of sample remaining. To find %, find *fraction* first.

         DATA: $0.04606\ \text{hr}^{-1} = k$
               $50.0\ \text{hr} = t$

         Strategy: The equation that relates these terms is $\boxed{\ln(\text{fraction remaining}) = -kt}$

         SOLVE: $\ln(\text{fraction remaining}) = -kt = -(0.04606\ \text{hr}^{-1})(50.0\ \text{hr}) = -2.303$

         Fraction remaining $= e^{\ln(\text{fraction remaining})} = e^{-2.303} = \textbf{0.100};$ remaining $= \textbf{10.0\%}$

    3. WANTED: $k = $ the rate constant

         DATA: Fraction remaining $= 50.0\% = \textbf{0.500}$ (% must be converted to fraction.)
               $t = 87.7\ \text{yr}$

         Equation: $\boxed{\ln(\text{fraction remaining}) = -kt}$

$$k = \dfrac{-\ln(\text{fraction remaining})}{t} = \dfrac{-\ln(0.500)}{87.7\ \text{yr}} = \dfrac{+0.693}{87.7\ \text{yr}} = \textbf{0.00790 yr}^{-1}$$

         The answer unit has a time^{-1} unit, which a k in decay rates must have.

Practice B 1a. WANTED: *Estimate* of % Sr-90 isotopes remaining after 40 yr. If one half-life is about 30 yr and 50% remains, and two half-lives is about 60 yr and 25% remains, then at 40 yr, an estimate might be that **about 40%** remains.

    1b. WANTED: % Sr-90 isotopes remaining at $t = 40.0$ yr

         $\boxed{\ln(\text{fraction remaining}) = -kt}$ and $\boxed{\ln(1/2) = -kt_{1/2}}$

         DATA: $28.8\ \text{yr} = t_{1/2}$

                $40.0\ \text{yr} = t$

         From half-life, k can be found using the second equation. From k and t, ln(fraction) and then fraction can be found.

         SOLVE: $k = \dfrac{-\ln(1/2)}{t_{1/2}} = \dfrac{-(-0.693)}{t_{1/2}} = \dfrac{0.693}{28.8\ \text{yr}} = \textbf{0.02406 yr}^{-1}$

         $\ln(\text{fraction remaining}) = -kt = -(0.02406\ \text{yr}^{-1})(40.0\ \text{yr}) = \textbf{−0.9624}$

         Fraction $= e^{\ln(\text{fraction})} = e^{-0.9624} = 0.382;$ **38.2%** Sr-90 remains after 40 yr

         Compare this to your estimate.

    2a. Estimate: We know that 20% remains after 321 days. We also know that 50% remains after one half-life, and 25% after two half-lives. 20% is close to 25%, and when 25% remains, it would be about 300 days, and 25% is two half-lives, so one half-life is **about 150 days**.

2b. WANTED: $t_{1/2}$

    DATA: 20.0% Po-210 remains, fraction remaining = **0.200**

               $t = 321$ days

See radioactive half-life and fraction and/or percentage? Write

$$\boxed{\ln(\text{fraction remaining}) = -kt} \qquad \text{and} \qquad \boxed{\ln(1/2) = -kt_{1/2}}$$

Knowing the fraction and t, k can be found from the first prompt equation. Half-life can then be found from the second equation. Solve for k in symbols first:

$$k = \frac{-\ln(\text{fraction})}{t} = \frac{-\ln(0.200)}{321 \text{ days}} = \frac{-(-1.61)}{321 \text{ days}} = \textbf{5.01} \times \textbf{10}^{-3} \textbf{ days}^{-1}$$

$$t_{1/2} = \frac{-\ln(1/2)}{k} = \frac{+0.693}{5.01 \times 10^{-3} \text{ days}^{-1}} = \boxed{\textbf{138 days} = \text{half-life of Po-210}}$$

Is this answer close to your estimate in part (a)?

3a. A radon-222 nucleus has **86 protons** and **136 neutrons**.

3b. Estimate: 10% remains after 12.6 days. 12.5% remains after three half-lives, which is close to 10%. If three half-lives is about 12 days, then one half-life would be **about 4 days**.

3c. WANTED: $t_{1/2}$

    DATA: 90.0% has decayed; 10.0% remains. Fraction remaining = **0.100**

               $t = 12.6$ days

    Equations: $\boxed{\ln(\text{fraction remaining}) = -kt} \qquad \text{and} \qquad \boxed{\ln(1/2) = -kt_{1/2}}$

    SOLVE: Knowing the fraction and t, k can be found from the first equation. Half-life can then be found from the second equation.

$$k = \frac{-\ln(\text{fraction})}{t} = \frac{-\ln(0.100)}{12.6 \text{ days}} = \frac{-(-2.30)}{12.6 \text{ days}} = \textbf{0.183 days}^{-1}$$

$$t_{1/2} = \frac{-\ln(1/2)}{k} = \frac{+0.693}{0.183 \text{ days}^{-1}} = \textbf{3.79 days}$$

Is this close to your estimate in part (b)?

Review Quiz

1. 6 protons and 8 neutrons.   2. $_{2}^{4}\text{He}$ 3a. Beta decay   3b. Alpha decay

4a. 0.875   4b. 12.5%   5. Exactly 200 particles   6. 4×10^{6} 7. 13.7   8. 3.73×10^{-6}

9. 1.25×10^{9} years = 1,250,000,000 years

Because the half-life is more than 1 billion years, ^{40}K decay does not occur very often, and a banana contributes only about 1% of the radiation dose that most of us receive each day from other sources of radiation in our environment.

10. Estimate: 0.50 is the fraction decayed after about 6,000 yr, so **about 0.15** is the fraction decayed in about one-third of that time?

    Calculate:

    WANTED: Fraction C-14 decayed = $1.000 - $ fraction remaining

DATA: $t_{1/2} = 5{,}730 \text{ yr}$

$t = 1{,}650 \text{ yr}$

Equations: $\boxed{\ln(\text{fraction remaining}) = -kt}$ and $\boxed{\ln(1/2) = -kt_{1/2}}$

From half-life, find k. From k and t, find $\ln(\text{fraction})$ then fraction.

SOLVE: $k = \dfrac{-\ln(1/2)}{t_{1/2}} = \dfrac{-(-0.693)}{t_{1/2}} = \dfrac{+0.693}{5{,}730 \text{ yr}} = \mathbf{1.21 \times 10^{-4} \ yr^{-1}}$

$\ln(\text{fraction remaining}) = -kt = -(1.21 \times 10^{-4} \ \text{yr}^{-1})(1{,}650 \text{ yr}) = \mathbf{-0.1996}$

Fraction remaining $= e^{\ln(\text{fraction remaining})} = e^{-0.1996} = \mathbf{0.819}$

Fraction *decayed* $= 1.000 - 0.819 = \mathbf{0.181}$ (Compare to your estimate.)

Table of Atomic Masses

Element	Symbol	Atomic Number[1]	Atomic Mass[2]	Element	Symbol	Atomic Number[1]	Atomic Mass[2]	Element	Symbol	Atomic Number[1]	Atomic Mass[2]
Actinium	Ac	89	(227)	Gold	Au	79	197.0	Promethium	Pm	61	(145)
Aluminum	Al	13	27.0	Hafnium	Hf	72	178.5	Protactinium	Pa	91	(231)
Americium	Am	95	(243)	Hassium	Hs	108	(277)	Radium	Ra	88	(226)
Antimony	Sb	51	121.8	Helium	He	2	4.00	Radon	Rn	86	(222)
Argon	Ar	18	39.9	Holmium	Ho	67	164.9	Rhenium	Re	75	186.2
Arsenic	As	33	74.9	Hydrogen	H	1	1.008	Rhodium	Rh	45	102.9
Astatine	At	85	(210)	Indium	In	49	114.8	Roentgenium	Rg	111	(272)
Barium	Ba	56	137.3	Iodine	I	53	126.9	Rubidium	Rb	37	85.5
Berkelium	Bk	97	(247)	Iridium	Ir	77	192.2	Ruthenium	Ru	44	101.1
Beryllium	Be	4	9.01	Iron	Fe	26	55.8	Rutherfordium	Rf	104	(261)
Bismuth	Bi	83	209.0	Krypton	Kr	36	83.8	Samarium	Sm	62	150.4
Bohrium	Bh	107	(264)	Lanthanum	La	57	138.9	Scandium	Sc	21	45.0
Boron	B	5	10.8	Lawrencium	Lr	103	(262)	Seaborgium	Sg	106	(266)
Bromine	Br	35	79.9	Lead	Pb	82	207.2	Selenium	Se	34	79.0
Cadmium	Cd	48	112.4	Lithium	Li	3	6.94	Silicon	Si	14	28.1
Calcium	Ca	20	40.1	Lutetium	Lu	71	175.0	Silver	Ag	47	107.9
Californium	Cf	98	(251)	Magnesium	Mg	12	24.3	Sodium	Na	11	23.0
Carbon	C	6	12.0	Manganese	Mn	25	54.9	Strontium	Sr	38	87.6
Cerium	Ce	58	140.1	Meitnerium	Mt	109	(268)	Sulfur	S	16	32.1
Cesium	Cs	55	132.9	Mendelevium	Md	101	(258)	Tantalum	Ta	73	180.9
Chlorine	Cl	17	35.5	Mercury	Hg	80	200.6	Technetium	Tc	43	(98)
Chromium	Cr	24	52.0	Molybdenum	Mo	42	95.9	Tellurium	Te	52	127.6
Cobalt	Co	27	58.9	Neodymium	Nd	60	144.2	Terbium	Tb	65	158.9
Copernicium	Cn	112	(285)	Neon	Ne	10	20.2	Thallium	Tl	81	204.4
Copper	Cu	29	63.5	Neptunium	Np	93	(237)	Thorium	Th	90	232.0
Curium	Cm	96	(247)	Nickel	Ni	28	58.7	Thulium	Tm	69	168.9
Darmstadtium	Ds	110	(271)	Niobium	Nb	41	92.9	Tin	Sn	50	118.7
Dubnium	Db	105	(262)	Nitrogen	N	7	14.0	Titanium	Ti	22	47.9
Dysprosium	Dy	66	162.5	Nobelium	No	102	(259)	Tungsten	W	74	183.8
Einsteinium	Es	99	(252)	Osmium	Os	76	190.2	Uranium	U	92	238.0
Erbium	Er	68	167.3	Oxygen	O	8	16.0	Vanadium	V	23	50.9
Europium	Eu	63	152.0	Palladium	Pd	46	106.4	Xenon	Xe	54	131.3
Fermium	Fm	100	(257)	Phosphorus	P	15	31.0	Ytterbium	Yb	70	173.0
Fluorine	F	9	19.0	Platinum	Pt	78	195.1	Yttrium	Y	39	88.9
Francium	Fr	87	(223)	Plutonium	Pu	94	(244)	Zinc	Zn	30	65.4
Gadolinium	Gd	64	157.3	Polonium	Po	84	(209)	Zirconium	Zr	40	91.2
Gallium	Ga	31	69.7	Potassium	K	19	39.1				
Germanium	Ge	32	72.6	Praseodymium	Pr	59	140.9				

[1]The atomic number is the number of **protons** in the nucleus of the atom.

[2]The atomic masses in this table use fewer significant figures than most similar tables in college textbooks. By keeping the numbers simple, it is hoped that you will use mental arithmetic to do easy numeric cancellations and simplifications before you use a calculator for arithmetic. For radioactive atoms, ( ) is the mass number of the most stable isotope.

The Periodic Table of the Elements

1 1A	2 2A	3 3B	4 4B	5 5B	6 6B	7 7B	8 8B	9 8B	10 8B	11 1B	12 2B	13 3A	14 4A	15 5A	16 6A	17 7A	18 8A
1 1.008 H Hydrogen																	2 4.00 He Helium
3 6.94 Li Lithium	4 9.01 Be Beryllium											5 10.8 B Boron	6 12.07 C Carbon	7 14.0 N Nitrogen	8 16.0 O Oxygen	9 19.0 F Fluorine	10 20.2 Ne Neon
11 23.0 Na Sodium	12 24.3 Mg Magnesium											13 27.0 Al Aluminum	14 28.1 Si Silicon	15 31.0 P Phosphorus	16 32.1 S Sulfur	17 35.5 Cl Chlorine	18 39.9 Ar Argon
19 39.1 K Potassium	20 40.1 Ca Calcium	21 45.0 Sc Scandium	22 47.9 Ti Titanium	23 50.9 V Vanadium	24 52.0 Cr Chromium	25 54.9 Mn Manganese	26 55.8 Fe Iron	27 58.9 Co Cobalt	28 58.7 Ni Nickel	29 63.5 Cu Copper	30 65.4 Zn Zinc	31 69.7 Ga Gallium	32 72.6 Ge Germanium	33 74.9 As Arsenic	34 79.0 Se Selenium	35 79.9 Br Bromine	36 83.8 Kr Krypton
37 85.5 Rb Rubidium	38 87.6 Sr Strontium	39 88.9 Y Yttrium	40 91.2 Zr Zirconium	41 92.9 Nb Niobium	42 95.9 Mo Molybdenum	43 (98) Tc Technitium	44 101.1 Ru Ruthenium	45 102.9 Rh Rhodium	46 106.4 Pd Palladium	47 107.9 Ag Silver	48 112.4 Cd Cadmium	49 114.8 In Indium	50 118.7 Sn Tin	51 121.8 Sb Antimony	52 127.6 Te Tellurium	53 126.9 I Iodine	54 131.3 Xe Xenon
55 132.9 Cs Cesium	56 137.3 Ba Barium	57 138.9 La Lanthanum	72 178.5 Hf Hafnium	73 180.9 Ta Tantalum	74 183.8 W Tungsten	75 186.2 Re Rhenium	76 190.2 Os Osmium	77 192.2 Ir Iridium	78 195.1 Pt Platinum	79 197.0 Au Gold	80 200.6 Hg Mercury	81 204.4 Tl Thallium	82 207.2 Pb Lead	83 209.0 Bi Bismuth	84 (209) Po Polonium	85 (210) At Astatine	86 (222) Rn Radon
87 (223) Fr Francium	88 (226) Ra Radium	89 (227) Ac Actinium	104 (261) Rf Rutherfordium	105 (262) Db Dubnium	106 (266) Sg Seaborgium	107 (264) Bh Bohrium	108 (277) Hs Hassium	109 (268) Mt Meitnerium	110 (271) Ds Darmstadtium	111 (272) Rg Roentgenium	112 (285) Cn Copernicium						

58 140.1 Ce Cerium	59 140.9 Pr Praseodymium	60 144.2 Nd Neodymium	61 (145) Pm Promethium	62 150.4 Sm Samarium	63 152.0 Eu Europium	64 157.3 Gd Gadolinium	65 158.9 Tb Terbium	66 162.5 Dy Dysprosium	67 164.9 Ho Holmium	68 167.3 Er Erbium	69 168.9 Tm Thulium	70 173.0 Yb Ytterbium	71 175.0 Lu Lutetium
90 232.0 Th Thorium	91 (231) Pa Protactinium	92 238.0 U Uranium	93 (237) Np Neptunium	94 (244) Pu Plutonium	95 (243) Am Americium	96 (247) Cm Curium	97 (247) Bk Berkelium	98 (251) Cf Californium	99 (252) Es Einsteinium	100 (257) Fm Fermium	101 (258) Md Mendelevium	102 (259) No Nobelium	103 (262) Lr Lawrencium

Index